Contemporary
Human Geography

Contemporary Human Geography

CULTURE, GLOBALIZATION, LANDSCAPE

SECOND EDITION

Roderick P. Neumann
Florida International University

Patricia L. Price
Baruch College, City University of New York

w. h. freeman
Macmillan Learning
New York

Vice President, STEM: Daryl Fox
Program Director, Earth Sciences: Andrew Dunaway
Senior Program Manager: Jennifer Edwards
Developmental Editor: Marion B. Castellucci
Marketing Manager: Leah Christians
Marketing Assistant: Madeleine Inskeep
Senior Media Editor: Amy Thorne
Lead Content Developer: Emily Marino
Media Project Manager: Jason Perkins
Director, Content Management Enhancement: Tracey Kuehn
Senior Managing Editor: Lisa Kinne
Workflow Manager: Lisa McDowell
Senior Content Project Manager: Edward J. Dionne
Project Manager: Linda DeMasi, Lumina Datamatics, Inc.
Director of Design, Content Management: Diana Blume
Design Services Manager: Natasha A. S. Wolfe
Interior Design: Patrice Sheridan
Cover Design Manager: John Callahan
Art Manager: Matthew McAdams
Illustrations: International Mapping
Senior Photo Editor: Sheena Goldstein
Photo Researcher: Lisa Passmore
Composition and Layout: Lumina Datamatics, Inc.
Printing and Binding: LSC Communications
Cover Photo: simon wong/Moment/Getty Images

Library of Congress Control Number: 2018960096
ISBN-13: 978-1-319-05981-1
ISBN-10: 1-319-05981-3

©2019, 2015 by W. H. Freeman and Company

Printed in the United States of America

1   2   3   4   5   6   22   21   20   19   18

W. H. Freeman and Company
One New York Plaza
Suite 4500
New York, NY 10004-1562
www.macmillanlearning.com

Contents in Brief

Contents

Dear Reader

We are very pleased to present the second edition of *Contemporary Human Geography.* With this edition we have maintained our dual commitments to keeping the textbook cutting edge with the latest advances in human geography while presenting concepts and examples in language that is clear and accessible. Throughout all chapters we draw directly from recent scholarly work to help convey a sense of human geography as engaged and relevant. We ourselves are conducting new research, publishing in geography journals, and serving on editorial boards. We believe our active scholarship provides *Contemporary Human Geography* with a distinct identity and best positions us to ensure that the textbook keeps pace with advances in the field. At the same time, we place great emphasis on writing that is crisp, lively, and peppered with contemporary examples to help students relate to abstract concepts. We work to ensure that the textbook actively aids student learning. Collectively we have over four decades of experience teaching and mentoring students, from lower-division undergraduate studies to the doctoral level. This experience informs our understanding of what is important for students of human geography to learn and the best ways to facilitate that learning.

As part of our efforts to keep the textbook current, for the second edition we conducted a thorough overhaul of all key vocabulary terms. We consulted many of the recently published human geography encyclopedias, handbooks, and dictionaries. Each of us has either edited or authored contributions to several of these reference works. We also consulted many foundational human geography publications to confirm that we conveyed the most accurate meanings. As a result, we refined many existing definitions, deleted some obsolete or idiosyncratic terms, and added recently introduced terminology. In addition, adopters of the previous edition will find new sections throughout that cover themes and practices that are fundamental to human geography and that introduce additional terminology.

We also redoubled our efforts to facilitate student learning. Educators know how much importance is placed today on documenting effective teaching and student learning. To that end we have thoroughly revised and expanded the materials related to learning outcomes. Among other sources, we consulted the Advanced Placement Curriculum for Human Geography and the education resources of the American Association of Geographers to develop the most appropriate learning outcomes. We have also placed greater emphasis throughout the textbook on actively stimulating student learning. We have included, for example, both active and experiential learning exercises in each chapter.

Reading through the table of contents and the following pages, you will find many other improvements and additions. We hope that the care and attention we gave to revising the second edition will be readily evident.

Sincerely,

Roderick P. Neumann
Patricia L. Price

Preface

Five themes provide a framework for human geography

Contemporary Human Geography is organized around five themes. These themes are introduced and explained in the first chapter and serve as the framework for the nine topical chapters that follow. Each theme is applied to a variety of human geography topics, such as language, ethnicity, politics, religion, and agriculture. This thematic organization allows students to relate to the most important aspects of human geography at every point in the text. As instructors, we have found that beginning students learn best when provided with a precise and useful framework, and the five-themes approach provides such a framework for understanding human geography. The themes are identified by colored banners that serve as a visual reminder to students when they appear throughout the book:

Region

Mobility

Globalization

Nature–Culture

Cultural Landscape

Our *region* theme appeals to students' curiosity about the differences among places. *Mobility* conveys the dynamic aspect of human geographical phenomena, a theme particularly relevant in a world of incessant and rapid movement. Students acquire an appreciation of how people, things, and ideas move (or do not move). The topics employed to illustrate the concepts of mobility include many popular cultural references familiar in students' daily lives. *Globalization* permits students to understand the complex processes that link the various economies, cultures, and societies around the world. An understanding of globalizing processes is necessary for explaining how those linkages can create economic and cultural similarities as well as disparities. *Nature–culture* addresses the complicated relationship between culture and the physical environment. With today's complex and often controversial relationship between the natural environment and our globalizing world, both the tensions and the alliances that arise with regard to this relationship are now at the forefront of this theme. Finally, the theme of *cultural landscape* heightens students' awareness of the visible and lived qualities of places and regions.

Engaging topics and features show students geography's relevance to their lives

Fascinating contemporary topics are presented, including how Google Maps work; refugees fleeing Syria, the Sudan, and Myanmar; barriers to the global movement of digital information; the purchasing of place names; driving while black; urban agriculture; and celebrities' role in development.

Who Speaks Wukchumni?
◐ Watch at 🍃 SaplingPlus

The Wukchumni are a Native American people resident in central California. Today, fewer than 200 Wukchumni are alive. Marie Wilcox, born in 1932, is the only remaining native speaker of the Wukchumni language. Together with family members, Marie has composed a Wukchumni dictionary and recorded herself speaking the native language. The predicament of the Wukchumni people and their language is mirrored throughout the United States, where 130 Native American languages are on the "endangered" list.

Thinking Geographically

1. As you watch this video, notice the poverty and isolation of the rural area where Marie and her fellow Wukchumni tribe members live. Do you think these factors have helped to preserve, or to endanger, the Wukchumni language?

2. Marie, her daughter Jennifer Malone, and her grandson Donovan Treglown have all contributed to the compilation and recording of the Wukchumni dictionary featured in this video. Their work has involved years of sacrifice. What factors have motivated them to pursue and complete this project?

3. As Marie relates in this video, the deaths of individual Wukchumni speakers have endangered the language to the point of extinction. Do you believe that, when Marie passes away, the Wukchumni language will die out entirely? Why or why not?

Available and assignable in SaplingPlus, each chapter includes a link to a relevant *New York Times* video, giving students a fresh way to observe and think about human geographic phenomena. Discussion questions in the text accompany each Video Connection in addition to the questions in SaplingPlus. Six videos are new to this edition.

Subject to Debate

This feature highlights a current controversy, asking students to think critically about all sides of an issue.

SUBJECT TO DEBATE Whither the Political Promise of the Internet?

In the early years of the Internet, many observers celebrated its potential to liberate people from political oppression. They believed it would allow citizens to bypass the state to directly link individuals with individuals, make geography irrelevant and sovereignty obsolete, and open a new era of personal freedoms and democratic governance. Many saw the role of social media in the Arab Spring of 2010–2011, discussed in this chapter, as exemplary. In Egypt and Tunisia, pro-democracy activists used Twitter, YouTube, and Facebook to organize street protests and ultimately topple two authoritarian governments. This was the promise of the Internet realized.

Recent political uses of social media reveal that the Internet brings perils as well. The Internet can endlessly fragment and customize information to match the interests of individual consumers. As a consequence, the shared worldview that binds a country's citizens together has eroded, a situation that is increasingly exploited by antidemocratic forces. Technology experts found that organizations linked to the Russian government use social media to promote political division in an effort to destabilize established democracies in the United States and Western Europe. Hate groups in the United States use social media to promote racial and ethnic divisions among citizens. The very social media used in support of the Arab Spring are now used by violent militants in the region to organize attacks on civilians and recruit allies, with the ultimate goal of imposing their own political authority. It appears the Internet can be a tool for liberation or repression, unity or discord.

Continuing the Debate

The Internet is largely self-regulated by the corporations—Google, Facebook, Twitter, etc.—that serve as neutral platforms providing information services to consumers. Specifically, service providers are not subject to the same disclosure regulations regarding political influence as are television, newspapers, and radio. With this knowledge and given the documented uses of the Internet for political purposes, consider the following discussion questions.

- Would you characterize the Internet as promoting primarily centrifugal forces, centripetal forces, or neither?
- How might we go about changing the Internet to reduce its more antidemocratic and divisive uses?
- Do you think that governments should regulate how information is disseminated and presented on the Internet, or should individual consumers be allowed to make their own judgments on its value?
- In what ways do your own social media habits narrow or expand your access to multiple political points of view?
- When you see political information on the Internet, what methods do you use to assess its validity and reliability?

Social media in the aid of political movements. Proponents initially touted social media as democratizing force, but it can also be used by antidemocratic and hate groups such as these white supremacist demonstrators in Charlottesville, Virginia, in 2017. (Samuel Corum/Anadolu Agency/Getty Images)

The Great Zimbabwe National Monument was constructed between 1100 and 1450 c.e., and encompasses over 1739 acres (700 hectares) in a region populated by the Shona people of Zimbabwe. Placed on the list of World Heritage Sites in 1986, Great Zimbabwe contains the best example of precolonial African architecture south of the Sahara.

- In the Great Enclosure, imposing dry masonry stone walls rise 36 feet (11 meters), enclosing numerous dwellings and other stone structures. In addition to the Great Enclosure, two other distinct zones comprise the complex: the Hill Ruins and the Valley Ruins.
- For nearly four centuries, the site served as the principal city of the kingdom of Zimbabwe, housing 10,000 to 20,000 residents and functioning as the center of an international trading network that stretched across the Indian Ocean and, ultimately, to China. The kingdom ruled over a gold-rich plateau covering a portion of present-day Zimbabwe and Mozambique. Locally mined gold was smelted into ingots for trade with coastal cities, which imported porcelain, glass beads, and other luxury goods.
- Racist nineteenth-century theories of history prevented Europeans from recognizing the ruins as the grand achievement of an African civilization. They speculated that ancient Phoenicians, Egyptians,

and even the Queen of Sheba had built it. Even in the 1970s, the white-minority government of Zimbabwe (then Southern Rhodesia) officially denied the site's black African origins. Twentieth-century archaeological studies have confirmed that the Shona's ancestors built and continuously occupied the site for nearly four centuries.

- During the twentieth-century pan-African struggle against European colonialism and white-minority rule, the site became a key source of cultural pride among black Africans across the continent. It helped form both Shona ethnic identity and postcolonial Zimbabwean national identity. Indeed, the present-day country of Zimbabwe took its name from the site after overthrowing white-minority rule in 1980.

THE GREAT ENCLOSURE: The most remarkable feature of the site is the Great Enclosure, an area dominated by a monumental, oval-shaped outer wall 820 feet (250 meters) long and 10 feet (3 meters) wide. The craftsmanship is stunning, beginning with the skillful splitting of granite blocks excavated from a nearby quarry. The granite split easily along fracture planes, resulting in cubelike blocks that craftsmen smoothed, stacked, and fitted together. This technique resulted in a finished quality that rivals modern brick walls and allowed the stone city to be built without mortar and yet remain intact for centuries.

- Cut stones set in herringbone and chevron patterns ascend the top of the outer wall. Within the wall are a smaller stone enclosure and a narrow stone passageway leading to the beehive-shaped conical tower. The tower is constructed of solid stone blocks and is assumed to have had a primarily symbolic or ceremonial purpose.
- The enclosure once contained houses built of clay and gravel, of which only traces remain. These were organized into walled-off community areas composed of a kitchen, dwelling huts, and a court. One theory suggests that the enclosure contained the royal residence of the kingdom's ruler. Thus, the Great Enclosure represents the seat of political and economic power for the expansive Zimbabwe kingdom.

TOURISM: The site is readily accessible and also relatively close to other attractions such as Victoria Falls (another World Heritage Site) and many of southern Africa's wildlife parks.

- International visitors freely roam the site and explore the very different perspectives and features offered by the Great Enclosure, the Hill Ruins, and Valley Ruins.
- Recent political and economic instability in Zimbabwe, however, has resulted in a significant drop-off in tourist activity. Although ruinous to the local economy, the lull in tourist traffic may be giving this archaeological wonder a needed respite from overuse.

- Built between 1100 and 1450 c.e. by the Shona people of Zimbabwe, the site is a unique example of African architecture.
- The site has three main areas: the Great Enclosure, the Hill Ruins, and the Valley Ruins.
- Placed on the list of World Heritage Sites in 1986, the site has been visited by thousands of tourists, but tourism has slowed due to political and economic instability in the country.

http://whc.unesco.org/en/list/364

World Heritage Site

These spectacular two-page features help students analyze the landscapes of some stunning UNESCO World Heritage Sites from a geographer's point of view.

GEOGRAPHY @ WORK

Matthew Toro

Director of Map, Imagery, and Geospatial Services, Arizona State University

Education: MA Geography, University of Miami

BA Geography and International Relations, Florida International University

Q. *Why did you major in geography and decide to pursue a career in the geography field?*

A. I wanted to understand the way the world works, and no other field seemed to offer the sort of comprehensive, multidimensional, multiscale answers afforded by the space-based perspective of geography. Other disciplines I explored seemed to look at one or a few isolated components of social or "natural" reality. Geography was able to interweave seemingly unrelated human and geophysical systems, helping me see more completely the way the world system operates. I'll never succeed at comprehending the world in its infinite complexity, but my geographic perspective gives me an insight that I'm confident no other field can match.

Q. *Please describe your job.*

A. I run a library-based mapping center. It's a place for learning about GIS and cartographic resources (data, maps, satellite and aerial imagery, etc.), as well as various geospatial software tools that can bring those resources to life in powerful ways. It's a place where people can explore how to integrate spatial analysis, map interpretation, and 21st century cartography into their work. We collect and process hundreds of geospatial datasets, digitize and georeference thousands of aerial photos and maps, create engaging educational programming, and conduct original research. Outside academia, I consult on various data analysis and geo-visualization projects.

Q. *How does your geographical background help you in your day-to-day work?*

A. Geography exposes the world as a set of processes rather than a fixed, immutable reality. That's empowering, as it provides a framework for changing those processes in more sustainable ways. My professional life involves multiple roles: from getting deep into the technicalities of data analysis, to GIS project management, to overseeing the operations of a geospatial technology center. Geography endowed me with indispensable critical thinking and information processing skills to effectively jump among those roles.

Q. *In terms of employment, what advice do you have for students considering a career in the geography field?*

A. Exploit geography's multi- and transdisciplinary nature to the fullest. Gain exposure to multiple theoretical approaches and applied methods, while specializing in a few that lend useful skills for the real economy. Be entrepreneurial and ambitious, and don't be afraid to market yourself. Appreciation for geographers—and how we think and what we can do—is growing tremendously across all sectors.

Geography @ Work

These interviews offer firsthand accounts of the working lives of geographers and other professionals who do work related to geography. Each person explains what they do, where they work, and how they apply their study of geography to their work.

Maps and Illustrations

Every map in the textbook has been revised to ensure that it is up to date with the latest data and visually appealing. Many new maps and illustrations have been added throughout.

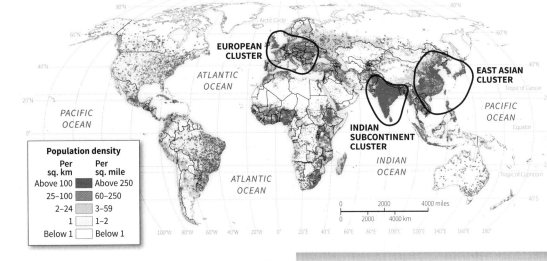

Population density

Per sq. km	Per sq. mile
Above 100	Above 250
25–100	60–250
2–24	3–59
1	1–2
Below 1	Below 1

FIGURE A1 Mercator projection. The poleward distortion of land area size is particularly noticable in the depiction of Antartica at the bottom of the map.

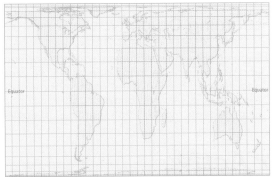

FIGURE A2 Peters projection. Here relative the relative size of land areas is accurate, but shapes are greatly distorted.

Map Reading Skills

An expanded appendix on how to read maps offers students instruction on map projections and guidance for reading maps and thinking critically about them.

Emphasis on evidence-based critical thinking pedagogy

Contemporary Human Geography presents geography as a dynamic academic pursuit based on observation and evidence, helping students develop informed ideas about geographic issues based on observation and scientific evidence.

Learning Objectives

3.1 Identify regional patterns of population characteristics and describe how these are distributed spatially and change over time.

3.2 Explain patterns, causes, and consequences of population migrations.

3.3 Understand theories of population growth and control and how these have changed over time.

3.4 Describe how the natural world shapes population characteristics, and how population characteristics in turn shape the natural world.

3.5 Analyze the imprint of demographic factors on the cultural landscape.

(Steve Raymer/Getty Images)

Learning Objectives

Each chapter now features updated Learning Objectives to aid in student learning and assessment. Learning objectives are presented at the beginning of each chapter, and then revisited at the chapter's end, with summary points, allowing students to review the chapter's main concepts. Practice assessments designed to help students master each learning objective are available in SaplingPlus.

the HUB *For Active Learning*

Additional instructor resources for these activities are available on The Hub for Active Learning.

Doing Geography:
Active Learning

These new activities provide instructors with ways to engage students in the classroom. With three to five activities per chapter, instructors can choose what works best for their course. Active Learning activities are available in SaplingPlus as an instructor supplement.

Doing Geography:
Experiential Learning

An Experiential Learning activity, including two new to this edition, appears at the end of each chapter. These activities put students in the role of a working geographer, giving them the information and resources they need to tackle a specific geographic exercise.

Doing Geography

ACTIVE LEARNING:
Listening to the Dialects of English

As we have seen, linguistic influences, historical patterns, and migratory flows all can play a role in shaping spoken languages. In this activity, you will use your phone or computer to listen to recordings of people speaking various dialects of English. As you listen, you'll try to understand what you are hearing and decide which dialects are easy or difficult to make out, keeping in mind the differences between dialects, pidgins, creoles, and "standard" languages, and how these differences sound.

More guidance at 🪧 Sapling Plus

EXPERIENTIAL LEARNING:
Exploring the Political Economy of Purchased Place-Names

In this activity, you will work individually or in small groups to identify places that bear the names of corporate or individual sponsors. As you learned in the section titled "The Political Economy of Toponyms," elements of the built environment—buildings, parks, sports venues, as well as vehicles and the clothing we wear—can bear the name of a wealthy person or family, or a corporation. Purchasing the right to name a place often confers a type of power on the purchaser. This experiential learning activity will encourage you to make these connections in your local landscape.

Steps to Exploring the Political Economy of Purchased Place-Names

Step 1: Work with your instructor to decide whether you will conduct this activity alone or in small groups.

Step 2: Determine the place you will explore. For some of you, the immediate college or university campus has ample places named by benefactors. For others, your city, town, or neighborhood will provide a better field research site.

Step 3: Set out to find places in the landscape that have been named by individuals or corporations. Photograph them with your cell phone and, if your instructor requests, note the location coordinates.

Step 4: Research the history of the named places that you find. Based on what you learn, answer the following questions.

* How much money did the naming rights cost?

* Who received the funds and what were the funds used for?

* Has there been any controversy over the naming, or has the naming of the place(s) you've identified changed over time from one sponsor to another?

Step 5: Report to your classmates and compare your findings.

The name of donors Joan and Sanford I. Weill is now part of the name of Cornell University's medical center. (Alessio Botticelli/GC Images/Getty Images)

SEEING GEOGRAPHY

Feisbuquear: "To Facebook"

In 2002, Hispanics surpassed African Americans as the nation's numerically most significant minority group. In some U.S. cities, Hispanics constitute more than half of the population, a fact that brings into question the designation "minority." For example, the population of Miami, Florida, is two-thirds Hispanic, and more than three-fourths of the residents of El Paso and San Antonio, Texas, are Hispanic. In the United States today, Spanish-speaking peoples from Latin America provide the largest flow of immigrants into the United States. Even midsize and smaller towns in the midwestern and southern United States are becoming destinations for Spanish-speaking immigrants, sharply changing the ethnic composition of cities such as Shelbyville, Tennessee; Dubuque, Iowa; and Siler City, North Carolina.

By the same token, English has burrowed its way deeply inside the contemporary Mexican linguistic landscape. The ubiquity of U.S. consumer goods, English-language media, and American popular culture in Mexico is notable. Many Mexicans travel back and forth to the United States for work- and family-related reasons. And, given that the United States and Mexico share a nearly 2000-mile border, these flows of people and ideas have a long history. It is logical to assume that the Spanish spoken by Mexicans will be influenced by American English, and vice versa.

Indeed, Spanish and English have combined in a rich, complex fashion, producing a hybrid language called Spanglish. The phrase "Vámonos al downtown a tomar una bironga after work hoy" is an excellent example. It translates into Standard English as "Let's go downtown and have a beer after work today." Vámonos ("let's go"), tomar ("to drink"), and hoy ("today") are Spanish words that are combined in the same sentence with the English words downtown and after work. Linguists refer to this tendency to shift between languages in the same sentence, a very common practice among bilingual speakers, as code-switching. However, the noun bironga, which means "beer" in English, is a Spanglish invention: it exists in neither English nor Spanish. This is quite common in Spanglish, and neologisms such as hanguear ("to hang out"), deiof ("day off"), and parquear ("to park" a vehicle) abound.

In the image that opens this chapter, the word "feisbuquear" is wholly invented, using the sound of the noun "Facebook" and a Spanish verb ending. Spanglish reflects the growing Spanish-English bilingualism of many U.S. residents, as well as residents of Mexico. It underscores the flexibility of language, and the enduring creativity of human beings as we attempt to communicate with one another.

Cell phone service provider advertisement in Mexico. Feisbuquear is a Spanglish verb meaning "to Facebook." (Courtesy of Patricia L. Price.)

Seeing Geography

Students begin their exploration of geographical concepts at the start of each chapter with a critical thinking question based on the chapter-opening photo. Students then contemplate this question throughout the chapter and explore it in greater depth at the end of the chapter with the Seeing Geography feature. Four Seeing Geography features have been revised with new photographs and content; many include new ideas and analyses.

SaplingPlus resources target the most challenging concepts and skills in the course

Contemporary Human Geography is accompanied by a media and supplements package that facilitates student learning and enhances the teaching experience. Fully loaded with our interactive e-book and all student and instructor resources, SaplingPlus is organized around a set of pre-built units available for each chapter in *Contemporary Human Geography*.

Created and supported by educators, SaplingPlus's instructional online homework drives student success and saves educators time. Every homework problem contains hints, answer-specific feedback, and solutions to ensure that students find the help they need.

LearningCurve
macmillan learning

Put "testing to learn" into action. Based on educational research, LearningCurve really works: game-like quizzing motivates students and adapts to their needs based on their performance. It is the perfect tool to get them to engage before class and review after class! Additional reporting tools and metrics help teachers get a handle on what their class knows and doesn't know. Available in SaplingPlus and Achieve Read and Practice.

Powered by ArcGIS Story Maps

One story map per chapter on a key topic has been designed as an extension activity for students.

Powered by ArcGIS Web Maps

Students use zoom and search tools, basemaps, and layers to answer questions and deepen their geographic understanding of key topics.

Practice at SaplingPlus

Read the interactive e-Text, review key concepts, and test your understanding.

 Story Map. Explore the five themes of the text by examining the geography of coffee.

 Web Map. Examine maps that illustrate issues in human geography.

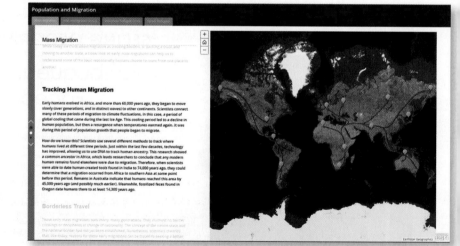

30 New Video Activities

A brief video related to key concepts from each chapter is followed by assessment questions, available in SaplingPlus.

Pre-Lecture Reading Quizzes

Available in SaplingPlus for each chapter, these quizzes help ensure that students have read the material before attending class.

Instructor Resources

The Hub for Active Learning

For the second edition, we've developed a suite of active learning resources to help instructors make their classes more engaging. Each chapter has a variety of activities based on the content in the text. The resources include handouts for students, instructor guides for each activity, and PowerPoint slide shows.

Lecture Outlines for PowerPoint

Adjunct professors and instructors who are new to the discipline will appreciate these detailed companion lectures, perfect for walking students through the key ideas in each chapter. These rich, pre-built lectures make it easy for instructors to transition to the book.

Test Bank

The Test Bank provides a wide range of questions appropriate for assessing your students' comprehension, interpretation, analysis, and synthesis skills. Each question is tagged for difficulty, level on Bloom's Taxonomy, book sections and page numbers, as well as learning objectives.

Clicker Questions

Clicker questions allow instructors to jump start discussions, illuminate important points, and promote better conceptual understanding during lectures.

Acknowledgments

An introductory text covering a wide range of topics must draw heavily on the research and help of others. Many geographers contributed advice, comments, ideas, and assistance as this book moved from draft to publication. We would like to thank those colleagues who contributed their helpful feedback and opinions:

Mohammad Alam, University of Winnipeg

Reuben Allen, Ball State University

Geordie Armstrong, Santa Barbara City College

Jennifer Bernstein, Santa Barbara City College

Michael Boester, Monroe Community College

John Conley, Santa Ana College

Thomas R. Craig, Oklahoma State University

Brent Doberstein, University of Waterloo

Matt Dyce, University of Winnipeg

Michael Farrell, Los Angeles City College

Lisa Ford, University of Manitoba

Brian Gilson, Oklahoma State University

Joseph P. Henderson, Georgia Gwinnett College

Todd Lindley, Georgia Gwinnett College

Elizabeth Lobb, Mount San Antonio College

Thomas Loder, Texas A&M University

Jose Javier Lopez, Minnesota State University

Neusa Hidalgo Monroy, University of Toledo

Monica Milburn, San Jacinto College

Shawn Mitchell, University of South Alabama

Michael Niedzielski, University of North Dakota

Bimal Kanti Paul, Kansas State University

Baishali Ray, Young Harris College

Jason Rhodes, Kennesaw State University

Thomas A. Smucker, Ohio University

Michael Steinberg, University of Alabama

Ray Sumner, Long Beach City College

Angie E. Wood, Chattanooga State Community College

Ramin Zamanian, Houston Community College

We would also like to thank those colleagues who offered helpful comments during the preparation of the first edition of *Contemporary Human Geography*:

Michael Cote, University of Southern Maine

Seth Giertz, University of Nebraska, Lincoln

Shawn Knabb, Western Washington University

Marc Law, University of Vermont

Michael Makowsky, Johns Hopkins University

Robert McComb, Texas Tech University

Paul Menchik, Michigan State University

Erin Moody, University of Maryland

Matthew Rutledge, Boston College

Daniel Sacks, Indiana University, Bloomington

Finally, our thanks go to the staff at Macmillan Learning, whose encouragement, skills, and suggestions created a special working environment and to whom we express our deepest gratitude. In particular, we thank Jennifer Edwards, program manager, for her patience with us, her dedication to this project, and her many contributions to it; Marion Castellucci, developmental editor, for her professionalism and invaluable advice; Shannon Howard, marketing manager; Ed Dionne, senior content project manager; Linda DeMasi, project manager; Amy Thorne, senior media editor; Patrice Sheridan, interior designer; John Callahan, cover design manager; Matt McAdams, art manager; Sheena Goldstein, photo editor; and Lisa Passmore, photo researcher. The fine work of all these people can be detected throughout the book.

About the Authors

Courtesy of Roderick P. Neumann

Roderick P. Neumann is professor of geography in the Department of Global and Sociocultural Studies in the School of International and Public Affairs at Florida International University. He earned his PhD at the University of California, Berkeley. Neumann's research is organized around three intersecting themes: the tensions between social justice and biodiversity conservation; political economy of natural resources; and the co-constitution of identity, nature, and landscape. He has conducted ethnographic fieldwork and archival research in East Africa, western Europe, and the U.S. West. This research was funded by the National Science Foundation, the National Endowment for the Humanities, and the Social Science Research Council and published in the *Annals of the American Association of Geographers, Progress in Human Geography, Cultural Geographies,* and *Political Geography.* His books include *Imposing Wilderness: Struggles over Livelihoods and Nature Preservation in Africa* (1998), *Making Political Ecology* (2005), and *The Commercialization of Non-Timber Forest Products* with Eric Hirsch (2000). His latest research incorporates insights from science and technology studies and the posthuman turn in the humanities and social sciences.

Courtesy of American Council on Education

Patricia L. Price is associate provost for academic administration and faculty development at Baruch College, one of the City University of New York (CUNY) senior colleges. She teaches and holds faculty appointments at CUNY's Graduate Center and in Baruch's Department of Sociology and Anthropology. She previously served as the dean of academic affairs at Guttman Community College, also a CUNY school. Before transitioning to full-time academic administration, Price was a professor of geography in the Department of Global and Sociocultural Studies at Florida International University in Miami, Florida. Her research encompasses critical geographies of race and ethnicity, diversity in higher education, and Latinos in the United States. She is currently researching the Cuban exile experience in Miami to explore how exiles creatively connect place and identity. This research was funded by the National Endowment for the Humanities. In an earlier research project, she and a team of students and faculty surveyed residents of Latino neighborhoods in Chicago, Phoenix, and Miami to compare the civic engagement of Latinos in different U.S. cities. This research was funded by the National Science Foundation. She has authored many journal articles, chapters, encyclopedia entries, and two books: *Dry Place: Landscapes of Belonging and Exclusion* (2004), and *The Cultural Geography Reader* with Tim Oakes (2008).

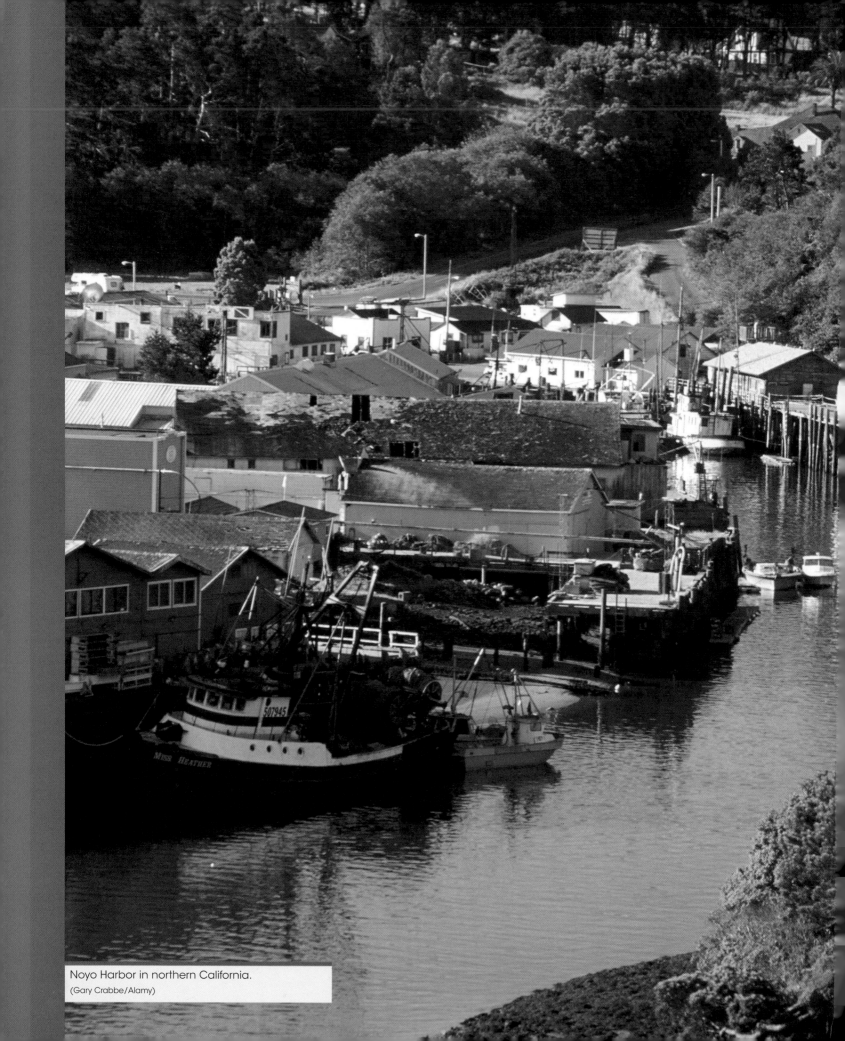

Noyo Harbor in northern California.

(Gary Crabbe/Alamy)

chapter 1

Human Geography

A CULTURAL APPROACH

Why do some locations have particularly strong identities?

Examine the photo and think about this question as you read. We will revisit them in Seeing Geography on page 43.

Learning Objectives

1.1 Explain the concept of culture and how culture interacts with the physical environment.

1.2 Explain the basic classification of cultures, including folk, indigenous, and popular cultures, and the processes of cultural diversification.

1.3 Explain the three basic perspectives on human geography: spatial modeling, humanistic, and social theoretical.

1.4 Apply key visualization tools used in human geography—including the longitude and latitude system and geospatial technologies—to everyday consumer devices.

1.5 Compare and contrast formal, functional, and vernacular regions.

1.6 Explain the different types of diffusion and migration.

1.7 Explain the concept of globalization and how it shapes interdependence among regions.

1.8 Explain the historical development of different approaches to nature–culture relations in human geography.

1.9 Explain the concept of cultural landscape and how geographers analyze it.

(Geoffrobidoe/Alamy)

We are born geographers. From the beginning of time, we humans have had to learn our way in the world to find food and shelter, to avoid danger, and to be with family and friends. Geography defines us through the places we inhabit and identify with and the journeys we make over the course of a day or a lifetime. It is no wonder that we are curious about the world and its peoples. Our curiosity about the world—about observable patterns (both fixed and in motion) and how we explain them— is at the heart of human geography. You are already geographers. We hope that our book will make you better ones.

This chapter introduces our approach to human geography and some of the essential terminology and methods of the discipline. We begin by elaborating on what we mean by a cultural approach to human geography and by presenting the broad cultural categories that we use throughout the textbook. We then examine the three main traditions in human geography: the spatial modeling, humanistic, and social theoretical perspectives. We describe geography's visualization tools, including the latest technologies in geographic information systems and remote sensing. Finally, we introduce five themes in human geography that we use to structure all subsequent chapters: region, mobility, globalization, nature–culture, and cultural landscape.

Our Cultural Approach to Human Geography

1.1   Explain the concept of culture and how culture interacts with the physical environment.

The discipline of geography is divided into two major strands: physical geography and human geography. Physical geographers study natural phenomena such as climates, river systems, and the distribution of wild plants and animals. Human geographers study social phenomena such as the formation of cities, the location of manufacturing, and the geographic distribution of languages. Thus **human geography** is concerned with the relationships between people and the spaces in which they live. Human geographers explore how these relationships create the diverse spatial arrangements that we see around us, arrangements that include homes, neighborhoods, cities, and countries.

In *Contemporary Human Geography*, we adopt a cultural approach to human geography. To understand the full scope of a cultural approach, we must first be clear about what we mean by culture. Culture is one of the most complex words in the English language, carrying different meanings in different contexts. At one level, we think of **culture** as being comprised of a variety of human behavioral traits and beliefs, which are learned as opposed to innate or inborn. Culture, therefore, includes a means of communicating the learned beliefs, practices, meanings, and values that influence individual and collective behaviors. Collectively, these acquired, learned behavioral traits are one meaning of culture. A cultural approach to the study of human geography implies, therefore, an emphasis on the meanings, values, attitudes, and beliefs that different groups of people lend to and derive from the spaces they inhabit and move through.

Furthermore, culture is not a static, fixed phenomenon, and it does not explicitly govern or control people's behavior. Rather, as geographers Kay Anderson and Fay Gale put it, "Culture is a *process* in which people are actively engaged." Individuals can and do change **cultural practices**, those social activities and interactions—ranging from religious rituals to architectural styles to food preferences to clothing—that collectively distinguish group identity. This understanding of culture recognizes that ways of life constantly change and that tensions over change commonly exist both within and between groups.

Throughout the book, we also use the word *culture* when referring to distinctive group identity. All people express cultural patterns, but not all in the same way, and these variations give rise to distinguishable human groupings. Collectively shared beliefs, practices, languages, technologies, and so forth produce characteristic ways of life. We therefore can also speak of cultures in the plural. Geographers have historically thought in terms of both generic categories of cultures (e.g., tribal, urban) and also specific cultures associated with specific places and regions (e.g., Tuareg camel pastoralists, Navajo Indians). There are an almost infinite number of ways of defining human groups in cultural terms or of creating categories of cultures. Today we are accustomed to hearing about not only Tuareg culture and Navaho culture, but also hip-hop culture and gay culture. The now-familiar appeals to multiculturalism and cultural diversity reflect a growing appreciation of the complex, ongoing processes of cultural differentiation that are characteristic of human society.

Cultures, however we may categorize and label them, are internally heterogeneous because people's perceptions, beliefs, and behaviors may vary according to age, class, or sexuality. In other words, cultures, while often distinctively identifiable, usually have internal variations. One of the most pervasive and universal internal cultural divisions is based on gender. Cultural geography studies have shown, for example, that women and men often experience the same spaces in different ways. A certain street corner or tavern might be a comfortable and familiar hangout for men but a threatening or uncomfortable space that women avoid (see Seeing Geography, Chapter 3). Therefore, geographers may refer to **gendered spaces**, spaces that are classified by social norms and practices as predominantly male or female.

Sometimes internal cultural differences can be so great that groups break apart and form new and distinct cultures. Historically, such fissions have often been associated with mass human migrations to new lands and environments. Nor are cultural groups consistent across time and space. Cultures are defined as much by dynamism and change as by fixity and stability. In short, we follow Anderson and Gale in taking the view of culture as a process rather than as a thing.

A cultural approach to human geography, then, studies the relationships among space, place, environment, and culture. It examines the ways in which culture is expressed and symbolized in the landscapes we see around us, including homes, commercial buildings, roads, farms, gardens, trash dumps, parks, and urban slums. It analyzes the ways in which language, religion, the economy, government, and other cultural phenomena vary or remain constant from one place to another, and provides a perspective for understanding the geography of human life (**Figure 1.1**).

In seeking explanations for cultural-geographic diversity, geographers consider a wide array of factors that influence this diversity. Some of these involve the **physical environment**, which includes the terrain, climate, natural vegetation, wildlife, variations in soil, and the pattern of land and water. Indeed,

there is solid scientific evidence that a major shift in climate triggered the initial migrations of *Homo sapiens* out of Africa. Because we cannot understand a culture removed from its physical setting, human geography offers not only a spatial perspective but also an ecological one.

Many forces, interacting in very complex ways, are at work in shaping cultural phenomena. The complexity of the forces that affect culture can be illustrated by an example drawn from agricultural geography: the distribution of wheat cultivation in the world. If you look at **Figure 1.2**, you can see important areas of wheat cultivation in Australia but not in Africa, in the United States but not in Chile, in China but not in Southeast Asia. Geographers recognize such consistent or characteristic spatial arrangements of objects or phenomena as a **pattern**. Identifying a spatial pattern is a first step toward explanation in geography. Why does this particular geographic pattern of wheat cultivation exist? Partly, it results from environmental factors such as climate, terrain, and soils. Some regions have always been too dry for wheat cultivation. The land in others is too steep or infertile. Indeed, a strong correlation exists between wheat cultivation and midlatitude climates, level terrain, and fertile soil.

Still, we should not place too much importance on such physical factors. People can modify the effects of climate on

FIGURE 1.1 Two traditional houses of worship. Geographers seek to learn how and why cultures differ, or are similar, from one place to another. Often the differences and similarities have a visual expression, such as these two nineteenth-century Catholic churches embedded in distinct cultural traditions, one from Spanish colonial New Mexico and one from New England. (Left: Roderick Neumann; Right: Nancy Catherine Walker/Alamy)

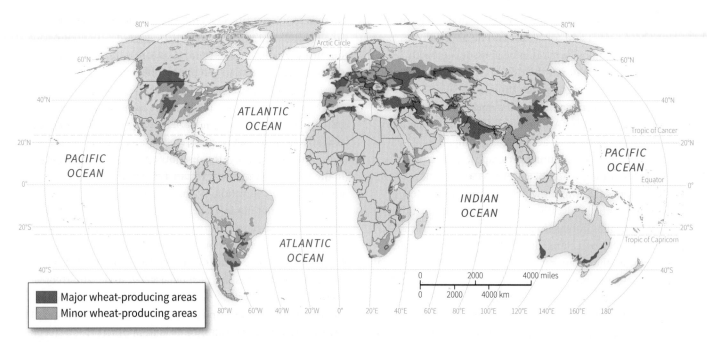

FIGURE 1.2 Areas of wheat production in the world today. These regions are based on a single trait: the importance of wheat in the agricultural system. This map tells us what and where. It raises the question of why.

agriculture through irrigation; the use of hothouses; or the development of new, specialized strains of wheat. We can conquer slopes by terracing, and we can make poor soils productive by adding fertilizers. For example, farmers in mountainous parts of Greece traditionally wrested an annual harvest of wheat from tiny terraced plots where soil had been trapped behind hand-built stone retaining walls. Even in the United States, environmental factors alone cannot explain the curious fact that major wheat cultivation is concentrated in the semiarid Great Plains, some distance from states such as Ohio and Illinois, where the climate for growing wheat is better. Wheat grows in a cultural environment as well as a physical one.

Ultimately, agricultural patterns cannot be explained by the characteristics of the land and climate alone. Many factors complicate the distribution of wheat, including people's tastes and traditions. Food preferences and taboos, often backed by religious beliefs, strongly influence the choice of crops to plant. Where wheat bread is preferred, people are willing to put great effort into overcoming physical surroundings hostile to growing wheat. They have even created new strains of wheat, thereby decreasing the environment's influence on the distribution of wheat cultivation. Other factors, such as government policies, can also encourage or discourage wheat cultivation. For example, import tariffs protect the wheat farmers of France and other European countries from competition with more efficient American and Canadian producers.

This is by no means a complete list of the forces that affect the geographical distribution of wheat cultivation. Like the distribution of wheat cultivation, the distribution of all cultural elements is a result of the constant interplay of diverse factors. Human geography is the discipline that seeks such explanations and understandings.

Many Cultures

1.2 Explain the basic classification of cultures, including folk, indigenous, and popular cultures, and the processes of cultural diversification.

Culture is a complex concept, and there are many ways to assess it, including both material and nonmaterial measures. Using these measures, we can classify cultures into types, including folk, indigenous, and popular cultures. These types are not rigid or static, however; they are always subject to processes of cultural diversification.

We classify cultures using many different criteria, both material and nonmaterial:

- **Material culture** includes the physical objects made and/or used by members of a cultural group: buildings,

Discovery (when Europeans began to explore the world by sea), they started thinking about "European culture" in relation to other cultures they encountered around the world. Although now discredited as ethnocentric and racist, hierarchical models—in which European culture was ranked above other cultures—were invented to categorize and rank cultures. Religion and technology were key criteria in constructing these hierarchies, with Christian and technologically developed cultures ranked at the top and non-Christian, nonindustrial cultures ranked at the bottom.

Folk Culture

As Europe and North America industrialized and urbanized in the nineteenth century, people began thinking about the historic cultural traditions that were being lost. Urban city dwellers started to view rural country spaces as inhabited by distinct, traditional cultures. Thus, a new term, **folk culture**, was introduced to distinguish traditional ways of life in rural and agricultural spaces from ways of life in the new urban and industrial spaces. Folk cultures are considered rural, unified, largely self-sufficient groups that share similar customs, material practices, and ethnicities.

When Europe's kingdoms and empires fragmented into new countries throughout the nineteenth century and into the twentieth, the urbanized, cosmopolitan elites looked to the rural folk as the source of distinctive national identities (**Figure 1.4**). Similarly, as the United States urbanized and expanded westward, rural folk were seen as the source of an American national character defined by self-reliance, perseverance, and

FIGURE 1.3 Would you eat these? Food preferences are strong markers of culture difference. These water beetles are a common food source in Southeast Asia. (Graham Day/Getty Images)

furniture, clothing, food, artwork, musical instruments, and so forth. The elements of material culture are visible.

- **Nonmaterial culture** includes the wide range of beliefs, values, myths, and symbolic meanings passed from generation to generation of a given society.

Often it is difficult to draw clear lines between material and nonmaterial features of culture. Food, for example, is material, but it is also charged with symbolic meaning and valued differently in different cultures (**Figure 1.3**). Many Westerners would gag at the mere thought of eating the worms and insects that are the daily food staples and even much-valued delicacies in non-Western, rural cultures around the world. Therefore, cultures can be thought of in terms of both material and nonmaterial characteristics.

Let's look briefly at how these material and nonmaterial characteristics were first used to identify, categorize, and map cultures. In eighteenth-century Europe and North America, people began to speak of "cultures" in the plural form. Specifically, following Europe's Age of Exploration and

FIGURE 1.4 Young men in Bavaria dance the "Schuhplattler," one of the traditional folk dances of Germany. The revival of such traditional folk dances by urban, educated classes has been important in the formation of modern national identities. (Imagno/ Getty Images)

FIGURE 1.5 National Folk Museum of South Korea, Seoul.
While folk cultures have largely disappeared in industrialized and urbanized countries, they remain preserved in museums as a source of national pride. (Jane Sweeney/AWL Images/Getty Images)

a pioneering spirit. Hence, the cowboy—an economically minor and short-lived rural occupation in the nineteenth-century western United States—came to symbolize an imagined American national character. Folk culture and national culture thus have been closely associated. The term **national culture**, sometimes regarded as a controversial idea, suggests that a country's population possesses a set of recognizable characteristics or traits—often including the same ethnic and linguistic traits—that express the core culture of each modern nation. Hence, many countries have museums and institutes dedicated to the celebration of folk culture and its relevance to contemporary national identity (**Figure 1.5**).

In terms of material life in folk cultures, many items of daily use such as clothing, furniture, and housing are handmade, often from raw materials found locally. Most food is grown and consumed locally, which has often resulted in the formation of distinctive regional cuisines, such as the Japanese diet of rice and raw fish or the Mesoamerican diet of maize, beans, and spicy peppers. In terms of nonmaterial life, folk cultures typically have strong family or clan structures and highly localized rituals, such as an annual blessing of water wells and springs or harvest dances. Generations of gifted individuals within folk communities create songs, musical styles, and storytelling that are unique to their regions. As with food, folk cultures often produce distinctive, highly regionalized musical genres, Appalachian bluegrass music being one well-known example.

In Europe and North America, folk cultures, strictly defined, no longer exist, though many material and nonmaterial traces can be found in the landscape. Moreover, the reasons for the creation of folk cultures —human inventiveness, curiosity, and need, uniquely expressed in specific places—endure, as we will illustrate throughout this textbook.

Indigenous Culture

Today, geographers frequently use terms such as *local, place-based,* or *indigenous* when referring to some of the distinctive cultural traits typically associated with folk culture. Indeed, the concept of **indigenous culture** has emerged as a focus of study in cultural geography. A simple definition of *indigenous* is "native" or "of native origin." In the modern world of independent countries, the word has acquired much greater cultural and political meanings. The International Labour Organization's (ILO) Indigenous and Tribal Peoples Convention 169 (Article 1.1) presents a legal definition that recognizes indigenous peoples as comprising a distinct culture. According to the ILO, indigenous peoples are self-identified tribal peoples whose social, cultural, and economic conditions distinguish them from the national society of the country in which they live (**Figure 1.6**). As with folk cultures, indigenous cultures are generally associated with rural regions, particularly geographically isolated arid lands, forests, and mountain ranges worldwide.

Indigenous peoples are regarded as descending from peoples occupying a territory at the time of **colonization**, which is defined as the appropriation, control, and occupation of a territory by a geographically distant power. In effect, indigenous peoples have been displaced from their lands to

FIGURE 1.6 A Karen of northern Thailand works on roof thatching. The Karen and other indigenous, tribal peoples around the world have maintained cultures distinct from the dominant national cultures of their home countries. (Nicholas J. Reid/The Image Bank/Getty Images)

make way for colonial settlers. Although indigenous peoples may share some of the material and nonmaterial characteristics that define folk cultures, their histories (and geographies) are quite different. Indigenous cultures are, in effect, those peoples who were colonized—mostly, but not exclusively—by European cultures. Thus they are now minorities in their homelands. Whereas folk culture was seen as providing the foundation for a national culture, indigenous culture is usually positioned in opposition to national culture.

Popular Culture

The notion of folk cultures emerged historically when urban elites began to fear that the ways of traditional, rural life were being erased and replaced by the rapid spread of **popular culture**. Popular culture refers to modern material and symbolic practices associated with the rise of mass-produced, machine-made goods and the invention of long-distance communication technologies, such as the telegraph, radio, and television (**Figure 1.7**). People sometimes use the terms mass culture and consumer culture to mean the same thing as popular culture. "Mass" emphasizes that popular culture is widespread, undifferentiated, and placeless. Mass media such as the Internet, social media, film, print, television, and radio shape cultural preferences and identities. "Consumer" emphasizes that popular cultural identity is shaped through the purchase and consumption of mass-produced commodities.

In popular culture, people are more mobile and less likely to live their lives in a single place. Secular institutions of authority—such as schools and courts—become increasingly important for maintaining social order. Individual choice is highly valued. Social relationships multiply, but fewer of them are conducted face to face. Rather, relationships become "stretched" across time and space. Think of online credit card purchases in contrast to the bartering that took place in the local village markets of historic folk cultures. Whereas the latter exchange was conducted in person between two people, think of how many individuals and corporations are involved in an online credit card purchase and home delivery of a simple pair of shoes, and how geographically distant from one another they all are.

Finally, geographer Edward Relph suggested that popular culture is characterized by a profound **placelessness**, a feeling resulting from the standardization of the built environment, increasingly on a global scale, that diminishes regional variation and eliminates the unique meanings associated with specific locations (**Figure 1.8**). Shopping malls, airports, and global hotel and fast-food chains are examples of placeless locations that could be anywhere. One can travel, for instance, to Beijing, China, stay overnight at an airport Hilton much like any other Hilton across the world, eat steak and mashed potatoes for dinner much like one would order in any major city, shop for shoes at the global chain stores in the airport mall while waiting for one's flight, and experience little that is Chinese or unique to Beijing.

FIGURE 1.7 Popular culture is reflected in every aspect of life, from the clothes we wear to the recreational activities that occupy our leisure time. Musician Kanye West performs with Jay-Z at American Airlines Arena in Miami. (Left: Scott Olson/Getty Images; Right: Kevin Mazur/WireImage/Getty Images)

FIGURE 1.8 Placelessness in urban landscapes: scenes almost anywhere. One of these photos was taken in the United States and the other two in the United Kingdom and Japan, respectively. It is difficult to detect many local distinctions in these popular urban landscapes. (Left to right: Emile Wamsteker/Bloomberg via Getty Images; Adina Tovy/Lonely Planet Images/Getty Images; Photo Japan/Alamy)

Cultural Diversification

There are so many ways through which cultural distinctions can be created, blurred, transformed, and erased. Cultural difference is evident everywhere—not only in the geographic distribution of different cultures but also in the way that difference is created or reinforced by geography. For example, in the United States,

the historical legacy of segregating "whites" from "blacks" has been important in establishing and maintaining cultural differences between these groups (**Figure 1.9**).

Alternatively, segregation can happen through the choices people make when relocating. Greater mobility in the modern era means that people from local, ethnically distinct cultures are frequently transplanted to faraway regions and cities where they

FIGURE 1.9 Greyhound bus station in 1943 Memphis, Tennessee, with separate facilities for blacks and whites. It exemplifies the racial segregation found historically throughout much of the United States. Though such practices have long been outlawed, historic patterns of segregation may be found in the landscape today. (Buyenlarge/Getty Images)

FIGURE 1.10 New York City's "Little Italy." Many major cities in the United States have such neighborhoods named for the predominant (or once-predominant) immigrant population. (Picture Partners/Alamy)

form recognizable enclaves, distinct and separate from the spaces of the predominant culture. As a result, cities around the world are dotted with neighborhoods called "Little Italy," "Chinatown," "Little Haiti," "Germantown," and so on (**Figure 1.10**).

Geographers increasingly refer to such transplanted ethnic, racial, and national populations as **diaspora cultures**. Historically, diaspora referred to a people (especially ancient Jewish culture) that had been displaced and geographically scattered. In modern times, however, international mobility has increased the size and number of diaspora cultures around the globe. In addition to settling in distinct enclaves, diaspora cultures often maintain strong social and economic ties to their homelands. In such cases, people may exhibit hybrid or multiple cultural identities, depending on the geographical and social settings.

Sometimes groups within a dominant culture become distinctive enough in norms, values, and material practices that we label them as a **subculture**. A subculture can be the result of resistance to the dominant culture or of a separate religious, ethnic, or national group forming its own distinct community within a larger culture. Subcultures may create distinct identities through the reuse and reinterpretation of mass-produced goods. Think of the importance of safety pins—a common household item originally associated with motherhood and diapers—to the material expression of punk culture (**Figure 1.11**). The list of subcultures emerging today is nearly infinite in length. Gay, hip-hop, NASCAR, and grunge

FIGURE 1.12 McKinley Morganfield, better known as Muddy Waters. Muddy Waters was born in the Mississippi Delta and moved to Chicago, where he became a recording star after electrifying his blues guitar. He was a key artist in the transformation of Delta folk music traditions into rock and roll. (Hulton Archive/Getty Images)

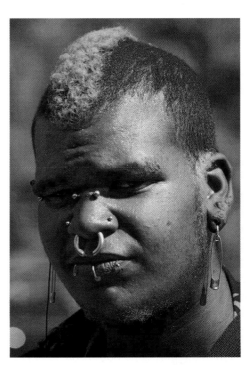

FIGURE 1.11 Punk culture in Massachusetts. The use of safety pins—designed originally for cloth diapers—in body piercings became an iconic symbol of punk culture. (Education Images/UIG/Getty Images)

are just a few of the adjectives that we place before "culture" to talk about popular cultural identities, each with its particular geographic expression. Culture is as dynamic and limitless as the human imagination.

Finally, just as the line between the material and the nonmaterial elements of culture can be blurred, so too can the line between elements of folk, indigenous, and popular cultures. The trafficking of ideas and things between different culture groups can give rise to further diversification. For example, blues music had its origins in the African American folk culture in the U.S. South, particularly the Mississippi Delta region (**Figure 1.12**). Later, in the 1950s, blues evolved into rock and roll, a musical genre that has come to epitomize global popular culture. There are many more examples of folk or indigenous cultural characteristics that have been incorporated into popular culture, including foods, textiles, and architecture. Conversely, items of popular culture have been incorporated into rural, folk-oriented cultures in surprising new ways. For example, the introduction of cheap, mass-produced glass beads from Europe provided the basis for vibrant new art forms in many Native American cultures (**Figure 1.13**).

FIGURE 1.13 Native American artistry with beads imported from Europe. A Lakota Sioux craftsperson of the mid-nineteenth century produced this animal-hide bag decorated in an exquisitely detailed floral design using glass beads. Indigenous and folk cultures have often used such mass-produced objects to create new local art forms. (Courtesy of Roderick Neumann.)

Three Perspectives on Human Geography

1.3 Explain the three basic perspectives on human geography: spatial modeling, humanistic, and social theoretical.

Geographers have developed three basic perspectives on the study and understanding of the complexity of human differences and variations. We refer to them as the spatial modeling, humanistic, and social theoretical perspectives. Each of these perspectives emphasizes different questions about human geographic patterns and uses different research methods to study those questions.

Spatial Modeling Perspective

Some geographers use mathematics to construct **spatial models** to understand and explain human-geographic patterns. This perspective, also called spatial science, asks questions that seek out universal laws and regularities, emphasizes the abstract over the concrete, and treats geographic patterns as problems of geometry. It draws its methodological inspiration from physics and geometry. **Space**, in this tradition, is an abstract, geometric property that can be scientifically modeled through the application of universal laws.

This perspective is not as prevalent as it once was, yet it made and continues to make important contributions to explaining human-geographic patterns. Spatial modeling has contributed to our understanding of transportation systems, urban settlement patterns, and social services delivery, to name just a few successful investigations. Aware that many causal forces are at work in the real world, model builders create simulations that focus on one or more potential causal force. That is, they abstract a limited set of possibilities from the complexities of social life in an effort to uncover the fundamental forces behind spatial patterns and forms.

Some model builders devise culture-specific models to describe and explain certain facets of spatial behavior within specific cultures. They still seek regularities and spatial principles, but only within the bounds of individual cultures, rather than universal. For example, several geographers proposed a model for Latin American cities in an effort to stress similarities among them and to understand why cities are formed the way they are (**Figure 1.14**). Obviously, no actual city in Latin America conforms precisely to their uncomplicated geometric plan. Instead, as in all model building, they deliberately generalized and simplified so that an urban type could be recognized and studied. The model will look strange to a person living in a city in the United States or Canada, for it describes a very different kind of urban environment, based in another culture.

No matter what sort of model one is building, it is essential to determine the scale of analysis. Scale of analysis, sometimes referred to as **spatial scale**, is the geographic extent of the area that is under investigation. We are used to talking about spatial scale in an everyday way when we talk about events at the "local scale," "national scale," or "global scale." We know from everyday experience that different events are occurring at different scales, as in an election when a candidate wins at the local scale, but loses at the national scale. Did the candidate win? The answer depends on our scale of analysis. When we use scale to analyze a geographic problem, it becomes a methodological issue linked to the nature of our inquiry. Let us use the previous model of the Latin American city. The investigators were trying to understand a common urban form. The scale of analysis is therefore defined by the area of a single urban settlement. But what if we wanted to

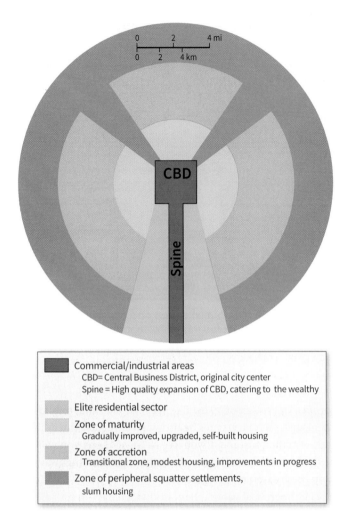

FIGURE 1.14 A generalized model of the Latin American city.
Urban structure differs from one culture to another, and in many ways the cities of Latin America are distinctive and have much in common. Geographers Ernst Griffin and Larry Ford developed the model diagrammed here to help describe and explain the processes at work shaping the cities of Latin America. (After Griffin and Ford, 1980, p. 406.)

Legend (from figure):
- Commercial/industrial areas
  CBD= Central Business District, original city center
  Spine = High quality expansion of CBD, catering to the wealthy
- Elite residential sector
- Zone of maturity
  Gradually improved, upgraded, self-built housing
- Zone of accretion
  Transitional zone, modest housing, improvements in progress
- Zone of peripheral squatter settlements,
  slum housing

CBD

Spine

Humanistic Perspective

Humanistic geographers ask questions that quantitative modelers do not. Where spatial modeling seeks universal mathematical solutions, humanistic geography focuses on human agency and creativity and subjective human experiences. Humanistic geographers seek to understand how people give geography meaning and value, and how humans' actions shape the geography of their lives and give it significance. Humanistic geographers, therefore, draw methodological inspiration from literature, the visual arts, and history, rather than math and science.

Just as space identifies the perspective of the model-building geographer, place is the key word connoting this more humanistic view of geography. **Place** emphasizes the way locations take on meaning and value through people's subjective experiences of them. Worldwide, people commonly develop strong emotional reactions to certain locations, which we refer to as a **sense of place**. Commonly, a key part of what defines a culture is a shared sense of place. Geographer Edward Relph suggests that "to be human is to have and know your place" in the geographical sense. This perspective on human geography highlights subjective meanings and experiences rather than objective geometric abstraction. Humanistic geographers pose questions about how and why certain places continue to evoke strong emotions in people, both positive and negative, even though those people may have little direct or daily connection with those places. Places that memorialize past events, such as the U.S. Civil War battlefields or Holocaust sites in Europe that draw millions of visitors annually, provide cases in point (**Figure 1.15**).

People's strong emotional and psychological attachment to ordinary locations means that a shared sense of place may influence social, political, and economic decision making. Geographers are therefore interested in how and why people

create a model showing how cities are located in relation to one another? We would need to expand our scale of analysis to an area large enough to include two or more cities. Spatial scale, as will become clear throughout the text, is a concept whose importance is not limited to model building. It is critical to any geographical investigation and to the answers that are produced by it.

FIGURE 1.15. Sense of place at a World War II memorial site.
Places can evoke a wide range of emotional responses. In the case of preserved memorial sites, such as the Nazi concentration camp of Auschwitz-Birkenau, individuals may find those emotions can be quite strong and difficult to process. Nevertheless, so many people want to visit Auschwitz-Birkenau annually that reservations are required well in advance. (Courtesy of Roderick Neumann.)

become attached to and derive meaning from their local neighborhoods or communities, and explore how those meanings may come into conflict with each other. Many of the debates in the news or in social media—debates on the construction of a new high-rise building or the location of a highway, for example—can only be understood by examining the meanings and values different groups of people give to and derive from particular places. Similarly, in the world of politics place is often a critical factor determining how and whether people vote in elections.

Social Theoretical Perspective

The third perspective aims to balance the humanistic focus on individual human agency with an attention toward the way social life is structured beyond individual human control. Contemporary social theory is notoriously diverse, and therefore the social theoretical perspective, which makes up the bulk of contemporary human geography, cannot be easily defined. At the most general level, social theoretical geographers ask questions about the uneven distribution of power in society and of the inequalities defined by gender, race, ethnicity, class, and sexuality. The perspective relies on a mixed-methods approach—survey questionnaires, interviews, participant observation—commonly found in social sciences such as anthropology, sociology, and political economy.

Cultures are rarely, if ever, homogeneous. Often certain groups of people have more power in society, and their beliefs and ways of life dominate and are considered the norm, whereas other groups of people with less power may participate in alternative cultures. These divisions are often based on gender, economic class, racial categories, ethnicity, or sexual orientation. The social hierarchies that result are maintained, reinforced, and challenged through many means. Those means can include such things as physical violence, but often social hierarchies are maintained in ways far subtler.

For example, some geographers are concerned with the role of economic power in structuring public access to space. An illustration comes from the familiar situation of the wealthiest families owning property with the best views of and having the most direct access to water, whether rivers, lakes, or oceans. This is often the case with city parks as well. Parks and many water bodies are **public spaces**, geographical areas that are owned and controlled by a government agency and theoretically accessible to all citizens. Through private property laws, however, wealthy homeowners can monopolize access, a process geographers call the privatization of public space. If the shoreline is bordered by private property and rows of private homes, it is hard for the public to reach the public beaches, and therefore access is essentially privatized. California state legislators recognized this problem and derived a regulatory solution known as the

FIGURE 1.16 Protecting public space and access. In California, the state legislature mandated that public coastal lands be made accessible to the public through the establishment of easements between private properties. (Duncan Selby/Alamy)

California Coastal Act of 1976. The act mandates that public easements between properties be provided at regular intervals along the Pacific coastline. If you live or travel there, you will notice the "coastal access" signs along the roads and highways marking these easements (**Figure 1.16**).

Social theoretical geographers study ideology—a set of dominant ideas and beliefs—in relationship to place, environment, and landscape in order to understand how power works culturally. For instance, most nations maintain a set of powerful beliefs about their relationships to the land, some holding to the idea that a deep and natural connection exists between a particular territory and the people who have inhabited it. These ideas often form part of a national identity (discussed in Chapter 6) and are expressed so routinely in poems, music, laws, and rituals that people accept these ideas as truths. Many American patriotic songs, for example, express the idea that the country naturally spreads from "sea to sea." Yet immigrants to a given culture and country, and people who have been marginalized by that culture, may hold very different ideas of identity with the land. Native Americans have claims to land that long predate those of the U.S. government, and they would argue against an American national identity that includes a so-called natural connection to all the land between the Atlantic and the Pacific. Uncovering and analyzing the connections between ideology and power, then, are often integral to the geographer's task of understanding the diversity within a culture.

These different approaches to thinking critically about human geography are both necessary and healthy. Moreover, they are not mutually exclusive. For example, one could construct a model of differential access to health care or exposure to industrial pollution that is based on social theories of racial inequality. All lines of inquiry yield valuable findings. We present all these perspectives throughout *Contemporary Human Geography.*

The Visualization Tools of Human Geography

1.4 Apply key visualization tools used in human geography—including the longitude and latitude system and geospatial technologies—to everyday consumer devices.

The use of visualization techniques to analyze and explain geographical relationships is fundamental to the discipline. It is therefore important that we introduce a few basic geography concepts and tools early on, so that we can use them to better understand the material in the chapters to come. We will use the example of a popular navigation software, Google Maps, to introduce the latitude and longitude system and geospatial technologies like the satellite-based Global Positioning System, remote sensing, and geographic information systems.

Longitude and Latitude

Increasingly, the computerized visualization of geographical relationships is part of our everyday lives, from the map displays on our vehicle dashboards to the Uber apps on our smartphones. We use these tools to help us visualize and therefore understand where we are located relative to other locations, how to get from one location to another, and how long it will take us to move across the space separating locations. Let us use one popular digital mapping software used by the majority of smartphone owners worldwide, Google Maps, to illustrate key visualization concepts and practices in geography. Google Maps is a web-based mapping service using software originally developed by a geospatial data visualization company. That means the

GEOGRAPHY @ WORK

Courtesy of Matt Rosenberg

Matt Rosenberg

Geography Expert for About.com

Education: BA Geography, University of California, Davis

MA Geography, California State University, Northridge

Q. *Why did you major in geography and decide to pursue a career in the field?*

A. I "stumbled" into the field after taking a course on urban and economic geography my freshman year in college and loving it. I like that geography is a wide-ranging discipline that requires students to be well versed in a variety of topics; ultimately the study produces well-rounded individuals (which I think employers especially appreciate).

Q. *Please describe your job.*

A. I think of myself as a "geographical research library." I distill research, offer guidance, and answer questions related to all aspects of the geography field. I also serve as editor-in-chief for the About.com Geography site and write a regular blog featured on the site.

Q. *What types of people contact you via the web site?*

A. I get a lot of high school and college students from around the world that pose questions based on their studies. I also hear from plenty of "armchair geographers" who ask a variety of questions, such as "Which countries are not competing in the Olympic games?" "What is the newest nation in the world?" and "Why are there climate extremes in the world?"

Q. *What advice do you have for college students considering a career in geography?*

A. I think it is very important to do one (or more) internships while in college to help you figure out what area of geography you want to work within. When I was in college, I interned at a city planning department and a GIS department, which helped me decide what I wanted to do in the field.

company specialized in software that could take data that had some geographic or spatial quality (e.g., location on the Earth's surface) and use it to create digitized images (e.g., road maps on a computer screen). The Google corporation bought the software company and converted the program to a web-based system.

How exactly was Google Maps constructed and how does it work? Let us start with the most basic geospatial datum point, location on the Earth's surface. How do we find the **absolute location** or precise position of anything on the surface of the Earth? Geographers invented the latitude and longitude system to solve that problem. The system starts by visualizing a mathematical grid spread over the Earth's surface. Latitude lines are the horizontal lines of the grid. These encircle the globe running parallel to the Earth's equator, so they are also called parallels (**Figure 1.17**). In this system, locations are identified not by linear distance measures, but by angular measures, denoted in degrees, minutes, and seconds. **Latitude** lines are therefore measured as the degree of distance north or south from the equator, which is at 0 degrees, as far as the poles, which are at 90 degrees.

Longitude lines are the vertical lines on the Earth's surface that mark imaginary great circles connecting the poles. (Great circles divide a sphere—in this case the Earth—exactly in half.) The first problem that arises is that since all longitude lines converge at the poles, they are not parallel, and therefore the distance between them is not constant. In effect, they are continually arcing poleward from the equator across the Earth's surface. The second problem is, where do we mark 0 degrees longitude? There is no geometrically determined way to mark the starting point for longitude along the equator. At an international conference in 1884 the world's most powerful countries reached a political agreement that the zero-degree longitude line would run through Greenwich, England. This line is called the **prime meridian** or Greenwich Meridian, meridian being another term for a longitude line (**Figure 1.18**). Longitude lines are therefore measured as the degree of distance east or west of the Greenwich Meridian, which is 0 degrees, to the line exactly on the opposite side of the Earth, which is 180 degrees. Remember, latitude is measured north or south between 0 and 90 degrees and longitude is measured east or west between 0 and 180 degrees. One of the authors is at the moment of writing sitting at 25° 41′ 60″ north latitude and 80° 17′ 60″ west longitude. Can you locate the author on a map?

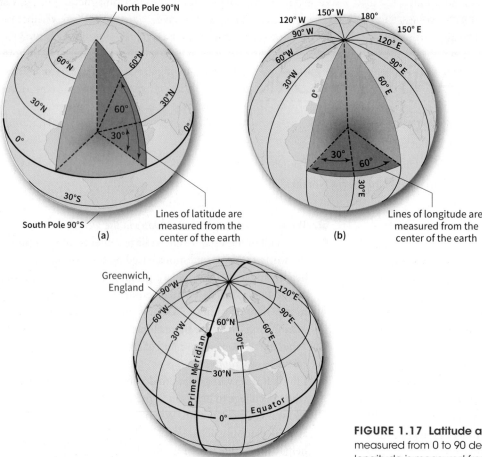

FIGURE 1.17 Latitude and longitude system. Latitude is measured from 0 to 90 degrees north or south of the equator and longitude is measured from 0 to 180 degrees east or west of the prime meridian.

FIGURE 1.18 A foot in two hemispheres. In 1884 the prime meridian, or 0 degrees longitude, was established by international agreement at the Royal Observatory in Greenwich, England. The site is now a museum and a popular tourist destination. (Oli Scarff/Getty Images)

FIGURE 1.19 Global positioning system satellites in orbit. The GPS satellite fleet needs to be continually updated with the newest communications technology. These represent the latest generation of satellites, which began launching in 2010. (United States Government)

Geospatial Technologies

All modern map-making, including Google Maps, is built on the foundation of the latitude and longitude system. Indeed, the grid system underlies all of the latest geospatial technologies, those modern tools we use for mapping and analyzing human and biophysical systems. Map apps, for instance, are not merely static representations, but function as interactive locational systems. How does that work? How does Google Maps know how to guide you to the nearest ice cream shop?

Map apps need a space-age technology called the **Global Positioning System (GPS)**. When Google Maps seeks permission "to use your current location," it is asking to link with your device's GPS receiver. The GPS used by most devices is a U.S. government–owned, space-based global navigation system controlled and maintained by the U.S. Air Force. The GPS consists of a constellation of 24 or more satellites that orbit the Earth twice daily and transmit radio signals earthward (**Figure 1.19**). If the signals of at least four orbiting satellites reach a receiver, it can be located precisely in time and three-dimensional space. It is this geospatial technology that allows our interactive engagement with the electronic devices that we increasingly rely on tell us where we are—that is, our absolute location—and how to get to where we want to go—that is, our relative location. **Relative location** simply means the position of one place (or person!) in relation to the position of another place. It can be measured in time or distance, which is one reason many map apps ask you which measure and mode of transportation you prefer.

If you are a Google Maps user who prefers a "bird's-eye" view of locations as they actually appear from above, then you are employing another common visualization technique in

geography's tool kit known as remote sensing. **Remote sensing** refers to a method of collecting information about the Earth's surface using devices located high above. The devices may record in the visible portion of the electromagnetic spectrum or they may use radar or radio waves. The "satellite" view function of Google Maps employs two basic types of remotely sensed imagery. For major cities, it uses aerial photographs taken from directly overhead by fixed-wing aircraft flying at 790 to 1510 feet (240 to 460 meters) altitude. **Aerial photography** produces fine-grained, high-resolution images that show the smallest detail. Recently, drone aircraft have begun to replace fixed-wing aircraft for this operation (**Figure 1.20**). For most other areas,

FIGURE 1.20 Aerial photography from remote-piloted drones. At one time, aerial photography required piloted aircraft with highly skilled navigators. Increasingly, drones are taking over the tasks, guided by GPS satellites. (Anton Gvozdikov/Alamy)

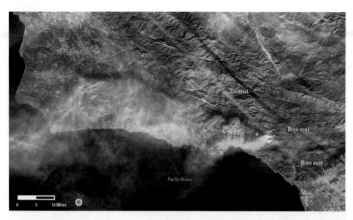

FIGURE 1.21 Satellite imagery. Remotely sensed images from low-orbiting satellites have become indispensable in tracking environmental change. This photo documents the extent of deforestation in Cambodia. (USGS)

Google Maps uses satellite imagery. **Satellite imagery** refers to information about the Earth's surface gathered from sensors mounted on orbiting satellites that record in both the visible and non-visible portions of the electromagnetic spectrum (**Figure 1.21**).

Collating all of the data obtained by the techniques discussed above required Google Maps to use another technique common to geographers, a **geographic information system (GIS)**. GIS is a means of collecting, storing, retrieving, and visually presenting geospatial data. Although various forms of GIS predated the arrival of computers, it has come to refer specifically to a software application that allows for the rapid manipulation of geospatial data for problem solving and research.

Creating user-friendly interactive maps on a hand-held electronic device is only one of thousands of possible applications of a geographer's visualization tools. They are used routinely in civilian, commercial, and military navigation, natural resource management, urban planning, and many uses yet to be invented. These visualization tools are the twenty-first-century answer to some age-old questions: Where am I? Where can I find what I am seeking? How do I get there? There is more information on how to read and interpret the maps used in *Contemporary Human Geography* in the Appendix. We recommend you become familiar with map reading early on and refer to the Appendix periodically so that you can be sure you are interpreting the maps correctly.

One final note on visualization tools that reflects this book's cultural approach. As we stated at the beginning, finding one's way and wondering about our place is a defining human characteristic. We should not be surprised, then, to learn that technologies of geographic visualization have been invented many times over in many different places by many different peoples. The cultures of Africa, Asia, the Americas, and the Pacific islands all have unique traditions of visualizing geographic relationships. Just as with foreign languages and

scripts, such geographic visualizations require translation. Throughout the textbook we hope to inspire an appreciation of the multiple cultural geographies produced by the human imagination.

Organizing Themes for *Contemporary Human Geography*

We have organized this book around five geographical themes: region, mobility, globalization, nature–culture, and cultural landscape. We use these themes both because they help to organize the diverse issues of the discipline and because they represent major concepts important to human geographers. Each theme stresses one particular aspect of the discipline and, while we have separated them for purposes of clarity, it is important to remember that the themes are interrelated. When discussing the theme of mobility, for example, we will inevitably bring up issues related to globalization, and vice versa. These themes give a common structure to every chapter and are stressed throughout the book.

Region

1.5 Compare and contrast formal, functional, and vernacular regions.

We all recognize that portions of the Earth's surface, including those modified by humans, take on identifiable characteristics that make them different from one another. We refer to these as regions. Most broadly, geographers think of regions as a geographic grouping of similar places or of places with similar characteristics. Regions are at the core of much geographic analysis. How and why are places alike or different? How do they interact together into functioning networks? How do their inhabitants perceive them and identify with them? These are central geographical questions. A **region**, then, is a geographical unit based on a set of common characteristics or functions. There is no single way to objectively identify regions, as their boundaries will be determined by the characteristics or functions under study. Geographers therefore recognize three types of regions: formal, functional, and vernacular.

Formal Regions

A **formal region** is a geographical area inhabited by people who have one or more traits in common, such as language, religion, or a system of livelihood. It is an area, therefore, that is relatively homogeneous with regard to one or more cultural traits. Geographers use the idea of formal regions to map spatial differences throughout the world. For example, an Arabic-language formal region can be drawn on a map of languages and would include the areas where Arabic is the native language of the majority of the population. Similarly, a wheat-farming

formal region would include the parts of the world where wheat is a predominant crop (look again at Figure 1.2).

The examples of Arabic language and of wheat cultivation represent the concept of formal region at its simplest level. Each is based on a single cultural trait: language and crop type. More commonly, formal regions depend on multiple related traits (**Figure 1.22**). Thus, an Inuit (Eskimo) culture region might be based on language, religion, economy, social organization, and types of dwellings. The region would reflect the spatial distribution of these five Inuit cultural traits. Areas in which all five of these traits are present would be part of the culture region.

Formal regions are the geographer's somewhat subjective creations. No two cultural traits have the same distribution, and the territorial extent of a culture region depends on what and how many defining traits are used. Why *five* Inuit traits, not four or six? Why not *foods* instead of (or in addition to) dwelling types? Consider, for example, Greeks and Turks, who differ in language and religion. Formal regions defined on the basis of speech and religious faith would separate these two groups. However, Greeks and Turks hold many other cultural traits in common. Both groups are monotheistic, worshipping a single god. Both groups enjoy certain traditional foods, such as shish kebab. Whether or not Greeks and Turks are placed in the same formal region or in different ones depends entirely on how the

FIGURE 1.22 An Inuit hunter with his dogsled team. Various facets of a multitrait formal region can be seen here, including the clothing, the use of a dogsled as transportation, and hunting as a livelihood system. (Louise Murray/Alamy)

geographer chooses to define the region. That choice, in turn, depends on the specific purpose of research or teaching that the region is designed to serve. Thus, an infinite number of formal regions can be created.

The geographer who identifies a formal region must locate borders. Because cultures overlap and mix, such boundaries are rarely sharp, even if only a single cultural trait is mapped. For this reason, we find border zones where different regions meet and sometimes overlap, rather than lines. These zones broaden with each additional trait considered because no two traits have the same spatial distribution. The U.S.–Mexican border is a good example of a border zone, where foods, languages, and other cultural markers overlap and blend into a recognizable border culture. *Tex-Mex* is a term frequently used to describe food and music in the Texas–Mexico border zone.

Functional Regions

The hallmark of a formal region is cultural homogeneity. By contrast, a **functional region** need not be culturally homogeneous; instead, it is a geographic area that has been organized to function politically, socially, culturally, or economically as one unit. A city, a precinct, a church diocese or parish, a trade area, and a Federal Reserve Bank district are all examples of functional regions. Functional regions have **nodes**, or central points where the functions are coordinated and directed. Examples of such nodes are city halls, national capitals, precinct voting places, parish churches, factories, and banks.

FIGURE 1.23 Aerial view of Denver. This image clearly illustrates the node of a functional region—here, the dense cluster of commercial buildings—that coordinates activities throughout the area surrounding it. (Jim Wark/Airphoto)

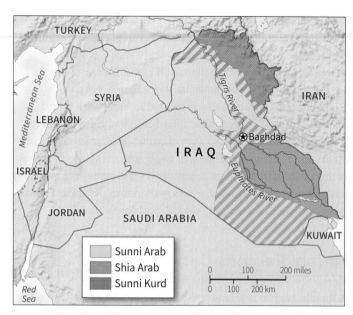

FIGURE 1.24 The three Iraqs. European empires established Iraq's modern state boundaries as a functional region for the purposes of colonial administration. The boundaries, however, encompass three different formal regions, which are sometimes in violent conflict with one another and which often pursue interests different from those of the functional region.

Some functional regions have clearly defined borders. A metropolitan area is a functional region that includes all the land under the jurisdiction of a particular urban government (**Figure 1.23**). The borders of this functional region may not be so apparent from a car window, but they will be clearly delineated on a regional map by a line distinguishing one jurisdiction from another. Similarly, each state in the United States and each Canadian province is a functional region, coordinated and directed from a capital, with government control extended over a fixed area with clearly defined borders.

Not all functional regions have fixed, precise borders, however. A good example is a daily newspaper's circulation area. The node for the paper would be the plant where it is produced. Every morning, trucks move out of the plant to distribute the paper throughout the city and surrounding suburbs and countryside. The sales territory of one newspaper may thus overlap with the sales territories of other city newspapers, making boundaries difficult to define. Take, for example, the *Kansas City Star,* which is distributed in 33 counties in Kansas and Missouri. To the east, the paper's distribution overlaps with that of the *St. Louis Post-Dispatch,* which has a much larger distribution area including Kansas City itself.

Functional regions often do not coincide spatially with formal regions, and this disjuncture may create problems for the effective performance of some functional regions. Iraq provides an example (**Figure 1.24**). As an independent state, Iraq forms a functional region, with political power and a transportation and energy infrastructure centered on the capital city, Baghdad. The boundaries of the modern state largely reflect the interests of nineteenth-century European colonial powers rather than the common cultural traits of its modern-day inhabitants. Modern Iraq is culturally divided into three major formal regions characterized by differences in language, religion, and ethnic identity. One formal region is dominated by people of Arab descent adhering to Sunni Islam, a second of Arab descent adhering to Shia Islam, and a third of Kurdish descent adhering to Sunni Islam. These differences make the functioning of the Iraqi state more difficult and help explain why political disunity continues to plague the government and its citizens.

Vernacular Regions

A **vernacular region** is one that is *perceived* to exist by its inhabitants, as evidenced by the widespread acceptance and use of a special regional name. We might think of vernacular regions as the most democratic type of all. Vernacular regions are named and their boundaries imagined through popular consensus. (Indeed, the term "popular region" is sometimes used in a similar way.) Like old gospel or folk songs passed down over the years from one musician to another, it is often difficult to trace the precise origins of a vernacular regional designation. Typically, the name of a vernacular region first becomes widely known through a local newspaper column, popular song, blog site, or some other form of popular mass communication. The vernacular name then "sticks." That is, it becomes widely adopted in everyday speech, if not in official documents.

Some vernacular regions are based on physical environmental features, but have local cultural meanings. For example, there are many regions called simply "the valley" in conversation among local residents. Wikipedia lists more than 30 different regions in the United States and Canada referred to as such, places as varied as the Sudbury Basin in Ontario and the Lehigh Valley in Pennsylvania. "The valley" in Southern California is synonymous with the San Fernando Valley, which is associated with a type of landscape (suburban), type of person (a white, teenage girl, called the "valley girl"), and way of speaking ("valspeak").

Other vernacular regions are so named because of economic, political, or historical characteristics. "Redneck Riviera" is a popular designation for the Gulf of Mexico coastal area of the Florida panhandle, running from Panama City in the east to Pensacola and beyond in the west. "Redneck" is a term referring to politically conservative, working-class, rural culture in the United States. "Riviera" refers, ironically, to the affluent resort area of the French Mediterranean coast (**Figure 1.25**). Country-and-western singer Tom T. Hall released

FIGURE 1.25 The Redneck Riviera. The coastline of Florida's panhandle region is popularly known in Florida and beyond as the Redneck Riviera, though local city governments in the region prefer to call it the Emerald Coast. (nik wheeler/Corbis via Getty Images)

a song entitled "Redneck Riviera" that included the lyrics: "They got beaches of the whitest sand / Nobody cares if gramma's got a tattoo." Unhappy with the negative connotations of the term "Redneck," the region's local governments and chambers of commerce created an alternative designation for it, "Emerald Coast." They have also worked hard to eliminate all references to the Redneck Riviera. In Gulf Shores, Alabama, the city government even passed a prohibitive ordinance when a television production company scouted the town as a location for a proposed reality TV show to be called *Redneck Riviera*. City ordinances aside, we would be willing to wager a week's salary that far more Floridians and Alabamans can identify the vernacular Redneck Rivera than can identify the officially sanctioned Emerald Coast.

Vernacular regions, like most regions, generally lack distinct borders, and the inhabitants of any given area may claim residence in more than one such region. Vernacular regions often lack the organization necessary for functional regions, although they may be centered on a single urban node, and they frequently do not display the cultural homogeneity that characterizes formal regions. They vary in size from city neighborhoods to sizable parts of continents. At a basic level, a vernacular region grows out of people's sense of belonging to and identification with a particular area. By contrast, many formal or functional regions lack this attribute and, as a result, are often far less meaningful for people.

Mobility

1.6 Explain the different types of diffusion and migration.

The concept of regions helps us see that similar or related sets of elements are often grouped together in space. Equally important in geography is understanding how and why these different cultural elements move through space and locate in particular settings. Regions themselves, as we have seen, are not stable but are constantly changing as people, ideas, practices, and technologies move around in space. Are there some patterns to these movements? These questions define our second theme, **mobility**, and two of its aspects, diffusion and migration.

Diffusion

One important way to study mobility is through the concept of diffusion. **Diffusion** can be defined as the pattern by which a phenomenon—people, ideas, technologies, or preferences—spreads from a particular location through space and time. Through the study of diffusion, the human geographer can trace how spatial patterns in culture emerged and evolved. After all, any culture is the product of almost countless innovations that

spread from their points of origin to cover a wider area. Some innovations or inventions occur only once, and geographers can sometimes trace them back to a single place of origin. In other cases, **independent invention** occurs: the same or a very similar innovation is separately developed at different places by different people working independently.

Using spatial models developed by geographer Torsten Hägerstrand, we can recognize several different kinds of diffusion (**Figure 1.26**). **Relocation diffusion** occurs when individuals or groups with a particular idea or practice migrate from one location to another, thereby bringing the idea or practice to their new homeland. Religions frequently spread this way. An example is the migration of Christianity with European settlers who came to America. In **expansion diffusion**, ideas or practices spread throughout a population, from area to area, in a snowballing process, so that the total number of knowers or users and the areas of occurrence increase.

Expansion diffusion can be further divided into three subtypes. In **hierarchical diffusion**, ideas leapfrog from one important person to another or from one urban center to another, temporarily bypassing other persons or rural territories. We can see hierarchical diffusion at work in everyday life by observing the acceptance of new modes of dress or foods. For example, sushi originally diffused to the United States from Japan in the 1970s, but very slowly because few people ate raw

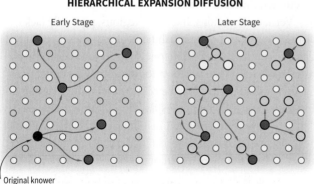

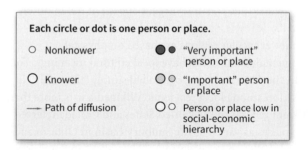

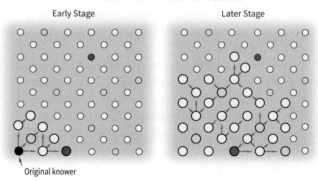

FIGURE 1.26 Types of cultural diffusion. These diagrams are merely suggestive; in reality, spatial diffusion is far more complex. In hierarchical diffusion, different scales can be used, so that, for example, the category "very important person" could be replaced by "large city."

fish. In the United States, the first sushi restaurants appeared in the global cities of Los Angeles and New York. Only gradually throughout the 1980s and 1990s did eating sushi become more common in less cosmopolitan parts of the country. By contrast, **contagious diffusion** involves the wavelike spread of ideas in the manner of a contagious disease, moving throughout space without regard to hierarchies. In agricultural regions, the use of a new crop or seed variety may spread in a pattern of contagious diffusion.

Hierarchical and contagious diffusion often work together. The 2009 influenza pandemic is a sobering example of how these two types of diffusion can reinforce each other. A pandemic is a worldwide disease epidemic. In March 2009, Mexico announced that a new influenza virus strain, known as H1N1, had reached epidemic proportions in Mexico City. Many countries cut off airline traffic to Mexico in an attempt to inhibit the spread of H1N1. However, the virus had already been carried out of the country by air travelers. Through hierarchical diffusion it began to appear in major cities worldwide, and from there it spread by contagious diffusion to the countryside. China offers a clear example of this process. Only two months after the disease was detected in Mexico, Chinese authorities detected the virus in their cities with international airports. The virus remained confined for the next four to six months, but thereafter H1N1 experienced its greatest geographic expansion as it spread across all of China (**Figure 1.27**).

Sometimes a specific trait is rejected but the underlying idea is accepted, resulting in **stimulus diffusion**. For example, early Siberian peoples domesticated reindeer only after exposure to the domesticated cattle raised by cultures to their south. The Siberians had no use for cattle, but the idea of domesticated herds appealed to them, and they began domesticating reindeer, an animal they had long hunted. Stimulus diffusion therefore typically involves some combination of cultural imitation and innovation to produce a new form or variation.

If you throw a rock into a pond and watch the spreading ripples, you can see them weaken as they move away from the point of impact. In the same way, diffusion weakens as a cultural innovation moves away from its point of origin. That is, diffusion decreases with distance. An innovation will usually be accepted most thoroughly in the areas closest to where it originates. Because innovations also take time to diffuse, time is also a factor in the spread of an innovation. Acceptance of a cultural innovation generally decreases with distance and time, producing what is called **time–distance decay**. Geographers say that innovations and other social phenomena are subject to the **friction of distance**, the inhibiting effect of distance on the intensity and volume of most forms of human interaction. Modern mass media and new communication and transport technologies, however, have greatly accelerated diffusion, overcoming the friction of distance and diminishing the impact of time–distance decay.

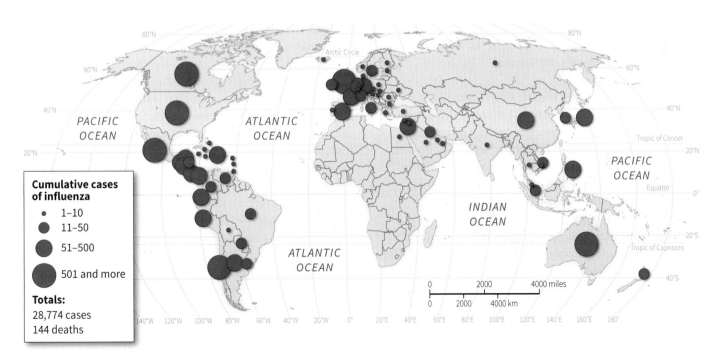

Cumulative cases of influenza

- 1–10
- 11–50
- 51–500
- 501 and more

Totals:
28,774 cases
144 deaths

FIGURE 1.27 Diffusion of influenza infections during the H1N1 pandemic of 2009. The influenza outbreak started in Mexico City. Global air travel facilitated hierarchical diffusion to international airports around the world. From those points, the virus spread within countries through contagious diffusion.

In addition to the gradual weakening or decay of an innovation through time and distance, barriers can retard its spread. **Absorbing barriers** completely halt diffusion, allowing no further progress. For example, many of us have become accustomed to using Google's "Street View" mapping function, which provides panoramic and zoom-in views from positions along many streets in the world. This technology allows users to plan everything from dining out to purchasing real estate. Street View, however, has raised privacy questions worldwide. Austria and Greece have chosen to prohibit the use of Street View technology within their borders because of privacy concerns. Among other effects on these countries, the ban will likely slow or halt the innovative uses of Street View now being explored in the visual arts.

Few absorbing barriers are effective or persist for long. More commonly, barriers are **permeable barriers**, allowing innovations to diffuse but oftentimes partially and in a weakened way. When a school board objects to students with tattoos or body piercings, they may set limits by mandating that these markings be covered by clothing. However, over time, those mandates may change as people get used to the idea of body markings. In this way, the school board acts as a permeable barrier to cultural innovations.

Although all places and communities hypothetically have equal potential to adopt a new idea or practice, diffusion typically produces a core–periphery spatial arrangement. Hägerstrand offered an explanation of how diffusion produces such a regional configuration. The distribution of innovations can be random, but the overlap of new ideas and cultural traits as they diffuse through space and time is greatest toward the center of the region and least at the peripheries. As a result of this overlap, more innovations are adopted in the center, or core, of the region.

Other geographers, most notably James Blaut and Richard Ormrod, regarded the Hägerstrandian concept of diffusion as too narrow and mechanical because it does not give enough emphasis to cultural and environmental variables and because it assumes that information automatically produces diffusion. They argue that nondiffusion—the failure of innovations to spread—is more prevalent than diffusion, a condition Hägerstrand's system cannot accommodate. Similarly, the Hägerstrandian system assumes that innovations are equally beneficial to all people throughout geographical space. In reality, susceptibility to an innovation is far more crucial, especially in a world where communication is so rapid and pervasive that it renders the friction of distance almost meaningless. The inhabitants of two regions will not respond identically to an innovation, and we seek to understand this spatial variation in receptiveness in order to explain diffusion or the failure to diffuse. Within the context of their culture, people must perceive some advantage before they will adopt an innovation. As with most human-geographic models, diffusion helps us recognize persistent or recurring spatial patterns, but does not offer universally valid explanations.

Increasingly, we see many other examples of mobility, such as the almost-instant global communication about new ideas and social trends through social media and the web, the rapid movement of goods from the place of production to that of consumption, and the seemingly nonstop movement of money worldwide through digital financial networks. These types of movements through space do not necessarily follow the pattern of core–periphery but instead create new and different types of patterns. The term **circulation** might better fit many of these forms of mobility because this term implies an ongoing set of multidirectional movements of people, goods, and ideas with no particular center or periphery.

Migration

The mass movement of people between different regions is a type of mobility best defined as **migration** from one region or country to another through particular routes. The spatial characteristics of human migrations have long been of interest to geographers, some of whom have attempted to establish quantitative models. Hägerstrand, for example, used regression analysis to quantify the effects of distance decay on migration patterns. More recently, interest has turned to the social and economic causes and consequences of migration—for example, how migrations are linked to changing conditions in labor markets, showing a general pattern of migration from areas of low employment or low wages to areas of higher employment or higher wages (see The Video Connection). Geographers are also increasingly interested in probing the subjective experience of migrants, in both places of origin and places of destination. Geographer Jeremy Foster, for example, studied the experiences of European migrants in colonial South Africa to reveal how their painting, photography, poetry, and other forms of artistic expression helped transform a colonized foreign land into a homeland. Foster was interested in how migrant settlers come to feel "at home" (or not) in a foreign land through engagement with certain kinds of cultural practices.

Migration sometimes involves crossing boundaries of some sort. Migration that crosses country borders is considered **international migration**. All residents of the United States, with the exception of Native Americans, can trace their presence in the country to international migration, often multiple migrations from different points of origin, via different routes, and at different times (**Figure 1.28**). One's "American" ancestry might easily involve early eighteenth-century migration from western Europe, nineteenth-century migration from North Africa, and twentieth-century migration from East Asia.

▶ The Video Connection

West Virginia, Still Home
◉ Watch at 🍁 SaplingPlus

This video focuses on McDowell County, West Virginia, describing the recent decline of the region's coal industry and the resulting outmigration of most of its population. Despite economic hardship and social upheaval, many remaining residents cannot imagine living elsewhere and remain deeply attached to the landscape. One resident tells the camera, "We are very connected to these mountains," a sentiment that reveals a strong sense of place, belonging, and regional identity. As we have learned in this chapter, the relationship between people and the places and landscapes in which they live is a main topic of study for human geographers. Quite a few other basic ideas in human geography are expressed by the video as well.

Thinking Geographically

1. The video shows how social and economic forces operate across geographic scales—the global slump in the demand for coal and mechanization in the coal industry nationwide have led to the decline of once prosperous and well-populated towns in a southern corner of West Virginia. Can you think of similar examples in the area where you live or with which you are familiar?

2. One of the residents in the video spoke of a "brain drain" and another of her hopes that some of those people will "have the desire to come back to this community." How do these and other statements about the town's population relate to the discussion of migration in this textbook? How many types of migration are illustrated in the video?

3. Recall the various ways that the concept of region is discussed in the chapter. How many of those concepts are illustrated by the case of McDowell County?

4. How does this video illustrate the concepts of place and sense of place? Think about the personal histories of some of the residents interviewed and their onscreen statements about the town and surrounding countryside. What sorts of emotions and sentiments do people express when discussing their home?

FIGURE 1.28 Becoming American. New citizens take the Oath of Allegiance to the United States' laws and Constitution. (CJ GUNTHER/EPA-EFE/REX/Shutterstock)

This example illustrates some of the primary characteristics geographers use to categorize migration, including spatial scale (regional/global), time (temporary/permanent), and distance (long/short).

Migration that occurs *within* the borders of a country is classified as **internal migration**. A classic example of internal migration in the United States is the **Great Migration**, the twentieth-century movement of 6 million African Americans from the rural southern states to the cities of the midwestern and northeastern states (**Figure 1.29**). Such mass migrations have deep, long-lasting impacts on social, economic, and cultural conditions in both the places of origin and of destination. We can view the Great Migration as a significant, but relatively small, part of a global-scale pattern of mass migration from rural to urban areas sparked by the Industrial Revolution. In 2007 the United Nations determined that for the first time in human history, the majority of the world's population lived in cities. This historic urbanization is primarily the result of rural-to-urban migration now experienced by nearly every country in the world.

Whether international or internal, geographers categorize types of migration as stepwise, return, seasonal, or forced. **Stepwise migration** is migration conducted in a series of stages. Migration might begin with a move from the countryside to a provincial capital. From there the migration continues to a country's capital city and then might proceed across international

FIGURE 1.29 The Great Migration. African Americans in the early twentieth century filled the railway stations of the U.S. South, waiting for trains to take them to new lives in the northern cities. (Courtesy of the State Archives of Florida.)

FIGURE 1.30 Demonstrations in Munich, Germany, against the deportation of immigrants. Many migrant-worker groups are fighting to maintain the rights of transnational migrants who would like to be able to move freely between their home countries and those that currently provide work for them. (Sachelle Babbar via ZUMA Wire)

borders to another country. Such stepwise migrations can take place over the course of one or more generations. **Return migration** refers to migrants going back, or returning, to their place of origin after long-term residency elsewhere. For example, it is not uncommon for migrants to return to their home countries upon reaching retirement age, even after spending most of their adult lives working abroad. Finally, **seasonal migration** refers to cycles of movement based on the time of year. Agricultural workers are often seasonal migrants because labor demand is seasonally variable and highly concentrated during the harvest periods of particular crops in particular places. So-called "snowbirds," people from the northern United States who move to their second homes in Florida during the winter months, are another type of seasonal migrant.

Another form of migration that occurs both internally and internationally is **forced migration**, which occurs when people are involuntarily compelled by disasters, social conflicts, or development projects to leave their homes. Disasters include both extreme natural events such as earthquakes, floods, and droughts, and human-caused environmental disasters such as chemical spills and radioactive contamination. Social conflicts include war; political repression based on ethnicity, race, religion, or other identity markers; and generalized violence. Development projects are initiatives, such as dams, urban clearance, and deforestation, that require population displacements. Forced migration is, unfortunately, on the increase worldwide. The United Nations High Commission for

Refugees announced that an all-time record high 65.6 million people were displaced from their homes at the end of 2016.

In today's globalizing world, with better and faster communication and transportation technologies, many migrants more easily maintain ties to their homelands even after they have migrated, and some may move back and forth between their home countries and those to which they have migrated. Scholars refer to this kind of migration as **transnational migration** (**Figure 1.30**). We will discuss these and other patterns of mobility in greater detail throughout the rest of the book.

Globalization

1.7 Explain the concept of globalization and how it shapes interdependence among regions.

The modern technological age allows an almost instantaneous diffusion of ideas and innovations, accelerating a phenomenon called **globalization**. Geographer Matthew Sparke provides a succinct definition of this very complex phenomenon. He defines it as "processes of economic, political, and social integration that have collectively created ties that make a difference to lives around the planet." Simply put, actions and conditions in one place are increasingly linked to actions and conditions in other places around the globe.

This interconnected and interdependent world has been created from a set of factors, primarily the accessibility and speed of travel and the Internet. Yet before our current period of rapid globalization, different countries and different parts of the world were linked. For example, in early medieval times, overland trade routes connected China with other parts of Asia. The British East India Company maintained maritime trading routes between England and large portions of South Asia as early as the seventeenth century, thereby creating interdependencies between the regions that persisted for centuries.

Interdependence refers to the reciprocal ties established between regions and countries that over time collectively create a global economic system. Interdependence does not mean equality. It simply means that there is a two-way flow of goods, money, people, and ideas that creates dependencies at both ends of the connection, but not of equal consequence. Spain, Portugal, and England all established interdependent global empires rooted in Europe's Age of Exploration and Discovery. Some geographers refer to such moments as early global encounters and suggest that they set the stage for contemporary globalization. That is, they trace today's interdependencies and inequalities under globalization to those under yesterday's colonial empires (see Chapter 9).

Some scholars believe that the historical roots of interdependency in European empires have been the primary driver of globalization. Other scholars believe that current communication technology, as well as multinational corporations, play a greater role in globalization, sometimes exceeding the effects of the wealth and power of national governments. Collectively, these trends have significantly diminished the relevance of national borders to the global circulation of people, goods, and ideas. According to geographers R. J. Johnston, Peter Taylor, and Michael Watts, with globalization, activities and outcomes "do not merely cross borders, these processes operate as if borders were not there." Most importantly from a human-geography perspective, globalization means that an ever-greater proportion of culture is organized on a global scale.

These increasingly linked and interdependent economic, political, and cultural networks around the world might lead many to believe that different groups of people around the globe are becoming more and more alike. In some ways this is true, but what these new global encounters have enabled is an increasing recognition of the differences between groups of people. And some of those differences have been caused by globalization itself. Some groups of people have access to advanced technologies, better health care, and education, whereas others do not. Even within our own neighborhoods and cities, we know that there are people who are less able to afford these things. We will discuss uneven socioeconomic development in depth in Chapter 9.

Globalization both helps make us more aware of uneven socioeconomic development and contributes to it. How does this happen? Globalization may be thought of as both a set of processes that are economic, political, and cultural in nature, and as the effects of those processes. For example, economic globalization refers to the interlinked networks of money, production, transportation, labor, and consumption that allow, say, the parts of an automobile to be manufactured in several countries, assembled in yet another, and then sold throughout large portions of the world. These economic networks and processes, in turn, have significant and often uneven effects on the economies of different countries and regions. Some regions gain employment, whereas others lose jobs; some consumers are able to afford these cars because they are less expensive, whereas other potential consumers who have become unemployed cannot afford to purchase them. These global economic processes and effects are, in turn, linked to politics. For example, those countries chosen by the automobile manufacturer as sites of production might see their standards of living improve, leading to larger consumer markets, better communications and media, and often changing political sensibilities. And these changing political ideas, in turn, will shape economic decisions and other factors. Globalization, therefore, involves looking at complex interconnections between a set of related processes and their effects.

Culture, of course, is a key variable in these interactions and interconnections. If we consider culture as a way of life, then globalization is a key shaper of culture and is shaped by it. We have seen the power of globalization occurring through Internet-based cultural media and communication. Some scholars have suggested that globalizing processes and an increase in mobility will work to homogenize different peoples, breaking down culture regions and eventually producing a single global culture. Other scholars see a different picture, one where new forms of media and communication will allow local cultures to maintain their distinct identities, reinforcing the diversity of cultures around the world. Throughout *Contemporary Human Geography,* we will return to these issues, considering the complex role of culture and cultures in an increasingly global world.

Nature–Culture

1.8 Explain the historical development of different approaches to nature–culture relations in human geography.

The themes of region, mobility, and globalization help us understand patterns, movements, and interconnections that characterize the ways in which people create spatial patterns on the Earth. Our fourth theme, **nature–culture**, adds a different dimension to this analysis; it focuses our attention on how people inhabit the Earth, and their relationships to the physical environment and nonhuman species. This theme helps us investigate how groups of people interact with the Earth's biophysical environment and examine how the culture, politics, and economies of those groups affect their ecological situation and resource use. The relationship between people and nature is viewed as a two-way interaction. People's values, beliefs, perceptions, and practices have ecological impacts, and ecological conditions, in turn, influence perceptions and practices. In labeling this theme "nature–culture" rather than "nature and culture," we acknowledge that "nature" and "culture" are defined relationally—that is, they are defined principally in relation to one another. Throughout the chapters we will therefore directly confront the modern paradox of humans as both *a part of* nature and *apart from* nature. That said, the interaction between nature and culture is sometimes blurry.

The term **ecology** was coined in the nineteenth century to refer to a new biological science concerned with studying the complex relationships among living organisms and their physical environments. Later, people also used the term when referring to the biophysical conditions of a particular region or place, as in "the ecology of Lake Erie." Geographers borrowed this term in the mid-twentieth century and joined it with the term culture to identify a field of study—**cultural ecology**—that deals with the interaction between culture and local ecologies. Later, another concept, the *ecosystem,* was introduced to describe a territorially bounded system consisting of interacting organic and inorganic components. Plant and animal species were said to be adapted to specific conditions in the ecosystem and to function so as to keep the system stable over time.

It soon became clear, however, that human cultural interactions with the environment were far too complex to be analyzed with concepts borrowed from biology. Furthermore, the idea that people interact with their ecosystems in isolation from larger-scale political, economic, and social forces is difficult to defend. We can readily observe, for example, that even in the most remote rural areas of the world goods are imported, agricultural commodities are grown and exported to a global market, money circulates, taxes are collected, and people migrate in and out for wage labor. So the term nature–culture has evolved to refer to the complex interactions among all these variables and to reflect the fact that studies of local human–environment relations need to include political, economic, and social forces operating on both national and global scales.

Through the years, human geographers have developed various perspectives on the interaction between humans and the land. Four schools of thought have developed: environmental determinism, possibilism, environmental perception, and humans as modifiers of the Earth.

Environmental Determinism

During the late nineteenth and early twentieth centuries, many geographers accepted **environmental determinism**: the belief that the physical environment is the dominant force shaping cultures and that humankind is essentially a passive product of its physical surroundings. Humans are clay to be molded by nature. Similar physical environments produce similar cultures.

For example, environmental determinists believed that peoples of the mountains were predestined by the rugged terrain to be simple, backward, conservative, unimaginative, and freedom loving. Desert dwellers were likely to believe in one god but to live under the rule of tyrants. Temperate climates produced inventiveness, industriousness, and democracy. Coastlands pitted with fjords produced great navigators and fishers. Environmental determinism had serious consequences, particularly during the time of European colonization in the late nineteenth century. For example, many Europeans saw Latin American native inhabitants as lazy because living in a tropical climate supposedly ensured that people didn't have to work very hard for their food. Europeans were able to rationalize their colonization of large portions of the world in part along climatic lines. Because the natives were "naturally" lazy and slow, the European reasoning went, they would benefit from the presence of the "naturally" more industrious Europeans who came from more temperate lands.

Determinists overemphasize the role of environment in human affairs, and their conclusions were often either very positive or very negative overgeneralizations of entire cultures and races. This does not mean that environmental influences are inconsequential or that geographers should not study such influences. Rather, the physical environment is only one of many forces affecting human culture and is never the sole determinant of behavior and beliefs.

Possibilism

After the 1920s, **possibilism**, the view that any physical environment offers a number of possible ways for a culture to develop, became prominent among geographers. Possibilists claim that while the local environment helps shape its resident culture, the way of life ultimately depends on the choices people make among the possibilities offered by the environment. These choices are guided by cultural heritage and are shaped by a particular political and economic system. Possibilists see the physical environment as offering opportunities and limitations; people make choices among these to satisfy their needs. **Figure 1.31** provides an interesting example: the cities of San Francisco and Chongqing were both built on similar physical terrains that dictated an overall form, but differing cultures lead to very different street patterns, architecture, and land use. In short, local

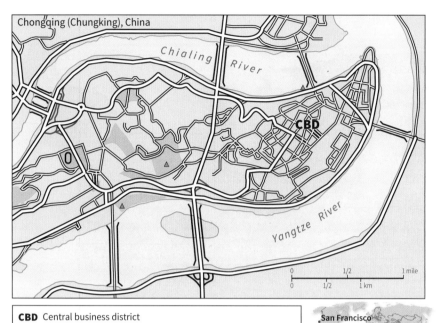

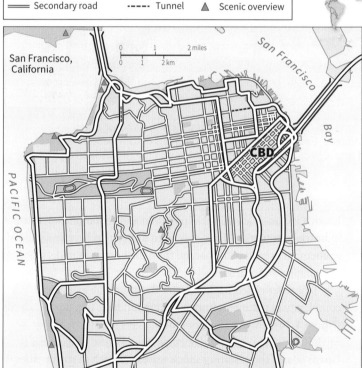

FIGURE 1.31 Chongqing (Chungking) and San Francisco. Both cities developed on elongated, hilly sites flanked on all but one side by water, and both were connected in the twentieth century by bridges leading to adjacent land across the water. In certain other respects, too—such as the use of tunnels for arterial roads—the cities are similar. Note, however, the contrast in street patterns. In Chongqing, the streets were laid out to accommodate the rugged terrain, but in San Francisco, relatively little deviation from a gridiron pattern was permitted. Note, too, that although San Francisco has a much smaller population than Chongqing, it covers a far larger area.

traits of culture and economy are the products of culturally based decisions made within the limits of possibilities offered by the environment.

Most possibilists believe that the more technologically advanced a culture is, the greater the number of possibilities and the weaker the influence of the physical environment. Technologically advanced cultures, in this view, have achieved some mastery over their physical surroundings. Geographers Jim Norwine and Thomas Anderson, however, warn that even in these advanced societies, "the quantity and quality of human life are still strongly influenced by the natural environment," especially climate. They argue that humankind's control of nature is anything but supreme, and is perhaps even illusory. One only has to think of the devastation caused by the September 2017 Hurricane Maria in Puerto Rico or the March 2011 earthquake and tsunami in Japan to underscore the often-illusory character of the control humans think they have over their physical surroundings.

Environmental Perception

Another approach to the theme of nature–culture is **environmental perception**, which describes how humans *perceive* nature. Each person and cultural group has mental images of the physical environment, shaped by knowledge, experience, values, and emotions. Whereas the possibilist sees humankind as having a choice of different possibilities in a given physical setting, the environmental perceptionist declares that the choices people make will depend more on what they perceive the environment to be than on the actual character of the environment. Perception, in turn, is colored by cultural contexts (**Figure 1.32**).

Some of the most productive research done on environmental perception has been on **natural hazards**, such as floods, hurricanes, volcanic eruptions, earthquakes, insect infestations, and droughts. Natural hazards are sometimes called "acts of God," highlighting that such disasters are beyond human control. A closer look, however, reveals that there is sometimes some human input into the making of disasters. For example, towns in the upper Mississippi River drainage basin were built close to the riverbanks to facilitate commerce, but were then vulnerable to the flooding common in the region. Sometimes hazards are known, but collectively forgotten because the time between disastrous occurrences is great and people become complacent (**Figure 1.33**). And sometimes human-made hazards combine with natural hazards to create even worse disasters. Human-made

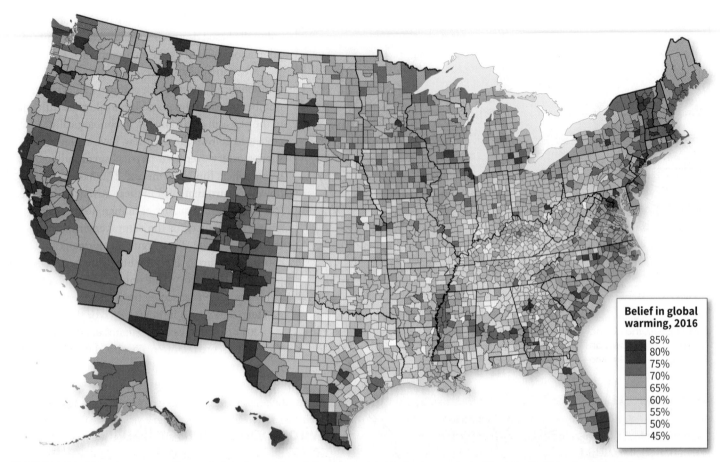

FIGURE 1.32 Mapping perceptions of global warming. Scientists worldwide have concluded that the Earth's atmosphere is warming dangerously, though many people in the United States do not agree. This map shows how perceptions of climate warming vary geographically, with the strongest levels of acceptance in coastal and arid regions, and some of the weakest in the midwestern and interior southern states. (Source: Adapted from Yale Climate Opinion Maps, 2016.)

Belief in global warming, 2016

85%
80%
75%
70%
65%
60%
55%
50%
45%

FIGURE 1.33 Record-setting rains from Hurricane Harvey flood Houston. For several decades, Houston, Texas, has allowed housing developments to spread into once-open prairie and wetlands. Consequently, when Hurricane Harvey hit in 2017, the record rainfall had no place to go, and the result was the worst flooding in the city's history and one of the most costly U.S. disasters ever. (Polaris/Newscom)

hazards include such things as pollution and contamination from industrial production. Thus, it is often difficult and even misguided to attempt to separate the natural from the human in understanding vulnerability to hazards.

Japan's so-called triple disaster illustrates the extent and complexity of nature–culture coupling inherent in natural hazards (**Figure 1.34**). On the afternoon of March 11, 2011, a strong earthquake occurred in the northwestern Pacific Ocean near Japan's coast. The resulting shock waves were felt in major cities across Japan, but government planning minimized the impacts. Because the country is prone to earthquakes, the government had invested in new construction technologies, enforced some of the world's strictest building codes, and installed an early earthquake warning system. These initiatives prevented property damage and saved many lives in Tokyo and other cities.

The earthquake, however, produced a 26-foot (8-meter) displacement in the seafloor, resulting in a tsunami (a wave produced by the displacement of a large volume of water). Within an hour of the initial earthquake, the tsunami began flooding coastal areas at heights ranging from 10 to 23 feet

FIGURE 1.34 Japan's 2011 "triple disaster." First, a very strong earthquake occurred off the coast, causing widespread structural damage (left). Second, the earthquake generated a tsunami that swept away entire communities, such as Minamisanriku (center). Third, the tsunami triggered a meltdown and radiation release at the Fukushima Nuclear Power Plant (right). (The Asahi Shimbury/Getty Images; TOSHIFUMI KITAMURA/AFP/Getty Images; Gamma-Rapho via Getty Images)

(3 to 7 meters). Though the government also had installed a tsunami warning system, it was inadequate to deal with the magnitude of the flooding. Indeed, dozens of government evacuation shelters, which were supposedly out of harm's way, were flooded and destroyed. It was later revealed that previous generations of residents had marked the coastline with stone tablets warning of the tsunami hazard. Some markers were hundreds of years old and advised residents to not build below certain elevations. The Japanese government and most present-day residents had not heeded these warnings, with the result that whole towns and ports were swept away by the 2011 tsunami.

The third disaster followed the tsunami, which destroyed the electrical power infrastructure and inundated nuclear power facilities along the coast. At the Fukushima Daiichi and Daini power plants, the waves swept over seawalls and destroyed the facilities' backup power systems. As a result, the plants were without electrical power and were unable to maintain control of radioactive materials on site. Several explosions and radioactivity leakage followed, leading to the evacuation of over 200,000 people. As the nuclear reactors melted down, it became clear that the company in charge had failed to adequately plan for the tsunami hazard. The releases of radioactivity left many areas uninhabitable, caused the closure of local fisheries, and spurred the government to abandon nuclear power as an energy source for the nation. Japan's triple disaster demonstrates how labeling a hazard "natural" or an "act of God" masks the human component present in all disasters.

In virtually all cultures, people knowingly inhabit hazard zones, especially floodplains, exposed coastal sites, drought-prone regions, and the environs of active volcanoes. In the United States, for example, every year more people live in areas subject to frequent hurricanes along the coast of the Gulf of Mexico and atop earthquake faults in California. How accurately do they perceive the hazard involved? Why have they chosen to live in such places? How might we minimize the eventual disasters? Human geographers seek the answer to these questions and aspire, with other experts, to mitigate the inevitable disasters through strategies such as land-use planning.

Humans as Modifiers of the Earth

Many human geographers, observing the environmental changes people have wrought, emphasize humans as modifiers of the habitat. This presents yet another facet of the theme of nature–culture. In a sense, the human-as-modifier school of thought is the opposite of environmental determinism. Whereas the determinists proclaim that nature molds humankind, and possibilists believe that nature presents possibilities for people, those geographers who emphasize the human impact on the land assert that humans mold the environment.

In addition to deliberate modifications of the environment through such activities as mining, logging, and irrigation, we now know that even seemingly harmless behavior, repeated for millennia, centuries, or in some cases mere decades, can have catastrophic effects on the environment. Plowing fields and grazing livestock can eventually denude regions (**Figure 1.35**). The use of certain types of air conditioners or spray cans

SUBJECT TO DEBATE Human Activities and Global Climate Change

One of today's most important and vexing scientific and political issues is global climate change—both grasping its causes and deciding what policies to pursue to lessen its effects. It's also an issue that human geographers, with their emphasis on understanding nature–culture relationships, are well prepared to discuss.

The Earth's temperatures are rising. Since the late nineteenth century, the surface temperature of the Earth has risen 2°F, with the greatest increase occurring since 1978. Global climate, of course, is always changing. What is critical today, however, is the degree to which scientists have been able to correlate global warming trends with the rise in the levels of carbon dioxide (CO_2) in the atmosphere. Since the start of the Industrial Revolution, atmospheric CO_2 levels have risen dramatically, and today they are higher than they have been in over 500,000 years. Carbon dioxide is one of several greenhouse gases that keep radiative energy from the sun (and therefore warmth) trapped in the Earth's lower atmosphere. Greenhouse gases occur naturally in the atmosphere but are also released when fossil fuels such as coal, oil, and natural gas are burned. Changes in the levels of carbon dioxide in the atmosphere, therefore, lead to changes in the Earth's temperature. Sixteen of the last 17 years had the hottest average global temperatures ever recorded. Changes in temperature, in turn, lead to other changes in the environment, such as melting glaciers, droughts, an increase in extreme weather events, and rising sea levels.

To what degree are our activities—particularly our energy demands that lead to the burning of fossil fuels—responsible for this climate change? This is where the real debate starts and where culture and politics can obscure scientific facts. A very few scientists are wary of pointing the finger at carbon dioxide emissions as the primary culprit, arguing that other factors or atmospheric changes are impacting the Earth's climate. A few are critical of the data that show increases in surface temperature. And a tiny minority believes that the recent fluctuation in climate is a natural occurrence rather than human induced.

Nevertheless, there is growing worldwide consensus that human activities are the primary factors responsible for recent global warming. The Intergovernmental Panel on Climate Change (IPCC), a group of scientists from many different countries, concluded that because of increased greenhouse gases, average global temperatures are likely to rise between 2.5°F and

apparently has the potential to alter the planet's very ability to support life.

Most critically, the increasing release of fossil-fuel emissions from vehicles, power plants, and factories has led to an increase

FIGURE 1.35 Human modification of the Earth includes severe soil erosion. This erosion could have been caused by road building or poor farming methods. The scene is in the Amazon Basin of Brazil. (Michael Nichols/National Geographic)

of greenhouse gases in the Earth's atmosphere. **Greenhouse gases** (GHG) are compounds in the atmosphere, such as carbon dioxide (CO_2) released from fossil fuel combustion, that absorb and trap heat energy close to the Earth's surface. Scientists now know that the rising levels of CO_2 in the atmosphere are causing a corresponding rise in global temperature. This global warming trend is referred to as the **greenhouse effect** (see Subject to Debate). Clearly, access to energy and technology is the key variable that controls the magnitude and speed of environmental alteration. Geographers seek to understand and explain the processes of environmental alteration as they vary from one culture to another and, through applied geography, to propose alternative, less destructive modes of behavior.

The cumulative global effects of human modification of the Earth have become so significant that some scientists believe our species has initiated a new geologic epoch. Atmospheric chemist and Nobel laureate Paul Crutzen argued in 2000 that the Holocene epoch, which began 11,700 years ago at the end of the last ice age, has now ended and that the Earth has entered a new epoch, which he labeled the *Anthropocene,* from *anthropo,* for "human," and *cene,* for "new." The **Anthropocene** designates a new geological epoch beginning around 1800 C.E. in which

10.4°F above 1990 levels by 2100. The question, then, is not whether the climate is warming, but what to do about it. The first step, scientists argue, is to find ways to decrease levels of carbon dioxide released by looking to new technologies and alternative energy sources. Yet, at the same time, we must deal with the effects of global warming that are already occurring.

Geography is a discipline well suited to confronting global warming effects because, as we have seen, one of its primary activities is to understand nature–culture relationships. Valued cultural landscapes, cities, and farmlands will all be affected, and human geographers will no doubt be on the front lines of mitigating those effects.

Continuing the Debate

As geographers, we know that various cultures interact with the environment differently and have varied beliefs and ideas about the role of science in explaining physical phenomena. Keeping all this in mind, consider these questions:

- How might such cultural differences affect people's conclusions about the causes and effects of global climate change?
- How are your ideas about global climate change affected by your position in the world?

Why is the Earth's climate warming? As our dependence on fossil fuel combustion has grown, atmospheric carbon dioxide and global temperatures have risen in tandem. (tainkm/iStock/Getty Images)

the dominant influence on the Earth's natural systems is human activity. Not all scientists have accepted this thesis, and even those who do continue to debate the starting date. An official ruling from the International Union of Geological Sciences is pending. Nevertheless, scholars and scientists, including geographers, are increasingly using the Anthropocene concept as a tool to understand humanity's role in modifying the Earth.

Cultural Landscape

1.9 Explain the concept of cultural landscape and how geographers analyze it.

Our fifth and final theme is **cultural landscape**, which describes how cultural groups transform the landscape and the symbols and values associated with those changes in landscape. Every inhabited area has a cultural landscape, fashioned from the natural landscape, and each uniquely reflects the culture or cultures that created it (**Figure 1.36**). Landscape changes,

FIGURE 1.36 Terraced cultural landscape of an irrigated rice district in Indonesia. In such areas, the artificial landscape made by people blends with nature and forms a human pattern on the land. (Nacivet/Photographer's Choice/Getty Images)

FIGURE 1.37 Chinatown, San Francisco. Gates are symbolically important in denoting the entry into sacred space in traditional Chinese culture. Today they are often used in American cities like San Francisco to mark for tourists a transition into a neighborhood dominated by immigrant Chinese. (© Ron Niebrugge/Alamy)

including roads, agricultural fields, houses, parks, cities, gardens, and commercial buildings, reflect a culture group's needs, values, and attitudes toward the Earth. We can learn much about a culture by observing and studying the landscape.

Some cultural landscapes are so central to understanding human history, identity, and beliefs that they become nationally and even globally significant. For example, the Renaissance-era landscape of Venice, Italy, is of such symbolic importance that in the 1960s a massive international campaign to save it from decay and flooding raised millions of dollars. Similar successful campaigns in other places led the United Nations in 1975 to establish a program to coordinate efforts to identify, catalog, and protect cultural landscapes and artifacts of worldwide significance. Known as **World Heritage Sites**, these are places (e.g., buildings, cities, forests, lakes, deserts, archaeological ruins) that the UN's International Heritage Programme judges to possess outstanding cultural or natural importance to the common heritage of humanity. Venice was one of the first additions to the UN's list, which by 2017 included 850 cultural sites. The designation has drawn even more international tourists to Venice, which now receives roughly 50 times as many visitors per year as the number of permanent residents in the historic city.

When the membership of the United Nations created the World Heritage Sites designation (see page 34), it was a clear signal that landscape is universally valued as an expression of culture and identity. World Heritage Sites simultaneously celebrate both the vast diversity of cultural achievements worldwide and also the singular impulse to creatively manipulate the material environment, which unites

all humanity. Since the program's establishment, countries have lobbied hard to receive World Heritage Site status for sites in their landscapes and take pride in the global recognition of local cultural achievement. Members pledge precious government resources to forever protect landscapes within their borders and to respond with emergency assistance when threats to sites arise in any other member state.

Why is such importance attached to the cultural landscape? Perhaps part of the answer is that it visually reflects the most basic strivings of humankind: shelter, food, clothing, and creative expression. In addition, the cultural landscape reveals people's different attitudes toward the modification of the Earth. The landscape also contains valuable evidence about the origin, spread, and development of cultures because it usually preserves various types of archaic forms. Dominant and alternative cultures use, alter, and manipulate landscapes to express their diverse identities (**Figure 1.37**).

Aside from containing archaic forms, landscapes also reveal messages about their present-day inhabitants and cultures. According to geographer Pierce Lewis, "The cultural landscape is our collective and revealing autobiography, reflecting our tastes, values, aspirations, and fears in tangible forms." Cultural landscapes offer "texts" that geographers read to discover dominant ideas and prevailing practices within a culture as well as less dominant and alternative forms within it. This "reading," however, is often a very difficult task, given the complexity of cultures, cultural change, and recent globalizing trends that can obscure local histories.

FIGURE 1.38 Landscape triptych panel by Fra Angelico, fifteenth century. Notice the depiction of the beautiful and orderly agricultural landscape outside the city walls but the absence of people actually doing the work to maintain that order. (The Deposition from Cross or Altarpiece of Holy Trinity by Giovanni da Fiesole known as Fra Angelico (1400–1455), tempera on wood, 176x185 cm, detail, ca 1432/DE AGOSTINI EDITORE/Bridgeman Images)

FIGURE 1.39 Yokohama at dusk. This skyline is a powerful symbol of the economic importance of the world's largest city, Tokyo-Yokohama. (Jose Fuste Raga/Corbis)

FIGURE 1.40 Prague's skyline is dominated by church spires. Here, St. Vitus Cathedral sits majestically overlooking the Vltava River and the Small Town. (Courtesy of Mona Domosh.)

Geographers have pushed the idea of "reading" the landscape further to focus on the symbolic and ideological qualities of landscape. In fact, as geographer Denis Cosgrove has suggested, the very idea of landscape itself was ideological, in that its development in the Renaissance served the interests of the new elite class for whom agricultural land was valued not for its productivity but for its use as a visual subject. Land, in other words, was important to look at as a scene, and the actual activities necessary for agriculture were thus hidden from these views. If you go to an art museum, for example, it will be difficult, if not impossible, to find in the Italian Renaissance room any landscape paintings that depict agricultural laborers (**Figure 1.38**).

Closer to home, we need only to look outside our windows to note the symbolic qualities of landscapes. One of the most familiar and obvious symbols is found in the modern urban skyline. Composed of tall buildings that normally house financial service industries, it represents the power and dominance of finance and economics within that culture (**Figure 1.39**). However, other cities are dominated by tall structures that have little to do with economics but more with religion. In medieval Europe, for example, cathedrals and churches rose high above other buildings, symbolizing the centrality and dominance of Catholicism in that culture (**Figure 1.40**). We can therefore recognize that **symbolic landscapes**, landscapes that express the values, beliefs, and meanings of a particular culture, are to be found everywhere.

Examining architectural style is often useful when trying to understand the particular values and beliefs that culture

FIGURE 1.41 The American ranch house. Its low-slung, rectangular architecture is a common house form in the United States and Canada. So generic and widespread is this form that only the vegetation in the photo suggests its location in a specific region. (Courtesy of Roderick Neumann.)

groups may hold, and is therefore of interest to human geographers. Even the most mundane architecture can be interpreted as symbolic and ideological. Take for example the common North American ranch house, invariant in its basic form from Vancouver, British Columbia, to Miami, Florida (**Figure 1.41**). Its rectangular walls hold a whole set of mid-twentieth-century assumptions about an ideal social structure.

World Heritage Site: The Great Zimbabwe National Monument

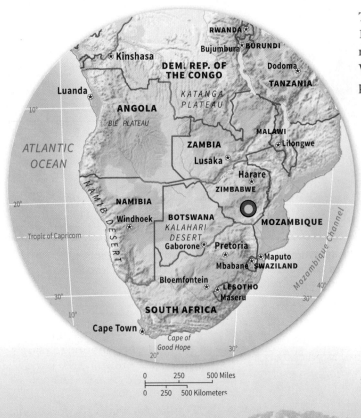

0   250   500 Miles

0   250   500 Kilometers

The Great Zimbabwe National Monument was constructed between 1100 and 1450 C.E. and encompasses over 1739 acres (700 hectares) in a region populated by the Shona people of Zimbabwe. Placed on the list of World Heritage Sites in 1986, Great Zimbabwe contains the best example of precolonial African architecture south of the Sahara.

- In the Great Enclosure, imposing dry masonry stone walls rise 36 feet (11 meters), enclosing numerous dwellings and other stone structures. In addition to the Great Enclosure, two other distinct zones comprise the complex: the Hill Ruins and the Valley Ruins.

- For nearly four centuries, the site served as the principal city of the kingdom of Zimbabwe, housing 10,000 to 20,000 residents and functioning as the center of an international trading network that stretched across the Indian Ocean and, ultimately, to China. The kingdom ruled over a gold-rich plateau covering a portion of present-day Zimbabwe and Mozambique. Locally mined gold was smelted into ingots for trade with coastal cities, which imported porcelain, glass beads, and other luxury goods.

- Racist nineteenth-century theories of history prevented Europeans from recognizing the ruins as the grand achievement of an African civilization. They speculated that ancient Phoenicians, Egyptians,

(Richard I'Anson/Lonely Planet Images/Getty Images)

(Danita Delimont/Gallo Images/Getty Images.)

and even the Queen of Sheba had built it. Even in the 1970s, the white-minority government of Zimbabwe (then Southern Rhodesia) officially denied the site's black African origins. Twentieth-century archaeological studies have confirmed that the Shona's ancestors built and continuously occupied the site for nearly four centuries.

- During the twentieth-century pan-African struggle against European colonialism and white-minority rule, the site became a key source of cultural pride among black Africans across the continent. It helped form both Shona ethnic identity and postcolonial Zimbabwean national identity. Indeed, the present-day country of Zimbabwe took its name from the site after overthrowing white-minority rule in 1980.

THE GREAT ENCLOSURE: The most remarkable feature of the site is the Great Enclosure, an area dominated by a monumental, oval-shaped outer wall 820 feet (250 meters) long and 10 feet (3 meters) wide. The craftsmanship is stunning, beginning with the skillful splitting of granite blocks excavated from a nearby quarry. The granite split easily along fracture planes, resulting in cubelike blocks that craftsmen smoothed, stacked, and fitted together. This technique resulted in a finished quality that rivals modern brick walls and allowed the stone city to be built without mortar and yet remain intact for centuries.

- Cut stones set in herringbone and chevron patterns accent the top of the outer wall. Within the wall are a smaller stone enclosure and a narrow stone passageway leading to the beehive-shaped conical tower. The tower is constructed of solid stone blocks and is assumed to have had a primarily symbolic or ceremonial purpose.

- The enclosure once contained houses built of clay and gravel, of which only traces remain. These were organized into walled-off community areas composed of a kitchen, dwelling huts, and a court. One theory suggests that the enclosure contained the royal residence of the kingdom's ruler. Thus, the Great Enclosure represents the seat of political and economic power for the expansive Zimbabwe kingdom.

TOURISM: The site is readily accessible and also relatively close to other attractions such as Victoria Falls (another World Heritage Site) and many of southern Africa's wildlife parks.

- International visitors freely roam the site and explore the very different perspectives and features offered by the Great Enclosure, the Hill Ruins, and Valley Ruins.

- Recent political and economic instability in Zimbabwe, however, has resulted in a significant drop-off in tourist activity. Although ruinous to the local economy, the lull in tourist traffic may be giving this archeological wonder a needed respite from overuse.

- Built between 1100 and 1450 C.E. by the Shona people of Zimbabwe, the site is a unique example of African architecture.
- The site has three main areas: the Great Enclosure, the Hill Ruins, and the Valley Ruins.
- Placed on the list of World Heritage Sites in 1986, the site has been visited by thousands of tourists, but tourism has slowed due to political and economic instability in the country.

http://whc.unesco.org/en/list/364

(Colin Hoskins/Alamy)

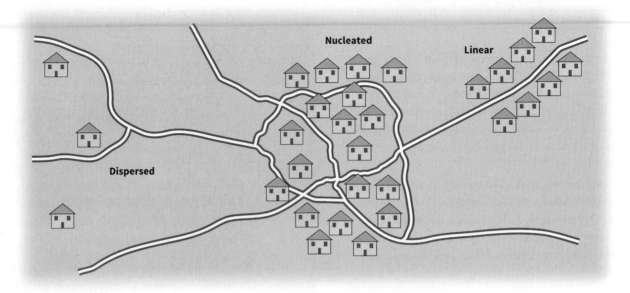

FIGURE 1.42 Types of settlement patterns. Population density provides only a crude measure of settlement. Geographers are interested in settlement patterns, the spatial arrangement of residences, which are important characteristics of the cultural landscape.

It is a highly ordered structure, materially and symbolically. The foundation of this social structure is the nuclear family: a heterosexual couple and their offspring. Each nuclear unit inhabits its own rectangular single-family house, which is positioned on a rectangular patch of ground for the family to groom and maintain, which is itself located within a larger rectangular pattern of land division. Cultural assumptions built into the division of the home's internal space include the value of individual privacy (everyone has his or her own bedroom); the idea that certain domestic functions should be spatially separate from others (cooking, eating, socializing, sleeping); and the notion that a family is composed of a single generation including a mother, father, and children (indicated by the "master" bedroom and smaller "children's" bedrooms). This example illustrates how even our most familiar cultural landscapes can be analyzed for their symbolic content and underlying assumptions.

As we have seen, the physical content of the cultural landscape is both varied and complex. To better study these complexities, geographers often focus on specific aspects of landscape, such as settlement pattern, land-division pattern, and architectural style. In the study of **settlement pattern**, human geographers describe and explain the spatial arrangement of buildings, roads, and other features that people construct while inhabiting an area. Towns and cities often have a pattern of buildings clustered around a center, called **nucleated settlement**. Many rural farming areas and some suburban developments have a **dispersed settlement** pattern, where individual buildings are scattered apart from one another.

Finally, buildings can be arranged in a **linear settlement pattern**, where buildings are arranged in a line, often along a road or river (**Figure 1.42**).

Another way to categorize settlement patterns is by degree of **density**, a term that refers to the number per unit area of a given landscape element, such as houses. Urban centers have a very dense settlement pattern as compared to rural areas. We can also think of settlement in terms of **concentration**, or the degree to which a given landscape element, such as a dwelling, is grouped. Take two settled areas of one square mile each and having equal numbers of houses. That is, the areas have the same density. However, in one area each house is equally distant from the next and in the other they are arranged, or *concentrated,* in groups of twenty. These basic settlement pattern concepts must be addressed in every new housing development and will affect the quality of daily life there.

The land-division pattern reveals the way people have divided the land for economic, social, and political uses. In cities, for example, you can see different patterns of land use. Some areas are devoted to economic uses, others to residential, governmental, cultural, or outdoor recreational uses. Each of these areas can be further subdivided. Economic uses may include offices for financial services, retail stores, warehouses, and factories. Residential areas are often divided into middle-class, upper-class, and working-class districts and/or are grouped by ethnicity and race (see Chapter 10). Such patterns, of course, vary a great deal from place to place and culture to culture, as we will see throughout this book.

Another way of thinking about land-division patterns is by examining property boundaries. Most countries have some form of **cadastral survey**, a systematic documentation of property ownership, shape, use, and boundaries. The maps produced by such surveys reveal much about the cultures settling those regions. For example, in some regions along rivers one can identify property lines in long, narrow shapes oriented perpendicular to the river shore. This pattern is an artifact of a farming community that organized settlement to provide as many families as possible direct access to the river for irrigation water (see Chapter 8 for further discussion). In other regions, one can recognize an abstract pattern of square parcels of uniform size. This pattern is evidence that the land was divided in a geometrical model prior to settlement, as is the case in parts of Canada and the United States (see Chapter 6 for further discussion).

Conclusion

We have now learned many of the most fundamental concepts and practices of human geography, which we will use throughout this book. We began by learning about the concept of culture, which serves as our overarching approach to the field. As we have seen and will continue to see, the interests of human geographers are diverse. Nowhere is this more evident than in our examination of three different methodological perspectives on human geography, categorized as spatial modeling, humanistic, and social theoretical. Each has its role in developing our understanding of human geography, and we will rely on the insights of all three in the chapters to come. We also learned some of the basic tools that geographers use for making sense of the world, including the longitude and latitude systems, remote sensing, GPS, and GIS. Whether we realize it or not, those same tools have become increasingly important in daily life as the use of hand-held electronic devices has spread. Finally, we became familiar with the five themes in human geography—region, mobility, globalization, nature–culture, and cultural landscape—that provide this book's organizing framework.

Chapter Summary

1.1 Explain the concept of culture and how culture interacts with the physical environment.

- Culture is a process that people actively engage in to create patterns of learned behavioral traits, beliefs, and material practices, such as food and clothing preferences.
- Culture can also be used to refer to distinguishable groups that share the same behavioral traits, beliefs, and material practices.

- The physical environment is an important influence on culture, but cultural practices are also capable of modifying the physical environment to suit culturally defined preferences.

1.2 Explain the basic classification of cultures, including folk, indigenous, and popular cultures, and the processes of cultural diversification.

- Folk culture is a label used to distinguish groups who follow traditional, self-sufficient ways of life in rural and agricultural regions.
- Indigenous culture is an important concept that distinguishes culture groups who were the original occupants of the land at the time of European colonization.
- Popular culture refers to modern material and symbolic practices associated with the rise of mass-produced, machine-made goods and the invention of long-distance communication technologies.
- The formation of new culture groups, including a wide range of subcultures and diaspora cultures, reflects a continuing process of cultural diversification.

1.3 Explain the three basic perspectives on human geography: spatial modeling, humanistic, and social theoretical.

- Spatial modeling uses mathematics to construct simulations of geographic situations in search of universal laws and regularities.
- Humanistic geography pursues questions about people's subjective experience of space and place, focusing on meanings, values, and emotions.
- The social theoretical approach seeks to balance humanistic geography's focus on individual

experiences with questions about how the uneven distribution of power in society operates to constrain or enable individual action.

1.4 Apply key visualization tools used in human geography—including the longitude and latitude system and geospatial technologies—to everyday consumer devices.

- The latitude and longitude system is a grid imposed on the globe, comprised of east–west and north–south lines, which provides the absolute location for any point on the surface of the Earth.
- Geospatial technologies such as GPS, GIS, remote sensing, and interactive mapping software are founded on the longitude and latitude system and are increasingly integral to many aspects of social life.

1.5 Compare and contrast formal, functional, and vernacular regions.

- Formal regions are geographical units defined by a set of one or more common traits, either material or nonmaterial.
- Functional regions are geographical units defined by a political, economic, religious, or other social function.
- Vernacular regions are defined informally by local inhabitants' subjective ideas about and meanings attributed to a geographic unit.

1.6 Explain the different types of diffusion and migration.

- There are five distinct ways that phenomena spread (or diffuse) through time and space: relocation diffusion, expansion diffusion, hierarchical diffusion, contagious diffusion, and stimulus diffusion.
- There are many forms of human migration, the most important of which are international migration, internal migration, stepwise migration, return migration, seasonal migration, forced migration, and transnational migration.

1.7 Explain the concept of globalization and how it shapes interdependence among regions.

- Globalization is complex set of political, economic, technological, and cultural phenomena that is increasingly integrating the local with the global.

- The historical roots of globalization can be traced to the era of European exploration and colonialism.
- Globalization has produced a pattern of interdependence among regions around the world that is expressed by a geographic unevenness in the level of socioeconomic development.

1.8 Explain the historical development of different approaches to nature–culture relations in human geography.

- Cultural ecology borrowed ideas from the field of biological ecology to understand nature–culture relations, but later evolved to include more political, economic, and social variables.
- Environmental determinism—the idea that physical environment shaped culture—was an early theory of nature–culture relations that was later rejected and replaced by the concept of possibilism.
- Environmental perception is an approach that recognizes that people's experience of nature is a subjective one, influenced greatly by cultural context.
- An important approach to nature–culture relations analyzes the role of humans in modifying the environment, even suggesting that we have entered a new geological epic defined by human influence, the "Anthropocene."

1.9 Explain the concept of cultural landscape and how geographers analyze it.

- In one sense, cultural landscapes are the visible outcomes of the material practices and symbolic meanings through which culture groups shape the Earth's surface.
- One prominent analytical approach is to read cultural landscapes for what they reveal about the cultures that created them, in part by giving attention to material elements such as architecture and settlement patterns.

Key Terms

human geography The study of the relationships between people and the places and spaces in which they live (page 2).

culture A dynamic process through which humans create, learn, and change shared behaviors, beliefs, values, meanings, and practices (page 2).

cultural practices The social activities and interactions—ranging from religious rituals to food preferences to clothing—that collectively distinguish group identity (page 2).

gendered space Spaces that are classified by social norms and practices as predominantly either male or female (page 2).

physical environment All aspects of the natural physical surroundings, such as climate, terrain, soils, vegetation, and wildlife (page 3).

pattern The consistent or characteristic spatial arrangements of objects or phenomena (page 3).

material culture All physical, tangible objects made and used by members of a cultural group, such as buildings, furniture, clothing, food, artwork, musical instruments, and so forth (page 4).

nonmaterial culture The wide range of beliefs, values, myths, and symbolic meanings passed from generation to generation of a given society (page 5).

folk culture Rural, unified, largely self-sufficient groups that share similar customs, material practices, and ethnicities (page 5).

national culture The controversial idea that citizens possess a set of recognizable values, behaviors, and beliefs—often including the same ethnic and linguistic traits—that express the core culture of each modern nation (page 6).

indigenous culture A culture group made up of the original inhabitants of a territory. The term commonly refers to subjugated cultures distinct from national cultures that are rooted in colonial domination (page 6).

colonization The appropriation, control, and occupation of a territory by a geographically distant power (page 6).

popular culture The modern material and symbolic practices associated with the rise of mass-produced, machine-made goods and the invention of long-distance communication technologies, such as radio and television (page 7).

placelessness A feeling resulting from the standardization of the built environment, increasingly on a global scale, that diminishes regional variation and eliminates the unique meanings associated with specific locations (page 7).

diaspora culture Ethnic, racial, and national population concentrations of people displaced and geographically scattered from their homelands. Such displaced groups often maintain strong social and economic ties to their homelands (page 9).

subculture A group of people with distinct norms, values, and material practices that differentiate them from the dominant culture surrounding them (page 9).

spatial model Simulations of social life, often represented in mathematical terms, that abstract a limited set of variables from real-world situations in search of universal laws and regularities in spatial patterns (page 10).

space In the spatial science tradition, an abstract, geometric property that can be scientifically modeled through the application of universal laws (page 10).

spatial scale The geographic extent of an area that is under investigation, generally determined by the problem being studied and therefore also called scale of analysis (page 10).

place Locations that acquire meaning and value through people's subjective experiences of them (page 11).

sense of place The subjective human emotions, both negative and positive, evoked by a particular place. A shared sense of place is commonly a key component in defining a culture (page 11).

public space Geographical areas that are owned and controlled by a government agency and legally accessible to all citizens (page 12).

absolute location The precise position of anything on (or below) the surface of the Earth (page 14).

latitude The degree of distance north or south from the equator, which is 0 degrees, as far as the poles, which are at 90 degrees (page 14).

longitude The degree of distance east or west of the prime meridian (Greenwich Meridian), which is 0 degrees, to the line exactly on the opposite side of the Earth, which is 180 degrees (page 14).

prime meridian The line of 0 degrees longitude running through Greenwich, England. Also called the Greenwich Meridian (page 14).

Global Positioning System or GPS A constellation of satellites orbiting the Earth, which transmit radio signals earthward to receiving devices, such as smartphones, locating them precisely in time and three-dimensional space (page 15).

relative location The position of one place, person, or object in relation to the position of another place, person, or object (page 15).

remote sensing The collection of information about the Earth's surface using devices located at distance, generally high above (page 15).

aerial photography A remote sensing technique using cameras mounted on fixed-wing aircraft to systematically photograph the Earth's surface from directly overhead (page 15).

satellite imagery Information about the Earth's surface gathered from sensors mounted on orbiting satellites that record in both the visible and non-visible portions of the electromagnetic spectrum (page 16).

geographic information system (GIS) A computer software application that is used to collect, store, retrieve, and visually present geospatial data (page 16).

region A geographical unit based on a set of common characteristics or functions (page 17).

formal region A geographical area inhabited by people who have one or more cultural traits in common (page 17).

functional region A geographic area that is defined by some economic, political, cultural, or social function (page 18).

node A central location in a functional region where functions are coordinated and directed (page 18).

vernacular region A culture region perceived to exist by its inhabitants, based in the collective spatial perception of the population at large and bearing a generally accepted name or nickname (such as "Dixie" for the southern United States) (page 19).

mobility The relative ability of people, ideas, or things to move freely across space (page 20).

diffusion The pattern by which a phenomenon—people, ideas, technologies, or preferences—spreads from one location through space and time (page 20).

independent invention A cultural innovation developed in two or more locations by individuals or groups working independently (page 20).

relocation diffusion The spread of an innovation or other phenomenon that occurs with the bodily relocation (migration) of the individual or group responsible for the innovation (page 20).

expansion diffusion The spread of innovations within an area in a snowballing process, so that the total number of knowers or users becomes greater and the area of occurrence grows (page 20).

hierarchical diffusion A type of expansion diffusion in which innovations spread from one important person to another or from one urban center to another, bypassing other persons or rural areas (page 20).

contagious diffusion A type of expansion diffusion in which cultural innovation spreads by person-to-person contact, moving wavelike through an area and population without regard to social status (page 21).

stimulus diffusion A type of expansion diffusion in which a specific trait fails to spread but the underlying idea or concept is accepted (page 21).

time–distance decay The decrease in acceptance of a cultural innovation with increasing time and distance from its origin (page 21).

friction of distance The inhibiting effect of distance on the intensity and volume of most forms of human interaction (page 21).

absorbing barrier A barrier that completely halts diffusion of cultural innovations (page 22).

permeable barrier A barrier that permits some aspects of an innovation to diffuse through it but weakens and retards continued spread; an innovation can be modified in passing through a permeable barrier (page 22).

circulation An ongoing set of multidirectional movements of people, ideas, or things that have no particular center or periphery (page 22).

migration The mass movement of a population between different areas of the Earth's surface (page 22).

international migration Human migration across country borders (page 22).

internal migration Human migration that occurs within the borders of a country (page 23).

Great Migration The twentieth-century movement of 6 million African Americans from the rural southern United States to the cities of the midwestern and northeastern states (page 23).

stepwise migration Human migration conducted in a series of stages (page 23).

return migration The phenomenon of migrants returning to their place of origin after long-term residency elsewhere (page 24).

seasonal migration Often associated with crop harvest periods; migrants move according seasonal changes in weather (page 24).

forced migration Situations where people are involuntarily compelled, by either disasters or social conditions, to leave their homes (page 24).

transnational migration The movement of groups of people who migrate continuously between countries, thereby maintaining ties to both their countries of origin and their countries of destination (page 24).

globalization Processes of economic, political, and social integration that operate on a global scale and have collectively created ties that make a difference to lives around the planet (page 24).

interdependence Relations between regions or countries of mutual, but not necessarily equal, dependence (page 25).

nature–culture How values, beliefs, perceptions, and practices have ecological impacts, and how, in turn, ecological conditions influence perceptions and practices (page 26).

ecology A biological science that studies the complex relationships among living organisms and their physical environments. Also used when referring to the biophysical conditions of a particular region or place (page 26).

cultural ecology A field of study that uses concepts borrowed from biological ecology to study culture as a system that facilitates human adaptation to the environment (page 26).

environmental determinism The belief that the physical environment is the dominant force shaping cultures and that humankind is essentially a passive product of its physical surroundings (page 26).

possibilism The theory that the environment presents constraints and opportunities for cultural innovation, but the culture that develops in a particular environment ultimately depends on human agency (page 26).

environmental perception The process whereby people act based on how they perceive their environment, rather than on how it actually is (page 27).

natural hazard A physical danger present in the environment, such as floods, hurricanes, volcanic eruptions, or earthquakes, that may be perceived differently by different peoples (page 27).

greenhouse gas (GHG) A compound in the atmosphere, such as the carbon dioxide (CO_2) released from fossil fuel combustion, that absorbs and traps heat energy near the Earth's surface (page 30).

greenhouse effect The global warming trend caused by the increase of greenhouse gases such as carbon dioxide (CO_2) (page 30).

Anthropocene The proposed designation for a new geological epoch beginning around 1800 C.E. in which the dominant influence on the Earth's natural systems is human activities (page 30).

cultural landscape The built forms that cultural groups create in inhabiting the Earth—farm fields, cities, houses, and so on—and the meanings, values, representations, and experiences associated with those forms (page 31).

World Heritage Sites Places, including buildings, cities, forests, lakes, deserts, and archaeological ruins, that the UN's International Heritage Programme judges to possess outstanding cultural or natural importance to the common heritage of humanity (page 32).

symbolic landscapes Landscapes that express the values, beliefs, and meanings of a particular culture (page 33).

settlement pattern The spatial arrangement of buildings, roads, towns, and other features that people construct while inhabiting an area (page 36).

nucleated settlement A pattern of buildings clustered around a center (page 36).

dispersed settlement A pattern where individual buildings are scattered apart from one another (page 36).

linear settlement A pattern where buildings are arranged in a line, often along a road or river (page 36).

density The amount per unit area of a given landscape element, such as houses (page 36).

concentration The degree to which landscape elements are grouped (page 36).

cadastral survey A systematic documentation of property ownership, shape, use, and boundaries (page 37).

Practice at ◎ SaplingPlus

Read the interactive e-Text, review key concepts, and test your understanding.

 Story Map. Explore the five themes of the text by examining the geography of coffee.

 Web Map. Examine maps that illustrate issues in human geography.

Doing Geography

ACTIVE LEARNING:
Space, Place, and Knowing Your Way Around

We started this chapter by saying that most of us are born geographers, with a sense of curiosity about the places and spaces around us. Think, for example, of the place you call home. You are probably familiar enough with its streets and buildings and greenspaces, and with the people who inhabit these spaces, to make connections among them—you know how to "read" the place. There are many other places, however, where this is not the case. Most of you have had the experience of going somewhere new, where it is difficult to find your way around, literally and metaphorically. How did you find your way? How did the abstract space become a familiar place?

This activity requires you to draw on your past experiences to understand two fundamental concepts in human geography: space and place. The concept of *space* is somewhat more abstract than the concept of *place*. Space describes a two-dimensional location on a map. Place is used to describe a location that has meaning. Your college campus, for example, may have been simply an abstract space presented on a map when you applied to the school, yet now it is a place because you have filled it with your own meanings.

More guidance at SaplingPlus

EXPERIENTIAL LEARNING:
The Privatization of Public Space

In this chapter we discussed perspectives on human geography, including the perspective that focuses on power and ideology. Geographers working from this perspective are interested in the ways that some groups or individuals are able to control access to space. These might include physical barriers such as gates or walls, legal mechanisms such as local zoning regulations and property laws, and surreptitious practices such as selective policing that deny or hinder public space access to certain social groups.

This activity requires you to go out into your community and identify a significant public space. It could be a city playground or a state park or a shoreline, or even a sidewalk. For example, sidewalks are public spaces with free access to anyone who obeys the social norms and legal requirements of public behavior. Many cities rent out that space to restaurant owners who use it for outdoor seating. Where once there might have been a city bench free for anyone to use,

there are now tables and chairs that require payment to the restaurant to use.

Steps to Recognizing the Privatization of Public Space

Step 1: Identify a public space with which you are familiar.

Step 2: Closely examine all of the ways access is controlled (points and times of entrance, policing, legal mechanisms, road layouts, gates, fences, walls, and so forth). Web-based maps and remote sensing techniques will help.

Step 3: Write a description of where and how access to the public is facilitated or inhibited. For instance, do homeowners have more privileged access to public amenities than the general public does?

Step 4: Discuss in your findings what types of social groups might be included or excluded. For example, do you need money or a vehicle to gain access? Would poorer citizens have less access to this space than more affluent citizens would?

Use your experiences to answer the following questions:

• What criteria must be considered to ensure wide public access to public spaces?

• How do you think privileged access to public space might benefit private property owners?

U.S. Representative Pete McCloskey walks with surfer Joao Demacedo on an access road to Martin's Beach, California, which had been illegally blocked by a private land owner. (ROBERT GALBRAITH/REUTERS/Newscom)

SEEING GEOGRAPHY

Noyo Harbor in Northern California

Why do some locations have particularly strong identities?

The terms *place* and *sense of place* help us understand how people give meaning to particular locations and how some locations can evoke strong emotions in people and help shape cultural identities. A sense of place, however, may be strong or weak. Indeed, it can be so weak that people can experience a sense of placelessness. How do we explain such variation? Why is a strong sense of place associated with some locations?

Let's take a closer look at this photo and think about how the central activity featured in it, commercial fishing, might create a strong sense of place. First, look at the natural setting, which is quite distinctive. It features steep, forested slopes meeting the waters of a relatively small Pacific coastal harbor. Second, carefully examine the human-made features in the photo. The preponderance of commercial fishing vessels is immediately evident. The buildings lining the harbor are distinctive as well, made almost exclusively of local timber and clearly designed for function rather than aesthetic appearance. Imagine the smells and sounds of an active fishing port. Now try to imagine the daily lives of people in the surrounding community. What sort of jobs would they have? If they are not directly involved in catching, processing, or selling fish,

they are likely working in a supporting business, such as a marine supply store. Utilitarian vehicles, particularly pickup trucks, will likely be more common than imported luxury sedans. In short, the daily lives of a great proportion of community members will either directly or indirectly center on the activities in Noyo Harbor.

Putting all of this together, we might speculate that residents of communities who are dependent on a single natural resource industry, such as logging, mining, or fishing, will likely feel a strong sense of place. Such industries require distinctive natural settings—forests, underground mineral deposits, bodies of water—that set them apart from other industries. The associated jobs create distinctive cultural identities—logger, miner, fisher— that set the residents apart from other wage workers. Since most people's livelihoods are dependent on one industry, there is likely to be a shared interest in that industry that creates a bond among residents and a shared emotional attachment to the location. Even if our speculations here are valid, these are not the only situations that can produce a strong sense of place. Perhaps you can think about other situations as you read through the textbook.

Commercial fishing boats and industry-related buildings in Noyo Harbor, Fort Bragg, California. (Gary Crabbe/Alamy)

Enjoying nature in a national park campground.
(Courtesy of Roderick Neumann.)

chapter 2

Geographies of Cultural Difference

ONE WORLD OR MANY?

What can this scene tell us about nature–culture relations in North American popular culture?

Examine the photo and think about this question as you read. We will revisit them in Seeing Geography on page 79.

Learning Objectives

2.1 Explain how cultural traits and patterns, such as leisure preferences, ethnic identities, and gender and sexuality, vary by region.

2.2 Recognize the mobility of cultural traits and interactions through the examples of tourism, diaspora cultures, and the Internet.

2.3 Explain how globalizing processes are influencing cultural interactions and change, such as the experience of space-time and cultural homogenization.

2.4 Explain how culture and the physical environment interact in indigenous, folk, and popular cultures and the influences of gender and the global economy on those interactions.

2.5 Compare and contrast folk and popular cultural landscapes.

(Courtesy of Roderick Neumann)

No matter where we live, if you look carefully, you will see how important cultural identity is to our daily lives. Cultural difference is evident everywhere—not only in the geographic distribution of different cultures but also in the way that difference is created or reinforced by geography. In this chapter, we will begin to explore the vast range of geographies of cultural difference. The term *difference* implies a relationship and a set of criteria for comparison and assessment. That is, cultures are defined as they relate to each other using a predetermined set of characteristics. But what does it mean to speak of *geographies* in the plural? Isn't there only one geography? The plural form emphasizes that there is no single way of seeing and experiencing place and space.

Region

2.1 Explain how cultural traits and patterns, such as leisure preferences, ethnic identities, and gender and sexuality, vary by region.

Regions are central to the creation, identification, and maintenance of cultural differences. In this section, we will explore three examples of the many ways that culture finds regional expression. We take in-depth looks at college sports, indigenous peoples, and LGBT communities.

U.S. College Sports

When we think of college sports, chances are football comes to mind first. Successful college football programs tend to define universities and draw national publicity unmatched by that earned from other sports or academic achievements. Universities, however, compete with one another in dozens of other sports. Most of these sports do not have the broad national appeal of football. Rather, the vast majority of sports sponsored by the National Collegiate Athletic Association (NCAA) have distinctive regional associations. Frequently, these regional associations can be traced back in time to the folk traditions of specific places.

Let's look at just three of the dozens of sports that the NCAA sponsors for men in its Division I category: ice hockey, lacrosse, and volleyball. For each of these sports, a regional pattern soon emerges.

Ice Hockey Beginning with ice hockey, there is a clear concentration of teams in the northern states east of the Rocky Mountains (**Figure 2.1**). Michigan has 7 of the 60 college teams, while only one team, the University of Alabama–Huntsville, is located in the South. Only four teams can be found in the western United States, most of which are clustered around Denver, Colorado.

A clearly defined collegiate ice hockey region thus exists in the northern United States. Furthermore, the number of ice

FIGURE 2.1 The distribution of NCAA Division I men's ice hockey teams. There is a discernible concentration of teams in the northern, midwestern, and northeastern states, suggesting a core ice hockey region.

hockey teams in the region—and the absence of the sport in other regions—is closely associated with the geographic origins and development of the sport. Modern ice hockey emerged in eastern Canada in the nineteenth century. Its invention appears to be rooted in the stick and ball games of the indigenous peoples of the region as well as winter folk traditions imported from Europe. Student enthusiasts formed the first ice hockey club with formal rules at Canada's McGill University in 1877. The sport spread from there, restricted almost exclusively to areas that experienced winters long and cold enough to form outdoor ice hockey rinks, as the perspective of possibilism would suggest. Neighborhood kids organized games whenever and wherever it was cold enough for a local pond or creek to freeze over (**Figure 2.2**). By the time technology made indoor ice

rinks feasible anywhere, the regional cultural preferences had been already established.

Lacrosse The NCAA's 71 college lacrosse teams also exhibit a distinctive regional clustering, with the heaviest concentration found in the northeastern United States (**Figure 2.3**). As with ice hockey, lacrosse has flourished close to its point of geographic origin. It is derived from a game played by the indigenous peoples of the northeastern United States and eastern Canada. European Canadians created the formal rules for modern lacrosse in 1867, and intercollegiate competition followed soon thereafter. The Ivy League universities and private liberal arts colleges of the northeastern United States established the first teams, followed by a few large public universities, such as the University of Michigan (**Figure 2.4**). At the college level, lacrosse is virtually unknown west of the Rocky Mountains.

Volleyball Our final example, volleyball, has a history and regional pattern different from those of lacrosse and ice hockey. There are far fewer Division I college volleyball teams—only 23—and they are concentrated in three separate regions: the West Coast, the Midwest, and the Northeast (**Figure 2.5**). The game itself was invented by an educator/physician working for the Young Men's Christian Association (YMCA) in Massachusetts in 1896 as part of an adult health and fitness program.

Although the game has since spread throughout the United States and beyond, college-level competition developed a separate and unanticipated geographic pattern. Volleyball was increasingly associated with a particular environment and culture far from its Massachusetts origin and with a particular set of regional cultural

FIGURE 2.2 A backyard game of ice hockey. (Radius Images/Alamy)

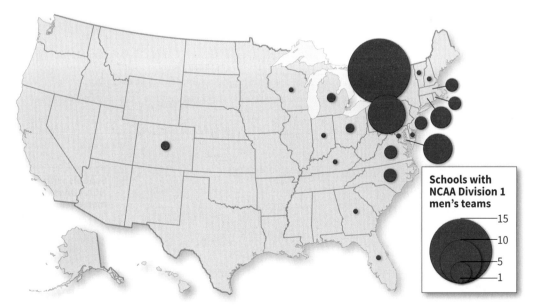

Schools with NCAA Division 1 men's teams

15
10
5
1

FIGURE 2.3 The distribution of NCAA Division I men's lacrosse teams. There is a discernible concentration of teams in the northeastern states, suggesting a core lacrosse region.

FIGURE 2.4 NCAA Division I men's lacrosse match between Syracuse and Duke universities. Syracuse, which has won 10 NCAA championships, has a history of recruiting Native Americans from the reservations throughout upstate New York. The game originated among Native American tribes in the region. (Drew Hallowell/Getty Images)

FIGURE 2.6 Pickup volleyball game on a southern California beach. Though the game was invented in nineteenth-century Massachusetts as an indoor sport, it became a dominant symbol of California beach culture after the 1960s. (Arthur Tilley/Getty Images)

values very different from those of the nineteenth-century YMCA. By the 1960s, volleyball, like surfing, had become symbolically associated with Southern California's beach culture (**Figure 2.6**). As with hockey on winter days in the North, kids in Southern California spent their summer days playing pickup volleyball on the beach or at neighborhood outdoor sand courts. Today, there is a strong concentration of college teams in California, where 9 of the 23 Division I teams are located. While three separate regional concentrations exist, California universities dominate the sport. In 48 years of Division I championships, California teams have won all but 12 championships, and all but one of those winning teams

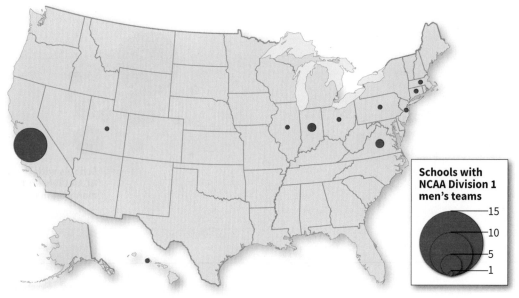

Schools with NCAA Division 1 men's teams

15
10
5
1

FIGURE 2.5 The distribution of NCAA Division I men's volleyball teams. There is a discernible concentration of teams on the Pacific coast, suggesting a core volleyball region.

is located in the southern half of the state. College sports are clearly useful for thinking about and identifying regional cultural patterns.

Indigenous Culture Regions

As we have seen in the example of lacrosse, the cultural practices of indigenous people are sometimes incorporated into popular culture. Yet most indigenous communities continue to be relatively isolated. Indigenous peoples generally live in areas with few roads or modern communications systems, such as mountainous areas, vast arid and semiarid regions, or large expanses of forest or wetlands. These concentrations of people constitute indigenous culture regions. Worldwide, large concentrations of indigenous populations exist outside the strong influence of national cultures and the control of governments, which are located in faraway capital cities. Control of indigenous regions by a central government is often weakened by the minimal infrastructure, rough topography, or harsh environmental conditions of the region. In many cases, indigenous peoples either fled or were subjected to forced migration by central governments to environmentally marginal regions, such as arid lands.

North American Indigenous Culture Regions In the United States and Canada, for example, this process of **geographic marginalization**, whereby a population is driven to remote, often rugged and unproductive lands, is particularly evident. In the United States, the central government has had a complex and often contradictory relationship with indigenous peoples, sometimes treating them as independent nations and at other times as second-class citizens. The way in which indigenous peoples are distributed in the United States reflects both the history of the east-to-west movement of European settlers and nineteenth-century government policies (**Figure 2.7**). For example, the Indian Removal Act of 1830 was intended to make way for European settlers by forcing eastern Native American tribes to migrate west of the Mississippi River, many of them to Oklahoma. Western Native American tribes were eventually forced onto government-created reservations, generally on the most unproductive and arid lands. Some of the larger reservation complexes in the arid West constitute an indigenous culture region today (**Figure 2.8**).

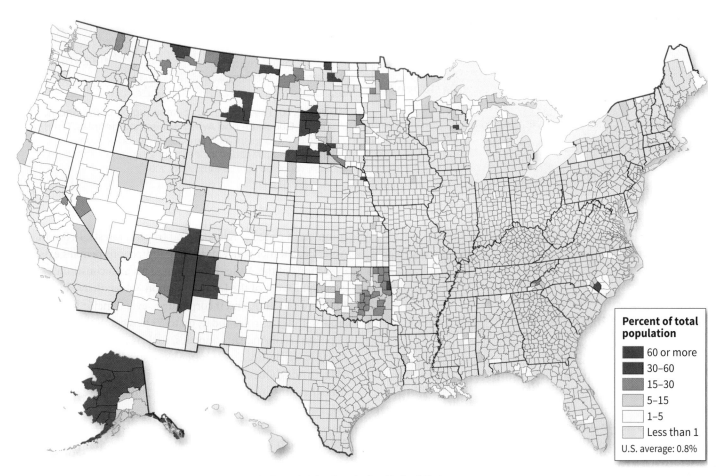

Percent of total population

- 60 or more
- 30–60
- 15–30
- 5–15
- 1–5
- Less than 1

U.S. average: 0.8%

FIGURE 2.7 Indigenous Native American population distribution in the United States. (U.S. Census Bureau)

FIGURE 2.8 An indigenous culture region in the United States. (U.S. Geological Survey)

FIGURE 2.9 The Mayan culture region in Middle America. The ancient Mayan Empire collapsed centuries ago, but its Mayan-speaking descendants continue to occupy the region. In many cases Mayan communities, after centuries of political and economic marginalization, are today actively struggling to have their land rights recognized by their respective governments.

Central and South American Indigenous Culture Regions

Indigenous culture regions also exist in Central and South America. There is a distinct Mayan culture region that encompasses parts of Mexico, Belize, Guatemala, and Honduras (**Figure 2.9**). Concentrations of Mayan speakers are especially

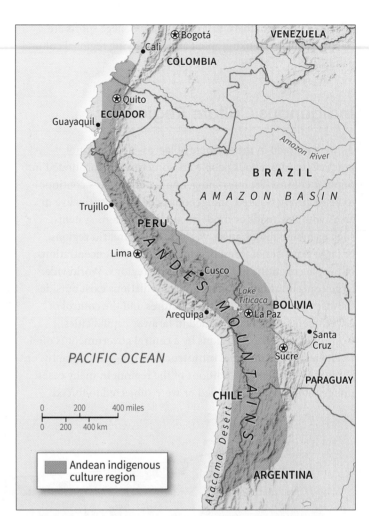

FIGURE 2.10 The indigenous culture region of the Andes, including Quechua- and Aymara-speaking peoples. This is the core of the ancient Inca Empire, where today many of the indigenous people speak their mother language rather than Spanish. (Adapted from de Blij & Muller, 2004.)

common in rugged highlands and tropical forests. In South America, another concentration of indigenous peoples exists in sections of the Andes Mountains. This area constituted the geographic core of the Inca civilization, which thrived between 1300 and 1533 C.E. and incorporated several major linguistic groups under its rule. Today, up to 55 percent of the national populations of Andean countries, such as Bolivia, Peru, and Ecuador, are indigenous. On the slopes and in the high valleys of the Andes, Quechua and Aymara speakers constitute an overwhelming majority, signifying an indigenous culture region (**Figure 2.10**).

South Asian Indigenous Culture Regions

The so-called Hill Tribes of South Asia are another good example of an indigenous culture region. Mountain ranges, including the Chittagong Hills, the Assam Hills, and the Himalayas,

FIGURE 2.11 A Murong grandmother with her grandchildren. The Murong and other indigenous peoples form the indigenous culture region of Bangladesh's Chittagong Hill Tracts. (Rashed Hasan/ AGE Fotostock)

surround the fertile valleys and deltas in which the ancient South Asian Hindu and Islamic civilizations were centered. Various indigenous peoples occupy these highland regions, which are remote from the lowland centers of authority and culturally distinct from them (**Figure 2.11**).

A series of indigenous culture regions ringing the valleys of South Asia thus exists, occupied by what the British colonial authorities referred to as Hill Tribes. Most of these peoples practice some version of swidden agriculture, which involves multiyear cycles of forest clearing, planting, and fallowing (see Chapter 8). Most hold Christian or animist beliefs (see Chapter 7) and speak languages distinct from those spoken in the lowlands. A similar pattern of highland indigenous culture regions can also be identified in the countries of Southeast Asia, such as Myanmar and Thailand, where indigenous peoples such as the Shan and Karen have populations in the millions.

LGBT Districts

Like sports, indigenous identity, and other aspects of human culture, sexuality can be expressed geographically. One of the most dynamic developments in cultural geography over the past two decades has been the emergence of a new field of study on sexuality and space, referred to as *queer geography* (or sometimes *gay geography*). The emergence of queer geography as a field of study reflects the increasing acceptance of lesbian, bisexual, gay, and transgendered (LGBT) identities in popular culture.

As geographers Larry Knopp and Michael Brown have noted, from the 1990s onward there has been an expanding

---

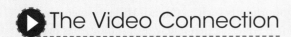

The Video Connection

An Attack on Equality

 Watch at 🌿 Sapling Plus

This video centers on an interview with two young victims of a hate crime against same-sex partners. The attack took place in daylight on a busy public street near Madison Square Garden. Such hate crimes, the video reports, are on the rise in New York City, despite the city's reputation for embracing diversity. The victims explain that the circumstances of the attack have caused them to reevaluate their ideas about New York City and its residents' readiness to accommodate and defend cultural diversity. They conclude that the State of New York's legalization of same-sex marriage has not meant the end of discrimination and violence against openly gay citizens.

Thinking Geographically

1. Besides the hate crimes, what are some examples of unequal treatment of same-sex couples mentioned in the video? In particular, think about how same-sex couples may experience public space differently than heterosexual couples as a result of unequal treatment.

2. Sexuality, as one mode of cultural difference, is expressed geographically. How do the violent attacks discussed in the video help explain why so many cities have defined LGBT districts?

3. This chapter discusses the idea of place images. In the video, one person describes his childhood image of New York as "this dream metropolis" and "basically the Emerald City that I've always wanted to go to." What do you think he meant by that? How do you think the attacks have altered the place image that fueled his move to New York?

FIGURE 2.12 Stonewall riots. Protestors attempt to impede police arrests outside the Stonewall Inn in Greenwich Village in 1969. These protests are viewed as the catalyst of the gay rights movement, which continues to work to end all forms of repression and discrimination based on sexual preferences. (New York Daily News Archive/Getty Images)

FIGURE 2.13 Seattle's gay pride parade. Gay pride parades in the United States trace their origins to the 1970 marches in several cities to commemorate the first anniversary of the Stonewall riots. To understand the huge cultural shift that has occurred over five decades, compare the police in Figure 2.12 with the police in this figure. (Elaine Thompson/AP Photo)

interest in documenting the LGBT experience, most notably through a series of case studies of North American cities. These studies have coincided with a dramatic turnaround in popular attitudes toward sexualities that historically had been kept "in the closet." That is, human sexuality that was different from a narrowly prescribed heterosexual norm was until recently repressed in public discourse and made invisible in the landscape (**Figure 2.12**).

Mapping LGBT culture is one way that "hidden" lives are made visible and unrecognized contributions to urban social life are acknowledged. Using GIS mapping software, archived materials, and interviews with LGBT activists, Knopp and Brown were able to reconstruct the geographic experience of LGBT residents of Seattle. They documented the location of a wide range of "gay-friendly" places, including bookstores, bars, theaters, and restaurants as well as the public spaces where LGBT residents felt comfortable socializing. In addition, they used their findings to produce a map and a poster that were exhibited at the city's Gay Pride Festival, thereby highlighting the many LGBT contributions to the cultural, political, and economic development of the city (**Figure 2.13**). Among Knopp and Brown's other findings was that LGBT influence was strong throughout the city

and not just confined to the widely recognized gay-friendly neighborhood of Capitol Hill.

Neighborhoods such as Capitol Hill are so common across North America that a new terminology has evolved to describe them. Known as a "gayborhood," "gay village," or simply "**LGBT district**," such urban spaces may be found in many major cities. As with numerous urban subcultures, LGBT residents have tended to cluster in neighborhoods where they feel comfortable among accepting or like-minded people and are able to live without hiding their sexual identity. Today, rather than try to suppress or hide such districts, local governments are celebrating them as part of the cultural diversity of their cities. Chicago's "Boystown" is a good example. Boystown is a new addition to Chicago's diverse patchwork of neighborhoods. It is the first (and one of the largest) officially recognized LGBT district in the United States. It is featured on the city's web site "Choose Chicago," and signified by the city's installation of rainbow-colored Art Deco pillars to mark its borders (**Figure 2.14**). Choose Chicago's mission is to "sell" the city to tourists, business investors, and prospective homeowners. It is interesting to note how an LGBT district has suddenly gone from being hidden to being marketed on the city's web site.

FIGURE 2.14 Sidewalk towers with rainbow color bands mark Chicago's Boystown neighborhood. Rather than suppress and hide the presence of LGBT groups, as was done in the recent past, Chicago has today chosen to celebrate this neighborhood. (Charles Cook/Getty Images)

Mobility

2.2 Recognize the mobility of cultural traits and interactions through the examples of tourism, diaspora cultures, and the Internet.

The movement of things, ideas, and people shapes and is shaped by geography. For example, things and people in motion can sometimes minimize the relevance of nation-state borders. At other times, nation-state borders can greatly restrict mobility. The interactions of culture and geographic mobility are complex and not altogether predictable. We will examine a range of examples to illustrate the complexities of geographic mobility, including international tourism, diasporas, and digital information.

Vampire Tourism

No other practice better exemplifies mobility in popular culture than tourist travel. **Tourism**, a modern activity that arose with the introduction of air, sea, and land mass transit systems, involves people traveling from their homes for purposes of business, leisure, entertainment, and recreation. There is an entire industry and infrastructure (e.g., theme parks, cruise-ship ports) that caters to an endless variety of tourist interests and destinations. The opportunity to travel to different cultures, places, and landscapes is now a leading form of leisure, one that plays a key role in the way cultural difference is experienced and understood. Let us explore the complex mobilities of tourism with a look at vampires.

Vampire tourism offers a surprising blend of reality and fiction, and of folk and popular cultures. According to geographer Duncan Light, vampire tourism is a form of "media tourism," a practice whereby fans of a fictional story or character—whether portrayed in a novel, film, or television program—travel to key places and landscapes in the plot. Vampires are popping up everywhere in popular culture. They appear in novels and cinema (the *Twilight* series), animated feature films (*Hotel Transylvania*), graphic novels (*Interview with the Vampire: Claudia's Story*), video games (*Vampyr*), and television (*The Vampire Diaries*). Between 2010 and 2017, an average of four vampire-themed films were released each year in the United States. So saturated is popular culture with vampire stories that one would be hard pressed to find a mass entertainment medium in which they do not materialize. Interestingly, the undead lives of these ubiquitous blood suckers are anchored in real geography, resulting in the rapid growth in vampire tourism nationally and globally. The Lonely Planet, a well-known publisher of travel guides, features a list of the "world's best vampire-spotting locations" on its web site.

Fictional vampire characters appear in the most unexpected places, including Forks, Washington; New Orleans, Louisiana; Volterra, Italy; and Mystic Falls, Virginia. Actually, Mystic Falls, the setting for the *The Vampire Diaries,* is fictional, too, but the television show is shot in real-life Covington, Georgia. Entrepreneurs in Covington have wasted no time creating "Mystic Falls" tour packages. As one web site advertises, "The Mystic Falls Tour is the perfect destination for travelers looking for a memorable Vampire Diaries adventure." Forks, Washington, where the *Twilight* series is based, has been overrun by hundreds of tourists daily—many of them young

FIGURE 2.15 Transylvania, Count Dracula's imagined home. Bran Castle, on the edge of Romania's Transylvania region, has become the mecca of vampire tourism.

FIGURE 2.16 Souvenir shops at Bran Castle in Romania's Transylvania region. The building is popularly known as "Dracula's Castle" due to its resemblance to the description in Bram Stoker's novel *Dracula*. (Robert B. Fishman/picture-alliance/dpa/AP Images)

women around the age of Bella's character—a phenomenon that has significantly improved the fortunes of this economically depressed logging town.

The Curse of Dracula, Revisited Forks and Covington, however, are mere way stations on the pilgrimage route to the mecca of vampire tourism, the Transylvania region of Romania (**Figure 2.15**). Transylvania is home to the world's most famous vampire, Count Dracula, subject of Bram Stoker's 1897 novel *Dracula* and of the numerous subsequent film versions. Stoker loosely based Dracula on vampire folktales that circulated throughout southeastern Europe in the eighteenth century. In his novel, Stoker located the count's home in Transylvania's Dran Castle, which today makes it a global destination for vampire tourists (**Figure 2.16**). Ironically, no folk tradition of vampirism exists in the region's culture. This inconvenient fact did not prevent Stoker from loosely modeling Count Dracula on Vlad III (or Vlad Ṭepeș), a real-life Transylvanian ruler from the 1450s whom the Romanians revere as a national hero for his role in repelling an Ottoman Turkish invasion.

In his study *The Dracula Dilemma*, Light shows how vampire tourism raises difficult questions for the Romanian government and its citizens about national and cultural identity. The Dracula story, Light points out, positions Romania as a backward, superstitious region on the eastern margins of modern, civilized western Europe. Moreover, Stoker links the evil beast Dracula, who is threatening the English homeland,

to Vlad III, a heroic figure in the creation of Romanian cultural and national identity. Imagine the world outside the United States believing George Washington was a vampire. In short, Stoker's character casts Romania in a negative light and challenges its citizens' national identity at its deepest level. Dracula is an awkward figure around which to build a Romanian tourist industry.

Cashing In on the Count Until 1989, Romania was a communist country where government censorship of popular media, including the Dracula stories, was widespread. When the country abandoned communism, it lifted censorship and unleashed private entrepreneurship. Stoker's novel and the Dracula films began to circulate widely in Romania. The borders

FIGURE 2.17 Dracula mugs in a souvenir shop at Bran Castle. Businesses in Romania are cashing in on the Dracula name and image. (Sean Gallup/Getty Images)

were opened up to the global tourist industry. Businesses started trading on the Dracula name, which is used to market guesthouses, restaurants, and even beer and cigarettes to foreign tourists (**Figure 2.17**).

The Romanian government and many citizens were not as quick as entrepreneurs to embrace Stoker's portrayal of Transylvania. As Light points out, the Romanian nation-state wishes "to project a sense of its own political and cultural identity to the wider world on its own terms." Those terms are defined by a desire to be politically and economically integrated into the European Union (see Chapter 6) while maintaining a distinct and proud Romanian cultural heritage. The ironies and frustrations of being the nation and culture most identified with Dracula are manifold. Vampires made no appearance in Romanian folk culture, but the whole world thinks of Transylvania as the home of the most infamous vampire of all.

Vampire tourism illustrates a number of features of mobility in popular culture. There is the mobility of the vampire myth; the ways it is transported and transplanted from folk to popular culture; the ways it circulates globally, linking unlikely places as distant and dissimilar as Forks, Washington, and Transylvania, Romania; and the ways it has set people in motion in search of direct experiences with these places. Fictional characters move through factual landscapes and consequently affect real-life political, economic, and cultural relations through media tourism. The mobility of vampire tourists following fictional plot lines, as Light emphasizes for Romania, "illustrates global inequalities and asymmetries in cultural power." The Romanian government and citizenry have fought a losing battle for control over representation of their cultural heritage. In popular culture,

Transylvania appears doomed to an undying association with a superstition transported from folk traditions rooted in other places.

Sudanese Diaspora Culture

Diaspora cultures are, by definition, born of mobility. Civil war and international armed conflicts over the past few generations have set millions of people in motion around the world, creating many new diaspora cultures. The list is long and includes the Cuban diaspora (following the 1959 communist revolution in Cuba), the Hmong diaspora (following the end of the U.S.–Vietnam War in 1975), and the Sudanese diaspora (following civil wars ongoing since the 1980s).

Decades of civil wars in Sudan have displaced at least 4 million people, creating a new diaspora located on three different continents (**Figure 2.18**). Major armed conflict in Sudan developed around the economic, religious, cultural, and racial differences distinguishing the Arabic/Islamic northern region, where the military and government were centered, from the African/Christian southern region, where rich oil deposits are found. Thus, a significant portion of the diaspora are southerners fleeing the fighting between the northern military and southern insurgents. The United States has been among the largest receiver countries, taking in approximately 150,000 southern Sudanese. Major Sudanese diaspora concentrations are located in Minnesota, New York, and Texas.

FIGURE 2.18 Refugees from the Darfur region of Sudan awaiting international aid in northeastern Chad. Civil wars in Sudan spanning decades have displaced millions of people and produced a Sudanese diaspora. (Scott Nelson/Getty Images)

FIGURE 2.19 Southern Sudanese celebrate their first independence day in the capital city of Juba on Saturday, July 9, 2011. Members of the Sudanese diaspora watched the festivities on television and conducted their own celebrations in cities around the world. (Pete Muller/AP Photo)

Geographer Caroline Faria conducted research among the Sudanese diaspora communities in the United States to investigate how members forge cultural and national identities under conditions of displacement. Faria's study of the Sudanese diaspora took place during a critical historical juncture between the 2005 signing of a peace agreement and the 2011 establishment of South Sudan, the world's newest nation-state (**Figure 2.19**). During this period the diaspora community vigorously debated what a new nation should be and what it would mean, culturally, to be a South Sudanese citizen.

Faria analyzed the first and second "Miss South Sudan" contests held in Washington, D.C., in 2006 and Kansas City in 2007 to demonstrate the importance of diasporic beauty pageants for "promoting and imagining a distant home and nation." Beauty pageants, in general, are fruitful sites for studying debates around gender, race, and national identity, because the contest winner is assumed to represent, even embody, the essential qualities of the nation. Thus in the Miss South Sudan contests, organizers, participants, and spectators struggled to identify an ideal type of woman who would signify the new nation.

As Faria discovered after analyzing interviews and public commentary and debate, the ideal South Sudanese woman must conform to certain racial, religious, and gender identities. First, differences in religious beliefs both supported and were accentuated by the north–south armed conflict in Sudan. Christianity has become an important marker of difference

to distinguish southern Sudanese cultural identity from that of the north. Miss South Sudan pageant commentators and judges thus emphasized "a strong Christian Faith as a vital part of any perspective winner." Second, southern Sudan is imagined as "culturally and phenotypically oriented toward Sub-Saharan Africa while the North is associated with an Arab influence and a lighter skin tone." Consequently, participants and judges in the Miss South Sudan contests linked the notion of feminine beauty to "blackness," with the 2006 winner opting for the "natural" look of a simple hair weave that projected a more "African" identity for the new nation (**Figure 2.20**). Finally, the pageant exposed an unresolved contradiction in the Sudanese diaspora's ideas of womanhood and women's ideal role in the new nation of South Sudan. Contestants were expected on the one hand to delay marriage and aspire to higher education and professional careers, while on the other to maintain a traditional role as mothers who would reproduce the new nation.

The case of the Miss South Sudan pageants highlights the challenge facing many diaspora cultures around the world: how to establish a coherent identity under conditions of forced migration and uncertainty (see Subject to Debate). In imagining national and cultural identities from a distance, diaspora cultures must confront basic questions about national ideals of masculinity and femininity and the relationship of race and religion to nationhood. Showcases of national culture, such as beauty pageants, are not merely sources of popular entertainment. Rather, they are cultural cauldrons in which national identities are cemented, reproduced, and reimagined.

FIGURE 2.20 Contestants at the inaugural Miss South Sudan beauty contest. Rita Magoh, left, and Grace Bok, both South Sudanese from Kansas City. Such cultural events are important to diaspora communities for defining national identities. (Jahi Chikwendiu/The Washington Post/Getty Images)

SUBJECT TO DEBATE Mobile Identities: Questions of Culture and Citizenship

One of the most dynamic features of cultural difference in the world today is mobility. Diaspora communities have sprung up around the globe. Some of these communities have arisen quite recently and are historically unprecedented, such as the movement of Southeast Asians into U.S. cities in the late twentieth century. Some are artifacts of European empires, such as the large populations of South Asians in England or West and North Africans in France. Others are deeply historical and express a centuries-old interregional linkage, such as the contemporary movement of North Africans into southern Spain.

The movement and settlement of large populations of migrants have raised questions of belonging and exclusion. How do transplanted populations become English, French, or Spanish, not only in terms of citizenship but also in terms of belonging to that culture? Some observers have argued that diaspora populations find ways to blend symbols from their cultures of origin with those of their host cultures. Thus, people develop a sense of belonging through a process of cultural hybridization. Other observers point to long-standing situations of cultural exclusion. In 2005 in France, for example, riots broke out in more than a dozen cities in suburban enclaves of West African and North African populations. The reasons for the riots were complex. However, many observers pointed out that underprivileged youth of African descent feel excluded from mainstream French culture, even though many are second- and third-generation French citizens. Another round of riots occurred in 2017, which many observers say was rooted in the same exclusions and inequalities that underlay the 2005 riots.

The debate over how to address questions of citizenship and cultural belonging is played out in many venues. In terms of policy, the French emphasize cultural integration; for example, the French government approved a law in 2010 banning full-face veils (burqas) in public, whereas Britain and the United States acknowledge multiculturalism. But many questions remain about the effectiveness of state policies toward diaspora cultures.

Continuing the Debate

Based on the discussion presented above, consider these questions:

- Do the geographic enclaves of Asian and African diaspora populations in the former colonial capitals of Europe reflect an effort by migrants to retain a distinct cultural identity? Or do they reflect persisting racial and ethnic prejudices and efforts to segregate "foreigners"? Or is it a combination of factors?

- How long does it take an Asian immigrant community to become "English" or a West African immigrant community to become "French"? One generation? Two? Never?

- Does the presence of Asians and Africans make the landscape of England and France appear less "English" or less "French"? Why or why not?

Riots and protests, such as the Paris suburb of Bobigny, spread across several cities in France in 2017. Protesters' complaints centered on the discriminatory treatment of French citizens of African descent. (GEOFFROY VAN DER HASSELT/Getty Images)

Barriers to the Movement of Digital Information

Although mass media create the potential for the almost instant diffusion of information over very large areas, this potential can be greatly impeded and slowed if access is limited or denied. Access to the Internet is a case in point. Though older forms of mass communication, such as television and newspapers, are unlikely to disappear, an increasing proportion of people worldwide turn to the Internet for information. The shift in technology has created a so-called **digital divide** between those who have ready access to the Internet information stream and those who have limited or no access (**Figure 2.21**). Many factors contribute to the digital divide, but one of the most basic is lack of Internet infrastructure in many parts of the world (see Chapter 9). Thus, while information may, in seconds, diffuse globally on the Internet, it is irrelevant to those without a device and the infrastructure with which to connect. A historical

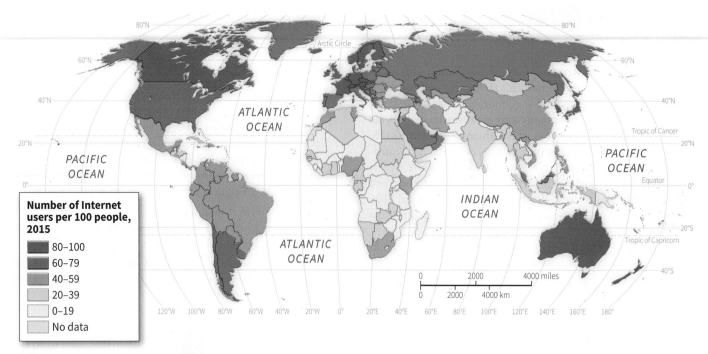

FIGURE 2.21 One way to assess the digital divide on a global scale. While the gaps among regions are closing, major hurdles to more equitable access remain.

analogy might be the invention of the printing press when most of the world's population was illiterate.

Recognition of the digital divide and potential solutions has arisen and spread quickly, precisely because information moves around the globe at such great speed. New initiatives meant to address the gap spring up continually. For example, the Digital Divide Institute operates as a policy think tank to advise companies and governments on solutions, and the Pew Research Center conducts surveys on who does and does not use the Internet and why. The Internet has both exacerbated socioeconomic inequalities and allowed for a rapid response to try to correct them.

In addition to the existing inequalities in Internet access, governments around the world periodically implement technology bans that have the effect of limiting the diffusion of information and ideas (**Figure 2.22**). For years Cuba banned cell phones, fearing that citizens would have access to information unfavorable to the government. A huge black market in cell phones forced the government to lift the ban in 2008. The government of the Czech Republic in 2010 banned the use of Google's Street View technology over privacy concerns. It has since lifted the ban after negotiating very strict rules on how Google is allowed to use the technology. The United Arab Emirates briefly banned the BlackBerry because the device's encryption makes it difficult for governments to monitor content for what they consider objectionable material.

Though all these bans were ultimately lifted, governments will continue to try to limit or control the diffusion of

FIGURE 2.22 Some countries have attempted to ban communication technologies in an effort to control citizens' access to information. For years, the Cuban government tried to ban cell phones. The ban ultimately failed because people were able turn to an illegal market in technology. (STR/AFP/Getty Images)

information across nation-state borders. In attempts to combat global terrorism, governments are focusing on restricting the use of encryption software in smartphones and other devices. Responding to revelations of governments' mass surveillance of personal communications, in 2014 Apple Corporation added a default encryption to their iPhones' operating software. The government of the United Kingdom has proposed legislation to ban such encryption, while the United States government is seeking a ban through the federal court system. The unresolved tensions between individual privacy rights and national security will keep government technology bans in public debate for years to come.

Globalization

2.3 Explain how globalizing processes are influencing cultural interactions and change, such as the experience of space-time and cultural homogenization.

The effects of globalization on cultural difference and vice versa are a much-discussed and debated issue for geographers and other social scientists. Rapid transportation systems (e.g., intercontinental jets) and new communications technologies (e.g., smartphones, GPS) are accelerating interactions among people and places, resulting in cultural change. In this section we will examine the interactions of globalization and cultural practices, beginning with a look at how globalized communication and transportation have altered people's sense of time and space, and what that might mean for processes of cultural homogenization and efforts to retain distinctive indigenous identities.

The Differing Experiences of Time and Space

As we noted in Chapter 1, globalization is deeply rooted in the history of increasing interdependence and integration of places and cultures, facilitated by faster and faster forms of transportation and communication. Geographers have observed that this speeding up of life and movement produces a uniquely modern experience of time and space, which they call **time-space convergence**. Time-space convergence is the phenomenon whereby the time taken to travel between places is progressively reduced through the introduction of new transportation technologies. This process is very ancient—think of the invention of the wheel—but it has accelerated greatly since the Industrial Revolution (see Chapter 9) and so is usually associated with modern times. From the horse carriage, to steamships and railroads, to jet planes, the world feels increasingly smaller **(Figure 2.23)**.

Time-space convergence provides another tool for thinking in terms of multiple geographies, rather than a single, universal geography. To begin, we know that transportation changes do not actually "shrink" the globe. The Earth remains unchanged in terms of **absolute distance**, which is the precise measurement

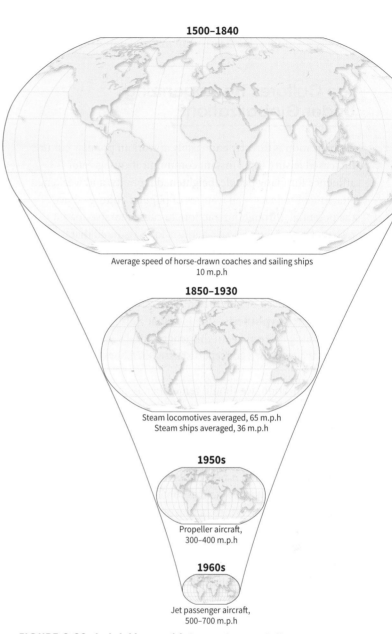

1500–1840
Average speed of horse-drawn coaches and sailing ships 10 m.p.h

1850–1930
Steam locomotives averaged, 65 m.p.h
Steam ships averaged, 36 m.p.h

1950s
Propeller aircraft, 300–400 m.p.h

1960s
Jet passenger aircraft, 500–700 m.p.h

FIGURE 2.23 A shrinking world. As new transportation technologies shorten the time between places, the more affluent among us experience the feeling that the world is becoming smaller.

of physical space between one point on the Earth's surface and another (as with the longitude and latitude system). However, the **relative distance**, measured by changes in time and cost of movement between points on the surface, has been greatly altered by technological change. Consequently, one's sense of time and space reflects one's cultural, social, and economic circumstances. Life moves much more slowly, and the world seems unfathomably larger to a poor subsistence farmer in Nicaragua (see later discussion on page 63) than to a geography professor in London. The differing experiences of time and space is a defining characteristic of folk and popular cultures.

Are Cultures Homogenizing Under Globalization?

Globalization is directly and visibly at work in popular culture. Increased leisure time, instant communications, greater affluence for many people, heightened mobility, and weakened attachment to place—all features of popular culture—have the potential, through interaction, to cause massive spatial restructuring. Many social scientists long assumed that the result of such globalizing forces and trends, especially mobility and electronic media, would be the homogenization of formerly diverse cultures, wherein the differences among places are reduced or eliminated. This assumption is called the **convergence hypothesis**—the idea that cultures are converging, or becoming more alike. In the geographical sense, this would yield placelessness, a concept discussed in Chapter 1 (see Figure 1.8).

Globalization, we should remember, is an ongoing process or, more accurately, a set of processes. It is incomplete and its outcome is far from predetermined. Geographer Peter Jackson is a strong proponent of the position that cultural differences are not simply obliterated under the wave of globalization. For Jackson, globalization is best understood as a "site of struggle." He means that cultural practices rooted in place shape the effects of globalization through resistance, transformation, and hybridization. In other words, globalization is not an all-powerful force. People in different places respond in different ways, rejecting outright some of what globalization brings while transforming and absorbing other aspects into local culture. Rather than one homogeneous globalized culture, Jackson sees multiple local consumption cultures.

Jackson was referring to consumption practices and preferences—in food, clothing, music, and so on—formed in specific places and historical moments. These local consumption cultures often shape globalization and its effects. In some ways, globalization revitalizes local difference. That is, people reject or incorporate into their cultural practices the ideas and artifacts of globalization and in the process reassert place-based identities. For example, global fast-food chains such as KFC and McDonald's often alter menus to match local food tastes. In China, KFC offers local favorites such as pumpkin porridge and *youtiao* (deep-fried dough), egg soup, and Beijing chicken rolls.

In some places, resistance to globalization may take the form of **consumer nationalism**. This occurs when local consumers avoid foreign companies or imported products and favor domestic businesses and products (**Figure 2.24**). It appears

FIGURE 2.24 Residents protest against Pepsi in Ahmedabad, Gujarat, India. Highly visible global brands, such as Pepsi, are often targeted outside the United States as symbols of foreign intrusion. (Shailesh Raval/The India Today Group/Getty Images)

that for every homogenizing tendency under globalization, we can find examples of place-based resistance to it.

Local Indigenous Cultures Go Global

Global-local tensions also exist for the world's indigenous peoples, who often interact with globalization in interesting ways. On the one hand, new global communications systems, institutions of global governance, and international nongovernmental organizations (NGOs) are providing indigenous peoples with extraordinary networking possibilities. Local indigenous peoples around the world are now linked in global networks that allow them to share strategies, rally international support for local causes, and create a united front to defend cultural survival. On the other hand, globalization brings the world to formerly isolated cultures. Global mass communications introduce new values, and multinational corporations' search for new markets and new sources of gas, oil, genetic, forest, and other resources can threaten local economies and environments.

Both aspects of indigenous peoples' interactions with globalization were evident at the World Trade Organization's (WTO) Ministerial Conference in Cancún, Mexico, in 2003. Indigenous peoples' organizations from around the world gathered for the conference, hosted by the Mayan community in nearby Quintana Roo. Though not officially part of the conference, they came together there to strategize ways to forward their collective cause of cultural survival and

FIGURE 2.25 A group of indigenous Filipinos participate in the opening of the Forum for Indigenous People at the Casa de la Cultura in Cancún, Quintana Roo, Mexico, during the World Trade Organization ministerial meetings. Indigenous peoples' groups from around the world organized the forum as a counterpoint to the WTO talks. (Jack Kurtz/The Image Works)

self-determination, gain worldwide publicity, and protest the WTO's vision of globalization (**Figure 2.25**). One outcome of this meeting was the International Cancún Declaration of Indigenous Peoples (ICDIP), a document highly critical of current trends in globalization.

Globalization is clearly a critical issue for indigenous cultures. Some argue that because globalization facilitates the creation of global networks that provide strength in numbers, it may ultimately improve indigenous peoples' efforts to control their own destinies. The future of indigenous cultural survival will depend on how globalization is structured and for whose benefit.

Nature-Culture

2.4 Explain how culture and the physical environment interact in indigenous, folk, and popular cultures and the influences of gender and the global economy on those interactions.

People who depend directly on the land for their livelihoods—farmers, hunters, loggers, and ranchers, for example—often have a different view of nature from those who work in the offices, schools, factories, and shopping malls of the cities. Let's now look at the relationship between nature and culture and how different cultures and subcultures differ in their interactions with the physical environment. In this section, we will examine a variety of relationships using the examples of indigenous ecology and property rights, folk ecology, gendered nature, and nature in popular culture.

Indigenous Ecology and Property Rights

Many observers believe that indigenous peoples possess a close relationship with and a great deal of knowledge about their physical environment. In many cases, indigenous cultures have developed sustainable land-use practices over generations of experimentation in a particular environmental setting. As a consequence, academics, journalists, and even corporate advertisers frequently portray indigenous peoples as defenders of endangered environments, such as tropical rain forests. It was not always this way. Historically, Europeans often accused indigenous populations of destroying the environment. In hindsight, it is easy to see that this claim was related to now-discredited European ideas about the racial inferiority of nonwhite peoples.

Debate continues today, with some observing that although indigenous cultures once may have lived sustainably, globalization is making their knowledge and practices less useful. That is, globalization introduces new markets, new types of crops, and new technologies that displace existing land-use practices. Others note that it is impossible to generalize about sustainability in indigenous cultures because the way indigenous peoples use their environments varies from place to place and the indigenous societies are internally heterogeneous. A key discussion centers on the role of indigenous peoples in conserving the environment. The discussion is important because indigenous peoples often occupy territories of high **biodiversity** (**Figure 2.26**). Biodiversity—a blending of biology and diversity—refers to the variety and variability of life on Earth as measured at various scales, including diversity among individuals, populations, species, communities, and ecosystems. Biodiversity thus includes genetic diversity, species diversity, and habitat diversity.

For example, many of the most biologically diverse areas in Latin America are under government protection in national parks and reserves. Eighty-five percent of these protected areas in Central America and 80 percent in South America contain resident indigenous populations. There is also a close geographic correspondence between indigenous territories and tropical rain forests, not only in Latin America, but also in Africa and Southeast Asia. Tropical rain forests, although they cover only 6 percent of the Earth's surface, are estimated to contain 60 percent of the world's biodiversity. Thus, some see indigenous

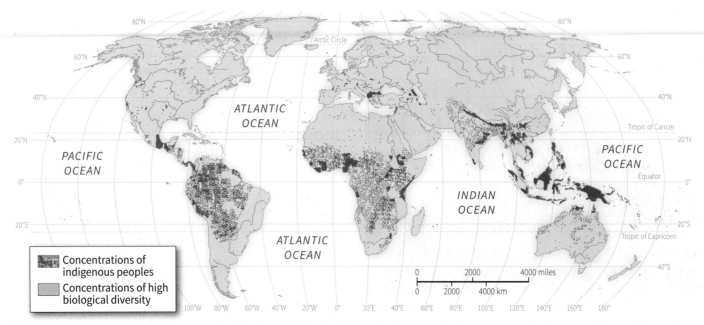

FIGURE 2.26 The global congruence of cultural and biological diversity. Conservationists are aware that many of the world's most biologically diverse regions are occupied by indigenous cultures. Many suggest, therefore, that cultural preservation and biological preservation should go hand in hand. (Adapted from IDRC, 2004.)

cultures as traditional caretakers of a precious resource. With the rise of genetic engineering, conservationists and corporations alike view tropical forests as in situ gene banks. That is, genetic diversity can be held in place in standing tropical forests as a reserve to be tapped in the future.

As multinational biotechnology companies look to the tropics for genetic resources for use in developing new medicines or crop seeds, however, indigenous peoples are increasingly vocal about their proprietary rights over the biodiversity of their homelands. Indigenous rights organizations claim that "our territories and resources, our indigenous knowledge, cultures and identities are grossly violated" by international trade and investment rules. The ICDIP document discussed on the previous page urges governments worldwide to make no further agreements under the WTO and to reconsider previous agreements. The control of plant genetic resources is a particularly important concern. Many indigenous peoples argue that generations worth of their labor and cumulative knowledge have gone into producing the genetic resources that transnational corporations are trying to privatize for their own profit.

Indigenous Technical Knowledge In light of such conflicts, cultural geographers emphasize the value of both the knowledge indigenous peoples bring to environmental management and their land-use practices for sustainable development. Initially, geographers focused on how indigenous cultures adapted to ecological conditions. For example, some studied the social

norms and land-use practices that helped certain cultures adapt to periodic drought. Later, they came to realize that external political and economic forces were just as important in shaping nature–culture relationships. Much of the interest in indigenous perceptions and practices falls under the category of **indigenous technical knowledge (ITK)**. This is a concept that anthropologists and geographers developed to describe the detailed local knowledge about the environment and land use that is part of many indigenous cultures. Some geographers have argued that ITK is sometimes superior to Western scientific knowledge and, therefore, should be incorporated into local environmental management and agricultural planning. Many indigenous peoples would also argue that ITK is proprietary knowledge that must be recognized and compensated.

ITK is place based. It is produced in particular places and environments through a process of trial and error that often spans generations. Thus, these local systems of knowledge are highly adapted to local conditions. Increasingly, however, the power of ITK is weakened through exposure to the economic forces of globalization. Sometimes the global economy applies such pressure to local **subsistence economies** that they become ecologically unsustainable. Subsistence economies are those that import little if any food and are oriented primarily toward production of food for local consumption, rather than for sale on the market. When an indigenous society organized for subsistence production begins producing for an external market, social, ecological, and economic difficulties often ensue.

Indigenous Encounters with the Global Economy

Geographer Bernard Nietschmann's classic study of the indigenous Miskito communities living along the Caribbean coast of Nicaragua showed how external markets can undermine local subsistence economies. Miskito communities had developed a subsistence economy founded on land-based gardening and the harvesting of marine resources, including green turtles. Marine resources were harvested in seasons in which agriculture was less demanding. The value of green turtles increased dramatically when companies moved in to process and export turtle products (meat, shells, leather). They paid cash and extended credit so that the Miskito could harvest turtles year-round instead of seasonally. Subsistence production in other areas suffered as labor was directed to harvesting turtles. Turtles became scarce, so more labor time was required to hunt them in a desperate effort to pay debts and buy food. Ultimately, the turtle population was decimated and the subsistence production system collapsed.

This study might sound like yet another tragic story of "disappearing peoples" or a "vanished way of life," but it did not end there. Nietschmann continued his research with Miskito communities, discovering, among other things, that the Miskito people continued to defend their cultural independence and also responded to the turtle scarcity. In 1991, with the cooperation of the Nicaraguan government, they created a protected area as part of a local environmental management plan. They were supported in this endeavor by academics and international conservation NGOs. Known as the Cayos Miskitos and Franja Costera Marine Biological Reserve, it encompasses 5019 square miles (13,000 square kilometers) of coastal area and offshore keys with 38 Miskito communities. Through this program, the Miskito were able to regulate and control their own exploitation of marine resources while reducing pressures from outsiders. While this is an ongoing experiment, it provides a good example of indigenous peoples' efforts to maintain a distinct cultural identity and way of life.

The Miskito case demonstrates the resiliency of indigenous cultures, the limits of ITK, and the strength of global economic forces. More recent studies of the cultural and political ecology of indigenous peoples reflect these themes. For example, geographer Anthony Bebbington conducted research among the indigenous Quechua populations in the Ecuadorian Andes to assess how they interact with modernizing institutions and practices. He found that although the Quechua people often possess extensive ITK about local farming and resource management, having that type of knowledge alone is not sufficient to allow them to prosper in a global economy. For instance, there is little indigenous knowledge about the way international markets work and thus little understanding of how to price and market produce. As a consequence, Quechua farmers have sought the support and knowledge of government agencies, the Catholic Church, and NGOs. Bebbington further found that indigenous Quechua communities use outside ideas and technologies to promote their own cultural survival, attempting, in essence, to negotiate their interactions with globalization on their own terms.

Folk Ecology

As with indigenous cultures, ideas persist about the particular abilities of folk cultures to sustainably manage the environment. Although the attention to conservation varies from culture to culture, folk cultures' close ties to the land often produce detailed local ecological knowledge within folk groups. This becomes particularly evident when they migrate. Typically, they seek new lands similar to those left behind. A good example can be seen in the migrations of Upland Southerners from the mountains of Appalachia between 1830 and 1930. As the Appalachians became increasingly populous, many Upland Southerners began looking elsewhere for similar areas to settle. Initially, they found an environmental twin of the Appalachians in the Ozark-Ouachita Mountains of Missouri and Arkansas. Somewhat later, others sought out the hollows, coves, and gaps of the central Texas Hill Country. The final migration of Appalachian hill people brought some 15,000 members of this folk culture to the Cascade and Coast mountain ranges of Washington State between 1880 and 1930 (**Figure 2.27**).

Although folk cultures have largely disappeared in the United States, the human inventiveness, curiosity, and need related to local ecologies remain, as we noted. Thus, there are many examples where local people derive what we might call folk solutions when confronted with a challenge of physical geography. The history of mountain bikes provides an example. Mountain bikes were invented when young men in the mountain regions of the western United States attempted to figure out how to adapt existing technology to a new enterprise. Bikes were cannibalized for parts and reassembled into something new and local. Enthusiasts rode these single-speed "clunkers" with big balloon tires, careening down mountain trails and dirt roads in a sport that did not yet have a name. Eventually, these locally imagined, hand-assembled clunkers became the rough prototypes of today's mass-produced mountain bikes.

Another good example is the swamp buggy, a folk solution to the problem of navigating in and around the Everglades and associated wetlands of southern Florida (**Figure 2.28**). Local hunters in southwestern Florida needed a way to access wildlife. Around 1920 they came up with a solution that involved installing huge balloon tires, oversized suspensions, and other modifications on pickup trucks. Jacked up high, these vehicles are able to move through standing water and over rough terrain

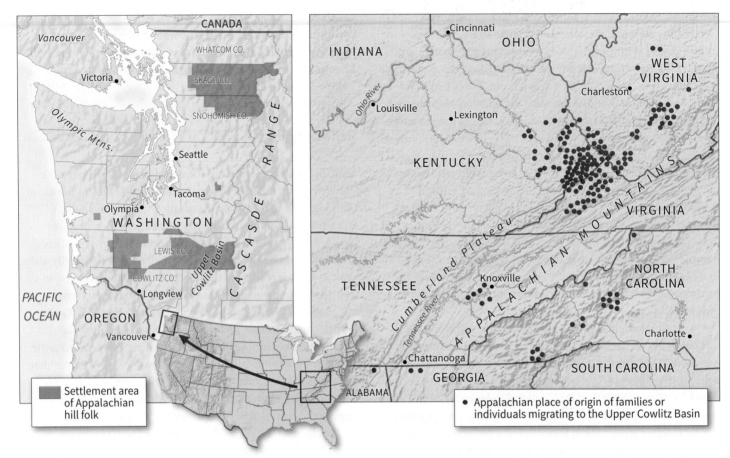

FIGURE 2.27 The relocation diffusion of Upland Southern hill folk from Appalachia to western Washington. Each dot represents the former home of an individual or family that migrated to the Upper Cowlitz River basin in the Cascade Mountains of Washington State between 1884 and 1937. Some 3000 descendants of these migrants lived in the Cowlitz area by 1940. (After Clevinger, 1938, p. 120; Clevinger, 1942, p. 4.)

FIGURE 2.28 A folk solution to navigating Florida's swamplands. In the 1920s, local residents of the greater Everglades region began modifying truck chassis to better navigate the swampy landscape. The oversized tires and suspensions allowed hunters to drive through inundated land. The solution was widely adopted and is used today by hunters and tour operators. (Lynn Pelham/The LIFE Picture Collection/Getty Images)

in a way that no mass-produced vehicle can. By the 1940s, they were a common sight in the south Florida wetlands landscape. They still are today, as they remain the best solution to the challenge of the local physical geography.

Nature and Gender

We noted in Chapter 1 and in this chapter's discussion of LGBT districts that cultures are heterogeneous and that gender is a key marker of difference within places and regions. Geographers and other social scientists have documented significant differences between men's and women's relationships with the environment. This observation holds for many different types of cultures. **Ecofeminism** is one way of thinking about how gender influences our interactions with nature (see also Chapter 7). Ecofeminism maintains that women—as traditional childbearers, gardeners, and gatherers—are more in touch with the rhythms of nature than men and therefore make better stewards of the Earth. This concept, however, might seem to suggest that there is something inherent or essential about men and women that makes them think about and behave differently toward the environment. Although cultural geographers would argue against such essentialism, few would disagree that gender is an important variable in nature–culture relations.

Diane Rocheleau's work on women's roles in the management of **agroforestry** systems makes this point. Agroforestry systems are farming systems that combine the growing of trees with the cultivation of agricultural crops. Agroforestry is practiced by folk and indigenous cultures across the tropical world, and has been shown to be a highly productive and ecologically sustainable practice. It is common in these production systems for men and women to have very distinct roles. Generally, for example, women are involved in seeding, weeding, and harvesting, whereas men take responsibility for clearing and cultivation. Gender differences also often exist in the types of crops men and women control and in the marketing of produce.

After conducting studies for many years—first in East Africa and later in the Dominican Republic—Rocheleau was able to see general themes regarding the way human–environment relations are different between men and women, not only in agroforestry systems but also in many rural and urban environments. Together with two colleagues, she identified three themes.

1. **Gendered knowledge.** Because women and men often have different tasks and move in different spaces, they possess different and even distinct sets of knowledge about the environment.

2. **Gendered environmental rights.** Men and women have different rights, especially with regard to the ownership and control of land and resources in agrarian communities.

3. **Gendered environmental politics.** For reasons having to do with their responsibilities in their families and communities, women are often the main leaders and activists in political movements concerned with issues of environment and health.

Taken together, these themes suggest that environmental planning or resource management projects that do not address issues of gender are likely to have unintended consequences, some of them negative, for both women and environmental quality.

Nature in Popular Culture

Popular culture is less directly tied to the physical environment than are folk and indigenous cultures, though it has enormous environmental impacts. City dwellers generally do not draw their livelihoods from the land. They have no direct experience with farming, mining, or logging activities, though they could not live without the products and materials produced by those

activities. Gone is the intimate association between people and land once known by past folk generations. Gone, too, is our direct vulnerability to many environmental forces, although this security is counterbalanced by new risks. Because popular culture is so tied to mass consumption, it can have significant impacts on the environment, such as producing air and water pollution and massive amounts of solid waste. Also, because popular culture fosters limited contact with and knowledge of the physical world, and that usually through recreational activities, environmental perceptions can become quite distorted.

Popular culture makes heavy demands on ecosystems. This is true even in the seemingly harmless activity of outdoor recreation. Recreational activities have increased greatly in the world's economically affluent regions. Many of these activities require machines, such as snowmobiles, off-road vehicles, and Jet Skis, that are powered by internal combustion engines and have numerous adverse ecological impacts ranging from air pollution to soil erosion. Even when outdoor recreation does not involve machinery directly, people use airplanes and cars to reach the "great outdoors." In national parks and protected areas worldwide, affluent tourists in search of nature have overtaxed protected environments and wildlife, and produced levels of congestion approaching those of urban areas (**Figure 2.29**).

FIGURE 2.29 Traffic jam in Yosemite National Park at the height of the summer tourist season. (Tom Meyers Photography)

Animal Geographies

Finally, some human geographers believe that studies of the modification of the Earth's surface are overly focused on human agency, neglecting the role of nonhuman species. In response to this criticism, nature–culture investigations in geography have experienced a "nonhuman turn," or "animal turn," in recent years. What does that mean? Look no further than the family dog, a member of the oldest domesticated species (see Chapter 8). For decades, the prevailing theory of canid domestication was a story of humans selecting less aggressive individuals from wild packs, breeding to minimize aggressiveness and accentuate desirable traits, and making individuals dependent on humans for food. All agency in this story belongs to humans. More recently, a convincing alternative theory supported by biological and archaeological evidence places canines at the center of the story alongside humans. In this story, those wild canids that were able to work and live cooperatively with humans sought a closer association that provided an evolutionary advantage to both species. The canids used humans, just as humans used canids. In summary, **animal geography** is a subfield of human geography that emphasizes the active participation of nonhuman species in nature–culture interactions, investigating how the human and nonhuman act together to transform space, place, and landscape.

Cultural Landscape

2.5 Compare and contrast folk and popular cultural landscapes.

Geographic patterns of cultural traits and practices enhance place making and produce distinctive cultural landscapes. Cultural landscapes reveal the important differences and commonalities within and between cultures. We'll explore a sampling of these by comparing and contrasting folk and popular cultural landscapes, and highlighting folk architecture, landscapes of popular consumption, leisure and amenity landscapes, and working landscapes.

Folk Architecture in North America

Every folk culture produces a highly distinctive landscape. One of the most visible aspects of these landscapes is **folk architecture**, buildings constructed in a local manner and style, without the assistance of professional architects or blueprints, using locally available raw materials. Folk buildings are important in creating distinctive folk cultural landscapes. Folk architecture springs not from the drafting tables of professional architects but from the collective memory of groups of a people (**Figure 2.30**). Material composition, floor plan, and layout are

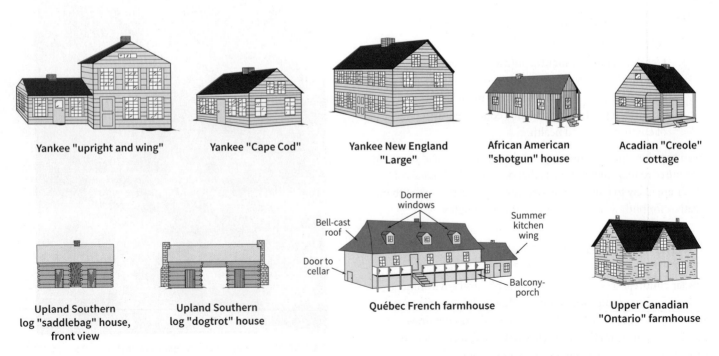

FIGURE 2.30 Selected folk houses. Many of these house types are associated with former folk regions. (After Glassie, 1968; Kniffen, 1965.)

important ingredients of folk architecture, but numerous other characteristics help classify farmsteads and dwellings. The form or shape of the roof, the placement of the chimney, and a building's location and orientation in the physical environment are important classifying criteria as well.

In North America, folk architecture today is mostly relic. Traces can still be found here and there, and they provide both a sense of past cultural landscapes and enduring symbols of cultural identity. One needs to know what to be on the lookout for in folk houses, however, in order to properly "read" the landscape. Yankee folk houses, for example, are of wooden frame construction, and shingle siding often covers the exterior walls. They were built with a variety of floor plans, including the New England "large" house, a huge two-and-a-half-story house built around a central chimney and two rooms deep. As the Yankee folk migrated westward, they developed the upright-and-wing dwelling. These particular Yankee houses are frequently massive, in part because the cold winters of the region forced most work to be done indoors.

By contrast, Upland Southern folk houses were smaller and built of notched logs. Many houses in this folk tradition consist of two log rooms, with either a double fireplace between, forming the saddlebag house, or an open, roofed breezeway separating the two rooms, a plan known as the dogtrot house. An example of an African American folk dwelling is the shotgun house, a narrow structure only one room wide but two, three, or even four rooms deep. Acadiana, a French-derived folk region in Louisiana, is characterized by the half-timbered Creole cottage, which has a central chimney and built-in porch. Scores of other folk house types survive in the American landscape. One can still find, for example, shotgun houses in the original centers of major cities, such as Miami and New Orleans. Increasingly, however, many exist only through special historic preservation efforts (see this chapter's World Heritage Site feature). Now that we have reviewed a few of the common styles of North American folk architecture, you can test your ability to identify them in the landscape (**Figure 2.31**).

FIGURE 2.31 Four folk houses in North America. Such distinctive house types often serve as visual clues to former folk regions. (Courtesy of Terry G. Jordan-Bychkov.)

farmhouses. The seven UNESCO-designated farmhouses are spread over an area stretching 62 by 31 miles (100 by 50 kilometers). Both they and the surrounding landscapes are protected under Swedish cultural heritage and environmental laws, reflecting their importance to that nation's cultural identity.

- The Decorated Farmhouses of Hälsingland represent the final flourishing of a centuries-old local tradition of timber construction. Beginning in the early nineteenth century, farmers in the region experienced a new period of prosperity based on forestry and flax production. Farmers invested their surplus earnings into building and decorating farmhouses. Most of these structures remain under the private ownership of farmers who continue to manage the surrounding lands for agriculture.

- The cultural landscape of Hälsingland has evolved over centuries and bears the imprint of occupation by a population of independent farmers. The decorated farmhouses are located on active farmsteads and within a larger agrarian landscape. The mixed agrarian livelihood practices of cattle breeding, crop cultivation, forestry, and hunting have visibly shaped the land. Historically, farmers used the pastures and woodlands communally and shared cultivated fields. Changes in Swedish law in the nineteenth century led to individual privatization of lands, which brought prosperity to some farmers and a flowering of folk architecture.

IN THE FARMHOUSES: The seven designated farmhouses are the Kristofers farm in Stene, Järvsö; the Gästgivars farm in Vallstabyn; the Pallars farm in Långhed; the Jon-Lars farm in Långhed; the Bortom åa farm in Gammelgården; the Bommars farm in Letsbo, Ljusdal; and the Erik-Anders farm in Askesta, Söderala. Each contains a number of decorated rooms for festivities, and some very prosperous farmsteads have entire houses, a *herrstuga*, dedicated solely to festive celebrations. The

Folk architecture has given way to professionally built and mass-produced housing. Recognizing that an important cultural heritage is vanishing, in 2012 UNESCO designated seven large farmhouses in northeastern Sweden's Hälsingland region as the Decorated Farmhouses of Hälsingland World Heritage Site. Out of more than 1000 possibilities, the committee chose these seven as outstanding examples of a regional folk architecture rooted in the Middle Ages. The timber farmhouses were constructed mostly between 1800 and 1870 and are richly decorated, inside and out, following a local folk-art tradition.

- Hälsingland, a small region in Sweden bordering the Gulf of Bothnia, contains a concentration of such farmsteads feature numerous intact buildings that were built and elaborately painted in a manner meant to display the prosperity and social status of the farmer.

- The houses are made entirely of wood, most constructed in a building tradition that may be traced to the Middle Ages, which features jointed horizontal timbers of pine or spruce from nearby forests. Prosperous farmers of the nineteenth century enlarged existing houses from one to two or two and a half stories or built entirely new houses on the farmstead. The emphasis was on creating exceptionally large structures using the most talented local carpenters and craftsmen.

- Each farmhouse is elaborately decorated. On the outside, local craftsmen carved decorative elements around the front porch and main entrance. Inside, the houses are decorated with textile paintings hung on the walls or with paintings applied directly to the wooden walls and ceilings. The rooms dedicated solely for festivities and special occasions are typically the most highly decorated.

- The paintings fuse local folk-art traditions with more universal styles, such as Baroque and Rococo, favored by the landed gentry of nineteenth-century Sweden. Wealthy farmers commissioned itinerant artists to paint their farmhouse interiors with depictions of landscapes and biblical characters in contemporary fashion or with detailed floral patterns. Ten of the painters who worked on the World Heritage farmhouses are known, but most remain anonymous.

TOURISM: The seven UNESCO-designated farmhouses are located in a scenic coastal region of northeastern Sweden, which has long been an attraction for both domestic and foreign tourists.

- Regional public and business groups, as well as individual farmers, actively collaborate to keep decorated farmhouses open to visitors. Museums in the region feature displays on the history of this folk tradition.

- The maintenance and conservation of Swedish cultural heritage is a popular movement in the country. Many craftsmen and women in Hälsingland carry on folk handicrafts, such as textiles, furniture making, and basketry, which draw many visitors to the region.

(J-P Lahall/Photolibrary/Getty Images)

- Established between 1800 and 1870 c.e., these farmsteads in northeastern Sweden represent the final flowering of a centuries-old tradition of folk architecture.

- In 2012, seven farmhouses were designated as World Heritage Sites, being noteworthy examples of the elaborately decorated interiors characteristic of the Hälsingland region's folk culture.

- Visitors are attracted to the region by its scenic beauty, agrarian landscapes, and celebration of local folk traditions and handicrafts.

http://whc.unesco.org/en/list/1282/

Folk Architecture Around the World

Although in North America folk architecture tends to be found only as a cultural relic of the past, it continues to thrive in less-industrialized countries where agriculture is the main occupation. Most construction is done with local materials, such as clay, stone, wood, bamboo, leaves, and grasses. Distinctive cultural landscapes emerge from the collective use of folk architecture, just two examples of which we present here, nipa huts and painted Ndebele houses.

Nipa Huts of the Philippines The characteristic folk housing of the Philippines is the nipa hut, still found in the countryside, but with many variations resulting from the mix of Malaysian, Spanish, Muslim, and American influences. Most commonly, it is constructed primarily of bamboo framing fastened and roofed with nipa leaves (**Figure 2.32**). Philippine folk architecture conveys a feeling of lightness (through, for example, the placement of large open-air windows) combined with a quality of warmth produced by the use of wood inside. The nipa hut is often raised on stilts, particularly in coastal areas and floodplains. The combination of large windows, wide eaves, and its elevated position results in the free flow of cooling breezes, even during rainy weather. The nipa hut is widely considered the Philippine national dwelling and thus is closely associated with cultural and national identity.

Ndebele Painted Houses of Southern Africa In southern Africa, one of the most distinctive house types are found

FIGURE 2.33 Ndebele village in South Africa. While the origins and meaning of Ndebele house painting remain matters of debate, there is no question that it creates a visually distinct cultural landscape. (Ariadne Van Zandbergen/Alamy)

in the Ndebele culture region, which stretches from South Africa north into southern Zimbabwe. In the rural parts of the Ndebele region, people are farmers and livestock keepers, and live in traditional compounds. What makes these houses distinctive is the Ndebele custom of painting brightly colored designs on the exterior house walls and sometimes on the walls and gates surrounding the grounds (**Figure 2.33**). The precise origins of this custom are unclear, but it seems to date to the mid-nineteenth century. Some suggest that it was an assertion of cultural identity in response to local residents' displacement and domination at the hands of white settlers. Others suggest a religious or sacred role.

What is clear is that the custom has always been handled by women, a skill and practice passed down from mother to daughter. Many of the symbols and patterns are associated with particular families or clans. Initially, women used natural pigments from clay, charcoal, and local plants, which restricted their palette to earth tones of brown, red, and black. Today, many women use commercial paints—expensive, but longer lasting—to apply a range of bright colors, limited only by the imagination. Another new development is the types of designs and symbols used. People are incorporating modern machines, such as automobiles, televisions, and airplanes into their designs. In many cases, traditional paints and symbols are blended with the modern to produce a synthetic design of old and new. Ndebele house painting is developing and evolving in new directions, all the while continuing to signal a persistent cultural identity to all who pass through the region.

FIGURE 2.32 Nipa hut in Baguio City, Philippines. Folk architecture, such as the nipa hut, typically uses local materials and is designed to adapt to local environmental conditions. (Photo by Ubo Pakes/Getty Images)

FIGURE 2.34 Indoor shopping mall at the foot of the Petronas Towers, Kuala Lumpur, Malaysia. Most of the world's largest malls are now located outside of North America. Malaysia boasts 2 of the 15 largest in the world. (Gavin Hellier/robertharding/Alamy)

Landscapes of Consumption

It is difficult to single out one type of popular landscape, just as it is difficult to distinguish one type of folk landscape. Popular culture, nonetheless, has been commonly associated with mass-produced suburban housing, commercial strips, and large indoor shopping malls. These latter retail districts have come to be known by geographer Robert Sack's term, "landscapes of consumption." Not only are such landscape features common in North America, they are now also seen in Brazil, China, India, and many other countries with rapidly growing economies increasingly linked through globalization (**Figure 2.34**). Indeed, not one of the world's 25 largest (in terms of leasable area) indoor shopping malls is located in North America, where they were initially introduced in the 1950s. Indoor malls are excellent examples of how the forces of globalization can rapidly promote the diffusion of a particular landscape feature and popular culture practice.

North America's largest indoor mall is West Edmonton Mall in the Canadian province of Alberta (**Figure 2.35**). Enclosing some 5.3 million square feet (493,000 square meters) and opened in 1986, it employs 23,500 people in more than 800 stores and services, and accounts for nearly one-fourth of the total retail space in all of greater Edmonton. Beyond its sheer size, West Edmonton Mall also boasts a water park, a sea aquarium, an ice-skating rink, a miniature golf course, a roller coaster, 21 movie theaters, and a 360-room hotel. Its "streets" feature motifs from such distant places as New Orleans, represented by a Bourbon Street complete with fiberglass ladies

FIGURE 2.35 The enormous West Edmonton Mall represents the growing infusion of spectacle and fantasy into the urban landscape. In addition to retail outlets, the mall includes a sea aquarium, an ice-skating rink, a miniature golf course, and the Submarine Ride shown here. (James Marshall/Getty Images)

of the evening. Jeffrey Hopkins, a geographer who studied this mall, refers to this as a "landscape of myth and elsewhereness," a "simulated landscape" that reveals the "growing intrusion of spectacle, fantasy, and escapism into the urban landscape." We would add that such malls blur the line between consumption and leisure landscapes, presenting shopping as a pastime rather than a necessity.

Leisure and Amenity Landscapes

Another common feature of popular culture is what geographer Karl Raitz labeled **leisure landscapes**. Leisure landscapes are designed to entertain people on weekends and vacations; often they are included as part of a larger tourist experience. Golf courses and theme parks such as Disney World are good examples of such landscapes (**Figure 2.36**). **Amenity landscapes** are a related landscape form. These are regions with attractive natural features such as forests, scenic mountains, or lakes and rivers that have become desirable locations for retirement or vacation homes. In many parts of the world, rural regions that can no longer economically support agriculture, logging, or ranching are increasingly viewed as amenity landscapes. There is a general shift in the perception of rural landscapes from one of production for livelihood and home for folk culture to one of consumption of scenery and recreation for popular culture. Such shifts have strong implications for the future of rural life in urbanized and industrialized countries, not least of which is the growing use of the countryside for vacation and retirement homes (**Figure 2.37**).

The past, reflected in surviving buildings, has also been incorporated into leisure and amenity landscapes. Most often, folk or historic buildings that cannot be preserved in place are relocated to create a sort of outdoor museum. Many small towns across North America feature such historical villages. Examples include the Heritage Historical Village of New London,

FIGURE 2.37 Vacation homes line the shore of Lake Tahoe, California. Lake Tahoe's natural amenities have long attracted summer vacationers. Initially, people flocked to a few motels, lodges, and campgrounds. In the past few decades, however, vacation home construction has boomed as increasing numbers of affluent urbanites were able to afford second homes. (Courtesy of Roderick Neumann.)

Wisconsin, which features five relocated and restored historic buildings, and the Historical Museum and Village of Sanibel, Florida, which includes seven buildings relocated from around the island (**Figure 2.38**). In cases where no historical buildings exist, history parks have been entirely reconstructed—for example, at Jamestown, Virginia, or Louisbourg on Cape Breton Island, Nova Scotia. Some of the larger re-created historical villages feature living history, which frequently includes both actors in period costume and skilled craftspeople practicing lost arts, such as blacksmithing or candle making. Such sites can play an important role in mythologizing the past to help create national and cultural identities.

Working Landscapes

As we stated earlier, it can be misleading to try to clearly distinguish between the features of folk and popular culture. The trafficking of ideas and materials back and forth across that supposed culture divide is quite common. Take, for example, the supposed opposite of leisure and amenity landscapes, the **working landscape**. A working landscape is a rural area that is shaped and maintained by the type of productive activity, such as ranching or winemaking, that predominates there.

FIGURE 2.36 A common leisure landscape. Golf courses are one of the most commonly encountered leisure landscapes in the United States. (nik wheeler/Alamy)

FIGURE 2.38 Sanibel Island, Florida, Historical Village and Museum. Heritage villages like this one are created by relocating historic buildings to a single site to facilitate preservation and tourism. (Courtesy of Roderick Neumann.)

FIGURE 2.39 A working landscape produced by generations of cattle ranchers. This area near the U.S. border with Mexico is being cooperatively managed to preserve both a way of life and a landscape. (Warner Glenn/Malpai Borderlands Group)

Ranching Landscapes of the West Geographer Nathan Sayre's study of cattle ranching in the American Southwest is illustrative. Ranching in the region of New Mexico and Arizona near the border with Mexico has been ongoing for nearly two centuries. It has produced a particular landscape consisting of ranch dwellings, livestock, roads, trails, wildlife, wildfire cycles, and patterns of vegetation (**Figure 2.39**). As pressure from real estate developers grew in the 1990s, residents recognized that the maintenance of this landscape was dependent on the continuance of ranching. A group of interested ranchers, scientists, public agencies, and private conservationists formed a not-for-profit organization called the Malpai Borderlands Group to try to ensure the continuance of both a rural way of life and the landscape (**Figure 2.40**). The group functions through a complex melding of the interests of urban-based conservationists, the local folk knowledge of rural-based ranchers acquired through five generations of living on the land, and the scientific knowledge of university-based researchers. Elements of both folk and popular culture are at work in this effort to maintain a working landscape.

The Wine Country of California Working landscapes are themselves increasingly perceived as amenities by people wealthy enough to buy large portions of rural land. The rural landscapes produced by generations working the land are often the main draw for many vacationers and owners of second homes (see Figure 2.39 and accompanying discussion). One of the clearest

examples of a working landscape that is simultaneously an amenity landscape is that of the "wine country" of Napa Valley, California (**Figure 2.41**). Wine grapes have been grown in the valley since the 1830s, and several wineries sprang up beginning

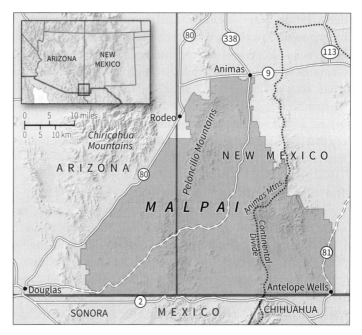

FIGURE 2.40 The Malpai Borderlands Group management area. This sprawling area dominated by cattle ranching is part of an experiment in land management involving ranchers, government agencies, conservationists, and scientists.

FIGURE 2.41 A working landscape filled with tourists. Napa Valley, California, is one of the world's premier wine-producing regions. It also serves as an amenity landscape for those wealthy enough to own a hobby vineyard. (leezsnow/E+/Getty Images)

in the 1860s. These early viticulturists and vintners tended to be from European immigrant families with histories in vineyard cultivation and wineries. By the 1970s, vineyards had become the dominant landscape feature. Around then, vineyard ownership began to shift and vineyards and wineries multiplied. Wealthy lawyers and physicians from the cities began buying "hobby" vineyards. Popular culture celebrities began producing their own wines, such as film director Francis Ford Coppola, who bought his first Napa Valley vineyard in 1975. Napa Valley had become, in sum, both a working landscape and the site of attractive scenery that appealed to the wealthy. Tourists now flock there by the tens of thousands to vacation in this working landscape.

The "Old West" Versus the "New West"   The cases of both the Malpai Borderland Group and Napa Valley are illustrative of the economic and cultural changes unfolding in the western United States. This phenomenon is popularly known as the "Old West" versus the "New West." The Old West is the historic regional economy and corresponding cultural landscapes based on cattle ranching, mining, and logging. Most Old Westerners worked as wage laborers in these extractive industries. The New West refers to an amenity-based economy that features luxury second homes, outdoor recreational resorts, and international art and film festivals. The New Westerners are affluent professionals who earned their money in distant

cities and are attracted to the scenic landscapes of the mountainous West.

Geographers John Harner and Bradley Benz conducted a study in southern Colorado that perfectly demonstrates how New West amenity landscapes emerge. They focused on the area surrounding Durango, Colorado, a former mining town that is now a tourist mecca featuring an antique steam-powered train, gourmet restaurants, ski shops, and boutique clothing stores (**Figure 2.42**). There, they measured the growth of "ranchettes," 35- to 70-acre private parcels of land too small to be economically productive or legally subdivided further. These new landholdings are created for wealthy New Westerners who vacation in or retire to their new luxury homes in the wide-open spaces.

Harner and Benz found that the number of ranchettes increased 67 percent between 1988 and 2008, signaling a huge influx of affluent migrants from the cities. Their study documented that the first ranchettes were constructed near the most desirable natural settings: streams and rivers, national parks and forests, and mountain vistas. The subdivision of large landholdings into ranchettes continued in a pattern of contiguous diffusion, gradually leveling out as the supply of amenity-rich spaces was exhausted.

Such fragmentation of landholdings and the corresponding creation of amenity landscapes have been repeated across the

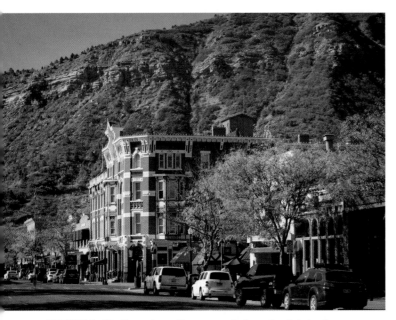

FIGURE 2.42 Downtown Durango, Colorado. Originating as an "Old West" mining town, today Durango's boutique stores and restaurants are emblematic of the "New West" economy based on amenity landscapes and tourism. (Inge Johnsson/Alamy)

FIGURE 2.43 The contested landscape of California's mountain ranges. To accommodate new residents, luxury retirement and vacation homes sprawl further into former ranch and farm lands. Newcomers and longtime residents have clashed politically over different values held for a shared landscape. (Courtesy of Roderick Neumann.)

western states many times over. (A main motivation behind the Malpai Borderlands Group's formation was combating this process.) Frequently, the process has led to clashes between newcomers and established residents. Geographers Peter Walker and Louise Fortmann studied the social conflicts emerging from the New West phenomenon in the foothills of California's Sierra Nevada range (**Figure 2.43**). They determined that the

most contentious political debates focused not on economic issues, but on cultural values expressed in landscape preferences. Affluent newcomers and established residents made different aesthetic judgments around landscape use, and this drove efforts to gain political control over land-use decisions. The New West phenomenon highlights the relevance of cultural landscape to regional politics and economies.

Conclusion

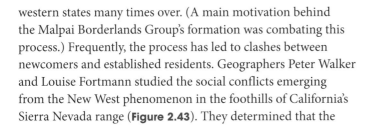

In this chapter, we have just scratched the surface of the complex nature of the geographies of cultural difference. Using just a few examples of cultural traits and practices, we learned that there are many ways that culture finds regional expression. We have explored the complex interactions of culture and mobility. We saw how globalization has shifted our perceptions of time and space and tended to homogenize cultural differences. Using our categories of popular, folk, and indigenous cultures, we learned how important nature and the physical environment are to culture. Finally, we learned how landscape is one of the key concepts in defining culture and cultural differences.

Chapter Summary

2.1 Explain how cultural traits and patterns, such as leisure preferences, ethnic identities, and gender and sexuality, vary by regions.

- Regional differences are an expression of cultural differences, which may be categorized in nearly unlimited ways.
- Different preferences for sports, for example, help define regional culture, which may be reflected in identifiable geographic patterns in colleges' support for athletics.

- We introduced a key marker of difference, indigenous culture, and showed how indigenous culture regions can be identified around the globe, often in the most mountainous, arid, and forested lands.
- Even subcultures based on sexuality can produce regional variation, most clearly expressed in the case of urban LGBT districts.

2.2 Recognize the mobility of cultural traits and interactions through the examples of tourism, diaspora cultures, and the Internet.

- Mobility is multidimensional, encompassing the movement across time and space of both the material and nonmaterial aspects of culture.
- Tourism provides an excellent example of the centrality of mobility to popular culture and also of the complex ways that tourism mobilizes cultural beliefs and practices.
- Major cultural groups, such as diaspora cultures, are defined by mobility, creating challenges for the establishment of coherent, stable national or cultural identities.
- Communications technology illustrates the presence of political and economic barriers to the free movement of and equitable access to information.

2.3 Explain how globalizing processes are influencing cultural interactions and change, such as the experience of space-time and cultural homogenization.

- Within popular culture, globalization is accompanied by the sense of a world that is shrinking, a sense that is not universally shared with folk cultures.

- The influences of globalization are varied and even contradictory, sometimes appearing to homogenize culture and at other times seeming to accentuate cultural differences.

2.4 Explain how culture and the physical environment interact in indigenous, folk, and popular cultures and the influences of gender and the global economy on those interactions.

- Indigenous cultures are commonly directly dependent on the local environment for their livelihoods, and often develop ways to conserve the resources that sustain them.
- Folk cultures depend directly on the local environment, and their knowledge of it is very detailed.
- Men and women often have different understandings of the environment because of gender roles.

2.5 Compare and contrast folk and popular cultural landscapes.

- Architecture is a common expression of difference in folk cultures and helps to visually define cultural landscapes.
- There are many categories of cultural landscape that help differentiate how they are valued and used.
- Some landscapes are dedicated to material consumption, some to production, and others to leisure and enjoyment.

Key Terms

geographic marginalization A government-led process whereby a population is forced to migrate to remote, often rugged and unproductive lands, far from political and economic centers (page 49).

LGBT district Urban neighborhoods inhabited or frequented by a disproportionately large number of lesbian, gay, bisexual, and transgendered people and identifiable by a geographic concentration of "gay-friendly" public spaces and establishments such as bookstores, bars, theaters, and restaurants (page 52).

tourism A modern activity that arose with the introduction of rail, sea, and air mass transportation and involves people traveling from their homes for the purposes of business, leisure, entertainment, or recreation (page 53).

digital divide A pattern of unequal access to advanced information technologies produced by socioeconomic inequalities and measured at scales ranging from the individual to countries and world regions (page 57).

time-space convergence The phenomenon whereby the time taken to travel between places is progressively reduced through the introduction of new transportation technologies (page 59).

absolute distance The precise measurement of physical space between one point on the Earth's surface and another, as in the case of the longitude and latitude system (page 59).

relative distance The space between one point on the Earth's surface and another measured by changes in time and cost of movement (page 59).

convergence hypothesis A hypothesis holding that cultural differences among places are being reduced by improved transportation and communications systems, leading to a homogenization of popular culture (page 60).

consumer nationalism A situation in which local consumers favor nationally produced goods over imported goods as part of a nationalist political agenda (page 60).

biodiversity Biodiversity—a blending of biology and diversity—refers to the variety and variability of life on Earth as measured at various scales, including diversity among individuals, populations, species, communities, and ecosystems (page 61).

indigenous technical knowledge (ITK) Highly localized knowledge possessed by indigenous culture groups about environmental conditions and sustainable land-use practices (page 62).

subsistence economies Rural economies that have few or no external inputs and are oriented primarily toward food production for local consumption, rather than production for sale on the market (page 62).

ecofeminism A philosophy that suggests that women—as traditional childbearers, gardeners, and nurturers of the family and home—are more in touch with the rhythms of nature than men and are therefore better stewards of the Earth (page 64).

agroforestry Farming systems that combine the growing of trees with the cultivation of agricultural crops (page 65).

animal geography A subfield of human geography that emphasizes the active participation of nonhuman species in nature–culture interactions, investigating how the human and nonhuman act together to transform space, place, and landscape (page 66).

folk architecture Building styles of folk cultures constructed in a local manner and style, without the assistance of professional architects or blueprints, using locally available raw materials (page 66).

leisure landscape Landscapes planned and designed primarily for entertainment and tourism, such as theme parks or ski and beach resorts (page 72).

amenity landscape Landscapes with attractive natural features such as forests, scenic mountains, or lakes and rivers that have become desirable locations for retirement or vacation homes (page 72).

working landscape Landscapes of rural areas that are shaped and maintained by the type of productive activity that predominates there, such as cattle ranching or various forms of agriculture (page 72).

Practice at SaplingPlus

Read the interactive e-Text, review key concepts, and test your understanding.

Story Map. Explore nature, culture, and sense of place in various regions of North America.

Web Map. Examine maps that illustrate cultural issues.

Doing Geography

ACTIVE LEARNING:
Place Image in the Media

In this chapter, we referred to the role of mass media in shaping popular understandings of place. Films and television programs often make place a central character in the plot. Los Angeles, San Francisco, Miami, and New York City have for years served filmmakers this way. An entire industry, media tourism, is dedicated to visiting places made popular by entertainment media.

For this exercise, you will work in class with a small group to analyze how place image is shaped by mass media.

More guidance at Sapling Plus

EXPERIENTIAL LEARNING:
Self-Representation of Indigenous Culture

Indigenous cultures are reasserting themselves after 500 years of marginalization. For centuries, the roar of dominant national cultures drowned out indigenous peoples' voices. Members of dominant national cultures—academics, missionaries, government officials, and journalists—have been largely responsible for writing about the histories and cultures of indigenous peoples. This is changing as some indigenous groups have prospered economically and as the indigenous rights movement gains increasing support worldwide. Today, more and more indigenous peoples are taking charge of representing themselves and their cultures to the outside world by building museums, producing films, hosting conferences, and creating web sites.

This exercise requires you to carefully study one of these platforms for cultural expression: *self-produced* indigenous peoples' web sites. To do so, follow the steps below.

Steps to Understanding Indigenous Culture

Step 1: Do some background research on the names and locations of major indigenous cultures. It is important to verify that the web sites you decide to study are self-produced. Confirm that the site is produced by a tribal or other indigenous organization, not by an external NGO, national government, corporation, or university.

Step 2: Think about how you are going to analyze the content of the web site in order to draw conclusions about the self-representation of indigenous cultures. Here are a few suggestions and possibilities: focus on questions of geography such as territorial claims, rights over natural resources, culturally significant relations with nature, and homeland self-rule.

Based on your research and analysis, systematically analyze how indigenous populations represent their cultural identities. Consider these questions:

* How do indigenous groups speak about their relationship to the land and the environment?
* What do they say about territorial claims and homelands? What roles do maps play on the web sites?
* What are their ideas on biodiversity conservation and bioprospecting (the search for genetic and other biological resources)?

In addition, look for discussions of conflict, cooperation, or disagreement with national governments or multinational corporations:

* Is there a project or policy (e.g., disposal of radioactive material) under dispute?
* What position is taken on the web site? How is the position framed in relation to indigenous rights and culture?
* What major issues and challenges does the site highlight and how do these relate to globalization?

Finally, think about possibilities for comparison:

* Are there regional (on either the U.S. national or global scale) differences in terms of the quantity and content of web sites?
* Did you find common themes across or within regions?
* Are there indigenous cultures that produce contrasting or competing representations?
* Do some indigenous cultures have more than one self-generated web site, and do those sites present different ideas?

Jim Enote, executive director of the A:shiwi A:wan Museum and Heritage Center in New Mexico, discusses a Zuni map art painting at an exhibition. Many Native American tribes build museums to represent their culture and way of life. (STAN HONDA/AFP/Getty Images)

SEEING GEOGRAPHY

Camping in the "Great Outdoors"

What can this scene tell us about nature–culture relations in North American popular culture?

How we think about and interact with nature reveals a great deal about how we understand our place in the world. For example, within contemporary popular culture, we often think of our relationship with nature as something outside the routine of daily life. We seek out nature on weekends, during vacations, or in retirement. Nature is equated with pretty scenery, which in popular culture typically means forests, mountains, and wide-open spaces. This scene from Everglades National Park in Florida reveals a great deal about one of the main ways within popular culture that we express this understanding of nature: camping in the "great outdoors."

Of course, there are many ways to camp. This particular form, recreational vehicle (RV) or motor home camping, is extremely energy intensive. Unseen in this photo is the supporting industrial manufacturing complex organized to produce a fleet of vehicles, trailers, and equipment, all of which are intended to be situated on an asphalt slab and connected to an electric power grid. A lot of natural resources have to be consumed to experience nature in this way. Such an experience is shaped less by direct physical interaction with nature than by interaction with mass-produced commodities. Nature serves mainly as background scenery.

Thinking of nature as background scenery suggests a stage set for actors to interact on. Look carefully at the photo and you will see that this campground is a stage for both extremely private and highly public interactions. On one hand, every set of "campers" has its own private home completely sealed from the outside (e.g., note the satellite dish and rooftop air conditioner). Each RV is self-sufficient, requiring interactions neither with nature nor with human neighbors. On the other hand, the RVs are extremely closely spaced. Interactions among campers are almost forced, for merely stepping out of the RV puts one in public view. Conversations with strangers—even sharing drinks and meals—become a cultural norm in such a setting. Indeed, meeting new people is one of the reasons campers give when explaining why they enjoy camping. Could it be that getting in touch with nature really means getting in touch with one another in ways that would be difficult in the daily routines of popular culture?

Enjoying nature in a national park campground.
(Courtesy of Roderick Neumann.)

A street scene in the large city of Kolkata, India.
(Steve Raymer/Getty Images)

chapter 3

Population Geography

A DEMOGRAPHIC PORTRAIT

Would you feel comfortable walking here? If not, why not?

Examine the photo and think about this question as you read. We will revisit them in Seeing Geography on page 115.

Learning Objectives

3.1 Identify regional patterns of population characteristics and describe how these are distributed spatially and change over time.

3.2 Explain patterns, causes, and consequences of population migrations.

3.3 Understand theories of population growth and control and how these have changed over time.

3.4 Describe how the natural world shapes population characteristics, and how population characteristics in turn shape the natural world.

3.5 Analyze the imprint of demographic factors on the cultural landscape.

(Steve Raymer/Getty Images)

The nineteenth-century French philosopher Auguste Comte famously asserted that "demography is destiny." Though this might be a bit of an exaggeration, one of the most important aspects of the world's human population is its demographic characteristics, such as age, gender, health, mortality, density, and mobility. In fact, many cultural geographers argue that familiarity with the spatial dimensions of demography provides a baseline for the discipline. Thus, **population geography** provides an ideal topic to launch our substantive discussion of the human mosaic.

The most essential demographic fact is that, as of 2017, the world's population reached the 7.5 billion mark. But numbers alone tell only part of the story of the delicate balance between human populations and the resources on which we depend for our survival, comfort, and enjoyment. Think of the sort of lifestyle you may now enjoy or aspire to in the future. Does it involve driving a car? Eating meat regularly? Owning a spacious house with central heating and air conditioning? If so, you are not alone. In fact, these "Western" consumption habits have become so widespread that they may now be more properly regarded as universal. Satisfying these lifestyle demands requires using a wide range of nonrenewable resources—fossil fuels, extensive farmlands, and fresh water among them—whose consumption ultimately limits the number of people the Earth can support. Indeed, it has been argued that if Western lifestyles are adopted by a significant number of the globe's inhabitants, then our current population of 7.5 billion is already excessive and will soon deplete or contaminate the Earth's life-support systems: the air, soil, and water we depend on for our very survival. Although we may think of our choices, such as how many children we will have or what to eat for dinner, as highly individual ones, when aggregated across whole groups, they can have truly global repercussions.

As we do throughout this book, we approach our study using the five themes of cultural geography. Under the regional theme, we consider the demographics of human populations: their size, age, gender composition, and spatial distribution. The mobility theme frames our discussion of population movements and the related spread of diseases. Although population density and impact are experienced locally, and policies affecting population are typically set at the national level, the debates over population size are most commonly pitched at the global scale. The ways in which human populations interpret and interact with the natural world are examined through the lens of the nature–culture theme. Finally, we consider the surprisingly diverse ways that places across the world respond to changing population dynamics in ways that shape—and are shaped by—the cultural landscape.

Region

3.1   Identify regional patterns of population characteristics, and describe how these are distributed spatially and change over time.

The principal characteristics of human populations—their densities, spatial distributions, age and gender structures, the ways they increase and decrease, and how rapidly population numbers change—vary enormously from place to place. Understanding the demographic characteristics of populations, and why and how they change over time, gives us important clues to their cultural characteristics.

Population Distribution and Density

If the 7,500,000,000 inhabitants of the Earth were evenly distributed across the land area, the **population density** would be about 129 persons per square mile (50 per square kilometer). However, people are very unevenly distributed, creating huge disparities in density. Australia, for example, has 8 persons per square mile (3 per square kilometer), whereas Bangladesh has 2927 persons per square mile (1130 per square kilometer) (**Figure 3.1**).

If we consider the distribution of people by continents, we find that 69.8 percent of the human race lives in Eurasia—Europe and Asia. The continent of North America is home to only 4.9 percent of all people, Africa to 16.2 percent, Latin America and the Caribbean to 8.6 percent, and Australia and the Pacific islands to 0.5 percent. When we consider population distribution by country, we find that 19 percent of all humans reside in China, 17 percent in India, and only 4.4 percent in the third-largest nation in the world, the United States (**Table 3.1**).

In analyzing data, it is convenient to divide population density into categories. For example, in Figure 3.1, one end of the spectrum contains thickly settled areas having 250 or more persons per square mile (100 or more per square kilometer); on the other end, largely unpopulated areas have fewer than 2 persons per square mile (less than 1 per square kilometer). Moderately settled areas, with 60 to 250 persons per square mile (25 to 100 per square kilometer), and thinly settled areas, inhabited by 2 to 60 persons per square mile (1 to 25 per square kilometer), fall between these two extremes. These categories create formal demographic regions based on the single trait of population density. As Figure 3.1 shows, a fragmented crescent of densely settled areas stretches along the western, southern, and eastern edges of the huge Eurasian

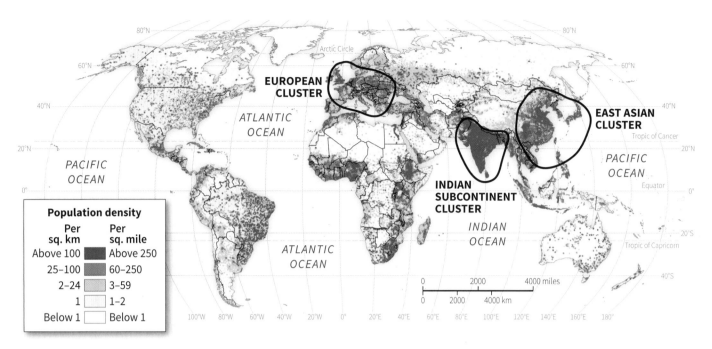

FIGURE 3.1 Population density in the world. Try to imagine the diverse shaping forces—physical, environmental, and cultural—that have been at work over the centuries to produce this complicated spatial pattern. It represents the most basic cultural geographical distribution of all. (Data from Population Reference Bureau.)

TABLE 3.1 The World's 10 Most Populous Countries, 2016 and 2050

Most Populated Countries in 2016 (estimated in millions)		Most Populated Countries in 2050 (estimated in millions)	
China	1,378	India	1,708
India	1,329	China	1,344
United States	324	United States	398
Indonesia	259	Nigeria	398
Brazil	206	Indonesia	360
Pakistan	203	Pakistan	344
Nigeria	187	Brazil	226
Bangladesh	163	Democratic Republic of Congo	214
Russia	144	Bangladesh	202
Mexico	129	Egypt	169

Source: Population Reference Bureau (2016).

landmass. Two-thirds of the human race is concentrated in this crescent, which contains three major population clusters: eastern Asia, the Indian subcontinent, and Europe. Outside of Eurasia, only scattered districts are so densely settled. Despite the image of a crowded world, thinly settled regions are much more extensive than thickly settled ones, and they appear on every continent. Thin settlement characterizes the northern sections of Eurasia and North America, the interior of South America, most of Australia, and a desert belt through North Africa and the Arabian Peninsula into the heart of Eurasia.

There is more that we as geographers want to know about population geography than simply population density. For example, what are people's standards of living and are these related to population density? Some of the most thickly populated areas in the world have the highest standards of living—and even suffer from labor shortages (for example, the major industrial areas of western Europe and Japan). The small principality of Monaco in southern Europe, for instance, is the most densely populated nation in the world; it also ranks among the highest incomes per capita, highest average life expectancies, and lowest unemployment rates. In other cases, thinly settled regions may actually be severely overpopulated relative to their ability to support their populations, a situation usually associated with marginal agricultural lands. Although 1000 persons per square mile (400 per square kilometer) is a "sparse" population for an industrial district, it is "dense" for a rural area. For this reason, **carrying capacity**—the maximum population size that can be supported in a given environment—is a far more meaningful index of overpopulation than density alone. Often, however, it is difficult to determine carrying capacity until the region under study is near or over its population size limit.

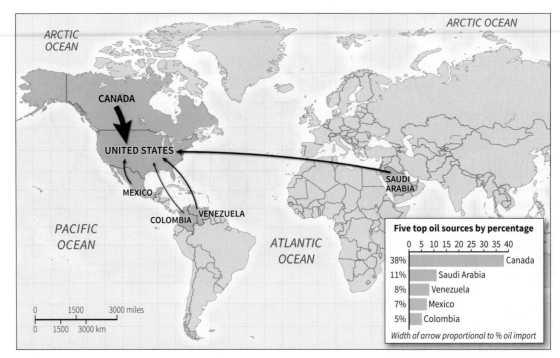

FIGURE 3.2 Where does U.S. oil come from? The United States has domestic oil resources, but they cannot fully supply the domestic demand, so oil is imported from other areas of the world. The United States consumes 20 percent of all the oil produced in the world. (Data from U.S. Energy Information Administration, 2016.)

Sometimes the carrying capacity of one place can be expanded by drawing on the resources of another place. Americans, for example, consume far more food, products, and natural resources than do most other people in the world: 20 percent of the entire world's petroleum production, for instance, is consumed in the United States (**Figure 3.2**). The carrying capacity of the United States would be exceeded if it did not annex the resources—including the labor—of much of the rest of the world.

We are also interested in demographic changes that occur over time. The population size of a place changes primarily in two ways: people are born and others die, and people move, or migrate, into and out of that location. Migration will be considered later in this chapter. For now we discuss births and deaths, which can be thought of as additions to and subtractions from a population. They provide what demographers refer to as natural increases and natural decreases.

Patterns of Natality

Births can be measured by several methods. The traditional way was simply to calculate the birth rate: the number of births per year per thousand people (**Figure 3.3**).

More revealing is the **total fertility rate (TFR)**, which is measured as the average number of children born per woman during her reproductive lifetime, considered to be from 15 to 49 years of age (with some countries having a younger upper-age limit of 44 years of age). The TFR is a more useful measure than the birth rate because it focuses on the female segment of the population, reveals average family size, and gives an indication of future changes in the population structure. A TFR of 2.1 is needed to produce a stabilized population over time, one that does not increase or decrease, a condition called **zero population growth**. A TFR of less than 2.1, if maintained, will cause a natural decline of population because deaths will exceed births.

Though the average global TFR is 2.5, the TFR varies markedly from one part of the world to another, revealing a vivid geographical pattern (**Figure 3.4**). In Europe, the average TFR is only 1.6. Every country with a TFR of 2.0 or lower will eventually experience population decline. Countries with very low TFRs are facing what have been termed "death spiral demographics." Bulgaria, for example, has a TFR of 1.5 and is expected to lose 28 percent of its population by 2050. In 2016, South Korea, Romania, Singapore, and Taiwan were tied at 1.2 for the lowest total fertility rate in the world.

By contrast, Africa has the highest TFR of any sizable part of the world at 4.7, led by Niger with 7.6 as of 2016. However, according to the World Bank, during the past two decades TFRs have fallen in all sub-Saharan African nations.

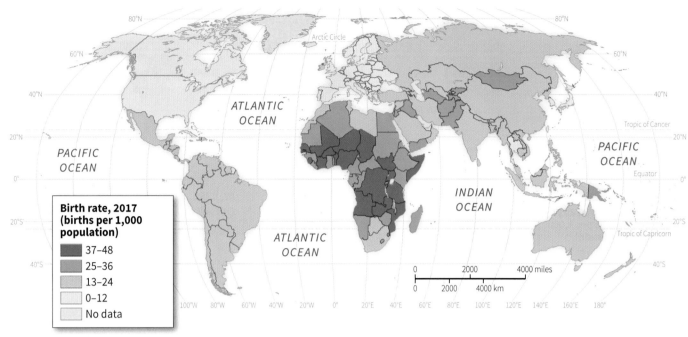

FIGURE 3.3 **Crude birth rate.** This map shows births per thousand population per year. (Data from Population Reference Bureau.)

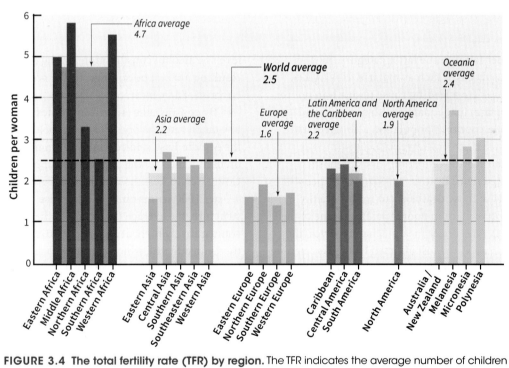

FIGURE 3.4 **The total fertility rate (TFR) by region.** The TFR indicates the average number of children born to women over their lifetimes. A rate of 2.1 is needed to produce a stable population over the long run; below that, population will decline. Fast growth is associated with a TFR of 5.0 or higher. (Data from United Nations.)

Patterns of Mortality

Another way to assess demographic change is to analyze **death rates**: the number of deaths per year per 1000 people (**Figure 3.5**). Of course, death is a natural part of the life cycle, and there is no way to achieve a death rate of zero. But geographically speaking, death comes in different forms. In the developed world, most people die of age-induced degenerative conditions, such as heart disease, or from maladies caused by industrial pollution of the environment.

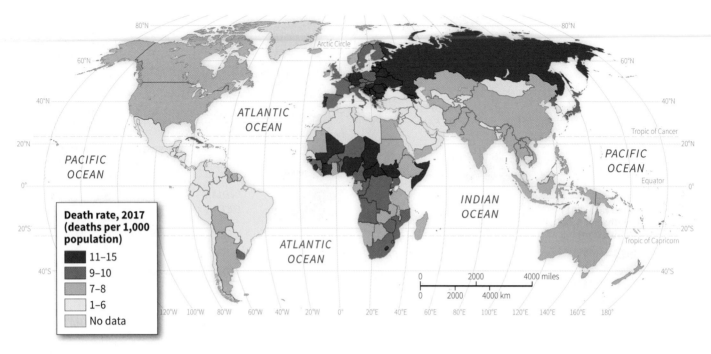

FIGURE 3.5 Crude death rate. This map shows deaths per thousand people per year. (Data from Population Reference Bureau.)

Many types of cancer fall into the latter category. By contrast, contagious diseases such as malaria, HIV/AIDS, and diarrheal diseases are a leading cause of death in poorer countries. Civil warfare, inadequate health services, and the age structure of a country's population also affect its death rate.

According to the World Population Clock, published by the Population Reference Bureau, worldwide 109 people die every minute; 24 of those deaths occur in the wealthy countries, while 85 take place in poorer nations. Some of the world's highest death rates occur in sub-Saharan Africa, the poorest world region and the most afflicted by life-threatening diseases and civil strife (**Figure 3.6** depicts the world regional distribution of new HIV cases). In general, death rates of more than 20 per 1000 people are uncommon today. The world's highest death rate as of 2016—just under 15 per 1000 people— was found in Lesotho, a small country nested within South Africa, and is primarily the result of a very high prevalence of HIV/AIDS. High death rates are also found in eastern European nations, six of which—Bulgaria, Lithuania, Ukraine, Latvia, Serbia, and Russia—rank among the world's ten highest death rates in 2016. In Russia, male life expectancy in 2015 was 64.7 years—3.3 years shorter than the global average of 68 years, and an astonishing 11.6 years less than the female life expectancy in Russia. This is due in no small measure to high levels of alcohol consumption among Russian men and the resulting alcohol poisonings, traffic accidents, homicides, liver cirrhosis, and cancer.

By contrast, the American tropics generally have rather low death rates, as does the desert belt across North Africa, the Middle East, and central Asia. In these regions, the predominantly young population suppresses the death rate. Compared to Lesotho or Bulgaria, Paraguay's death rate of only 4.7 per 1000 seems quite low. Because of its older population structure, the average death rate in the European Union is comparatively high, at 10.2 per 1000. Australia, Canada, and the United States, which continue to attract young immigrants, have lower death rates than most of Europe. Canada's death rate, for instance, is 8.5 people per 1000.

Various demographic traits, including infant mortality, can be used to assess standard of living and analyze it geographically. **Figure 3.7** is a simple attempt to map living standards using the **infant mortality rate**—a measure of how many children per 1000 live births die before reaching one year of age. Many experts believe that the infant mortality rate is the best single index of living standards because it is affected by many different factors: health, nutrition, sanitation, access to doctors, availability of clinics, education, ability to obtain medicines, and adequacy of housing. A striking geographical pattern is revealed by the infant mortality rate.

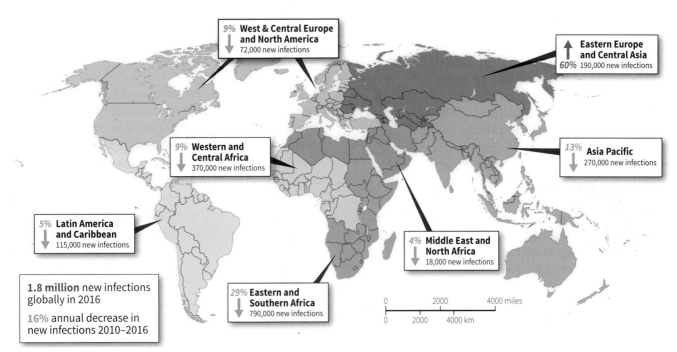

FIGURE 3.6 Number of new HIV infections in 2016 and change since 2010. There were 1.8 million people worldwide who were newly infected with HIV in 2016. While southern and eastern Africa are home to the majority—over 50 percent—of the world's existing HIV-positive individuals, there has been an alarming increase recently in central and eastern Europe. Russia alone accounts for 8 out of 10 new infections in the region. (Data from Avert.org and UNAIDS.)

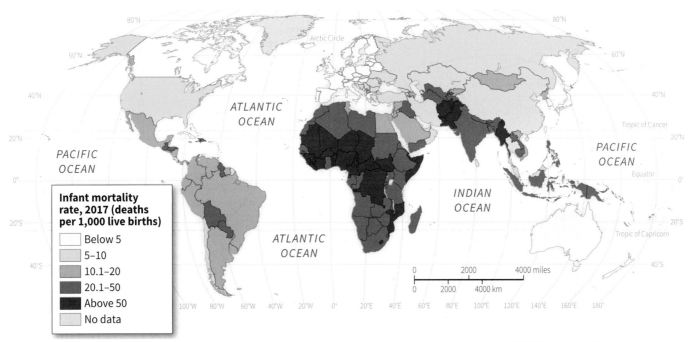

FIGURE 3.7 The world pattern of infant mortality rate. The key indicates the number of children per 1000 born who die before reaching one year of age. The world's infant mortality rate is 37. Many experts believe that this rate is the best single measure of living standards. (Data from Population Reference Bureau.)

The Demographic Transition

All industrialized, technologically advanced countries have low fertility rates and stable or declining populations, having passed through what is called the **demographic transition** (**Figure 3.8**). In preindustrial societies, birth and death rates were both high, resulting in almost no population growth. Because these were agrarian societies that depended on family labor, many children meant larger workforces, thus the high birth rates. But low levels of public health and limited access to health care, particularly for the very young, also meant high death rates. With the coming of the industrial era, medical advances and improvements in diet set the stage for a drop in death rates. Human life expectancy in industrialized countries soared from an average of 35 years in the eighteenth century to 75 years or more at present. Yet birth rates did not fall as quickly, leading to a population explosion as fertility outpaced mortality. In Figure 3.8, this is shown in late stage 2 and early stage 3 of the demographic transition model. Eventually, a decline in the birth rate followed the decline in the death rate, slowing population growth. An important reason leading to lower fertility levels involves the high cost of children in industrial societies, particularly because childhood itself involves a prolonged period of economic dependence on parents. Finally, in the postindustrial period, the demographic transition produced zero population growth or actual population decline (Figure 3.8 and **Figure 3.9**).

Achieving lower death rates is relatively cost effective, historically requiring little more than the provision of safe drinking water and vaccinations against common infectious diseases. Lowering death rates tends to be uncontroversial and quickly achieved, demographically speaking. Getting birth rates to fall, however, can be far more difficult, especially for a government official who wants to be reelected. Birth control, abortion, and challenging long-held beliefs about family size can prove quite controversial, and political leaders may be reluctant to legislate them. Indeed, the Chinese implementation of its one-child-per-couple policy probably would not have been possible in a country with a democratically elected government (see Subject to Debate later in this chapter, page 94). In addition, because it involves changing a cultural norm, the idea of smaller families can take three or four generations to become a reality. Increased educational levels for women are closely associated with falling fertility levels, as is access to various contraceptive devices (**Figure 3.10**).

The demographic transition is a model that predicts trends in birth rates, death rates, and overall population

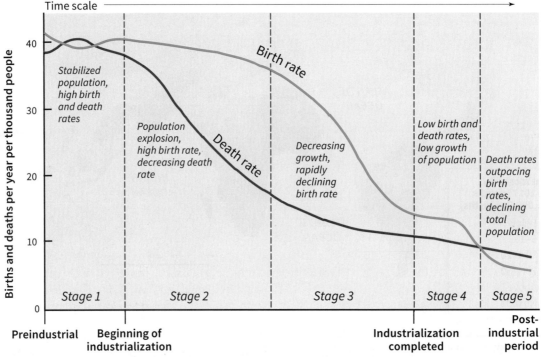

FIGURE 3.8 The demographic transition as a graph. The "transition" occurs in several stages as the industrialization of a country progresses. In the postindustrial stage, population decline eventually begins.

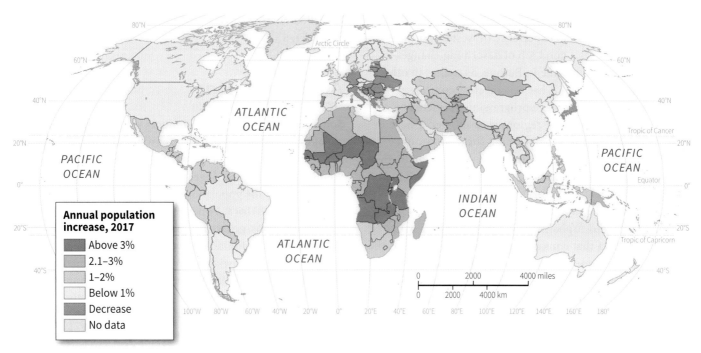

FIGURE 3.9 Annual population increase. The change is calculated as the difference between the number of births and deaths in a year, taken as a percentage of total population. Migration is not considered. Note the contrast between tropical areas and the middle and upper latitudes. In several places, countries with a very slow increase border areas with extremely high growth. (Data from Population Reference Bureau.)

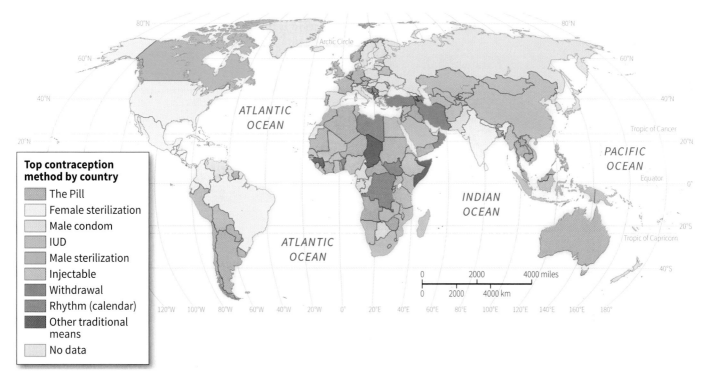

FIGURE 3.10 The geography of contraception. This map displays the top method of contraception used by married and cohabiting couples in each country. Contraception is much more widely practiced than abortion, but cultures differ greatly in their level of acceptance. (Data from United Nations.)

change levels. Yet, like many models used by demographers, it is based on the historical experience of western Europe: it is **Eurocentric**. It does a good job of describing changes in population patterns over time in Europe, as well as in other wealthy regions. However, it has several shortcomings. First is the inexorable stage-by-stage progression predicted by the model. Have countries or regions ever skipped a stage or regressed? Certainly. The case of China shows how policy, in this instance a government-imposed restriction on births, can fast-forward an entire nation to stage 4 (again, see the Subject to Debate feature). War, too, can cause a return to an earlier stage in the model by increasing death rates. For instance, Angola and Afghanistan are two countries with histories of conflict that continue today; both had topped the world's death rate list until recently. In other cases, wealth has not led to declining fertility. Thanks to oil exports, residents of Oman enjoy relatively high average incomes. But fertility, too, remains relatively high at nearly 3 children per woman in 2016. Indeed, the Population Reference Bureau has pointed to a "demographic divide" between countries where the demographic transition model applies well, and others—mostly poorer countries or those experiencing widespread conflict or disease—where birth and death rates do not necessarily follow the model's predictions. Even in Europe, there are countries where fertility has dropped precipitously, while at the same time death rates have escalated. In countries such as Russia, which, according to Peter Coclanis, has "somehow managed to reverse the so-called demographic transition," the model itself must be questioned.

Age Distributions

Some countries have overwhelmingly young populations. In a majority of countries in Africa, as well as some countries in Latin America and tropical Asia, close to half the population is younger than 15 years of age (**Figure 3.11**). In Niger, for example, 49.3 percent of the population is younger than 15 years of age. In sub-Saharan Africa, 41 percent of all people are younger than 15. Other regions, generally those that industrialized early, have a preponderance of middle-aged people in the over 15–under 65 age bracket. A growing number of affluent countries have remarkably aged populations. In Germany, for example, fully 21 percent of the people are over the traditional retirement age of 65. Many other European countries are not far behind. A sharp contrast emerges when Europe is compared with Africa, Latin America, or parts of Asia, where the average person never even lives to age 65. In Mauritania, Niger, Afghanistan, Guatemala, and many other countries, only 2 to 3 percent of the population has reached that age.

Countries with disproportionate numbers of old or young people often address these imbalances in innovative ways. Italy, for example, has one of the lowest birth rates in the world, with a TFR of only 1.4. In addition, Italy's population is one of the oldest in the world: 20.5 percent

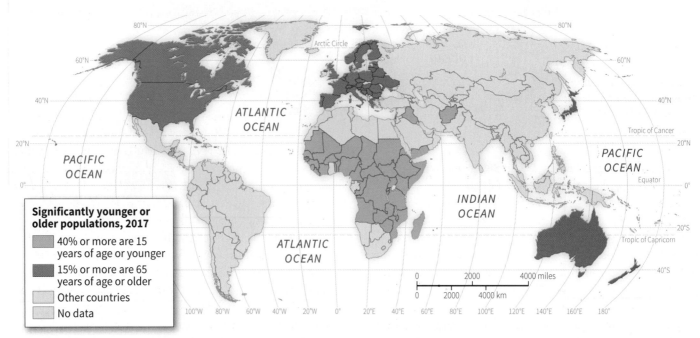

FIGURE 3.11 The world pattern of youth and old age. Some countries have populations with unusually large numbers of elderly people; others have preponderantly young populations. (Data from Population Reference Bureau.)

of its population is age 65 or older. As a result of both its TFR and age distribution, Italy's population is projected to shrink by 10 percent between 2010 and 2050. Of course, immigrants to Italy from other countries can and no doubt will counteract this trend, but that will lead to contentious political and cultural debates.

Given that the Italian culture does not embrace the institutionalization of the growing ranks of its elderly, and faced with the reality that more Italian women than ever work outside the home and thus are unwilling or unable to stay at home full-time to care for their elders, Italians have gotten creative. The northern Italian city of Milan, where 24 percent of the population is over 65 years of age, began an elder adoption program in 2010. In exchange for 200 euros (about 233 U.S. dollars) per month, a family can foster an elderly person, sharing a daily lunch and including them in leisure activities such as shopping or a trip to the hairdresser. Carla Romagnoni, age 102, was fostered by the Bonneti family. Maria Bonneti says, "Carla and I have fun together. We've been to La Scala to watch the ballet. Up till a few months ago, I used to take her to the cinema. Obviously, I cook as well. She's always asking me for Milan-style risotto. But mostly we talk. We tell each other the things that happen to us, our experiences." Is she a friend? "More than that. It's as if we were related. In fact, she's like a granny."

Within countries, the distribution of older populations also differs spatially. For example, rural populations in the United States and many other countries are usually older than those in urban areas. The flight of young people to the cities has left some rural counties in the midsection of the United States with populations whose median age is 45 or older. Some warm areas of the United States have become retirement havens for the elderly; parts of Arizona and Florida, for example, have populations far above the average age. Communities such as Sun City near Phoenix, Arizona, legally restrict residence to the elderly (**Figure 3.12**). In Great Britain, coastal districts have a much higher proportion of elderly than does the interior, suggesting that the aged often migrate to seaside locations when they retire.

A very useful graphic device for comparing age characteristics is the **population pyramid** (**Figure 3.13**, page 92). Careful study of such pyramids not only reveals the past acceptance of birth control but also allows geographers to predict future population trends. Youth-weighted pyramids, those that are broad at the base, suggest the rapid growth typical of the population explosion. What's more, broad-based population pyramids not only reflect past births but also show that future growth will rely on the momentum of all those young people growing into their reproductive years and having their own families, regardless of how small those families may be in contrast to earlier generations. Those population pyramids with

FIGURE 3.12 Residents of Sun City, Arizona, enjoy the many recreational opportunities provided in this planned retirement community, where the average age is 75. (A. Ramey/PhotoEdit)

more of a cylindrical shape represent countries approaching population stability or those in demographic decline. A quick look at a country's population pyramid can tell volumes about its past as well as its future. How many dependent people—the very old and the very young—live there? Has the country suffered the demographic effects of genocide or a massive disease epidemic (Figure 3.13)? Are significantly more boys than girls being born? These questions and more can be explored at a glance using a population pyramid.

The Geography of Gender

Although the human race is divided almost evenly between females and males, geographical differences do occur in the **sex ratio**—the ratio between men and women in a population (**Figure 3.14** and Subject to Debate, page 94). Slightly more boys than girls are born, but infant boys have slightly higher mortality rates than do infant girls. Recently settled areas typically have more males than females, as is evident in parts of Alaska, northern Canada, and tropical Australia. According to the U.S. Census Bureau, in 2016 there were 107 males for every 100 of Alaska's female inhabitants. Some poverty-stricken parts of South Africa are as much as 59 percent female. Prolonged wars reduce the male population. And, in general, women tend to outlive men. Returning to an earlier example, the gender gap in life expectancy has caused Russia's sex ratio to stand at 86 males for every 100 females. The population pyramid is also

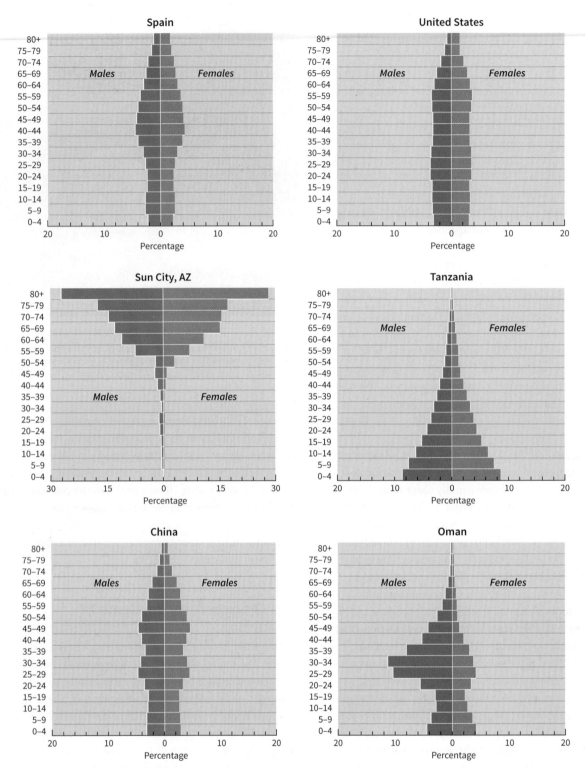

FIGURE 3.13 Population pyramids for selected countries and communities. Tanzania displays the classic stepped pyramid of a rapidly expanding population, whereas the U.S. pyramid looks more like a precariously balanced pillar. China's population pyramid reflects the lowered numbers of young people as a result of that country's one-child policy. Spain's population pyramid displays the visual impact of a low fertility rate. Sixty-five percent of the population of the Middle East is under 30 years of age. If provided educations and job opportunities, these individuals can constitute a valuable pool of human capital. If not, they can be the drivers of unrest. In addition, some Middle Eastern nations, as illustrated by Oman's population pyramid, have male-heavy populations, due to labor immigration from other, mostly poor Muslim, countries. (Data from Population Reference Bureau.)

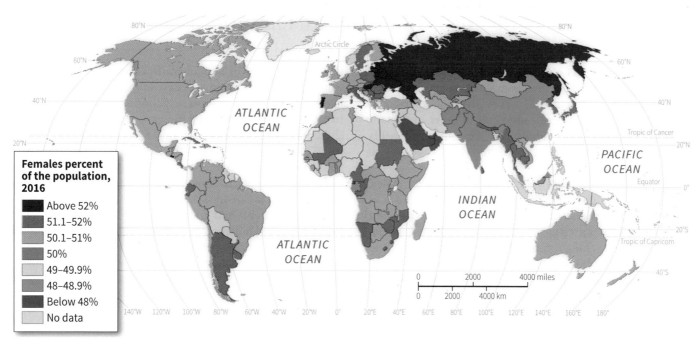

FIGURE 3.14 Females as a percentage of total population. Unbalanced sex ratios can occur in countries where people of one gender migrate to other countries for work, as has happened in Saudi Arabia, where male labor is imported to supplement a low-paid labor force. Sex ratios also become skewed when one gender outlives another, as women do in Russia and the former Soviet republics. India and China skew heavily male because of the preference for male offspring. (Data from Population Reference Bureau.)

useful in showing gender ratios. Note, for instance, the larger female populations in the upper bars for the United States in Figure 3.13.

Beyond such general patterns, gender often influences demographic traits in specific ways. Often **gender roles**—culturally specific notions of what it means to be a man and what it means to be a woman—are closely tied to how many children are produced by couples. In many cultures, women are considered more feminine when they produce many offspring. By the same token, men are seen as more masculine when they father many children. Because the raising of children often falls to women, the spaces that many cultures associate with women tend to be the private family spaces of the home. Public spaces such as streets, plazas, and the workplace, by contrast, are often associated with men. Some cultures go so far as to restrict where women and men may and may not go, resulting in a distinctive geography of gender, as **Figure 3.15** illustrates (take a look at Seeing Geography, page 115; see also Chapter 9).

Falling fertility levels that coincide with higher levels of education for women, however, have resulted in numerous challenges to these cultural ideas of male and female spaces. As more and more women enter the workplace, for instance, ideas of where women should and should not go slowly become modified. Clearly, gender is an important factor to consider in our exploration of population geography.

FIGURE 3.15 Segregated beach. This portion of the beachfront in the Israeli city of Tel Aviv allows women and men to swim on different days. Young children of both genders accompany their mothers, whereas older boys visit with their fathers. Gender-based segregation is important for Israel's orthodox Jews. (Courtesy of Patricia L. Price)

SUBJECT TO DEBATE   Female: An Endangered Gender?

Does the simple fact of being female expose a person to demographic peril? In most societies, women are viewed as valuable, even powerful, particularly as mothers, nurturers, teachers, and spiritual leaders. Yet in other important ways, to be female is to be endangered. We will consider this controversial idea with an eye to how demographics and culture closely shape each other.

Many cultures demonstrate a marked preference for males. The academic term describing this is *androcentrism;* you may know it as *patriarchy, male bias,* or simply *sexism.* Whether a preference for males is a feature of all societies has been disputed. Some societies pass along forms of their wealth, property, and prestige from mother to daughter, rather than from father to son. This is rare, however, and it is clear that the roots of the cultural preference for males are historically far reaching and widespread. In most societies, positions of economic, political, social, and cultural prestige and power are held largely by men. Men typically are considered to be the heads of households. Family names tend to pass from father to son, and with them, family honor and wealth. In traditional societies, when sons marry, they usually bring their wives to live in their parents' house and are expected to assume economic responsibility for aging parents.

For all of these reasons, in many places a cultural premium is placed on producing male children. Because couples often can choose to have more children, having a girl first is usually not a problem. However, in countries that have enacted strict population control programs, couples may not be given the chance to try again for a male child. This has resulted in severe pressures on couples to have boys in some countries, particularly in China and India. In both these nations, female-specific infanticide or abortion has resulted in a growing gender imbalance. Ultrasound devices that allow gender identification of fetuses are now available even to rural peasants in China. About 100,000 such devices were in use as early as 1990 in China, and by the middle of that decade there were 121 males for every 100 females among children two years of age or younger. The sex ratio in China is changing radically, and a profound gender imbalance exists there: in 2016 there were 870 girls for every 1000 boys, the most skewed sex ratio in the world. The naturally occurring sex ratio is 952 girls for every 1000 boys. In India, too, there were only 892 girls per 1000 boys in 2016. The cultural preference for boys may continue even when women migrate. In Manitoba, Canada, Indian-born women who already had two daughters were found to be almost twice as likely to have sons as their third child. The resulting sex ratio of 192 males for every 100 females is not a natural phenomenon; rather, it is the result of gender-selective abortion of female fetuses.

The cultural ramifications of such male-heavy populations are potentially profound. Men of marriageable age are increasingly unable to find female partners. Social analysts speculate that this will lead to human trafficking and violence against women, and this is already the case in India, where girls are kidnapped or trafficked in from neighboring Nepal and Myanmar. In China, the long-standing policy—now phased out—of one child per couple has resulted in the so-called four-two-one problem. This refers to the fact that the generational structure of many families now reflects four grandparents, two parents, and a single male child. This places enormous pressures on the shoulders of the male child to care for aging parents and grandparents. It also encourages parents and grandparents to lavish all their attention, wealth, and hopes on the only child. For some families, this has led to the "little emperor syndrome," whereby the male heir becomes spoiled, unable to cope independently, and even obese.

Continuing the Debate

As noted, most societies value females and males equally. For a number of reasons, however, some societies show a clear preference for males. Keeping all this in mind, consider the following questions:

- Are Chinese and Indian families somewhat justified in prioritizing the birth of a son?

- According to a recent report, Americans using technology to select their baby's gender are, unlike the Chinese, more likely to choose to have a girl. Why do you think there is a difference between male and female preference in these two societies?

This "little emperor" poses with his grandparents. (Lane Oatey/Blue Jean Images/Getty Images)

Mobility

3.2 Explain patterns, causes, and consequences of population migrations.

After natural increases and decreases in populations, the second of the two basic ways in which population numbers are altered is by the migration of people from one place to another, and it offers a straightforward illustration of relocation diffusion. When people migrate, they sometimes bring more than simply their culture; they can also bring disease. The introduction of diseases to new places can have dramatic and devastating consequences. Thus, the spread of diseases is also considered part of the theme of mobility because diseases move with people.

Migration

Humankind is not tied to one locale. *Homo sapiens* most likely evolved in Africa, and ever since, we have proved remarkably able to adapt to new and different physical environments. We have made ourselves at home in all but the most inhospitable climates, avoiding only such places as icy Antarctica and the Arabian Peninsula's arid "Empty Quarter." Our permanent habitat extends from the edge of the ice sheets to the seashores, from desert valleys below sea level to high mountain slopes. This far-flung distribution is the product of migration.

Early human groups moved in response to the migration of the animals they hunted for food and the ripening seasons of the plants they gathered. Indeed, the agricultural revolution, whereby humans domesticated crops and animals and accumulated surplus food supplies, allowed human groups to stop their seasonal migrations. But why did certain groups opt for long-distance relocation? Some migrated in response to environmental collapse, others in response to religious or ethnic persecution. Still others—probably the majority, in fact—migrated in search of better opportunities. For those who migrate, the process generally ranks as one of the most significant events of their lives.

Migration takes place when people decide that moving is preferable to staying and when the difficulties of moving seem to be more than offset by the expected rewards (see The Video Connection). In the nineteenth century, more than 50 million European emigrants left their homelands in search of better lives. Today, migration patterns are very different (**Figure 3.16**). Europe, for example, now predominantly receives immigrants rather than sending out emigrants. International migration stands at an all-time high, much of it labor migration associated with the process of globalization. According to the United Nations, about 244 million people today live outside the country of their birth.

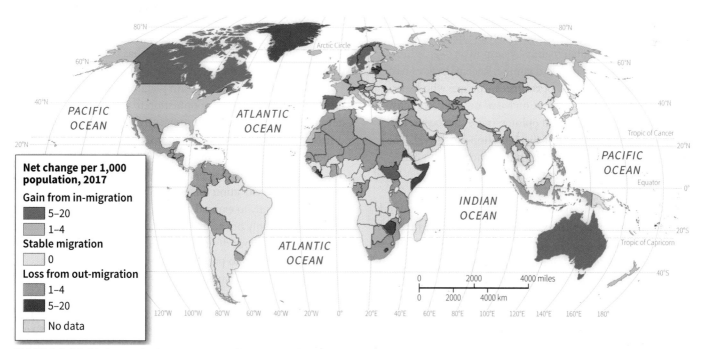

FIGURE 3.16 Major and minor migration flows today. Some nations are net recipients of immigrants, who go there in search of work or freedom from persecution. Other nations are net sources of emigrants, who leave their country of birth for a variety of reasons, including climate change and natural disasters. (Data from Population Reference Bureau.)

▶ The Video Connection

Paraíso

🔘 Watch at 🔵 SaplingPlus

Three Mexican immigrants who work as window washers on Chicago's downtown high-rise buildings are profiled in this video. The process of window washing is depicted, and the men speculate about the dangers—but also the beauty and craft—of the work they do. In addition, the men discuss more abstract notions: death, heaven, and the future that awaits them. Their conversations also bring up comparisons to the lives they might have lived had they stayed in Mexico. A great deal of nuance characterizes this video, and the three men's understandings of central contrasts—between danger and beauty, poverty and wealth, heaven and Earth—are not what you might expect.

Thinking Geographically

1. These immigrants are members of several communities. Identify these communities, and specify the contributions they make to them.

2. Window washing is a dangerous occupation. How do these three men rationalize that danger? What do they view as the positive aspects of their job?

3. The three men talk a lot about wealth and poverty. Give some examples from the video of their understandings of wealth and poverty. Does anything strike you as unexpected?

Historically, and to this day, forced migration also often occurs. The westward displacement of the Native American population of the United States; the dispersal of the Jews from Palestine in Roman times and from Europe in the mid-twentieth century; the export of Africans to the Americas as slaves; and the Clearances, or forced removal, of farmers from Scotland's Highlands to make way for large-scale sheep raising—all provide examples of forced migrations. Today, refugee movements are far too common, prompted mainly by despotism, war, ethnic persecution, climate change, and famine. Recent decades have witnessed a worldwide flood of **refugees**—people who leave their country because of persecution based on race, ethnicity, religion, nationality, or political opinion (note that economic persecution does not fall under the definition of *refugee*). The United Nations estimates that 22.5 million people who live outside their native country are refugees (see also Chapter 5 and Chapter 6).

Every migration, from the ancient dispersal of humankind out of Africa to the present-day movement toward urban areas, is governed by a host of **push and pull factors** that contribute to making the old home unattractive or unlivable and the new land attractive. Generally, push factors are the most central. After all, a basic dissatisfaction with the homeland is prerequisite to voluntary migration. The most important factor prompting migration throughout the thousands of years of human existence has been economic. More often than not, migrating people seek greater prosperity through better access to resources, especially land. Both forced migrations and refugee movements, however, challenge the basic assumption of the push-and-pull model, which posits that human movement is the result of choices and is primarily driven by economic factors.

Diseases on the Move

Throughout history, infectious disease has periodically decimated human populations. One has only to think of the vivid accounts of the Black Death plagues that killed one-third of medieval Europe's population to get a sense of how devastating disease can be. Importantly for human geographers, diseases both move in spatially specific ways and result in responses that are spatial in nature.

The spread of disease provides classic illustrations of both expansion and relocation diffusion. Some diseases are noted for their tendency to expand outward from their points of origin. Commonly borne by air or water, the viruses or bacteria that cause contagious diseases such as influenza and cholera spread from person to person throughout an affected area. Some diseases diffuse in a hierarchical fashion, whereby only certain social strata are exposed. The poor, for example, have always been much more exposed to the unsanitary conditions—rats, fleas, excrement, and crowding—that have led to disease epidemics. Other illnesses, particularly airborne diseases, affect all social and economic classes without discriminating. Indeed,

FIGURE 3.17 Fleeing from disease. This woodcut, which was printed in 1630, depicts Londoners leaving the pestilent city in a cart. In 1665–1666, London experienced the so-called Great Plague, an outbreak of bubonic plague that killed one-fifth of the city's population. (Bettmann/Getty Images)

disease and migration have long enjoyed a close relationship. Diseases have spread and relocated thanks to human movement. Likewise, widespread human migrations have occurred as a result of disease outbreaks.

When humans travel, their diseases move along with them. Thus, cholera—a waterborne disease that had been endemic to the Ganges River region of India for centuries—broke out in the early 1800s in Calcutta (now called Kolkata; see the photo and Seeing Geography on page 115). Because Calcutta was an important node in Britain's colonial empire, cholera quickly began to spread far beyond India's borders. Soldiers, pilgrims, traders, and travelers spread cholera throughout Europe, then from port to port across the world through relocation diffusion, making cholera the first disease of global proportions.

Early responses to contagious diseases were often spatial in nature. Isolation of infected people from the healthy population—known as quarantine—was practiced. Sometimes only the sick individuals were quarantined, whereas at other times households or entire villages were shut off from outside contact. In the late nineteenth century, whole shiploads of eastern European immigrants to New York City were routinely quarantined. Another early spatial response was to migrate away from the area where infection had occurred. During the time of the plague in fourteenth-century Europe, some healthy individuals abandoned their ill neighbors, spouses, and even children. Some of them formed altogether separate communities, whereas others sought merely to escape the city walls for the countryside (**Figure 3.17**).

Targeted spatial strategies were implemented once it became known that some diseases spread through specific means. The best-known example is John Snow's 1854 mapping of cholera

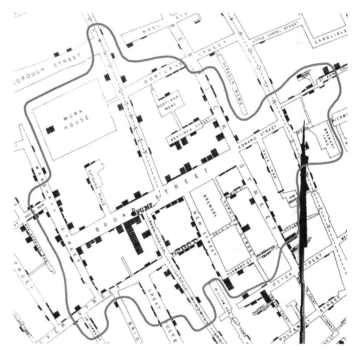

FIGURE 3.18 Mapping disease. London physician John Snow was skeptical of the notion that "bad air" somehow carried disease. He interviewed residents of the Soho neighborhood stricken by cholera to construct this map of cholera cases and used it to trace the outbreak to the contaminated Broad Street pump. Snow is considered to be a founder of modern epidemiology. (Courtesy of the John Snow Archive.)

outbreaks in London, illustrated in **Figure 3.18**, which allowed him to trace the source of the infection to one water pump. Thanks to his detective work, and the development of a broader medical understanding of the role of germs in the spread of disease, cholera was discovered to be a waterborne disease that could be controlled by improving the sanitary conditions of water delivery.

The threat of deadly disease is hardly a thing of the past. Today, geographers play vital roles in understanding the diffusion of HIV/AIDS, SARS (severe acute respiratory syndrome), and the so-called swine flu and bird flu in order to better address outbreaks and halt the spread of these global scourges. By mapping the spread of these relatively contemporary diseases, understanding the pathways traveled by their carriers, and developing appropriate spatial responses, it is hoped that mass epidemics and disease-related panics can be avoided.

Globalization

3.3 Understand theories of population growth and control and how these have changed over time.

One of the great debates today involves the population of the Earth. How many people can the Earth support? Should humans limit their reproduction, and who should decide? These debates occur at the global scale, rather than at the level of individual countries. These questions are so hotly contested because addressing them in a conscientious way may well determine the long-term fate of the human race.

Population Explosion?

One of the fundamental issues of the modern age is the **population explosion**—a dramatic increase in world population since 1900 (**Figure 3.19**). The crucial element triggering this explosion has been a steep decline in the death rate, particularly for infants and children, in most of the world, without an accompanying universal decline in fertility. At one time in traditional cultures, only two or three offspring in a family of six to eight children might live to adulthood, but when improved health conditions allowed more children to survive, the cultural norm encouraging large families persisted. This time lag—typically, several generations in duration—between falling infant mortality rates and reduced fertility rates is, in part, what gave rise to the large population increases depicted in stages two and three of the demographic transition model.

Until very recently, the number of people in the world has been increasing geometrically, doubling in shorter and shorter periods of time. It took from the beginning of human history

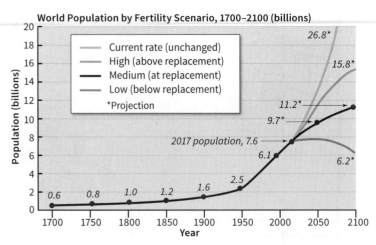

FIGURE 3.19 Projected world population in the year 2100. The United Nations projects that world population will reach 11.2 billion by 2100. This is based on a medium-high global fertility rate scenario. Notice, however, what a difference it makes if fertility is at or below replacement level.

until 1800 C.E. for the Earth's population to grow to 1 billion people. But from 1800 to 1930, it grew to 2 billion, and in only 45 more years it doubled again. The overall effect of even a few population doublings is astonishing.

As an illustration of a simple geometric progression, consider the following legend. A king was willing to grant any wish to the person who could supply a grain of wheat for the first square of his chessboard, two grains for the second square, four for the third, eight for the fourth, and so on. To cover all 64 squares and win, the candidate would have had to present a cache of wheat larger than today's worldwide wheat crop. Looking at the population explosion in another way, it is estimated that 107 billion humans have lived in the entire 200,000-year period since *Homo sapiens* originated. Of these, 7.5 billion are alive today.

Malthus Versus the Cornucopians Some scholars foresaw long ago that an ever-increasing global population would eventually present difficulties. The most famous pioneer observer of population growth, the English economist and cleric Thomas Malthus, published *An Essay on the Principle of Population*—known as the "dismal essay"—in 1798. He believed that the human ability to multiply far exceeded our ability to increase food production. Consequently, Malthus maintained that "a strong and constantly operating check on population" would necessarily act as a natural control on numbers. Malthus regarded famine, disease, and war as the inevitable outcome of the human population's outstripping the food supply (**Figure 3.20**). He wrote, "Population, when un-checked, increases in a geometrical ratio. Subsistence only increases in an arithmetical ratio. A slight acquaintance with numbers will show the immensity of the first power in comparison of the second."

The adjective **Malthusian** entered the English language to describe the dismal future Malthus foresaw. Being a cleric as well as an economist, however, Malthus believed that if humans could voluntarily restrain the "passion between the sexes," we might avoid their otherwise miserable fate.

But was Malthus right? From the very first, his ideas were controversial. The founders of communism, Karl Marx and Friedrich Engels, blamed poverty and starvation on the evils of capitalist society. Taking this latter view might lead one to believe that the miseries of starvation, warfare, and disease are more the result of unequal distribution of the world's wealth than of overpopulation. Indeed, severe food shortages in several regions point to precisely this issue of inequitable food distribution and its catastrophic consequences.

Malthus did not consider that when faced with complex problems such as scarce food supplies, human beings are highly creative. This has led critics of Malthus and his modern-day followers to point out that although the global population has doubled three times since Malthus wrote his essay, food supplies have doubled five times. Innovations such as the green revolution have led to food increases that have far outpaced population growth (see Chapter 8). Other measures of well-being, including life expectancy, air quality, and average

education levels, have all improved, too. Some of Malthus's critics, known as **cornucopians**, argue that human beings are, in fact, our greatest resource and that attempts to curb our numbers misguidedly cheat us out of geniuses who could devise creative solutions to our resource shortages. Modern-day followers of Malthus, known as **neo-Malthusians**, counter that the Earth's support systems are being strained beyond their capacity by the widespread adoption of wasteful Western lifestyles.

The Slowdown of World Population Growth The fact is that the world's population is growing more slowly than before. The world's TFR has fallen to 2.47; one demographer has declared that "the population explosion is over"; and Figure 3.19 projects a possible leveling off or even a drop in the global population by the year 2100, suggesting that the world's current "population explosion" may be merely a stage in a global demographic transition.

In reality, it is difficult to speculate about the state of the world's population beyond the year 2050. What is clear, however, is that the lifestyles we adopt will affect how many people the Earth can ultimately support. In particular, whether wealthy countries continue to use the amount of resources they currently do, and whether developing countries decide to follow a Western-style route toward more and more resource consumption, will significantly determine whether we have already overpopulated the planet.

The Rule of 72

A handy tool for calculating the **doubling time** of a population is called the Rule of 72: take a country's rate of annual increase, expressed as a percent, and divide it into the number 72. The result is the number of years a population, growing at a given rate, will take to double.

For example, the natural annual growth of the United States in 2016 was 0.43 percent. Dividing 72 by 0.43 yields 167, which means that the population of the United States will double every 167 years. This does not, however, factor in the relatively high levels of immigration experienced by the United States, which will cause its population to double faster than every 120 years.

What about countries with higher rates of growth? Consider Guatemala, which is growing at 2 percent per year. Doing the math, we find that Guatemala's population is doubling every 36 years.

Percentages such as 0.43 or 2 don't sound like such high rates. If we were discussing your bank account rather than the populations of countries, you would hope that your money would double more quickly than every 36 years! (Incidentally,

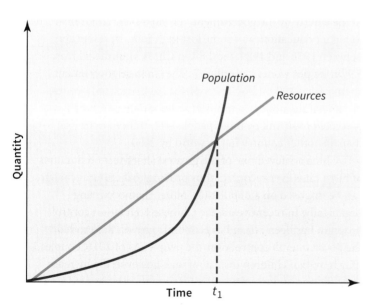

FIGURE 3.20 Malthus's dismal equation. Thomas Malthus based his theory of population growth on the simple notion that resources (food) increase in a linear fashion, whereas population grows geometrically. Beyond time t_1, population outstrips resources, resulting in famine, conflict, and disease.

you can apply the Rule of 72 to your bank account or to any other figure that grows at a steady annual rate.) You may be tempted to say, "Look, there are only 15 million people in Guatemala, so it doesn't really matter if its population is doubling quickly. What really matters is that at an annual increase of 1.2 percent, India's 1.3 billion people will double to 2.6 billion in 60 years. Or, at an annual increase of 0.47 percent, China's 1.38 billion will double almost as slowly as the U.S. population—every 153 and 167 years, respectively—but that will add another 1.38 billion to the world's population, more than four times what the United States will add by doubling!"

This assessment is partly right and partly wrong, depending on your vantage point. Viewed at the global scale, it does indeed make a significant difference when China's or India's population doubles. But if you are a resident of Guatemala or an official of the Guatemalan government, a doubling of your relatively small country's population every 36 years means that health care, education, jobs, fresh water, and housing must be supplied to twice as many people every 36 years. As mentioned in the introduction to this chapter, the scale at which population questions are asked is vital for the answers that are given.

Population Control Programs

Though the debates may occur at a global scale, most population control programs are devised and implemented at the national level. When faced with perceived national security threats, some governments respond by supporting pronatalist programs designed to increase the population. Labor shortages in the Nordic countries have led to incentives for couples to have larger families. Most population programs, however, are antinatalist: they seek to reduce fertility rates. Needless to say, trying to control this is an easier task for a nonelected government to carry out because limiting fertility rates challenges traditional gender roles and norms about family size.

Although China certainly is not the only country that has sought to limit its population growth, its so-called one-child policy provides the best-known modern example. Mao Zedong, the longtime leader of the People's Republic of China (1949–1976), did not initially discourage large families, in the belief that "every mouth comes with two hands." In other words, he believed that a large population would strengthen the country in the face of external political pressures. He reversed his position in the 1970s when it became clear that China faced resource shortages as a result of its burgeoning population. In 1980 the one-child-per-couple policy was adopted. With it,

FIGURE 3.21 China's one child per couple policy. This Chinese billboard encouraged families to use family planning in order to achieve the goal of one child per couple. (Iain Masterton/Alamy)

Chinese authorities sought not merely to halt population growth but, ultimately, to decrease the national population. All over China there were billboards and posters admonishing the citizens that "one couple, one child" is the ideal family (**Figure 3.21**). Violators faced huge monetary fines, could not request new housing, lost the rather generous benefits provided to the elderly by the government, forfeited their children's access to higher education, and some lost their jobs. In response, between 1970 and 1980, the TFR in China plummeted from 5.9 births per woman to only 2.7. The Chinese government claimed that 400 million fewer births have occurred over the lifetime of the policy. In terms of sheer numbers, China is predicted to shrink by more people—44 million, to be exact—than any other country in the world by 2050.

China achieved one of the greatest short-term reductions of birth rates ever recorded, thus proving that cultural changes can be imposed on a population, rather than occurring organically. In recent years, the Chinese population control program has been relaxed as economic growth has eroded the government's control over the people. As of 2016, couples may have two children instead of one; however, the increase in economic opportunity and migration to cities has led some couples to have smaller families voluntarily. Because of China's population size, its actions will greatly affect the global future as well as its own national one (see Subject to Debate on page 94 for a discussion of the gender implications of strict population control policies).

Nature-Culture

3.4 Describe how the natural world shapes population characteristics, and how population characteristics in turn shape the natural world.

The reasons that human populations have formed such diverse patterns are often cultural. Though the differences may have started out as adaptations to given physical conditions, when repeated from generation to generation, these patterns become woven tightly into the cultural fabric of places. Thus, demographic practices such as living in crowded settlements or having large families may well have deep roots in both nature and culture. In addition, perceptions of specific natural environments, whether positive or negative, vary by culture, and can change for a given group of people over time. Moreover, the adaptive strategies of human populations can modify nature, pushing it past the breaking point, with disastrous consequences both for the environment and for those inhabiting it.

The Environmental Influence on Population

Local population characteristics are often influenced by the availability of resources. In the middle latitudes, population densities tend to be greatest where the terrain is level, the climate is mild and humid, the soil is fertile, mineral resources are abundant, and the sea is accessible. Conversely, population tends to thin out with excessive elevation, aridity, coldness, ruggedness of terrain, and distance from the coast.

Climatic factors influence where people settle. Most of the sparsely populated zones in the world have, in some respect, difficult climates from the viewpoint of human habitation (see Figure 3.1, page 83). The thinly populated northern edges of Eurasia and North America are excessively cold, and the belt of sparse population extending from North Africa into the heart of Eurasia matches the pattern in major desert zones of the Eastern Hemisphere. Humans remain largely creatures of the humid tropics, subtropics, or midlatitudes and have not fared well in excessively cold or dry areas. Small populations of Inuit (Eskimo), Sami (Lapps), and other peoples live in some of the less hospitable areas of the Earth, but these regions do not support large populations, while Antarctica has no indigenous human population and is a temporary home to only a handful of researchers. Humans have proven remarkably adaptable, and our cultures include strategies that allow us to live in many different physical environments. As a species, however, perhaps we have not entirely moved beyond the **adaptive strategies** that suited us so well to the climatic features of sub-Saharan Africa, where we began.

Humankind's preference for lower elevations is especially true for the middle and higher latitudes. Living at lower altitudes allows the inhabitants to avoid some of the colder weather at these latitudes. Most mountain ranges in these latitudes stand out as sparsely populated regions. By contrast, inhabitants of the tropics often prefer to live at higher elevations, concentrating in dense clusters in mountain valleys and basins (see Figure 3.1). For example, in tropical portions of South America, more people live in the Andes Mountains than in the nearby Amazon lowlands. The capital cities of many tropical and subtropical nations lie in mountain areas above 3000 feet (900 meters) in elevation. Living at higher elevations allows residents to escape the hot, humid climate and diseases of the tropical lowlands. In addition, these areas were settled because the fertile volcanic soils of these mountain valleys and basins were able to support larger populations in agrarian societies.

Humans often live near the sea. The continents of Eurasia, Australia, and South America resemble hollow shells, with the majority of the population clustered around the rim of each continent (see Figure 3.1). In Australia, half the total population lives in only five port cities, and most of the remainder is spread out over nearby coastal areas. This preference for living by the sea stems partly from the trade and fishing opportunities the sea offers. At the same time, continental interiors tend to be regions of climatic extremes. For example, Australians speak of the "dead heart" of their continent, an interior land of excessive dryness and heat, and sparse resident human populations (**Figure 3.22**).

FIGURE 3.22 Australia's "dead heart." This shot was taken west of Alice Springs in Australia's Northern Territory, which is just about at the center of the continent. The climate here is inhospitable and settlements are sparse. (Auscape/UIG/Getty Images)

Disease also affects population distribution. Some diseases attack valuable domestic animals, depriving people of food and clothing resources. Such diseases have an indirect effect on population density. For example, in parts of East Africa, a form of sleeping sickness attacks livestock. This particular disease is almost invariably fatal to cattle but not humans. The people in this part of East Africa depend heavily on cattle, which provide food, represent wealth, and serve a religious function in some tribes. The spread of a disease fatal to cattle has caused entire tribes to migrate away from infested areas, leaving those areas unpopulated.

Environmental Perception and Its Impact on Settlement and Migration

Perception of the physical environment plays a major role in a group's decision about where to settle. Different cultural groups often perceive the same physical environment in different ways. These varied responses to a single environment influence settlement patterns. A good example appears in a part of the European Alps shared by German- and Italian-speaking peoples. The mountain ridges in that area—near the point where Switzerland, Italy, and Austria border one another—run in an east–west direction, so that each ridge has a sunny, south-facing slope and a shady, north-facing one. German-speaking people, who rely on dairy farming, long ago established permanent settlements on the shady slopes, while Italians are culturally tied to crops that grow best on the sunny slopes.

Sometimes a cultural group changes its perception of an environment over time, with a resulting redistribution of its population. The coalfields of western Europe are a good case in point. Before the industrial age, many coal-rich areas—such as the Midlands of England, southern Wales, and the lands between the headwaters of the Oder (or Odra) and Vistula rivers in Poland—were only sparsely or moderately settled. The development of steam-powered engines and the increased use of coal in the iron-smelting process, however, created a tremendous demand. Industries grew up near the European coalfields, and people flocked to these areas to take advantage of the new jobs. In other words, once a technological development gave a new cultural value to coal, many sparsely populated areas containing that resource acquired large concentrations of people.

Some migration within the United States today is prompted by a desire for a pleasant climate and other positive physical environmental traits, such as beautiful scenery. Surveys of immigrants to Arizona and Florida revealed that, for older Americans, those states' sunny, warm climate is a major attraction. However, young people—those in the 18–34 age bracket—are the most mobile Americans, as they relocate for college, careers, and later as families with young children in search of more space. Colorado attracted more of these migrants than any other U.S. state, according to a 2014 study of federal income tax returns. The attraction of Colorado for young adults is due, in part, to the mountainous physical environment, and the climbing, skiing, and hiking opportunities they provide. Thus, while different age and cultural groups may express different preferences, all are influenced by their perceptions of the physical environment in making decisions about migration.

Population Density and Environmental Alteration

People modify their habitats through their adaptive strategies. Particularly in areas where population density is high, radical environmental alterations occur often. This can also happen in fragile environments even at relatively low population densities because, as discussed earlier in this chapter, the Earth's carrying capacity varies greatly from one place to another and from one culture to another.

Many of our adaptive strategies are not sustainable. Population pressures and local ecological crises are closely related. For example, in Haiti, where rural population pressures have become particularly severe, the trees in previously forested areas have been felled for fuel, leaving the surrounding fields and pastures increasingly denuded and vulnerable to erosion (**Figure 3.23**). In short, overpopulation relative to resource availability can precipitate environmental destruction—which, in turn, results in a downward cycle of worsening poverty, leading to an eventual catastrophe that is both ecological and demographic (**Figure 3.24**).

FIGURE 3.23 Aerial view of Hispaniola. The border between Haiti, on the left, and the Dominican Republic, to the right, is clearly demarcated by the absence of forest cover on the Haitian side. (James P. Blair/National Geographic Creative)

FIGURE 3.24 Landslide on Whidbey Island, Washington State. This massive landslide in 2013 changed the topography of Whidbey Island by moving nearly 5.3 million square feet of earth and raising the elevation of the beach some 30 feet. Human settlements so close to the Island's edge may have triggered this landslide, when indigenous shrubs and trees were replaced with lawns. (Ted S. Warren/AP Photo)

The worldwide ecological crisis is not solely a function of overpopulation. A relatively small percentage of the Earth's population controls much of the industrial technology and consumes a disproportionate percentage of the world's resources each year. Americans, who make up less than 5 percent of the global population, account for about 25 percent of the natural resources consumed globally each year. New houses built in the United States in 2016 were, on average, 1000 square feet larger than those built in 1973, nearly doubling the living space per person. If everyone in the world had an average American standard of living, the Earth could support only about 500 million people—only 8 percent of the present population. As the economies of large countries such as India and China continue to surge, the resource consumption of their populations is likely to rise as well, because the human desire to consume appears to be limited only by the ability to pay for it. Indeed, in 2010, China surpassed the United States to become the world's largest consumer of energy resources.

Environmental Changes in Mexico City Sometimes environmental changes occur slowly over time. Depletion of the underground water table, for instance, has caused Mexico City to sink, on average, 2.5 inches each year. Some of the larger, heavier buildings in Mexico City, such as the city's main cathedral, have been sinking even more quickly than that. The Palace of Fine Arts, for instance, has sunk 13 feet in the last century, and its original ground floor has now become the building's basement.

Mexico City was originally the Aztec capital of Tenochtitlan. Building in the middle of a lake, the Aztecs made use of a series of canals and floating gardens to transport themselves and grow their food. The natural environment was thus minimally altered by its Aztec inhabitants. When the Spaniards conquered Mexico in the sixteenth century, they built Mexico City on top of the ruins of Tenochtitlan and drained the lake in an effort to control flooding. Since then, Mexico City has grown to one of the world's largest urban areas. Its 21 million inhabitants get their water from an aquifer—a natural underground reservoir. As the water is drained out over time, the underlying layers of soft clay and volcanic silt dry out and collapse unevenly.

Cracked foundations, window frames that are no longer square, and buildings leaning on each other present multiple inconveniences, costs, and hazards (**Figure 3.25**). According to one Mexico City–based engineer, "When a building tilts more than 1 degree, then I think it becomes very uncomfortable." He notes that people living in crooked buildings are unable to comfortably lie in bed or even wash dishes, with tap water flowing oddly. "Tables aren't stable. Liquids don't look right when they are in big containers. Windowpanes can break. Doors don't close right." Vertigo is a common complaint of people who live and work in tilted structures. Other urban features are also affected. Subway lines and sidewalks need extensive repairs due to sinking, tilting, and buckling. Liquids in drainage canals and pipelines flow uphill.

Some structures are in imminent danger of collapse, and 50 buildings have been condemned by the city since 2006, with another 5000 or so known to be unstable. These structures become vulnerable to natural disasters. Because Mexico City is located in an earthquake-prone area, frequent tremors further damage buildings, and a larger earthquake could prove disastrous. An earthquake that struck in 1985 caused 800 buildings to collapse and left thousands dead.

FIGURE 3.25 Sinking building in Mexico City. This line of rooftops was initially straight, but has shifted as the ground beneath became unstable over the years. (Albert Moldvay/National Geographic/Getty Images)

FIGURE 3.26 Environmental refugees fleeing wildfires. This image is from Ventura, California. Hundreds of thousands of Californians were displaced from their homes when central and southern portions of the state were devastated by wildfires in late 2017. Subsequently, mudslides resulting from rainfall in these newly deforested areas displaced thousands more. (Noah Berger/AP Photo)

Environmental Refugees Sometimes human adaptive strategies stress the natural environment past the breaking point and previous population densities can no longer be sustained. When that happens, people are forced to migrate. Sudden environmental disasters, such as floods, tornadoes, or forest fires, can give rise to a massive human exodus from a

destroyed place (**Figure 3.26**). People who are displaced from their homes due to severe environmental disruption are called **environmental refugees**.

The adaptive strategy of draining Mexico City's underground aquifer to provide water for its population has resulted in the slow collapse of the city. At what point will residents begin to migrate to other places because of the hazards posed? Another large earthquake like the one that struck in 1985 might bring about significant population movements out of the damaged city.

A number of environmental catastrophes are closely related to population pressures. Violent hurricanes and sea level rise are associated with global warming; land degradation occurs due to local pressures such as over-farming, contamination, or broader climate change resulting in desertification; and landslides result from deforestation and other effects of human settlement expansion. These are just some of the catastrophes resulting from unsustainable adaptive strategies (**Figure 3.27**). The Internal Displacement Monitoring Centre estimated that some 19 million people worldwide were forced to flee their homes due to weather-related events alone in 2015. These disasters are more human-made than they are "natural," and they are often the trigger for large-scale migrations of human populations in search of safer environments.

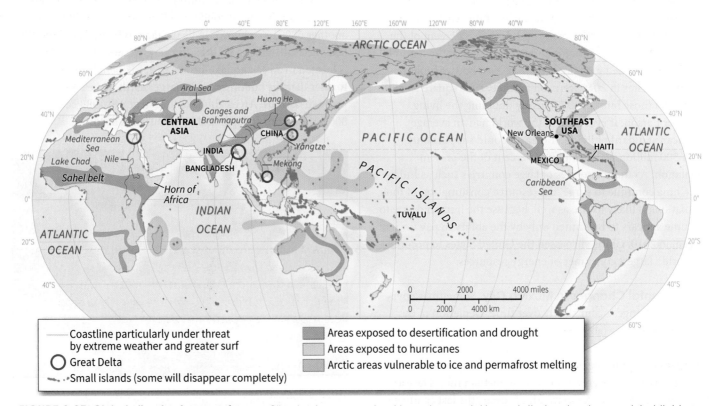

FIGURE 3.27 Global climate change refugees. Climate change can lead to environmental hazards that make places uninhabitable. This map depicts the types of hazards that are likely to occur. People in the developing world are most likely to become environmental refugees, because climate change is coupled with the effects of poverty and war.

GEOGRAPHY @ WORK

Aaron Hoard

Deputy Director, Office of Regional and Community Relations,
University of Washington
Education: BA Geography, University of Washington

Q. *Why did you major in geography and decide to pursue a career in the field?*

A. I was interested in urban planning as a career, but my university didn't offer an undergraduate planning degree at the time. The mix of subjects and technical skills taught in the geography department seemed closely aligned to what I wanted to do as a career.

Q. *Please describe your job.*

A. I act as a liaison between the university and local governments, businesses, and neighbors. In this capacity, I help manage projects related to transportation, land use, public safety, housing, and city regulations. I have helped organize the university's response to design and mitigation issues related to the construction of major transportation projects. I have worked on affordable housing strategies for university employees, including a home mortgage program and an affordable housing project. I also work closely with the community and city to address ongoing concerns about public safety and student behavior in the surrounding neighborhoods. And I represent the university on a variety of boards and committees in the community.

Q. *How does your geographical background help you in your day-to-day work?*

A. My projects cut across multiple fields of study, such as housing, transportation, public safety, social services, and economic development. Geography has a rich tradition of being an interdisciplinary degree, and having that experience working across different disciplines helped prepare me to work on complex challenges in my career. Many of my projects are also spatial in nature. Where do people live? Where can we develop new buildings? Where does crime occur? The ability to identify and answer these spatial questions helps to focus solutions.

Q. *What advice do you have for college students considering a career in geography?*

A. Develop a "toolbox" of analytical skills offered through your geography program. Skills in GIS, programming, cartography, statistics, demographics, remote sensing, and other areas can help make you a competitive candidate for your first job. The interdisciplinary nature of geography will serve you well throughout your career. Learning to fuse together different subjects to answer spatial questions in geography will help you to develop the intellectual flexibility you'll need to tackle a variety of complex problems.

Cultural Landscape

3.5 Analyze the imprint of demographic factors on the cultural landscape.

Population geographies are visible in the landscapes around us. The varied densities of human settlements and the shapes these take are clues to the mutually shaping interaction of demography and landscape. These clues are evident wherever people have settled, be they urban, suburban, or rural locations.

Less visible, but no less important, factors are also at work on and throughout the cultural landscape. In this final part of the chapter, we examine the diverse and often creative ways in which different places are organized spatially in order to accommodate the populations that live there.

Diverse Settlement Types

Human settlements range in density from the isolated farmsteads found in some rural areas to the teeming streets of megacities such as Kolkata, India (see Seeing Geography,

FIGURE 3.28 Nomadic family moving on. These Mongol herders are nomadic; they move often in search of the best pasture for their livestock. Here, a family packs their yurt, a traditional Mongolian tent structure, and their few household possessions. (Tuul & Bruno Morandi/Getty Images)

FIGURE 3.29 High-density dwelling in Amsterdam. The Borneo-Sporenburg development provides high-density, low-rise housing to accommodate Holland's small families in an urban setting. Instead of yards, these dwellings feature rooftop terraces. The waterfront also provides a "green space" for residents, who tie their boats up in front of their houses. (Iain Masterton/Alamy)

page 115). The size and pattern of human settlements depend, in part, on the size and needs of the population to be accommodated. Small rural populations and those engaged in subsistence farming don't inhabit large, dense cities. Nomadic peoples require dwellings that can be easily packed up and moved, and they tend to have few household possessions (**Figure 3.28**). On the other end of the spectrum, dense cities make more sense for populations employed in the service and information sectors. These people benefit from being as close as possible to their jobs, and higher densities—at least in theory—reduce commuting distances. People in postindustrial societies also tend to have smaller families, which can be more easily accommodated in apartments (**Figure 3.29**).

But there is also a significant cultural aspect at work, such that similar populations may live in settlements that look and feel vastly different from each other. Take rural settlements, for instance. In many parts of the world, farming people group themselves together in clustered settlements called **farm villages**. These tightly bunched settlements vary in size from a few dozen inhabitants to several thousand. Within the farm villages are individual **farmsteads** that include the house, barn, sheds, animal pens, and garden. The fields, pastures, and meadows lie out in the country beyond the limits of the village, and farmers must journey out from the village each day to work the land.

Farm villages are the most common form of agricultural settlement in much of Europe, in many parts of Latin America, in the densely settled farming regions of Asia (including much of India, China, and Japan), and among the sedentary farming peoples of Africa and the Middle East.

In many other parts of the world, the rural population lives on dispersed, isolated farmsteads, often some distance from the nearest neighbors (**Figure 3.30**). These dispersed rural settlements grew up mainly in Anglo America, Australia, New Zealand, and South Africa—that is, in the lands colonized by emigrating Europeans.

Why do so many farm people settle together in villages? Historically, the countryside was unsafe, threatened by roving bands of outlaws and raiders. Farmers could better defend themselves against such dangers by grouping together in villages. In many parts of the world, the populations of villages have grown larger during periods of insecurity and shrunk again when peace returned. Many farm villages occupy the most easily defended sites in their vicinity.

Various communal ties strongly bind villagers together. Farmers linked to one another by blood relationships, religious customs, communal landownership, or other similar bonds usually form clustered villages. Mormon farm villages in the United States provide an excellent example of the clustering force of religion. Communal or state ownership of the land—as in China and parts of Israel—encourages the formation of farm villages.

The conditions encouraging dispersed settlement are precisely the opposite of those favoring village development.

FIGURE 3.30 A truly isolated farmstead, in the Vesturland region of western Iceland. This type of rural settlement dominates almost all lands colonized by Europeans who migrated overseas. Iceland was settled by Norse Vikings a thousand years ago. (Jay Dickman/Getty Images)

These include peace and security in the countryside, eliminating the need for defense; colonization by individual pioneer families rather than by socially cohesive groups; private agricultural enterprise, as opposed to some form of communalism; and well-drained land where water is readily available. Most dispersed farmsteads originated rather recently, dating primarily from the colonization of new farmland in the past two or three centuries.

Landscapes and Demographic Change

Population change in a place can occur rapidly or more slowly over time. Rapid **depopulation** can come about as the result of sudden catastrophic events, such as natural disasters, disease epidemics, and warfare. At other times, depopulation occurs more slowly. For example, places may lose population over time because of the gradual out-migration of people in search of opportunities elsewhere, as a cumulative result of declining fertility rates, or, as we saw in the previous section, in response to climate change.

Populations can increase or decrease slowly or rapidly. A place might experience an abrupt influx of people who have been suddenly displaced from other areas, as in Uganda, which absorbed between 2000 and 6000 South Sudanese refugees per day in 2017 (**Figure 3.31**). However, population increases commonly occur more gradually, as the result of demographic improvements such as longer life spans or lower infant mortality rates, or the accumulation of steady streams of immigrants to attractive areas over time.

Depopulation in one area and population increase in another are often linked. For instance, from 2011 to the end of 2015, some 2.5 million Syrian refugees fled to neighboring Turkey. As a longer-term example, broad economic changes or technological innovations can lead to regional population shifts whereby some areas become depopulated and others gain population. Migration within the United States, away from economically moribund industrial cities in the Northeast and toward the thriving technology-based cities of the Pacific Northwest and northern California, provide a good example of a steady regional population shift over time.

Indeed, shifting economic and technological conditions have influenced the demographic landscape in profound ways. The process of industrialization during the past 200 years has caused the greatest voluntary relocation of people in world history. Within industrial nations, people moved from rural areas to cluster in manufacturing regions (see Chapter 10). Agricultural changes have also influenced population density. For example, the complete mechanization of cotton and wheat cultivation in mid-twentieth-century America allowed those crops to be raised by a much smaller labor force (see Chapter 8). As a result, profound depopulation occurred, to the extent that many small towns serving these rural inhabitants ceased to exist.

Regardless of the reasons, population changes must be accommodated, and these adaptations are invariably reflected in the cultural landscape.

FIGURE 3.31 Refugees flee South Sudan. Nearly 2.5 million individuals have fled South Sudan, with most going to neighboring Uganda. Drought and conflict have combined to create famine. (Paula Bronstein/Getty Images)

World Heritage Site: Pico Island Vineyard

Pico Island provides an excellent example of a cultural landscape shaped by geodemographic factors. Pico Island is the second largest island in the Azores, a chain of nine volcanic islands. Politically a part of Portugal, the islands are located some 950 miles west of Portugal, in the middle of the Atlantic Ocean. Pico Island is noted for its vineyard culture, established in the 1500s, which exemplifies the isolated farmstead settlement.

- Pico Island consists of 987 hectares (40,515 square feet) of volcanic rock whose man-made landscape has been shaped entirely for small-scale agriculture; namely, growing grapes and processing them into wine (viniculture).

- The island remained uninhabited until 1439, when the first Portuguese settlers arrived. Initially, they were cattle herders, but winemaking was introduced from the mainland a few decades later, with Portuguese Franciscan and Carmelite religious orders improving its techniques in the 1500s. Winemaking continued to be the island's mainstay until the mid-1800s, when the grape vines became afflicted by grape mildew and phyllorexa (plant diseases). Today, wine production has experienced a resurgence, although the sweet Verdelho wine exported to European nobility in past centuries is hard to find because the island's small size limits total production.

- With a population of just over 15,000 people, Pico Island counts some cities—the capital Madalena being the largest, with 6000 inhabitants—but the majority of its population lives in small farm villages and port towns.

(Thomas Stankiewicz/LOOK-foto/Getty Images)

(Yulia_B/Shutterstock)

THE WALLS: The viniculture landscape of Pico Island is unique and ingenious. The island has mild year-round temperatures and adequate rain, both of which are conducive to growing wine grapes. But the island is constantly buffeted by strong winds and sea surges that are harmful to grape vines. In response, islanders used volcanic basalt rock to construct walls along the coastline, which enclose small plots of land used to grow grapes. The walls protect the fragile plants from the wind and seawater, and support them as they grow. The resulting landscape is one of many small interlocking rectangular plots, called *currais*, bounded by dark-colored basalt rock walls. UNESCO has deemed this an "extraordinarily beautiful man-made landscape of small, stone-walled fields."

- Vine cultivation is done entirely by hand, rather than using farm machinery. This is possible because of the small scale of wine production on the island. In addition, the rocks used to construct the walls were stacked meticulously by hand: no mortar was used.

- Buildings constructed to support winemaking and the people engaged in it include cellars, warehouses, churches, ports, houses, and wells. The settlements are small and the architecture distinctive.

TOURISM: The island's natural features as well as its historic industries attract a small but growing number of tourists.

- Pico is "the mountain island" of the Azores, dominated by a volcanic cone rising nearly 8000 feet above sea level. Hiking to the crater at the top of the dormant volcano takes a full day and is quite challenging.

- Whaling, officially banned by the European Union in the 1980s, was Pico Island's main industry for a period of time. Today, whale watching is one of the island's primary tourist activities.

- Pico Island has received an international award for its efforts to promote sustainable tourism (tourism that results in low impact on the environment and local culture).

- Pico Island is a beautiful expression of the balance forged between nature and man in isolated conditions.
- The small, geometric basalt-walled plots are ingenious constructions that allow for the successful cultivation of wine grapes in a harsh environment.
- Placed on the list of World Heritage Sites in 2004, Pico Island has a small but growing tourism industry.

http://whc.unesco.org/en/list/1117

(Günter Lenz/imageBROKER/Alamy)

Depopulation in History: Ancient Rome There are as many theories accounting for the decline of the Roman Empire as there are historians theorizing about it. Lead poisoning, over-extension of the empire's reach, conquest by Germanic tribes, and at least 200 other hypotheses have been proposed.

Rome, the empire's capital city, suffered this process acutely. For nearly a millennium, Rome had been the world's wealthiest, most powerful, and most populous city (**Figure 3.32**). At the close of the first century C.E., Rome's population surpassed 1 million. As the Roman Empire's economic, political, and military might waned across the third, fourth, and fifth centuries, the demographic balance shifted as well. Under the emperor Constantine in 330 C.E., the empire was split into eastern and western halves, and its capital was relocated to Constantinople (now Istanbul, in modern-day Turkey).

As barbarian invasions further weakened the divided empire, Rome's physical infrastructure—its famed roads, bridges, aqueducts, and monumental buildings—crumbled, as did its impressive administrative infrastructure. With no access to former amenities, from running water to educational opportunities, Rome's population had shrunk to around 20,000 by 550 C.E. The Roman landscape had become a ghost of its former glory: dilapidated, uninhabited, and in ruins (see Figure 3.32).

Bogotá Rising Dilapidated-looking landscapes can result from population decline, as the example of ancient Rome illustrates, but they can also arise from rapid population influx. Many of the world's shantytowns exemplify the sort of chaotic landscape that can result from rapid population increases. Los Altos de Cazucá, a neighborhood in the Colombian capital of Bogotá, is but one example. In this case, the population influx comes mostly from people arriving in the capital after being displaced by armed conflict in the countryside.

The Internal Displacement Monitoring Centre estimates that, by the end of 2016, there were 40.4 million internally displaced persons (IDPs) due to armed conflict worldwide. Colombia is one of the globe's hotspots for IDPs, having the highest number of IDPs of any country in the world. Colombia's estimated 7.2 million IDPs have been uprooted largely as the result of ongoing civil warfare between the paramilitary group known as the FARC (Revolutionary Armed Forces of Colombia) and government forces bent on eradicating them. The United Nations High Commissioner for Refugees has identified this as the "worst humanitarian crisis in the western hemisphere."

Colombia's armed conflict has depopulated the rural areas where it occurs, forcing its victims into the cities. Terrorized, landless, and impoverished, the mainly indigenous and Afro-Colombian IDPs settle in the outskirts of cities like Colombia's capital, Bogotá.

Los Altos de Cazucá is one of the settlements inhabited by Colombia's internally displaced persons (**Figure 3.33**). Known as shantytowns, areas like Los Altos de Cazucá arise for different reasons, but they exist in all large cities throughout the developing world.

Housing is constructed by the residents themselves, using found materials like cardboard, tin panels, and old tires. Some shantytowns are located far away from downtown areas where wealthy people reside and where jobs are, while others are literally pressed up against wealthier neighborhoods (**Figure 3.34**).

Most shantytowns arise spontaneously, in order to address population influx to cities that lack the resources

FIGURE 3.32 Rome at its height and Rome in ruins. The image on the left, from a nineteenth-century engraving, shows Rome at its height, as seen from Mount Palatine, one of Rome's seven hills. The image on the right, from a nineteenth-century painting, shows the ruins of the Roman Forum from the Capitoline Hill, another of Rome's seven hills. (Both photos: The Granger Collection)

FIGURE 3.33 Los Altos de Cazucá. This is a shantytown on the outskirts of Colombia's capital city, Bogotá. Many of its approximately 50,000 residents are displaced people from other parts of the country. (imageBROKER/SuperStock)

FIGURE 3.34 Landscapes of poverty and wealth. This scene from Bogotá is a common Latin American urban landscape, with wealthy high-rises located next to impoverished shantytowns. (Florian Kopp/imageBROKER/Alamy Stock Photo)

to plan systematically for rapid growth. But are shantytowns temporary, disappearing as their residents become incorporated into city life? Hardly. Instead, shantytowns gradually become part of the urban fabric. Indeed, most of the spatial expansion of Latin American cities like Bogotá occurs precisely thanks to growth of the shantytowns surrounding them. Over time, the dwellings are constructed of more permanent materials such as concrete blocks. Roads are paved and running water installed. Power and phone lines are extended. The once-temporary areas slowly become visually, economically, and culturally integrated into the permanent fabric of the city. New shantytowns then arise beyond their borders, to accommodate recent arrivals.

Conclusion

Spatial aspects and expressions of human populations are fundamental geographic processes. For this reason, we began our topical discussion of human geography with this chapter on population geography. It provides the building blocks for understanding the coming chapters, helping us to understand how human populations settle, move, and flourish (or struggle) in concert with our environment. In this chapter, we have argued that the growth and decline of human populations is responsive to a variety of economic, political, environmental, and cultural factors. Decisions to move and settle are also shaped by these factors. In turn, contemporary as well as historical population changes have helped to mold economics, politics, environments, and cultures. Together, these form the main human geographic dimensions of the world around us.

Chapter Summary

3.1   Identify regional patterns of population characteristics and describe how these are distributed spatially and change over time.

- Population density varies from region to region, while the ability of a place to support a given population also varies due to environmental and economic factors.
- Fertility and mortality work together to increase or decrease populations over time.
- The demographic transition model posits that nations sequence through a series of stages, beginning with high birth and death rates, through subsequent high-growth stages where birth rates outstrip death rates, and finally settling in a no- or negative-growth stage with low birth and death rates.
- Young and old populations are unevenly distributed across the globe, as is the ratio of females to males; this can be visually displayed using a population pyramid.
- Demographic traits can reflect as well as shape variations in standards of living.

3.2   Explain patterns, causes, and consequences of population migrations.

- The movement of human populations through migration is another way that population size changes over time.
- When humans migrate, other factors—such as diseases—move along with them.

3.3   Understand theories of population growth and control and how these have changed over time.

- The world's population has been growing at an increasing rate over time, but is expected to level off at some point in the future as global fertility declines.
- There is no consensus on whether large populations are, on balance, a positive or negative thing.
- Though the "Rule of 72" can be used to calculate the doubling time of any population, the impact of population growth depends, in part, on the scale at which the issue is framed.
- Some countries have desired larger populations, while others have sought to reduce population size.

3.4   Describe how the natural world shapes population characteristics, and how population characteristics in turn shape the natural world.

- Favorable climate, location, and other natural resources support denser populations and are more attractive to people.
- The same physical environment can be interpreted in different ways by different cultures.
- In order to support larger populations, adaptive strategies are employed to modify the physical environment; over time, these can result in ecological disaster. When environments collapse due to sudden disaster or long-term climate change, people living there become displaced environmental refugees.

3.5   Analyze the imprint of demographic factors on the cultural landscape.

- Human settlement patterns vary from rural and isolated to high density.
- Over time, economic and political factors result in population changes that can be sudden or gradual.

Key Terms

population geography The study of the spatial and ecological aspects of population, including distribution, density per unit of land area, fertility, gender, health, age, mortality, and migration (page 82).

population density A measurement of population per unit area (e.g., people per square mile) (page 82).

carrying capacity The maximum number of people that can be supported in a given area (page 83).

total fertility rate (TFR) The number of children the average woman will bear during her reproductive lifetime (15–44 or 15–49 years of age) (page 84).

zero population growth A stabilized population created when an average of only two children per couple survive to adulthood. Eventually, the number of deaths equals the number of births (page 84).

death rate The number of deaths per year per 1000 people (page 85).

infant mortality rate The number of infants per 1000 live births who die before reaching one year of age (page 86).

demographic transition The movement from high birth and death rates to low birth and death rates (page 88).

Eurocentric Using the historical experience of Europe as the benchmark for all cases (page 90).

population pyramid A graph composed of back-to-back bars showing the age and sex composition of a population (page 91).

sex ratio The numerical ratio of females to males in a population (page 91).

gender roles The set of generally held expectations around the behaviors of men and women in different cultural and historical contexts (page 93).

refugees Those fleeing from persecution in their country of nationality. The persecution can be religious, political, racial, or ethnic (page 96).

push and pull factors Unfavorable, repelling conditions (push factors) and favorable, attractive conditions (pull factors) that interact to affect migration and other elements of diffusion (page 96).

population explosion The rapid, accelerating increase in world population since about 1650, and especially since 1900 (page 98).

Malthusian Those who hold the views of Thomas Malthus, who believed that overpopulation is the root cause of poverty, illness, and warfare (page 99).

cornucopians Those who believe that science and technology can solve resource shortages. In this view, human beings are our greatest resource rather than a burden to be limited (page 99).

neo-Malthusians Modern-day followers of Thomas Malthus (page 99).

doubling time The amount of time, in years, that it will take for a given population to double in size (page 99).

adaptive strategies The unique way in which each culture uses its particular physical environment to provide the necessities of life—food, clothing, shelter, and safety (page 101).

environmental refugees People who are displaced from their homes due to severe environmental disruption (page 104).

farm villages Clustered rural settlements of moderate size, inhabited by people who are engaged in farming (page 106).

farmstead The center of farm operations, containing the house, barn, sheds, and livestock pens (page 106).

depopulation A decrease in population that sometimes occurs as the result of sudden catastrophic events, such as natural disasters, disease epidemics, and warfare (page 107).

proxemics The study of the size and shape of people's envelopes of personal space (page 115).

 ## Practice at 🍂 SaplingPlus

Read the interactive e-Text, review key concepts, and test your understanding.

 Story Map. Explore migrations at various times, places, and scales.

 Web Map. Examine maps illustrating demographic issues.

Doing Geography

ACTIVE LEARNING:
Refugee Movements

As you have learned, people are sometimes forced to migrate by circumstances in their home countries. In this activity, you will use dynamic refugee maps to observe and analyze refugee flows from 2000 through 2015. Where are the majority of today's refugees coming from, and where are they going? What has changed over time? How does what you see in these maps align, or not, with news coverage of refugees? In other words, are there groups we hear a lot about in the news that are less significant, and others that are more numerically significant that we don't hear much about?

More guidance at 🔷 Sapling Plus

EXPERIENTIAL LEARNING:
Public Space, Personal Space: Too Close for Comfort?

Culture can condition people to accept or reject crowding. Personal space—the amount of space that individuals feel "belongs" to them as they move about their everyday business—varies from one cultural group to another. As the Seeing Geography section for this chapter notes, different people require different amounts of personal space. One's comfort zone varies with social class, gender, ethnicity, the situation at hand, and what one has grown accustomed to over one's life. Americans conducting business in Japan are often surprised at the level of physical closeness expected in their dealings. Such closeness might well be interpreted as overstepping one's bounds, literally, in the United States!

In this exercise, you will gather some data on the amount of personal space needed by those around you. Observe and record your findings, and discuss them in class as a group.

Steps to Understanding Personal Space

Step 1:   Observe your professors as they lecture in class. Is your class a large one that meets in a lecture hall? If so, where does your professor sit or stand in relation to the students? Does the professor have his or her own designated space in the classroom? Where is it located, and how big is it? Does the professor ever step outside of it? How does this professor's use of space compare to that of your other professors, and why do you think this is so? If you have a smaller class, compare the use of space by that professor. Is it different from the behavior of the professors in large lecture halls? Under what circumstances, if any, do your professors get close to students, and how close do they get?

Step 2:   How close can you get to friends? Strike up a conversation with a same-gender friend standing next to you. Discreetly move closer and closer to your friend until he or she moves away or says something about your proximity. How much space separated you when this happened? What do you think would happen if you tried this with a stranger? With a friend or a stranger of a different gender? With a friend or a stranger from a different culture?

Step 3:   Discuss your findings. Did all your classmates have similar experiences, or were your findings notably different? Are all students in your class from similar economic or ethnic backgrounds? If not, that may explain some of the differences that emerge.

Population densities in parks. These two images depict similarly designed public playgrounds featuring a large sandbox. The park on the left is located in Minnesota in the United States; the park on the right is located in Taipei, Taiwan. *Where would you rather play? Why?* (Top: James Shaffer/PhotoEdit; Bottom: Christian Klein/Alamy)

SEEING GEOGRAPHY

Kolkata, India

**Would you feel comfortable walking here?
If not, why not?**

Do you need your "personal space"? Most Americans and Canadians do. If so, Kolkata (formerly Calcutta), India, is a place you might want to avoid. West Bengal state, where Kolkata is located, has the highest population density in India.

Why do people form such dense clusters? Push factors encourage people to leave their farms and move to the city. Some can no longer make a living or feed their families with the food provided by the tiny plots of land they work. Others are forced off the land by landlords who want to convert their farms to use mechanized Western methods of agriculture that use far less labor (see Chapter 8). But there are pull factors exerted by cities such as Kolkata—the hope or promise of better-paying jobs, the encouragement of friends and relatives who came to the city earlier, or the greater availability of government services. And so, pushed and pulled, they come to the crowded city to jostle and elbow their way through the streets.

But it is not only cities in the developing world that become so dense. If you have ever visited, or lived in, Manhattan, New York, you are all too familiar with dense crowds of people. In fact, there are some who grow up in these environments and find that the relative solitude of rural areas verges on terrifying. They prefer the bustle of activity and the sounds of the city, and they feel at home in a crowd.

Being a woman is another reason that you might feel uncomfortable in this Kolkata street environment. Notice the relative absence of women in this crowd. Many societies have strict norms that dictate where women—and men—may and may not go. Harassment or even violence may be the result of violating these norms.

The study of the size and shape of people's envelopes of personal space is called **proxemics**. Anthropologist Edward T. Hall, who wrote *The Hidden Dimension,* is the founder of this science. Urban planners, architects, psychologists, and sociologists, as well as geographers, use proxemics to explain why some people need more space than others and how this varies culturally.

A street scene in the large city of Kolkata, India. (Steve Raymer/ Getty Images)

Cell phone service provider advertisement in Mexico. *Feisbuquear* is a Spanglish verb meaning "to Facebook." (Courtesy of Patricia L. Price.)

chapter 4

The Geography of Language

LOCATING THE SPOKEN WORD

Why make up a word that doesn't exist in Spanish or English, but instead is located between the two languages?

Think about this question and photo as you read. We will revisit them in Seeing Geography on page 147.

Learning Objectives

4.1	Identify the geographical patterns of languages.
4.2	Understand how languages and dialects have come to exist, move, and change.
4.3	Evaluate the relationship between technology and language.
4.4	Explain the relationships between language and the physical environment.
4.5	Characterize the ways languages are visibly part of the cultural landscape.

(Courtesy of Patricia L. Price.)

Language is one of the primary features that distinguish humans from other animals. Many animals, including dolphins, whales, and birds, do indeed communicate with one another through patterned systems of sounds, movements, or scents and other chemicals. Some nonhuman primates have been taught to use sign language to communicate with humans. The complexity of human language, its ability to convey nuanced emotions and ideas, and its importance for our existence as social beings sets it apart from the communication systems used by other animals.

In many ways, language is the essence of human culture. It provides the single most common variable by which different cultural groups are identified and by which groups assert their unique identity. Language not only facilitates the cultural diffusion of innovations; it also helps to shape the way we think about, perceive, and name our environment. **Language,** a mutually agreed-upon system of symbolic communication, is the main vehicle by which learned belief systems, customs, and skills pass from one generation to the next.

Region

4.1   Identify the geographical patterns of languages.

While there are relatively few linguistic families, the spatial variation of speech is remarkably complicated, in part because of the intricate regional patterns (**Figure 4.1**). Because language is such a central component of culture, understanding the

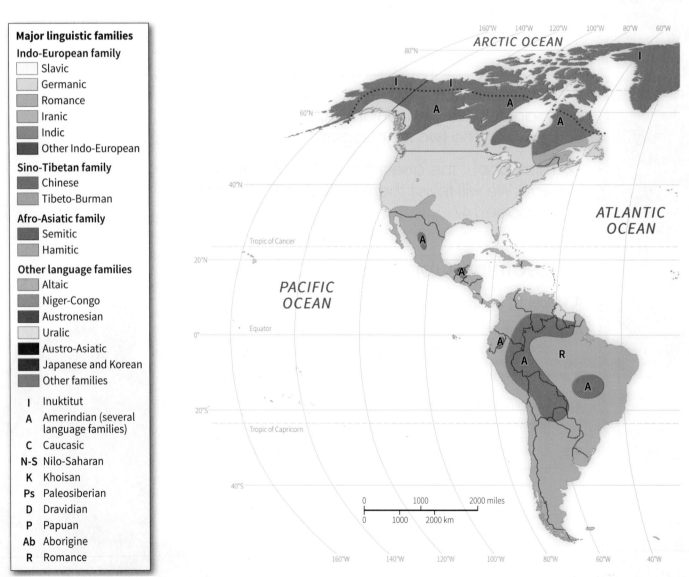

Major linguistic families

Indo-European family
- Slavic
- Germanic
- Romance
- Iranic
- Indic
- Other Indo-European

Sino-Tibetan family
- Chinese
- Tibeto-Burman

Afro-Asiatic family
- Semitic
- Hamitic

Other language families
- Altaic
- Niger-Congo
- Austronesian
- Uralic
- Austro-Asiatic
- Japanese and Korean
- Other families

- **I** Inuktitut
- **A** Amerindian (several language families)
- **C** Caucasic
- **N-S** Nilo-Saharan
- **K** Khoisan
- **Ps** Paleosiberian
- **D** Dravidian
- **P** Papuan
- **Ab** Aborigine
- **R** Romance

FIGURE 4.1 The major linguistic formal culture regions of the world. Although there are thousands of languages and dialects in the world, they can be grouped into a few linguistic families. The Indo-European language family represents about half of the world's population. It spread throughout the world, in part, through Europe's empire-building efforts.

spatiality of language, and how and why its patterns change over time, provides a valuable window into human geography. The logical place to begin is with the regional theme.

Language Classifications

Separate languages are those that cannot be mutually understood. In other words, a monolingual speaker of one language cannot comprehend the speaker of another (**Figure 4.2**, page 120). **Dialects,** by contrast, are variant forms of a language where mutual comprehension is possible. A speaker of English, for example, can generally understand that language's various dialects, regardless of whether the speaker comes from Australia, Scotland, or Mississippi. Nevertheless, a dialect is

distinctive enough to label its speaker as hailing from one place or another, or even from a particular city. About 7000 languages and many more dialects are spoken today.

When different linguistic groups come into contact, a **pidgin** language, characterized by a very small vocabulary derived from the languages of the groups in contact, often results. Pidgins primarily serve the purposes of trade and commerce: they facilitate exchange at a basic level but do not have complex vocabularies or grammatical structures. An example is Tok Pisin, which means, "talk business." Tok Pisin is a largely English-derived pidgin spoken in Papua New Guinea, where it has become the official national language in a country where many native Papuan tongues are spoken. Although New Guinea pidgin is not readily intelligible to a speaker of Standard English, certain common words such as *gut bai* ("good-bye"),

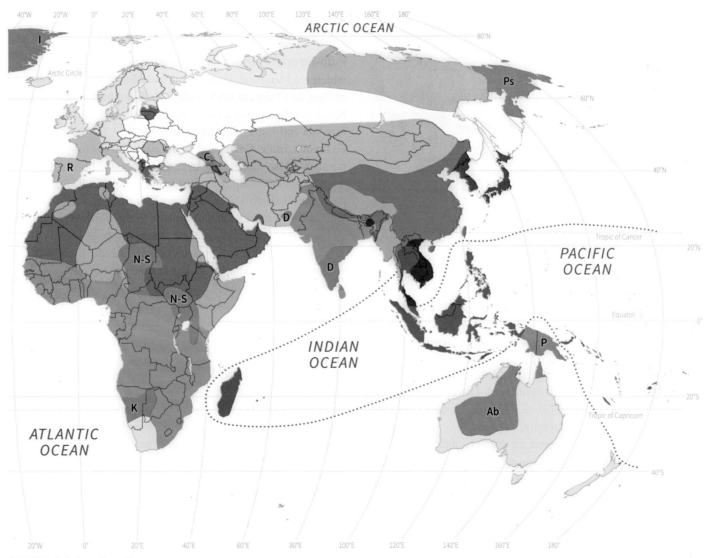

FIGURE 4.1 (continued)

FIGURE 4.2 **How to sound like a dog in 14 languages.** Although a dog obviously sounds the same in any part of the world, the speakers of various human languages render that sound very differently. (James Chapman)

tenkyu ("thank you"), and *haumas* ("how much") reflect the influence of English. When pidgin languages acquire fuller vocabularies and become native languages of their speakers, they are called **creole** languages. Obviously, deciding precisely *when* a pidgin becomes a creole language has at least as much to do with a group's political and social recognition as it does with what are, in practice, fuzzy boundaries between language forms (see also Doing Geography: Active Learning, on page 146).

Another response to the need for speakers of different languages to communicate with one another is the elevation of one existing language to the status of a **lingua franca.** A lingua franca is a language of communication and commerce spoken across a wide area where it is not a mother tongue. The Swahili language enjoys lingua franca status in much of East Africa, where inhabitants speak a number of other regional languages and dialects. English is fast becoming a global lingua franca. Finally, regions that have linguistically mixed populations may be characterized by **bilingualism,** which is the ability to speak

two languages with fluency. For example, along the U.S.–Mexico border, so many residents speak both English and Spanish (with varying degrees of fluency) that bilingualism in practice—even if not in policy—means there is no need for a lingua franca.

Language Families

One way in which geolinguists simplify the mapping of languages is by grouping them into **language families:** tongues that are related and share a common ancestry. Words are simply arbitrary sounds associated with certain meanings. Thus, when words in different languages are alike in both sound and meaning, they may well be related. Over time, languages interact with one another, borrowing words, imposing themselves through conquest, or organically diverging from a common ground. Languages and their interrelations can thus be graphically depicted as a tree with various branches (**Figure 4.3**).

Indo-European Language Family The largest and most widespread language family is the Indo-European, which is spoken on all the continents and is dominant in Europe, Russia, North and South America, Australia, and parts of southwestern Asia and India (see **Figure 4.4**, page 122). Romance, Slavic, Germanic, Indic, Celtic, and Iranic are all Indo-European subfamilies. These subfamilies, in turn, are divided into individual languages. For example, English is a Germanic Indo-European language. Six Indo-European tongues, including English, are among the 10 most spoken languages in the world as classified by the number of native speakers (**Table 4.1**, page 122).

Comparing the vocabularies of various Indo-European tongues reveals their kinship. For example, the English word *mother* is similar to the Polish *matka,* the Greek *meter,* the Spanish *madre,* the Farsi *madar* in Iran, and the Sinhalese *mava* in Sri Lanka. Such similarities demonstrate that these languages have a common ancestral tongue.

Sino-Tibetan Language Family Sino-Tibetan is another of the major language families of the world and is second only to Indo-European in numbers of native speakers. The Sino-Tibetan region extends throughout most of China and Southeast Asia (see Figure 4.3). The two language branches that make up this group, Sino- and Tibeto-Burman, are believed to have had a common origin some 6000 years ago in the Himalayan Plateau; speakers of the two language groups subsequently moved along the great Asian rivers that originate in this area. "Sino" refers to China and in this context indicates the various languages spoken by more than 1.4 billion people in China. Han Chinese (Mandarin) is spoken in a variety of dialects and serves as the official language of China. The nearly 400 languages and dialects that make up the Burmese and Tibetan branch of this language

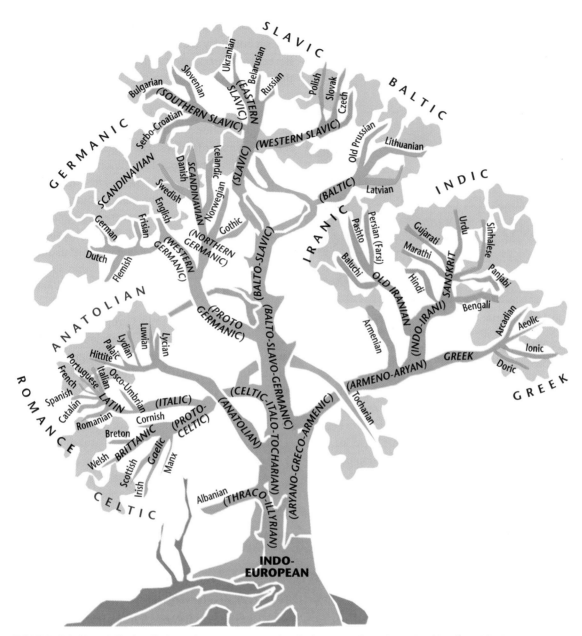

FIGURE 4.3 Linguistic family tree. Shown here is a detailed image of one branch of the linguistic family tree. (Source: Adapted from Ford.)

family border the Chinese language region on the south and west. Other East Asian languages, such as Vietnamese, have been heavily influenced by contact with the Chinese and their languages, although it is not clear that they are linguistically related to Chinese at all.

Afro-Asiatic Language Family The third major language family is the Afro-Asiatic. It consists of two major divisions: Semitic and Hamitic. The Semitic languages cover the area from the Arabian Peninsula and the Tigris-Euphrates river valley of Iraq westward through Syria and North Africa to the Atlantic Ocean. Despite the considerable size of this region, there are fewer speakers of the Semitic languages than you might expect because most of the areas that Semites inhabit are sparsely populated deserts. Arabic is by far the most widespread Semitic language and has the greatest number of native speakers, about 295 million. Although many different dialects of Arabic are spoken, there is only one written form.

Hebrew, which is closely related to Arabic, is another Semitic tongue spoken by 9 million people worldwide (but of those, only 7 million speak it fluently). For many centuries, Hebrew was a "dead" language, used only in religious ceremonies by millions of Jews throughout the world. With the creation of the state of Israel in 1948, a common language was

FIGURE 4.4 Sign in Arab quarter of Nazareth. Many of Israel's cities are home to diverse populations. This sign at a child-care center reflects Israel's polyglot population, with its English, Arabic, and Hebrew wording. (Courtesy of Patricia L. Price.)

needed to unite the immigrant Jews, who spoke the languages of their many different countries of origin. Hebrew was revived as the official national language of what otherwise would have been a **polyglot,** or multi-language, state (Figure 4.4). Amharic, a third major Semitic tongue, is spoken today by 22 million people in the mountains of East Africa.

Smaller numbers of people who speak Hamitic languages share North and East Africa with the speakers of Semitic languages. Like the Semitic languages, these tongues originated in Asia but today are spoken almost exclusively in Africa by the Berbers of Morocco and Algeria, the Tuaregs of the Sahara, and the Cushites of East Africa.

Other Major Language Families Most of the rest of the world's population speak languages belonging to one of six remaining major families. The Niger-Congo language family, which is spoken by about 400 million people, dominates Africa south of the Sahara Desert. The greater part of the Niger-Congo culture region belongs to the Bantu subgroup. Both Niger-Congo and its Bantu constituent are fragmented into a great many different languages and dialects, including Swahili. The Bantu and their many related languages spread from what is now southeastern Nigeria about 4000 years ago, first west and then south in response to climate change and new agricultural techniques.

Flanking the Slavic Indo-Europeans on the north and south in Asia are the speakers of the Altaic language family, including Turkic, Mongolic, and several other subgroups. The Altaic homeland lies largely in the inhospitable deserts, tundra, and coniferous forests of northern and central Asia. Also occupying tundra and grassland areas adjacent to the Slavs is the Uralic family. Finnish and Hungarian are the two most widely spoken Uralic tongues, and both enjoy the status of official languages in their respective countries.

As depicted in **Figure 4.5,** one of the most remarkable language families in terms of distribution is the Austronesian. Representatives of this group probably originated from

TABLE 4.1 The 10 Leading Languages in Numbers of Native Speakers*

Language	Family	Speakers (in millions)	Main Areas Where Spoken
Chinese	Sino-Tibetan	1284	China, Taiwan, Singapore
Spanish	Indo-European	437	Spain, Latin America, southwestern United States
English	Indo-European	372	British Isles, United States, Caribbean, Australia, New Zealand, South Africa, Philippines, former British colonies in tropical Asia and Africa
Arabic	Afro-Asiatic	295	Middle East, North Africa
Bengali	Indo-European	292	Bangladesh, eastern India
Hindi	Indo-European	260	Northern India, Pakistan
Portuguese	Indo-European	219	Portugal, Brazil, southern Africa
Russian	Indo-European	154	Russia, Kazakhstan, parts of Ukraine and other former Soviet republics
Japanese	Japanese and Korean	128	Japan
Lahnda	Indo-European	84.3	Pakistan

*"Native speakers" means mother tongue.
Source: Ethnologue, 2017, http://www.ethnologue.com/statistics/size.

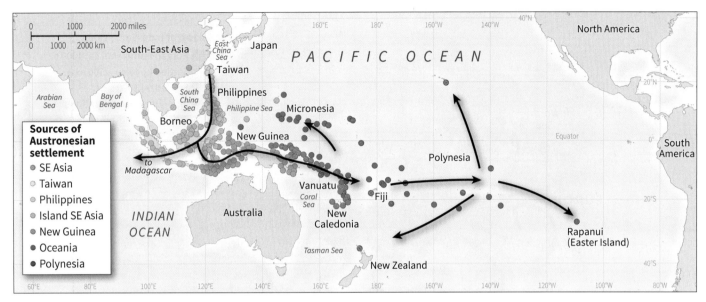

FIGURE 4.5 Map of the Pacific language family tree. This map shows the settlement of the Pacific by Austronesian language family speakers. By studying the relationships between these languages, the settlement history of the region is revealed. (Source: Adapted from https://www.sciencedaily.com/releases/2009/01/090122141146.htm.)

modern-day Taiwan. Today, speakers of the Austronesian languages live mainly on tropical islands stretching from Madagascar, off the east coast of Africa, through Indonesia and the Pacific islands, to Hawaii and Easter Island. This longitudinal span is more than half the distance around the world. The north–south, or latitudinal, range of this language area is bounded by Hawaii and Taiwan in the north and New Zealand in the south. The largest single language in this family is Javanese, with 75.5 million native speakers, but the most geographically widespread is Polynesian.

Japanese and Korean, with about 205 million speakers combined, probably form another Asian language family. The two perhaps have some link to the Altaic family, but even their kinship to each other remains controversial and unproven.

In Southeast Asia, the Vietnamese, Cambodians, Thais, and some tribal peoples of Malaysia and parts of India speak languages that constitute the Austro-Asiatic family. They occupy an area into which Sino-Tibetan, Indo-European, and Austronesian languages have all encroached.

Mobility

4.2 Understand how languages and dialects have come to exist, move, and change.

Different types of cultural diffusion have helped shape the linguistic map. Relocation diffusion has been extremely important because languages spread when groups, in whole or in part, migrate from one area to another. Some individual tongues or entire language families are no longer spoken in the regions where they originated, and in certain other cases the linguistic hearth is peripheral to the present distribution (compare Figures 4.3 and 4.7). Today, languages continue to evolve and change based on the shifting locations of peoples and on their needs as well as on outside forces.

Indo-European Diffusion

How did Indo-European languages arise and spread to become the largest language family on Earth? One theory suggests that the earliest speakers of the Indo-European languages lived in southern and southeastern Turkey, a region known as Anatolia, about 9000 years ago. According to the **Anatolian hypothesis,** the initial diffusion of these Indo-European speakers was facilitated by the innovation of plant domestication. As sedentary farming was adopted throughout Europe, a gradual and peaceful expansion diffusion of Indo-European languages occurred. As these people dispersed and lost contact with one another, different Indo-European groups gradually developed variant forms of the language, causing fragmentation of the language family.

The Anatolian hypothesis has been criticized by scholars who note that specific words used for animals (particularly horses), but not agriculture, appear to link Indo-European languages to a common origin. The **Kurgan hypothesis,** which is more widely accepted than the Anatolian hypothesis, places the rise of Indo-European languages in the central Asian steppes

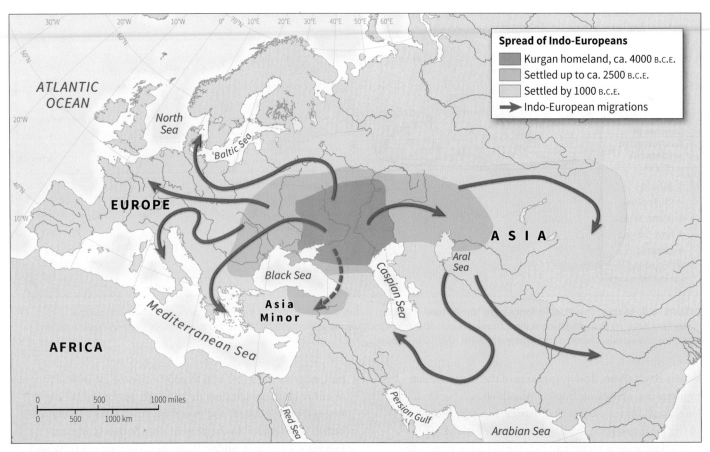

FIGURE 4.6 The spread of Indo-European language. This map depicts the so-called Kurgan hypothesis, named after the burial mounds (*kurgan*) characteristic of the warrior pastoralists who inhabited the area north of the Black and Caspian seas. Around 6000 B.C.E., they began to spread outward, conquering and imposing their language across Europe, central Asia, India, the Balkans, and Anatolia (modern-day Turkey).

only 6000 years ago (**Figure 4.6**). It asserts that the spread of Indo-European languages was both swifter and less peaceful than is maintained by those who subscribe to the Anatolian hypothesis. The domestication of horses by militaristic Kurgans allowed them to overtake the more peaceful agricultural societies and to rapidly spread their languages through imposition. No one theory has been definitively proven.

What is more certain is that in later millennia, the diffusion of certain Indo-European languages—in particular, Latin, English, and Russian—occurred in conjunction with the territorial spread of great political empires. In such cases of imperial conquest, relocation and expansion diffusion were not mutually exclusive. Relocation diffusion occurred as a small number of conquering elites came to rule an area. The language of the conqueror, implanted by relocation diffusion, often gained wider acceptance through expansion diffusion. Typically, the conqueror's language spread hierarchically—adopted first by the more important and influential persons and by city dwellers. The diffusion of Latin with Roman conquests, and Spanish with the conquest of Latin America, occurred in this manner.

Migration and the Survival of Language

As we have seen above, conquest can lead to the imposition of a new language and the abandonment or suppression of native tongues. However, these threatened languages may reappear and thrive in new places, as their speakers migrate for reasons of economic or cultural survival.

New York City is thought to be home to as many as 800 languages, making it the most linguistically dense place in the world. There are more speakers of Vlashki in Queens than in the Croatian mountain villages where the language originated, and roughly the same number of Garifuna speakers in the Bronx and Brooklyn as in Honduras and Belize. According to journalist Sam Roberts, these are but two of "a remarkable trove of endangered tongues that have taken root in New York."

How did New York City become home to such linguistic riches? These languages relocated there through the migration of their native speakers. While populations in the language source region may fall victim to ethnic conflict, disease, starvation,

SUBJECT TO DEBATE Imposing English

English-only laws are nothing new in the United States. Its history as a nation of immigrants has led to a population that, at any one point in time, has spoken a variety of languages besides English. In its early days as a colony, one could hear German, Dutch, French, and a multitude of Native American languages spoken alongside English. This prompted both Benjamin Franklin and John Adams to propose enforcing English as the sole acceptable language, and Theodore Roosevelt once said, "The one absolutely certain way of bringing this nation to ruin or preventing all possibility of its continuing as a nation at all would be to permit it to become a tangle of squabbling nationalities. We have but one flag. We must also learn one language, and that language is English."

Citing concerns that providing official documents and services in multiple languages would simply be too expensive, contemporary advocates of English-only legislation claim that mandating one language is one way to reduce the cost of government. Some proponents also believe that English-only laws encourage immigrants to assimilate by learning the official language of the United States. Opponents accuse the laws of being discriminatory and suggest that supporters of English-only legislation are threatened by cultural diversity. Linguistic unity, they say, does not lead to political or cultural unity. Furthermore, providing official documents and services only in English in effect denies these services and information to those who do not understand English. Debates such as these bring up questions of the legal, social, and political status of minority groups and their languages, debates that exist in many countries besides the United States.

Today, most of those who wish to legislate English as the official language of the country target Spanish-speaking immigrants as the object of their concern. Anxieties about being culturally "overwhelmed" by Spanish speakers who refuse to learn English culminate in claims that Latino immigrants are dividing the nation in two: one English-speaking and culturally "American," and the other Spanish-speaking and unwilling to assimilate into the mainstream.

Historically, most immigrants to the United States eventually abandon their native tongues. As late as 1910, one out of every four Americans could speak some language other than English with the skill of a native; today that number is significantly lower. This was a result of the mass immigrations from Germany, Poland, Italy, Russia, China, and many other foreign lands. Much of this linguistic diversity has given way to English, partly because these other languages lacked legal status, partly because of the monolingual educational system in the United States, and partly because of social pressures.

Continuing the Debate

As this discussion illustrates, many cultures do not want to assimilate into English-speaking society, choosing instead to actively assert pride in their language. Keeping this in mind, consider these questions:

- Do you think the wave of immigrants today, with their pride in language, is different from earlier waves? How so?

- Will monoglot English speakers become a dwindling minority as more and more people become bilingual through either choice or necessity?

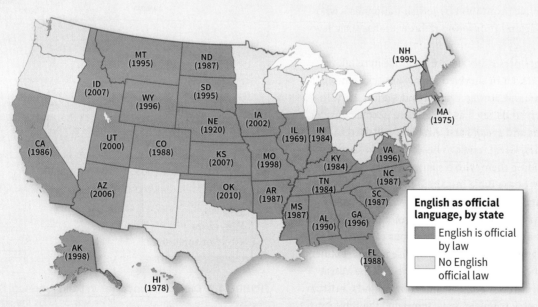

States that have some form of official English-only laws. Dates show the years English-only laws were enacted. In the 1980s, the influx of immigrants to the United States from Asia and Latin America prompted many of these laws. Typically, English-only laws require that state documents be published in English only. Some states' laws, however, also prohibit the state from doing business in a language other than English or providing services such as multilingual emergency medical hotlines.

compulsory schooling, or merely assimilation into dominant language groups, thereby losing their ability to speak their native language, migrant speakers from these places may have a better chance at keeping the tongue alive and well in their new homes.

Many of these relocated languages find themselves under new pressure from the dominant English language in the United States (see Subject to Debate). For this reason, members of the nonprofit Endangered Language Alliance canvass city neighborhoods in search of immigrant speakers of vulnerable languages. The speakers are videotaped, and the Alliance encourages the teaching and use of these languages. The reality, however, is that many of these languages will vanish from New York City's linguistic landscape when the children of these immigrants cease to speak them regularly or when parents stop teaching them to their children.

Religion and Linguistic Mobility

Cultural interaction creates situations in which language is linked to a particular religious faith or denomination, a linkage that greatly heightens cultural identity. Perhaps Arabic provides the best example of this cultural link. It spread from a core area on the Arabian Peninsula with the expansion of Islam. Had it not been for the evangelical success of the Muslims, Arabic would not have diffused so widely. The other Semitic languages also correspond to particular religious groups. Most Hebrew-speaking people are of the Jewish faith, and the Amharic speakers in Ethiopia tend to be Coptic, or Eastern, Christians. Indeed, we can attribute the preservation and recent revival of Hebrew to its active promotion by Jewish nationalists who believe that teaching and promoting Hebrew to diasporic Jews facilitate unity.

Many languages persist because of their use in religious ceremonies and texts. Latin survived mainly as the ceremonial language of the Roman Catholic Church and Vatican City. In non-Arabic Muslim lands, such as Iran, where people consider themselves Persians and speak Farsi, Arabic is still used in religious ceremonies. Great religious books can also shape languages by providing them with a standard form. Martin Luther's translation of the Bible in 1522 led to the standardization of the German language, and the Qur'an is the model for written Arabic. Because they act as common points of frequent cultural reference and interaction, great religious books can also aid in the survival of languages that would otherwise become extinct. The early appearance of a hymnal and the Bible in the Welsh language aided the survival of that Celtic tongue, and Christian missionaries in diverse countries have translated the Bible into local languages, helping to preserve them. In Fiji, the appearance of the Bible in one of the 15 local dialects elevated that dialect to the dominant native language of the islands.

Language's Shifting Boundaries

Dialects, as well as the language families discussed previously, reveal a vivid geography. Their boundaries—what separates them from other dialects and languages—shift over time, both spatially and in terms of what elements they contain or discard.

Geolinguists map dialects by using **isoglosses,** which indicate the spatial borders of individual words or pronunciations. For example, the dialect boundaries between Latin American Spanish speakers using *tú* and those using *vos* are clearly defined in some areas, as shown in **Figure 4.7.** The choice of *tú* or *vos* for the second-person singular carries with it a cultural indication. *Vos* represents a usage closer to the original Spanish but is considered by many in the Spanish-speaking world to be rather archaic and, in fact, has died out in Spain itself. In other regions, particularly throughout the countries of the Southern Cone, *vos* has long been used by the media; often, it is considered to reflect the "standard" dialect and usage for the area in which it is used. Because certain words or dialects can fall out of fashion or simply become overwhelmed by an influx of new speakers, isogloss boundaries are rarely

FIGURE 4.7 Dialect boundaries in Latin America. Spanish speakers in the Americas use either *vos* or *tú* as the second-person singular verb form. They represent dialects of Spanish: both are correct, linguistically speaking, but the *vos* form is older. Some regions use *vos* and *tú* interchangeably. (Source: Adapted from Pountain, 2005.)

clear or stable over time. Indeed, in Central America, the media are increasingly using *vos*—long used in conversation in the region—thus elevating *vos* to a more official status covering a larger territory. Because of this, geolinguists often disagree about how many dialects are present in an area or exactly where isogloss borders should be drawn. The language map of any place is constantly shifting.

The dialects of American English provide another good example. At least three major dialects, corresponding to major culture regions, had developed in the eastern United States by the time of the American Revolution: the Northern, Midland, and Southern dialects (**Figure 4.8**). As the three subcultures expanded westward, their dialects spread and fragmented. Nevertheless, they retained much of their basic character, even beyond the Mississippi River. These culture regions have unusually stable boundaries. Even today, the "*r*-less" pronunciation of words such as *car* ("cah") and *storm* ("stohm"), characteristic of the East Coast Midland regions, is readily discernible in the speech of its inhabitants.

Although we are sometimes led to believe that Americans are becoming more alike, as a national culture overwhelms regional ones, the current status of American English dialects suggests otherwise. Linguistic divergence is still under way, and dialects continue to mutate on a regional level, just as they always have. Local variations in grammar and pronunciation proliferate, confounding the proponents of standardized speech and defying the homogenizing influence of the Internet, television, and other mass media.

Shifting language boundaries involve content as well as spatial reach, and this, too, changes over time. Today, for example, some of the unique vocabulary of American English dialects is becoming old-fashioned. For instance, the term *icebox,* which was literally a wooden box with a compartment for ice that was used to cool food, was used widely throughout the United States before people had refrigerators. Although the modern electric refrigerator is ubiquitous in the United States today, some people, particularly those of older generations and in the South, still use the term *icebox.* Most speakers of American English, by contrast, no longer pause to say the entire word *refrigerator,* shortening it instead to *fridge.*

As illustrated by the birth of the new word *fridge,* slang terms are quite common in most languages, and American English

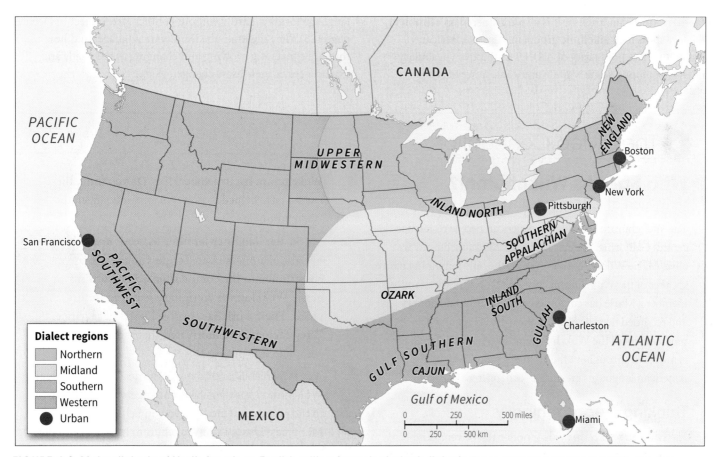

FIGURE 4.8 Major dialects of North American English, with a few selected subdialects. These dialects had developed by the time of the American Revolution and have remained remarkably stable over time. Some cities, identified on this map, have their own particular dialects.

is no exception. **Slang** refers to words and phrases that are not part of a standard, recognized vocabulary for a given language but are nonetheless used and understood by some or most of its speakers. Often, subcultures—for example, youth, drug dealers, and clubbers—have their own slang that is used within that community but is not readily understood by nonmembers. Slang words tend to be used for a period of time and then discarded as newer terms replace them. For example, *fresh, the bomb,* and *phat* were used to refer to desirable, attractive, or fashionable things in the 1990s. Although many people today still recognize these words, a new generation of young people is much more likely to use a new set of words. Their children, in turn, will more than likely use yet another set of words. Slang illustrates another way in which American English changes over time.

Some African Americans speak a distinctive form of English. African-American Vernacular English (AAVE) shares characteristics with the older Southern dialect and also displays considerable African influence in pitch, rhythm, and tone. Some linguists understand AAVE as a creole language that grew out of a pidgin that developed on the early slave plantations and is today spoken by some African Americans. Indeed, AAVE shares many characteristics with other English creole languages worldwide. Today, it is considered a dialect, or variation, of Standard American English. It is also considered an **ethnolect,** a dialect spoken by an ethnic group, in this case, African Americans. The popularity of AAVE's distinctive vocabulary and syntax among some white Americans, however, calls into question its ethnic exclusivity and serves to point out the instability of any language boundaries, including ethnic ones.

The grammar of AAVE is virtually uniform across the country, which has been attributed to the segregation of African American speakers or their relatively recent migration from the South (see Chapter 5, page 155). Some distinguishing characteristics of AAVE's speech patterns include the use of double negatives ("She don't like nothing"), omission of forms of the verb *to be* ("He my friend"), and non-conjugation of verbs ("She give him her paper yesterday"). There is some controversy over the place of AAVE in the U.S. educational system. Does AAVE constitute a distinctive language, rather than simply a dialect of English? Should it be taught—with its attendant grammar, structure, and literature—to American schoolchildren? Are those who speak AAVE and Standard English technically bilingual? This has led some linguists to refer to AAVE as African American Language (AAL), indicating the status of a separate language.

In the United States today, many descendants of Spanish speakers have adapted their speech to include words and variants of words in both Spanish and English, in a dialect known as "Spanglish" (see Seeing Geography, page 147). Although acceptance of AAVE and Spanglish as legitimate language forms is hardly without conflict (see Subject to Debate, page 125), they illustrate the fluidity of languages and how they are constantly evolving and changing as the needs and experiences of their users change.

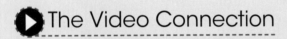 **The Video Connection**

Who Speaks Wukchumni?

 Watch at  SaplingPlus

The Wukchumni are a Native American people resident in central California. Today, fewer than 200 Wukchumni are alive. Marie Wilcox, born in 1932, is the only remaining native speaker of the Wukchumni language. Together with family members, Marie has composed a Wukchumni dictionary and recorded herself speaking the native language. The predicament of the Wukchumni people and their language is mirrored throughout the United States, where 130 Native American languages are on the "endangered" list.

Thinking Geographically

1.   As you watch this video, notice the poverty and isolation of the rural area where Marie and her fellow Wukchumni tribe members live. Do you think these factors have helped to preserve, or to endanger, the Wukchumni language?

2.   Marie, her daughter Jennifer Malone, and her grandson Donovan Treglown have all contributed to the compilation and recording of the Wukchumni dictionary featured in this video. Their work has involved years of sacrifice. What factors have motivated them to pursue and complete this project?

3.   As Marie relates in this video, the deaths of individual Wukchumni speakers have endangered the language to the point of extinction. Do you believe that, when Marie passes away, the Wukchumni language will die out entirely? Why or why not?

Globalization

4.3 Evaluate the relationship between technology and language.

More often than not, the diffusion of some languages has come at the expense of many others. Ten thousand years ago, the human race consisted of only 1 million people, speaking an estimated 15,000 languages. Today, a population 7000 times larger speaks only 47 percent as many tongues. Only 1 percent of all languages have as many as 500,000 speakers. It has been estimated that the world loses a language on average every two weeks. Some experts believe that all but 300 languages will be extinct or dying by the year 2100. Clearly, globalization has worked to favor some languages and eliminate others. There are, for instance, no children today who are learning any of California's nearly 100 native languages (see The Video Connection). Languages die out when their speakers do; often the entire cultural world associated with a language vanishes as well. Thus, globalization both presents the opportunity for more people to communicate directly with one another and, at the same time, threatens to extinguish the cultural diversity that goes hand in hand with linguistic diversity.

FIGURE 4.9 Multilingual sign. This sign, located in a New York City polling station, displays the world's three dominant languages: English, Spanish, and Chinese. (Patti McConville/Alamy)

Technology, Language, and Empire

Technological innovations affecting language range from the basic practice of writing down spoken languages to the sophisticated information superhighway provided by the Internet. Technological innovations have in the past facilitated the spread and proliferation of multiple languages, but more recently they have encouraged the tendency of only a few languages—especially English, but also Chinese and Spanish—to dominate all others (**Figure 4.9**). Particular language groups achieve cultural dominance over neighboring groups in a variety of ways, often with profound results for the linguistic map of the world. Technological superiority is usually involved (see the cartoon at right). Earlier, we saw how plant and animal domestication—the technology of the "agricultural revolution"—aided the early diffusion of the Indo-European language family.

An even more basic technology was the invention of writing, which appears to have developed as early as 5300 years ago in several hearth areas, including in Egypt, among the Sumerians in what is today Iraq, and in China. Writing helped civilizations develop and spread, giving written languages a major advantage over those that remained spoken only. Written

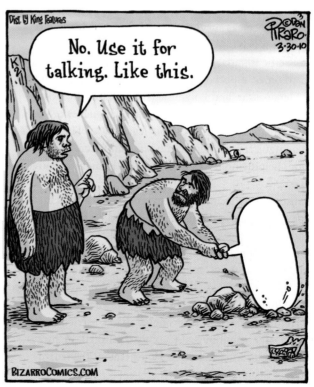

(Bizarro © 2010 Dan Piraro - Distributed by King Features Syndicate, Inc.)

languages can be published and distributed widely, and they carry with them the status of standard, official, and legal communication.

Written language facilitates record keeping, helping governments and bureaucracies to develop. Thus, the languages of conquerors tend to spread with imperial expansion. The imperial expansion of Britain, France, the Netherlands, Belgium, Portugal, Spain, and the United States across the globe altered the linguistic practices of millions of people. This empire building superimposed Indo-European tongues on the map of the tropics and subtropics. The areas most affected were Asia, Africa, Latin America, and the Austronesian island world. A parallel case from the ancient world is China, also a formidable imperial power that spread its language to those it conquered. During the Tang dynasty (618–907 C.E.), Chinese control extended to Tibet, Mongolia, Manchuria (in contemporary northeastern China), and Korea. The 4000-year-old written Chinese language proved essential for the cohesion and maintenance of its far-flung empire. Although people throughout the empire spoke different dialects or even different languages, a common writing system lent a measure of mutual intelligibility at the level of the written word. Today, however, the Chinese are concerned about the "Romanization" of their language, as Chinese schools prioritize learning English and youth are not familiar with as many Chinese characters as their parents or grandparents.

Even though imperial nations have, for the most part, given up their colonial empires, the languages they transplanted overseas survive. As a result, English still has a foothold in much of Africa, South Asia, the Philippines, and the Pacific islands. French persists in former French and Belgian colonies, especially in northern, western, and central Africa; Madagascar; and Polynesia (**Figure 4.10**). In most of these areas, English and French are the languages of the educated elite, often holding official legal status. They are also used as a lingua franca of government, commerce, and higher education, helping hold together states with multiple native languages.

Transportation technology also profoundly affects the geography of languages. Ships, railroads, and highways all serve to spread the languages of the culture groups that build them, sometimes spelling doom for the speech of less technologically advanced peoples whose lands are suddenly opened to outside contacts. The Trans-Siberian Railroad, built about a century ago, spread the Russian language eastward to the Pacific Ocean. The Alaska Highway, which runs through Canada, carried English into Native American refuges. The construction of highways in Brazil's remote Amazonian interior threatens the native languages of that region.

FIGURE 4.10 French, the colonial language of the empire, shares this sign on the isle of Bora Bora in French Polynesia with the native variant of the Polynesian tongue. Until recently, French rulers allowed no public display of the Polynesian language and tried to make the natives adopt French. (Courtesy of Terry G. Jordan-Bychkov.)

Another example is the predominance of English on the Internet, which can be understood as a contemporary information highway (**Figure 4.11**). It is unclear what will happen when other languages begin to challenge the dominance of English on the Internet, which is bound to happen sooner or later, although exactly when English will be surpassed by another language is anyone's guess. For example, from 2000 to 2017 there was a truly impressive 2263 percent growth in the number of Chinese speakers on the Internet. If this trend continues, and when—not if—the 47 percent of Chinese speakers who do not now use the Internet begin to log on, we can expect Chinese to surpass English as the most popular language on the Internet. Speakers of other languages grew even more impressively over this time period. For instance, Arabic-speaking Internet users grew 6806 percent from 2000 to 2017, while speakers of Russian grew 3273 percent over the same period.

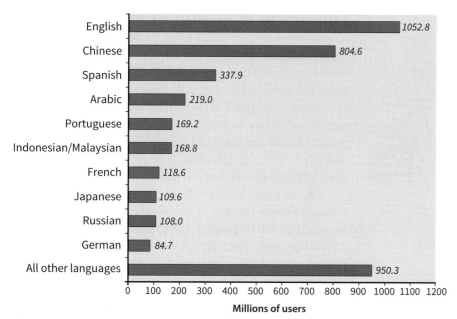

Language	Millions of users
English	1052.8
Chinese	804.6
Spanish	337.9
Arabic	219.0
Portuguese	169.2
Indonesian/Malaysian	168.8
French	118.6
Japanese	109.6
Russian	108.0
German	84.7
All other languages	950.3

FIGURE 4.11 The 10 most prevalent languages on the Internet, measured as a percentage of total users. English is the second most widely spoken language on Earth, after Mandarin Chinese. And English is the most widely spoken second language in the world. Mandarin Chinese, however, is an increasingly popular second language learned by individuals of non-Chinese ancestry. (Source: Adapted from http://www .internetworldstats.com/stats7.htm.)

Technology can also be used to preserve and revive endangered languages. Native American groups, for instance, are working with software developers to create language apps for the iPhone and iPad. They are also producing toys and video games that speak in native languages. Facebook, texting, video chats, and YouTube also help to connect speakers and promote the use of native speech, especially among younger people. Most speakers of Native American languages— around 90 percent—are middle-aged and older, so capturing the interest of younger speakers is vitally important to the survival of these languages. The U.S. government provides federal funds in the form of competitive grants to support language preservation efforts, and some tribes can utilize casino earnings for this purpose. Regardless of the funding source, most Native American groups are engaged in language preservation and instruction. Besides social media, apps, and video games, language immersion schools have been established (**Figure 4.12**), with mentor relationships forming between elders and youth.

Texting and Language Modification

Though English may dominate the Internet (see Figure 4.11), much of what comes across our computer and cell phone screens isn't a readily recognizable form of English.

As with the diffusion of spoken English to far-flung regions of the British Empire, the English language that is spread via electronic correspondence is subject to significant modification. E-mailing, instant messaging, and text messaging Standard English on cell phones requires a lot of typing, and text is notoriously deficient in conveying emotions when compared to the spoken word. For these reasons, abbreviations and symbols are used to shorten the number of keystrokes, to add emotional punctuation to correspondence, and to make electronic communication difficult to monitor by those who don't understand the language— particularly parents and teachers!

English is an alphabetic writing system, where letters represent discrete sounds that must be strung together to form a word's complete sound. A second major writing system is syllabic, where characters represent blocks of word sounds. This type of writing, prevalent throughout the Middle East and Southeast Asia, includes Arabic, Hebrew, Hindi, and Thai. The third form of writing is logographic, where characters represent entire words. Chinese is the only major language in this category.

English-speaking texters quickly learn to take shortcuts around the lengthiness inherent to alphabetic writing systems. The simplest and most commonly used shortcut involves

FIGURE 4.12 Learning Cherokee with technology. These two fifth-graders attend the Cherokee Nation Immersion School in Tahlequah, Oklahoma. They are working with their teacher to learn the Cherokee language with the assistance of an app. (Sue Ogrocki/ AP Photo)

"FIRST PAPER I'VE HAD TO GRADE WRITTEN ENTIRELY IN EMOTICONS."

(Sidney Harris/ScienceCartoonsPlus.com)

using acronyms instead of whole words to convey common phrases. *POS* (parent over shoulder), *VBG* (very big grin), *LOL* (laughing out loud), and *GMTA* (great minds think alike) are some examples of this technique. Another shortcut involves using characters that sound like words. For example, the number "8" can substitute for the word or sound "ate," so "h8" is "hate," and "i8" is "I ate." In the first example, a syllabic approach to writing is used because "8" condenses the syllable "ate" into one character. The second example substitutes a symbol for an entire word. Using "8" for the word "ate" is an example of this technique, called rebus writing, where a symbol is used for what it sounds like, as opposed to what it stands for. As your instructor for this class will likely confirm, such abbreviations and symbols have even worked their way into term papers at the college level, much to the consternation or delight of language scholars (see the cartoon above).

Texters sometimes use logographic writing, in which symbols such as ☺ and ♥, which are called pictograms, or word pictures, are used. Although the meaning of these symbols is understood by speakers of many languages because of their ubiquity, non-English languages also employ their own symbol combinations. In Korean, for instance, ^^ is used instead of :) to convey a smiling face and -_- is used instead of :(to depict a sad face. In Chinese, the number "5" is pronounced in a way that resembles crying, so "555" is the Chinese texter's way of conveying sadness.

Texting shortcuts are making their way into spoken and written English. For instance, LOL is now commonly used in speech. As speakers of non-English languages become more heavily involved in texting-based activities, their spoken and written languages also doubtlessly will become modified.

Language Proliferation: One or Many?

Could all the world's languages have derived from one single mother tongue? It may seem a large leap from the primordial tongue to a consideration of globalization and languages, but, in fact, the two are related. If we humans began with one language, why shouldn't we return to that condition? If one language can become 15,000 languages and the 7000 or so that remain will dwindle to 300 languages within a century, then is it possible to eventually end up with just one again?

Are the forces of modernization working to produce, through cultural diffusion, a single world language? And if so, what will that language be—English? Worldwide, about 335 million people speak English as their mother tongue and perhaps another 350 million speak it well as a second, learned language. Adding other reasonably competent speakers who can "get by" in English, the world total reaches about 1.5 billion, more than for any other language. What's more, the Internet is one of the most potent agents of diffusion, and its leading language is English.

English earlier diffused widely with the British Empire and U.S. imperialism, and today it has become the de facto language of globalization. Consider the case of India, where the English language imposed by British rulers was retained (after independence) as the country's language of business, government, and education. It provided some linguistic unity for India, which had 800 indigenous languages and dialects. This is why today many of India's 1.3 billion people speak English well enough to provide customer support services over the telephone for clients in the United States. Even so, many resent its use and wish India were rid of this linguistic colonial legacy once and for all (see Figure 4.22 on page 142). Although English is not likely to be driven out of India any time soon, it is true that the spoken English of India has drifted away from Standard British English. The same holds true for the English of Singapore, which is now a separate language called Singlish. Many other regional, English-based languages have developed—languages that could not be readily understood in London or Chicago.

But is the diffusion of English to the entire world population likely? Will globalization and cultural diffusion produce one world language? Probably not. More likely, the world will ultimately be divided largely among 5 to 10 major languages.

Language and Cultural Survival

Because language is the primary way of expressing culture, if a language dies out, there is a good chance that the culture of its speakers will, too. Languages, like animal species, can be

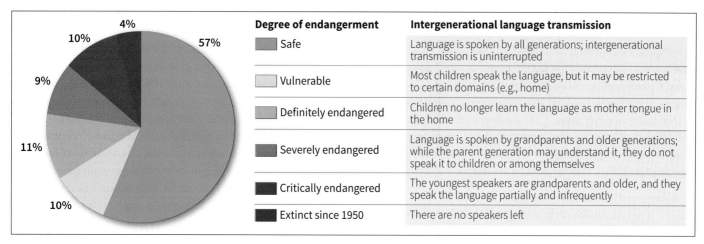

	Degree of endangerment	Intergenerational language transmission
	Safe	Language is spoken by all generations; intergenerational transmission is uninterrupted
	Vulnerable	Most children speak the language, but it may be restricted to certain domains (e.g., home)
	Definitely endangered	Children no longer learn the language as mother tongue in the home
	Severely endangered	Language is spoken by grandparents and older generations; while the parent generation may understand it, they do not speak it to children or among themselves
	Critically endangered	The youngest speakers are grandparents and older, and they speak the language partially and infrequently
	Extinct since 1950	There are no speakers left

FIGURE 4.13 Degrees of language endangerment. Forty-three percent of the world's languages are endangered to some degree. (Data from UNESCO Project: Atlas of the World's Languages in Danger, 2011.)

classified as endangered or extinct. Endangered languages are those that are not being taught to children by their parents and are not being used actively in everyday matters. Some linguists believe that more than half of the world's roughly 7000 languages are endangered. According to Ethnologue, an online language resource, 377 languages have become extinct since 1950, and another 906 are on the "dying" list. Languages that have only a few elderly speakers still living fall into this category (**Figure 4.13**). **Language hotspots**—places with the most unique, misunderstood, or endangered tongues—are located around the globe (**Figure 4.14**). Three of the world's most vulnerable regions are found in the United States. The

Americas and the Pacific regions together account for more than three-quarters of the world's current nearly extinct languages, thanks to their many and varied indigenous language groups. UNESCO estimates that, in the United States alone, 48 native languages are "critically endangered," as their youngest speakers are grandparents. When speakers die, it is most likely that their language will die out with them.

Languages can also be used to keep cultural traditions alive. Keith Basso, an anthropologist who has written an intriguing book titled *Wisdom Sits in Places*, discusses the landscape of distinctive place-names used by the Western Apache of New Mexico. The people Basso studied use place-names to

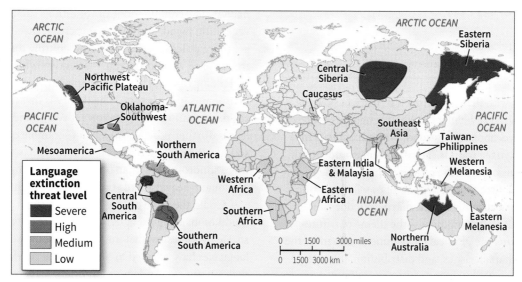

FIGURE 4.14 Global "language hotspots." The Enduring Voices Project and National Geographic Society have teamed up to document endangered languages and thereby attempt to prevent language extinction. One-quarter of the world's languages that are in trouble or dying may be found in the Americas. They represent a wealth of Native American languages slowly becoming suffocated by English, Spanish, and Portuguese. (Source: Adapted from http://travel.nationalgeographic.com/travel/enduring-voices/.)

invoke stories that help the Western Apache remember their collective history. According to Nick Thompson, one of Basso's interviewees, "White men need paper maps. . . . We have maps in our minds." Thompson goes on to assert that calling up the names of places can guard against forgetting the correct way of living, or adopting the bad habits of white men, once Western Apaches move to other areas. "The names of all these places are good. They make you remember how to live right, so you want to replace yourself again." One of the places Basso heard about is called Shades of Shit. Here is what he was told:

> It happened here at Shades of Shit.
>
> They had much corn, those people who lived here, and their relatives had only a little. They refused to share it. Their relatives begged them but still they refused to share it.
>
> Then their relatives got angry and forced them to stay at home. They wouldn't let them go anywhere, not even to defecate. So they had to do it at home. Their shades [shelters] filled up with it. There was more and more of it! It was very bad! Those people got sick and nearly died.
>
> Then their relatives said, "You have brought this on yourselves. Now you live in shades of shit!" Finally, they agreed to share their corn.
>
> It happened at Shades of Shit. (p. 24)

Today, merely standing at this place or speaking its graphic name reminds the Western Apache that stinginess is a vice that can threaten the survival of the entire community.

Nature-Culture

4.4   Explain the relationships between language and the physical environment.

Language interacts with the environment in three basic ways. First, the specific physical habitats in which languages evolve help to form their vocabularies. Second, physical habitats may shape the way a language sounds. Third, the environment can guide the migrations of linguistic groups or provide refuges for languages in retreat. From the viewpoint of possibilism—the notion that the physical environment shapes, but does not fully determine, cultural phenomena—the theme of nature–culture illustrates how the physical environment influences the vocabulary, tonal characteristics, and distribution of language.

Habitat and Vocabulary

Humankind's relationship to the land played a strong role in the emergence of linguistic differences, even at the level of vocabulary. For example, the Spanish language—which originated in Castile, Spain, a dry and relatively barren land rimmed by hills and high mountains—is especially rich in words describing rough terrain, allowing speakers of this tongue to distinguish even subtle differences in the shape and configuration of mountains. Similarly, Scottish Gaelic possesses a rich vocabulary to describe mountainous types of topography; this terrain-focused vocabulary is a common attribute of all the Celtic languages spoken by hill peoples. English, by contrast, which developed in the temperate wet coastal plains of northern Europe, is relatively deficient in words describing mountainous terrain. However, English abounds with words describing flowing streams and wetlands: typical physical features found in northern Europe. This vocabulary transferred well to the temperate East Coast of the United States. In the rural American South alone, one finds *river, creek, branch, fork, prong, run, bayou,* and *slough.* This vocabulary indicates that the area is a well-watered land with a dense network of streams.

Clearly, then, language serves an adaptive strategy. Vocabularies are highly developed for those features of the environment that involve livelihood. Without such detailed vocabularies, it would be difficult to communicate sophisticated information relevant to the community's livelihood, which in most places is closely bound to the physical landscape.

Natural Environment and Language Sounds

Recent research suggests that the sounds that birds use to communicate vary with the natural characteristics of their habitat. The combinations of conditions such as temperature, vegetation, wind, and mountains create specific habitats that are more or less conducive to transmitting different sound frequencies. Bioacousticians—scientists who study sound and biology—find that birds living in rain forests use fewer consonants in their song. The theory is that this is because the dense rainforest vegetation reflects sounds from trees and leaves; furthermore, warm air can scramble consonant-heavy sounds. Birds in such areas use more vowels, resulting in a more sonorous song than birds dwelling in open plains, where consonant-heavy sounds transmit more faithfully.

Could the same be true for human languages? Ian Maddieson, a linguist at the University of New Mexico, believes so. He hypothesizes that the linguistic soundscapes of human languages have adapted to the ecological conditions of the places they are spoken. As depicted in **Figure 4.15,** consonant-heavy languages, such as English, evolved in the cool open plains of northern Europe. Languages that are more focused on vowels, such as Hawaiian, were shaped by the dense vegetation of their natural habitat. In the geography of the spoken word, nature and culture appear to be intertwined in fascinating ways.

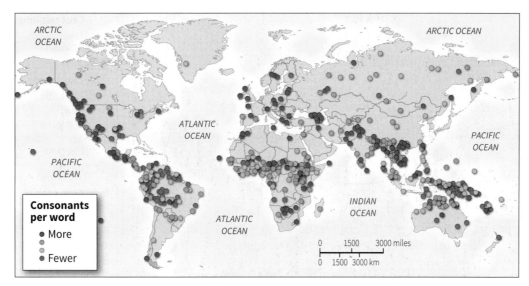

FIGURE 4.15 Did the language you speak evolve because of the environment? Places with more open spaces tend to have languages with lots of consonants. The sounds of such languages don't carry well in places that are windy or have dense forest cover. These places tend to have languages with more vowels. (Source: Based on an analysis by Ian Maddieson and Christophe Coupé.)

Habitat Helps Channel Language

Environmental barriers and natural routes have often guided linguistic groups onto certain paths. The wide distribution of the Austronesian language group, for instance, was profoundly affected by prevailing winds and water currents in the Pacific and Indian oceans (see Figure 4.6). The Himalayas and the barren Deccan Plateau deflected migrating Indo-Europeans entering the Indian subcontinent into the rich Ganga-Indus river plain. The northern and southern dialect boundaries in the United States are loosely limited by the Mississippi River.

Because such physical barriers as rivers and mountain ridges can discourage groups from migrating from one area to another, they often serve as linguistic borders as well. In parts of the Alps, speakers of German and Italian live on opposite sides of a major mountain ridge. Portions of the mountain rim along the northern edge of the Fertile Crescent in the Middle East form the border between Semitic and Indo-European tongues. Linguistic borders that follow such physical features generally tend to be stable, and they often endure for thousands of years. By contrast, language borders that cross plains and major routes of communication are often unstable, and shift over time.

Habitat Provides Refuge

The environment also influences language insofar as inhospitable areas provide protection and isolation. Such areas, referred to as **linguistic refuge areas,** provide minority linguistic groups protection from aggressive neighbors. Rugged hilly and mountainous areas, excessively cold or dry climates, dense forests, remote islands, and extensive marshes and swamps can all offer refuge to minority language groups. For one thing, unpleasant environments rarely attract conquerors. Also, mountains tend to isolate the inhabitants of one valley from those in adjacent ones, discouraging contact that might lead to linguistic diffusion.

Examples of these linguistic refuge areas are numerous. The rugged Caucasus Mountains and nearby ranges in central Eurasia provide protection for a large variety of peoples and their languages (**Figure 4.16**). In the Rocky Mountains of northern New Mexico, an archaic form of Spanish survives, largely as a result of isolation that ended only in the early 1900s. Similarly, the Alps, the Himalayas, and the highlands of Mexico form fine-grained linguistic mosaics, thanks to the mountains that provide both isolation and protection for multitudinous languages. Bitterly cold tundra climates of the far north have sheltered Uralic and Inuktitut speakers, and a desert has shielded Khoisan speakers from Bantu invaders. In short, rugged, hostile, or isolated environments protect linguistic groups that might otherwise be overtaken by more dominant languages.

Still, environmental isolation is no longer the vital linguistic force it once was. Fewer and fewer places are so isolated that they remain relatively untouched by outside influences. Today, inhospitable lands may offer linguistic refuge, but it is no longer certain that they will in the future. Even an island situated in the middle of the vast Pacific Ocean does not offer reliable refuge in the age of the Internet, social media, and global tourism. Similarly, marshes and forests provide refuge only if they are not drained and cleared by those who wish to use the land more intensively. The nearly 10,000 Gullah-speaking descendants

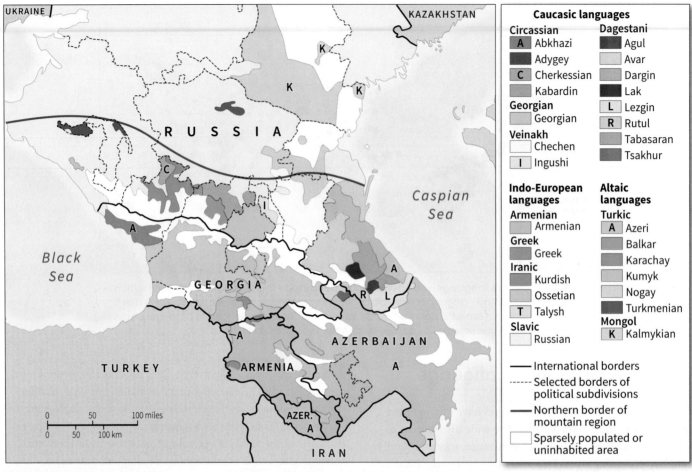

FIGURE 4.16 The environment provides a linguistic refuge in the Caucasus Mountains. The rugged mountainous region between the Black and Caspian seas—including parts of Armenia, Russia, Georgia, and Azerbaijan—is peopled by a great variety of linguistic groups, representing three major language families. Mountain areas are often linguistic mosaics because the rough terrain provides refuge and isolation.

of African slaves have long nurtured their distinctive African-influenced culture and language, in part because they reside on the Sea Islands of South Carolina, Georgia, and northern Florida. Today, the development of these islands for tourism and housing for wealthy nonlocals, as well as the out-migration of Gullah youth in search of better economic opportunities, threatens the survival of the Gullah culture and language. The reality of the world is no longer isolation, but contact.

Cultural Landscape

4.5 Characterize the ways languages are visibly part of the cultural landscape.

Road signs, billboards, graffiti, placards, and other publicly displayed writings not only reveal the locally dominant language but also can be a visual index to bilingualism,

linguistic oppression of minorities, and other facets of linguistic geography. Furthermore, differences in writing systems render some linguistic landscapes illegible to those not familiar with these forms of writing (**Figure 4.17**).

Linguistic Landscapes and Their Messages

Linguistic landscapes send messages, both friendly and hostile. Often these messages have a political content and deal with power, domination, subjugation, or freedom. In Turkey, for example, until recently Kurdish-speaking minorities were not allowed to broadcast music or television programs in Kurdish, to publish books in Kurdish, or even to give their children Kurdish names. People who spoke Kurdish were arrested and imprisoned. In 2002 Turkey reformed its legal restrictions to allow the Kurdish language to be used in daily life but not in public education. In 2012 Kurdish language instruction

FIGURE 4.17 Linguistic landscapes. This image, from Los Angeles's Koreatown, is difficult for non-Korean speakers to read. The Korean characters used are not the Latin alphabet with which most English speakers are familiar. (nik wheeler/Getty Images)

FIGURE 4.18 Road sign in Dublin, Ireland, shows the place-name in Irish Gaelic on top and English underneath. (Courtesy of Patricia L. Price.)

FIGURE 4.19 Graffiti in a parking lot. This wall in a parking lot in Cork, Ireland, is covered with graffiti. Gangs use stylized scripts often unintelligible to nonmembers to mark their territory. (Courtesy of Patricia L. Price.)

became an elective subject in Turkey's public schools. The Canadian province of Québec, similarly, has tried to eliminate English-language signs. French-speaking immigrants settled Québec, and its official language is French, in contrast to Canada's policy elsewhere of bilingualism in English and French. As **Figure 4.18** indicates, there is a practice in Ireland to supplement English-language place-name signs by adding the original Gaelic place-names. The suppression of minority languages and attempts to reinstate them in the landscape offer an indication of the social and political status of minority populations more generally.

Other types of writing, such as graffiti, can denote ownership of territory or send messages to others (**Figure 4.19**). Only those who understand the specific symbols used will be able to decipher the full meaning of the message. Misreading such writing can have dangerous consequences for those who stray into unfriendly territory. In this way, graffiti can be understood as a dialect that is particular to a subculture and transmitted through symbols or a highly stylized script.

Toponyms

Language and culture also intersect in the names that people place on the land, whether they are given to settlements, terrain features, streams, or various other aspects of their surroundings (**Figure 4.20**). These place-names, or **toponyms,** often reflect the spatial patterns of language, dialect, and ethnicity. Toponyms become part of the cultural landscape when they appear on signs and placards. Many place-names consist of two parts—the generic and the specific. For example, in the American place-names Huntsville, Harrisburg, Ohio River, Newfound Gap, and Cape Hatteras, the specific segments are *Hunts-, Harris-, Ohio, Newfound,* and *Hatteras.* The generic parts, called **generic toponyms,** which tell what *kind* of place is being described, are *-ville, -burg, River, Gap,* and *Cape.*

Generic toponyms are of greater potential value to the cultural geographer than specific names because they appear again and again throughout a culture region.

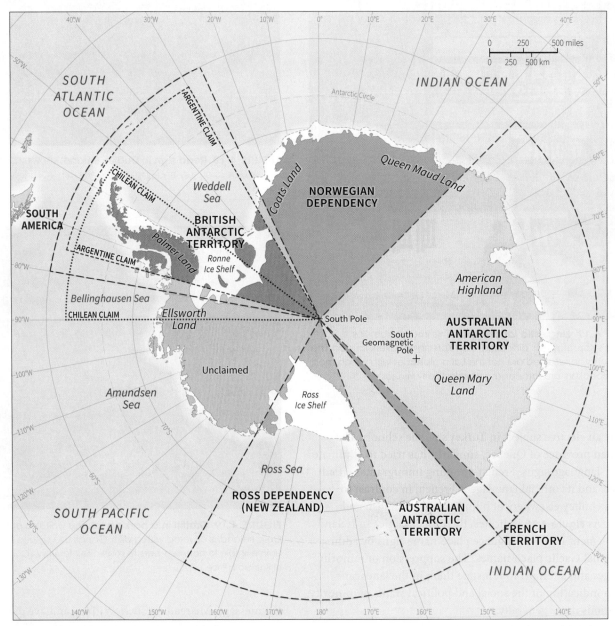

FIGURE 4.20 Naming place is closely related to claiming place. This map of Antarctica shows the pie slice–shaped land claims of various nations. Notice how place-names reflect the names of Antarctic explorers or European rulers. The gray portions are unclaimed. (Source: Adapted from Latrimer Clarke Corporation Pty. Ltd.)

There are literally thousands of generic place-names, and every culture or subculture has its own distinctive set of them. They are particularly valuable both in tracing the spread of a culture and in reconstructing culture regions of the past. Sometimes generic toponyms provide information on changes people brought about long ago in their physical surroundings.

Generic Toponyms of the United States The three dialects of the eastern United States (see Figure 4.9)—Northern, Midland, and Southern—illustrate the value of generic toponyms in cultural geographical detective work. For example, New Englanders, speakers of the Northern dialect, often used the terms *Center* and *Corner* in the names of the towns or hamlets. Outlying settlements frequently bear the prefix *East, West, North,* or *South,* with the specific name of the township as the suffix. Thus, in Randolph Township, Orange County, Vermont, we find settlements named Randolph Center, South Randolph, East Randolph, and North Randolph (**Figure 4.21**). Hewitts Corners is located a few miles away.

These generic usages and duplications are peculiar to New England, and we can locate areas settled by New Englanders as they migrated westward by looking for such place-names in other parts of the country. A trail of "Centers" and name duplications extending westward from New England through upstate New York and Ontario and into the upper Midwest clearly indicates their path of migration and settlement. Toponymic evidence of New England exists in areas as far away as Walworth County, Wisconsin, where Troy, Troy Center, East Troy, and Abels Corners are clustered; in Dufferin County, Ontario, where one finds places such as Mono Centre; and even in distant Alberta, near Edmonton, where the toponym Michigan Centre doubly suggests a particular cultural diffusion.

Similarly, we can identify Midland American areas by such terms as *Gap, Cove, Hollow, Knob* (a low, rounded hill), and *-burg,* as in Stone Gap, Cades Cove, Stillhouse Hollow, Bald Knob, and Fredericksburg. We can recognize Southern speech by such names as *Bayou, Gully,* and *Store* (for rural hamlets), as in Cypress Bayou, Gum Gully, and Halls Store.

Toponyms and Cultures of the Past Place-names often survive long after the culture that produced them vanishes from an area, thereby preserving traces of the past.

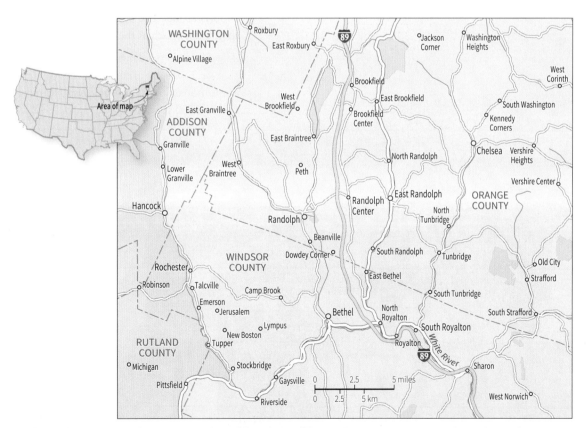

FIGURE 4.21 This map depicts a portion of the state of Vermont, displaying the generic toponyms that abound in New England.

The Alhambra fortress-palace complex, the gardens and rural estates of the Generalife, and the Albayzín residential quarter together form this World Heritage Site. All three are located in the city of Granada, in southern Spain. The Alhambra and the Albayzín are situated on hills above the modern lower city. Hills were important defense sites in medieval European cities (see Chapter 10). And the irrigated gardens of the Generalife were part of the rulers' rural estates.

- From the eleventh through fourteenth centuries, Muslim rulers, or *emirs,* oversaw the construction of the Alhambra fortress and residences, the gardens of the Generalife, and the Albayzín residential quarter. From the eighth through fifteenth centuries, Spain was ruled by the Moors, Muslims of North African descent.

- These sites are exceptional examples of Moorish imperial architecture. Though Spain became Christianized after the Moors were expelled in 1492, Spanish monarchs greatly admired and sought to safeguard the architectural achievements of the Moors in the Andalusian region of southern Spain. Through careful restoration, the architecture of these sites remains true to its Spanish-Moorish roots. The urban layout, colors, and materials look today much as they did over 700 years ago.

- The Albayzín residential quarter occupies a hill that, according to archaeological evidence, has been continuously occupied since early Roman times (about 2500 years).

(Karol Kozlowski/Alamy)

(caracterdesign/Getty Images)

WALL WRITING IN THE ALHAMBRA: The original Alhambra fortress-palace was built by Samuel Ha-Nagid, an eleventh-century Jewish grand vizier (a prime minister of sorts) to the Zirid sultans of Granada. In the thirteenth and fourteenth centuries, the Nazrid emirs constructed an entire complex of palaces and irrigated gardens at the site. The buildings are adorned with what appear at first glance to be elaborate, intricate designs. In reality, the walls, ceilings, and other architectural elements are covered with words. Over 10,000 Arabic inscriptions adorn the interior of the Alhambra complex. "There is probably no other place in the world where studying walls, columns and fountains is so similar to turning the pages of a book," according to Spanish researcher Juan Castilla.

- Islamic architecture is typically adorned with decorative writing called calligraphy and other geometric patterns, rather than representations of living beings. According to some Muslims, figural representation was a form of idolatry (the worship of a physical object as a god), while others saw the body as an imperfect covering for the soul.

- Because calligraphy was commonly regarded as the preeminent expression of the visual arts, calligraphers typically had a higher status than other artists and were housed in the Ministry of Writing.

- Efforts to digitally archive and transcribe the wall writing reveal that fewer than 10 percent of the inscriptions are Qur'anic (religious) verses or poetry. The rest offer praise to the Nasrid emirs or consist of other popular sayings, with "There is no victor but Allah" being the most common inscription.

TOURISM: This site attracts a huge number of international tourists, most of whom arrive in the summer months.

- The number of tourists to the Alhambra is limited to a daily quota of 6600.

- Unfortunately, the decorative surfaces of the site's buildings are not well protected, so they continue to be worn away by the touch of so many tourists.

- The picturesque city of Granada is situated on a series of hills against the backdrop of the Sierra Mountains. Tourists enjoy the mix of Mediterranean cultures, cuisines, and traditions there.

- This World Heritage Site was built by and for the Moorish rulers of southern Spain, and exemplifies the pinnacle of Andalusian architecture, a synthesis of classical Arabic and southern Spanish styles.
- The interior structures of the Alhambra palace are covered with more than 10,000 intricate inscriptions in Arabic calligraphy.
- The Alhambra and the Generalife were placed on the list of World Heritage Sites in 1984, and in 1994 the site was expanded to include the Albayzín quarter.

http://whc.unesco.org/en/list/314

(Geoff A Howard/Alamy)

Australia abounds in Aborigine toponyms, even in areas from which the native peoples disappeared long ago. No toponyms are more permanently established than those identifying physical geographical features, such as rivers and mountains. Even the most absolute conquest, exterminating an aboriginal people, usually does not entirely destroy such names. The abundance of Native American toponyms in the United States provides an example. The names of more than half of the states are of Native American origin. India, however, has recently decided to revert to traditional toponyms, after many Indian place-names had been Anglicized under British colonial rule (**Figure 4.22**).

In Spain and Portugal, seven centuries of Moorish rule left behind a great many Arabic place-names. An example is the prefix *guada-* on river names (as in Guadalquivir and Guadalupejo). The prefix is a corruption of the Arabic *wadi,* meaning "river" or "stream." Thus, Guadalquivir, corrupted from Wadi-al-Kabir, means "the great river." The frequent occurrence of Arabic names in any particular region or province of Spain reveals the remnants of Moorish cultural influence in that area. Many such names were brought to the Americas through Iberian conquest, so that Guadalajara, for example, appears on the map as an important Mexican city.

The Political Economy of Toponyms Without a doubt, you are familiar with places that bear the names of wealthy and influential individuals, politicians, or corporate sponsors. Sports stadiums, campus buildings, and museums are all specific spaces that can be named. The owners of these spaces typically use naming opportunities as a way to raise funds. For instance, universities provide naming rights for donors contributing anywhere from $5 million to over $300 million.

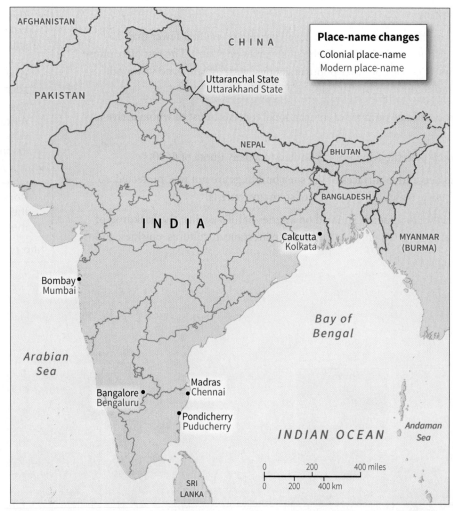

FIGURE 4.22 India's postcolonial toponym shift. More than 50 years after the English colonizers "quit" India, their colonial place-names are being swept from the map, too. (Source: Adapted from Sappenfield, 2006.)

FIGURE 4.23 Barclays Center Station. The New York City subway system is public. Yet this stop—just outside the Barclays Center arena—is named after the arena's corporate sponsor, a bank. (Richard Levine/Alamy)

But what happens when a place that has been built using public monies—say, a train station that was constructed using tax revenues—is given a corporate name (**Figure 4.23**)? Geographers Reuben Rose-Redwood and Derek Alderman examine this practice, using the example of the building that was constructed to replace the World Trade Center's Twin Towers, which were destroyed in the terrorist attacks of 9/11. The building was initially dubbed "The Freedom Tower," but once it acquired corporate tenants, the building began to be referred to by its legal address, One World Trade Center. According to Rose-Redwood and Alderman, this illustrates "how the naming of places has become . . . one of the next 'frontiers' in the neoliberalization of urban spaces."

What happens when the corporate sponsor of a place has a less-than-acceptable reputation? Florida Atlantic University, for instance, sold the naming rights for its sports stadium to GEO Group for $6 million in February 2013. GEO Group operates private prisons and has been implicated in the inhumane treatment of inmates at its facilities. Less than two months later, the agreement was cancelled. In this instance, the negative place image generated by a corporate sponsor wasn't worth the price paid (**Figure 4.24**).

Is no place safe from corporate ownership? Apparently not. People are paid to allow companies to use their cars as mobile billboards. What about the items of clothing and accessories that prominently bear the names of their designers? Is what poet Adrienne Rich called "the geography closest-in"— our bodies—a place that is also open for corporate naming opportunities?

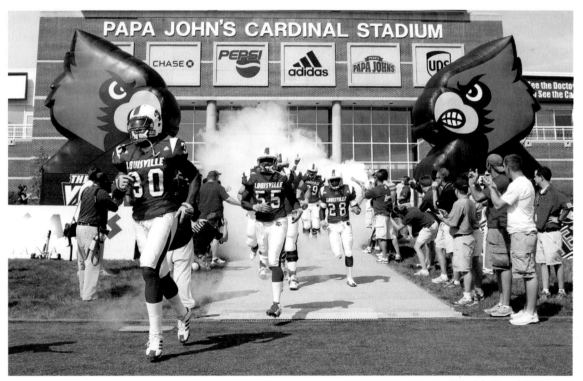

FIGURE 4.24 University of Louisville sports stadium. The home of the Cardinals was named after a pizza chain and supported by other corporate sponsors noted on the sign. Recently the university took Papa John's name off the stadium in the wake of a racial slur scandal. (Andy Lyons/Getty Images)

Conclusion

Human interactions, with one another and with the world around us, are primarily language-based. We shape our surroundings using language, and in turn, our surroundings shape the way different languages sound, their vocabularies, and their spatial expression.

Like human cultures, languages are fluid and adaptive to the needs of their users. Languages are also susceptible to the power dynamics at work in human societies more generally. Conquest, dominance, and repression, but also creativity and resilience, are at work in the rise and fall of languages and in their ever-changing expressions. The trend toward the dominance of a few "big" languages may afford opportunities to communicate on a global scale, but it also may signal the demise of much of the cultural richness across the Earth.

Chapter Summary

4.1   Identify the geographical patterns of languages.

- Language families, dialects, vocabulary, pronunciation, and toponyms display distinct spatial variations that can be identified on maps of linguistic culture regions.
- Pidgins, creoles, lingua franca, and bilingualism are all linguistic strategies that allow speakers of different languages to interact with each other; there is both creativity and power involved.
- Languages are connected to other languages in a tree-like fashion, with major branches—such as Indo-European or Sino-Tibetan—leafing into specific, related tongues.

4.2   Understand how languages and dialects have come to exist, move, and change.

- The rise and spread of Indo-European, the largest language family on Earth, has been explained by the Anatolian hypothesis and the Kurgan hypothesis.
- Relocation and expansion diffusion, both hierarchical and contagious, are apparent in the movement of language from one place to another.
- Political conquest and the spread of religions have been important for the diffusion of languages.
- Language is fluid and ever-changing along with the needs of users.

4.3   Evaluate the relationship between technology and language.

- The number of languages spoken on Earth is decreasing, while a few—notably Chinese, English, and Spanish—are spoken by an increasing number of people worldwide. Technological forces, ranging from the invention of writing to transportation innovations, help to spread languages across the globe.
- The rise of the Internet and texting have dramatically reshaped spoken and written language.
- When endangered languages become extinct, so too do the cultures of their speakers. Efforts to preserve and teach dying languages draw on the same cultural innovations that in other ways act to threaten the most vulnerable of tongues.

4.4   Explain the relationships between language and the physical environment.

- The natural habitat helps to shape linguistic elements, such as vocabulary and dominant sounds.
- Physical features, such as mountains and rivers, can channel the spread of language or provide barriers to its diffusion.
- Inhospitable physical environments can provide refuge and protection for speakers of minority languages.

4.5   Identify the ways languages are visibly part of the cultural landscape.

- Dominance of one group over another is often expressed in exclusion, or even extermination, from the linguistic cultural landscape.
- Toponyms provide clues to the history of places, political struggles over land, and display particular regional patterns in the United States.
- Corporations purchase the right to name places, and in so doing bring the controversy between public and private space into sharp focus.

Key Terms

language A mutually agreed-upon system of symbolic communication that has a spoken and usually a written expression (page 118).

dialect A distinctive local or regional variant of a language that remains mutually intelligible to speakers of other dialects of that language; a subtype of a language (page 119).

pidgin A composite language consisting of a small vocabulary borrowed from the linguistic groups involved in commerce (page 119).

creole A language derived from a pidgin language that has acquired a fuller vocabulary and become the native language of its speakers (page 120).

lingua franca An existing, well-established language of communication and commerce used widely where it is not a mother tongue (page 120).

bilingualism The ability to speak two languages fluently (page 120).

language family A group of related languages derived from a common ancestor (page 120).

polyglot Involving many languages (page 122).

Anatolian hypothesis A theory of language diffusion holding that the movement of Indo-European languages from the area in contemporary Turkey then known as Anatolia followed the spread of plant domestication technologies (page 123).

Kurgan hypothesis A theory of language diffusion holding that the spread of Indo-European languages originated with animal domestication in the central Asian steppes and grew more aggressively and swiftly than proonents of the Anatolian hypothesis maintain (page 123).

isogloss The spatial border of usage of an individual word or pronunciation (page 126).

slang Words and phrases that are not part of a standard, recognized vocabulary for a given language but that are nonetheless used and understood by some of its speakers (page 128).

ethnolect A dialect spoken by a particular ethnic group (page 128).

language hotspots Those places on Earth that are home to the most unique, misunderstood, or endangered languages (page 133).

linguistic refuge area An area protected by isolation or inhospitable environmental conditions in which a language or dialect has survived (page 135).

toponym A place-name, usually consisting of two parts—the generic and the specific (page 138).

generic toponym The descriptive part of many place-names, often repeated throughout a culture area (page 138).

 ## Practice at

Read the interactive e-Text, review key concepts, and test your understanding.

 Story Map. Explore the globalization of the English language.

 Web Map. Examine maps illustrating issues of linguistic geography.

Doing Geography

ACTIVE LEARNING:
Listening to the Dialects of English

As we have seen, linguistic influences, historical patterns, and migratory flows all can play a role in shaping spoken languages. In this activity, you will use your phone or computer to listen to recordings of people speaking various dialects of English. As you listen, you'll try to understand what you are hearing and decide which dialects are easy or difficult to make out, keeping in mind the differences between dialects, pidgins, creoles, and "standard" languages, and how these differences sound.

More guidance at 🍃 SaplingPlus

EXPERIENTIAL LEARNING:
Exploring the Political Economy of Purchased Place-Names

In this activity, you will work individually or in small groups to identify places that bear the names of corporate or individual sponsors. As you learned in the section titled "The Political Economy of Toponyms," elements of the built environment—buildings, parks, sports venues, as well as vehicles and the clothing we wear—can bear the name of a wealthy person or family, or a corporation. Purchasing the right to name a place often confers a type of power on the purchaser. This experiential learning activity will encourage you to make these connections in your local landscape.

Steps to Exploring the Political Economy of Purchased Place-Names

Step 1:   Work with your instructor to decide whether you will conduct this activity alone or in small groups.

Step 2:   Determine the place you will explore. For some of you, the immediate college or university campus has ample places named by benefactors. For others, your city, town, or neighborhood will provide a better field research site.

Step 3:   Set out to find places in the landscape that have been named by individuals or corporations. Photograph them with your cell phone and, if your instructor requests, note the location coordinates.

Step 4:   Research the history of the named places that you find. Based on what you learn, answer the following questions.

- How much money did the naming rights cost?

- Who received the funds and what were the funds used for?

- Has there been any controversy over the naming, or has the naming of the place(s) you've identified changed over time from one sponsor to another?

Step 5:   Report to your classmates and compare your findings.

The name of donors Joan and Sanford I. Weill is now part of the name of Cornell University's medical center. (Alessio Botticelli/GC Images/Getty Images)

SEEING GEOGRAPHY

Feisbuquear: "To Facebook"

In 2002, Hispanics surpassed African Americans as the nation's numerically most significant minority group. In some U.S. cities, Hispanics constitute more than half of the population, a fact that brings into question the designation "minority." For example, the population of Miami, Florida, is two-thirds Hispanic, and more than three-fourths of the residents of El Paso and San Antonio, Texas, are Hispanic. In the United States today, Spanish-speaking peoples from Latin America provide the largest flow of immigrants into the United States. Even midsize and smaller towns in the midwestern and southern United States are becoming destinations for Spanish-speaking immigrants, sharply changing the ethnic composition of cities such as Shelbyville, Tennessee; Dubuque, Iowa; and Siler City, North Carolina.

By the same token, English has burrowed its way deeply inside the contemporary Mexican linguistic landscape. The ubiquity of U.S. consumer goods, English-language media, and American popular culture in Mexico is notable. Many Mexicans travel back and forth to the United States for work- and family-related reasons. And, given that the United States and Mexico share a nearly 2000-mile border, these flows of people and ideas have a long history. It is logical to assume that the Spanish spoken by Mexicans will be influenced by American English, and vice versa.

Indeed, Spanish and English have combined in a rich, complex fashion, producing a hybrid language called Spanglish. The phrase "Vámonos al downtown a tomar una bironga after work hoy" is an excellent example. It translates into Standard English as "Let's go downtown and have a beer after work today." *Vámonos* ("let's go"), *tomar* ("to drink"), and *hoy* ("today") are Spanish words that are combined in the same sentence with the English words *downtown* and *after work*. Linguists refer to this tendency to shift between languages in the same sentence, a very common practice among bilingual speakers, as code-switching. However, the noun *bironga,* which means "beer" in English, is a Spanglish invention: it exists in neither English nor Spanish. This is quite common in Spanglish, and neologisms such as *hanguear* ("to hang out"), *deioff* ("day off"), and *parquear* ("to park" a vehicle) abound.

In the image that opens this chapter, the word "feisbuquear" is wholly invented, using the sound of the noun "Facebook" and a Spanish verb ending. Spanglish reflects the growing Spanish-English bilingualism of many U.S. residents, as well as residents of Mexico. It underscores the flexibility of language, and the enduring creativity of human beings as we attempt to communicate with one another.

Cell phone service provider advertisement in Mexico. Feisbuquear is a Spanglish verb meaning "to Facebook." (Courtesy of Patricia L. Price.)

Ethnic restaurants abound in Manhattan's Greenwich Village neighborhood. (Steve Lewis Stock/ Getty Images)

chapter 5

Geographies of Race and Ethnicity

MELTING POT OR SALAD BOWL?

When does cuisine cease to be ethnic and become simply "American"? What roles do mobility and globalization play in the process?

Examine the photo and think about these questions as you read. We will revisit them on page 183.

Learning Objectives

5.1 Differentiate between race and ethnicity, and describe how ethnic groups are distributed geographically.

5.2 Explain the ways that migration and mobility shape, and are shaped by, race and ethnicity.

5.3 Interpret how globalization has deepened, transformed, or erased ethnicity, race, and racism.

5.4 Analyze the diverse ways in which ethnic and racial groups interact with the natural environment, and discuss how these groups may be protected or made more vulnerable.

5.5 Identify the ways that ethnicity and race leave imprints on the cultural landscape.

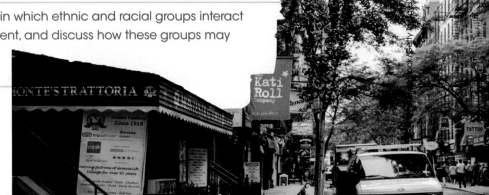

Ethnic geography is the study of the spatial aspects of ethnicity. Ethnic groups are the keepers of distinctive cultural traditions and the focal points of various kinds of social interaction. They are the basis not only of group identity but also of friendships, marriage partners, recreational outlets, business success, and political power. These interactions can offer cultural security and reinforcement of tradition. Ethnic groups often practice unique adaptive strategies and usually occupy clearly defined areas, whether rural or urban. In other words, the study of ethnicity has built-in geographical dimensions. The geography of race is a related field of study that focuses on the spatial aspects of how race is socially constructed and negotiated. Cultural geographers who study race and ethnicity tend to always have an eye on the larger economic, political, environmental, and social power relations at work.

This chapter draws on insights and examples from both ethnic geography and the geography of race. Race and ethnicity are related but distinct concepts, underscoring historically specific understandings of what constitutes meaningful human differences and solidarities. Being a member of an ethnic or racialized minority can impact many aspects of an individual's life, constraining day-to-day mobility as well as shaping long-term migratory patterns. Indeed, ethnic minority populations are often created through migration. Globalization has intensified both tensions around race and ethnicity and the pride expressed in belonging to racial or ethnic groups. The natural environment interacts in key ways with ethnic and racial groups, channeling their migratory patterns, protecting the cultural traits of distinct groups, and encouraging migrants to selectively relocate some aspects of their cultures while adopting aspects of their new host culture. Finally, race and

ethnicity leave a distinct mark on the cultural landscape, one that is at times the locus of bitter struggles over the past.

Region

5.1 Differentiate between race and ethnicity, and describe how ethnic groups are distributed geographically.

One of the enduring stories that the people of the United States proudly tell themselves is that "ours is an immigrant nation." This story is on display during annual festivals celebrating the variety of ethnic traditions in countless cities, towns, and villages across the nation. For example, the midwestern town of Wilber, settled by Bohemian immigrants beginning about 1865, bills itself as "The Czech Capital of Nebraska" and annually invites visitors to attend a "National Czech Festival." Celebrants are promised Czech foods, such as *koláce, jaternice,* poppy seed cake, and *jelita;* Czech folk dancing; Czech postcards and souvenirs imported from Europe; and handicraft items made by Nebraska Czechs (bearing an official seal and trademark to prove their authenticity). Thousands of visitors attend the festival each year. Without leaving Nebraska, these tourists can move on to "Norwegian Days" at Newman Grove, the "Greek Festival" at Bridgeport, the Danish "Grundlovs Fest" in Dannebrog, "German Heritage Days" at McCook, the "Swedish Festival" at Stromsburg, the "St. Patrick's Day Celebration" at O'Neill, several Native American powwows, and assorted other ethnic celebrations (**Figure 5.1**).

FIGURE 5.1 The town of Lindsborg, Kansas (*left*). Proud of its Swedish heritage, Lindsborg holds a Swedish Festival each year in June. **Hispanic Heritage Festival in Nebraska (*right*).** Hispanic immigrants have expanded the range of ethnic pride festivals throughout the Midwest. (Left: Betts Anderson Loman/PhotoEdit; Right: Steve Skjold/Skjold Photographs)

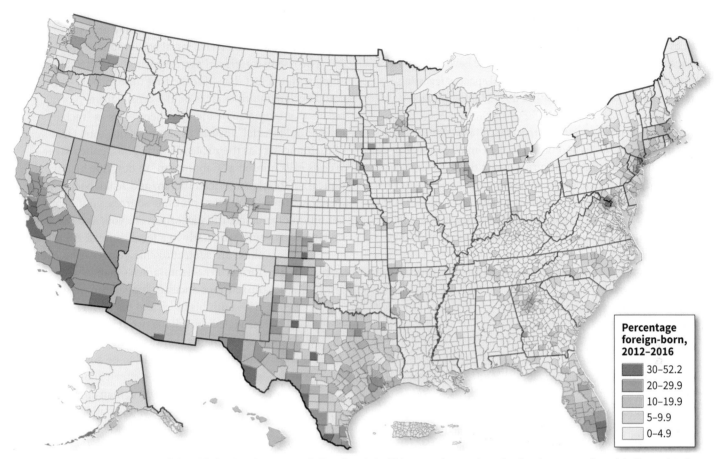

FIGURE 5.2 The distribution of the U.S. foreign-born population by state. This map shows where foreign-born people reside throughout the United States and Puerto Rico. The Southwest in general, and large cities throughout the country, are home to higher proportions of foreign-born. (Data from U.S. Census Bureau.)

Today, Nebraska is still a magnet for immigrants, but since the 1990s, the state's new arrivals have been overwhelmingly non-European. In particular, Mexican immigrants employed in Nebraska's meat-processing industry find destinations such as Nebraska and other upper midwestern states attractive (**Figure 5.2**). In general, immigrants to the United States today are more likely to come from Asia or Latin America than from Europe, and they are changing the face of ethnicity in the United States (**Figure 5.3**). Indeed, ethnicity is a central aspect of the cultural geography of most places.

Race or Ethnicity: What's the Difference?

Race is often used interchangeably with ethnicity, but the two terms have very different meanings, and we must be careful in choosing between them.

Race Race can be understood as a genetically significant difference among human populations. A few biologists today

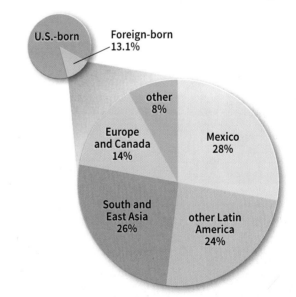

FIGURE 5.3 The foreign-born population of the United States. The smaller pie chart shows that 13.1 percent, or 41.3 million, of the total U.S. population in 2013 was born abroad. The larger pie chart shows that the majority of these people came from Latin America and Asia. (Data from Pew Research Center, 2015.)

support the view that human populations do form racially distinct groupings, arguing that race explains phenomena such as the susceptibility to certain diseases. In contrast, many social scientists (and many biologists as well) have noted how definitions of race change across time and space, suggesting that race is a social construct rather than a biological fact (**Figure 5.4**).

Because race is a social construction, it takes different forms in different places and times. In the United States, for example, the so-called one-drop rule meant that anyone with any African American ancestry at all was considered black. This law was intended to prevent interracial marriage. It also meant that moving out of the category "black" was, and still is today, extremely difficult because one's racial status is determined by

ancestry. Yet the notion that "black blood" somehow makes a person completely black is challenged by the growing numbers of young people with diverse racial backgrounds.

The rise in interracial marriages in the United States, along with the fact that—for the first time in 2000—one could declare multiple races on the census form, means that more people identify themselves as "racially mixed" or "biracial" instead of choosing only one facet of their ancestry as their sole identity. Thus, golfer Tiger Woods calls himself a "Cablinasian" to describe his mixed Caucasian, black, American Indian, and Asian background. Racial identities arise from multiple sources: your own definition of yourself; the way others see you; and the way society treats you, particularly in a legal context. All three of these are subject to change over time. Still, former President Barack Obama, whose mixed-race ancestry has led him to identify with both his black and his white relatives, struggled during his two terms in office against the claims that he was at once "too black" and "not black enough" to be a representative president.

In Brazil, by contrast, many physiognomic features, such as skin pigmentation, eye color, and hair texture, are used to identify a person racially, with the result that many racial categories exist in Brazil. Siblings are frequently classified in significantly different racial terms depending on their appearance. The same person can be put into multiple racial categories by different people, and individuals' own racial self-designations can depend on such variables as their mood at the time they are asked. Increased economic or educational status can "whiten" individuals formerly classified as black. Racial fluidity notwithstanding, one must keep in mind that Brazil

FIGURE 5.4 Racial classifications in colonial Mexico. These images, called *castas* paintings, were common in colonial Mexico. They depict the myriad racial and ethnic combinations perceived to arise from intermixing among Europeans, indigenous peoples, and African slaves in the New World. They provided a way for those in power to keep track of the confusing racial hierarchy they had created. (Schalkwijk/Art Resource, NY)

was also the last country in the Americas to abolish slavery, and discrimination against darker-skinned Brazilians is still common today.

Studies of genetic variation have demonstrated that there is far more variability within racial categories than between them, which has led most scholars to believe that all human beings are, genetically speaking, members of only one race: *Homo sapiens sapiens.* In fact, some social scientists have dropped the term race altogether. This is not to say, however, that **racism,** the belief that human capabilities are determined by racial classification and that some races are superior to others, does not exist. In this chapter, we use the verb **racialize** to refer to the processes whereby differences are understood—usually, but not always, by the powerful majority—through the lens of race. Across the world, hatreds based in racism are at the root of the most incendiary conflicts imaginable (see Subject to Debate, page 154).

Ethnicity What exactly is an ethnic group? The word *ethnic* is derived from the Greek word *ethnos,* meaning a "people" or "nation," but that definition is too broad. For our purposes, an **ethnic group** consists of people of common ancestry and cultural tradition. A strong feeling of group identity characterizes ethnicity. Membership in an ethnic group is largely involuntary, in the sense that a person cannot simply decide to join; instead, he or she must be born into the group. In some cases, outsiders can join an ethnic group by marriage or adoption.

Different ethnic groups may base their identities on different traits. For some, such as Jews, ethnicity primarily means religion; for the Amish, it is both folk culture and religion; for Swiss Americans, it is national origin; for German Americans, it is ancestral language; for African Americans, it is a shared history stemming from slavery. Religion, language, folk culture, history, and place of origin can all help provide the basis of the sense of "we-ness" that underlies ethnicity.

As with race, ethnicity is a notion that is at once vexingly vague yet hugely powerful. The boundaries of ethnicity are often fuzzy and shift over time. Moreover, ethnicity often serves to mark minority groups as different, yet majority groups also have an ethnicity, which may be based on a common heritage, language, religion, or culture. Indeed, an important aspect of being a member of the majority is the ability to decide how, when, and even if one's ethnicity forms an overt aspect of one's identity. Some scholars question whether ethnicity even exists as intrinsic qualities of a group, suggesting instead that the perception of ethnic difference arises only through contact and interaction.

Many, if not most, ethnic groups around the world originated when they migrated from their native lands and settled in a new country. In their old home, they often belonged to the host culture and were not ethnic, but when they were transplanted by relocation diffusion to a foreign land, they simultaneously became a minority and ethnic. At over 90 percent of the population, Han Chinese are not considered an ethnic minority in China, but if they come to North America, they are. Indigenous ethnic groups that continue to live in their ancient homes become ethnic when they are absorbed into larger political states. The Navajo, for example, reside on their traditional and ancient lands and became ethnic only when the United States annexed their territory. The same is true of Mexicans, who, long resident in what is today the southwestern United States, found themselves labeled ethnic minorities when the border was moved northward after the 1848 U.S.–Mexican War. "We did not jump the border, the border jumped us!" is a common local saying in response to this sudden shift in ethnic status.

Ethnic minorities typically adapt, in some measure, to the predominant ethnic culture. **Acculturation** often occurs, meaning that the ethnic group adopts enough of the ways of the host society to be able to function economically and socially. Stronger still is **assimilation,** which implies a blending-in with the host culture and may involve the loss of many distinctive ethnic traits. Intermarriage is perhaps the most effective way of encouraging assimilation. Many students of North American culture long assumed that all ethnic groups would eventually be assimilated into the **melting pot,** but relatively few have been; instead, they use acculturation as their way of survival. The past three decades, in fact, have witnessed a resurgence of ethnic identity across the globe. Perhaps, then, **transculturation,** which is the notion that people adopt elements of other cultures as well as contributing elements of their own culture, thereby transforming both cultures, is a better way to view the interaction among cultures. This is why demographers and others scholars studying contemporary immigration in North America have largely abandoned the analogy of the melting pot for the metaphor of a **salad bowl,** where racial and ethnic differences mix but never entirely dissolve.

Formal ethnic culture regions exist in most countries. Ethnic and racialized minority populations tend to cluster in particular spatial patterns. There are four types of ethnic culture regions: rural ethnic homelands and islands, and urban ethnic neighborhoods and ghettos. These tend to shift over time in response to changes in policies, attitudes, and the changing cultural boundaries of racial and ethnic difference.

Ethnic Homelands and Islands

The difference between ethnic homelands and islands is their size, in terms of both area and population. Rural **ethnic homelands** include sizable populations, cover large areas, and often have overlapping borders. Because of their size, the age

SUBJECT TO DEBATE Racism: An Embarrassment of the Past, or Here to Stay?

Most modern societies are quick to claim that they have moved beyond the scourge of racism. Brazil, for example, bills itself as a "racial democracy." Swedes see theirs as a nation that is particularly tolerant in matters of race. And in the "land of opportunity"—the United States—privileges are supposedly extended to all, regardless of race, creed, color, and so forth.

Yet we know through our day-to-day experiences of studying, working, and socializing in specific places that although "post-racism" may be the ideal of modern societies, it is not always the fact in real life. Racist incidents on the streets of North American cities appear in the news headlines with startling frequency. And the subtle, indirect, or unintentional racism that people of color face as they go about their daily lives—so-called racial **microaggressions**—shape how people experience and move through places.

Indeed, place is never simply a passive background against which the drama of race plays out. Take, for instance, violent deaths of unarmed black men in the United States: Trayvon Martin in Sanford, Florida, in 2012; Michael Brown in Ferguson, Missouri, in 2014; Freddie Gray in Baltimore, Maryland, in 2015; and Terrence Crutcher in Tulsa, Oklahoma, are just four of the more familiar names. In 2016 alone, *The Guardian,* a news organization that maintains a database called "The Counted," estimated that 258 black men and women in the United States were killed in encounters with the police.

A study conducted by Hehman et al. reveals that some regions of the United States are much more likely than others to experience a disproportionate number of black individuals being shot and killed by the police. Rather than stemming from racist attitudes on the part of individual police, scholars found that communities where there was more prejudice against blacks were the same communities where blacks were being killed by police at a higher rate than would be warranted by their share of the population. In other words, the bias held by a place rather than by an individual shaped life-and-death actions.

Race-related tensions are not only about ancestry or skin pigmentation, but also about religious differences, immigration, and social opportunities. The backlash against immigrant Latinos and Muslims, escalating in the wake of the election of Donald Trump as president of the United States, illustrates this claim. Indeed, as anxieties arise over ever-higher levels of contact in a globalizing world, racialized intolerance is all too often the response. A. Sivanandan has argued that "poverty is the new black" in a globalizing world that has become increasingly hostile toward involuntary migrants regardless of their skin color.

Continuing the Debate

Possible responses to racism lie along a continuum, from violent protest to waiting for a time to erase racism. Scale is an important factor in how racism is manifested and therefore addressed. Consider these questions:

- In the study mentioned here, racist community attitudes as a whole were found to be the most important factor in the disproportionate number of deaths of black individuals at the hands of the police. How can such "place bias" be addressed and reduced?

- Hispanics also have a problematic relationship with local police forces and have suffered disproportionate levels of police discrimination and violence. Yet there is no broad national movement among Latinos to address this issue. Why do you think this is the case?

- Will racism finally go away by itself as societies modernize and move beyond intolerance?

---

of their inhabitants, and their geographical segregation, they tend to reinforce ethnic identities. The residents of homelands typically seek or enjoy some measure of political autonomy or self-rule. Homeland populations usually exhibit a strong sense of attachment to the region. Most homelands belong to indigenous ethnic groups and include special, venerated places that serve to symbolize and celebrate the ethnic character or heritage of the region. In its fully developed form, the homeland represents that most powerful of geographical entities, one combining both formal and functional culture regions. By contrast, **ethnic islands** are small dots in the countryside, usually occupying an area smaller than a county and serving as home to several hundred to several thousand people of the same ethnic group

(at most). Because of their small size and isolation, they do not exert as powerful an influence as homelands do.

North America includes a number of existing ethnic homelands (**Figure 5.5** on page 156), including Acadiana, the Louisiana French homeland now increasingly identified with the Cajun people and also recognized as a vernacular region; the Hispano or Spanish American homeland of highland New Mexico and Colorado; the Tejano homeland of south Texas; the Navajo reservation homeland in Arizona, Utah, and New Mexico; and the French Canadian homeland centered on the valley of the lower St. Lawrence River in Québec. Some geographers include Deseret, a Mormon homeland in the Great Basin of the intermontane West.

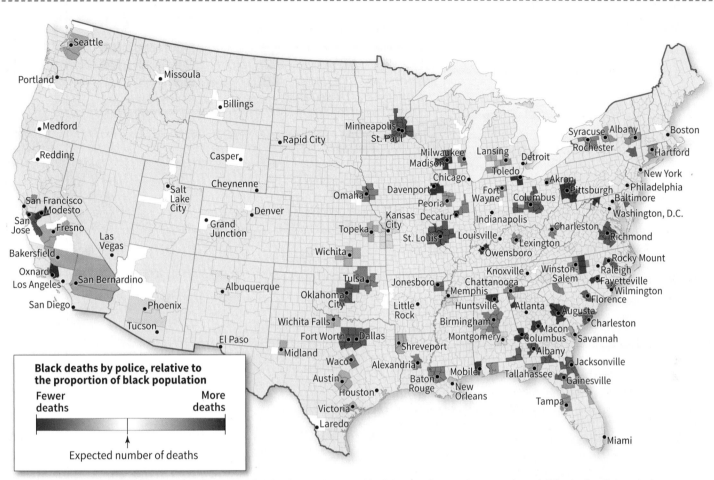

Disproportionate numbers of deaths of black individuals are displayed in this map. Researchers used models including 14 sociodemographic variables to explain these variations, but it was only the psychological factor—regional implicit biases toward blacks or the implicit stereotypical association between blacks and weapons—that predicted blacks being killed by police. Where these biases were stronger, more African Americans were killed by police. (Data from Eric Heyman et al., 2018.)

Some ethnic homelands have experienced decline and decay while others are thriving. For example, the Pennsylvania Dutch homeland has been weakened to the point of extinction by assimilation. Another homeland, the southern Black Belt, was once moribund, having been diminished by the collapse of the plantation-sharecrop system and the resulting African American relocation to northern and western urban areas. But it has experienced a revitalization through the recent return migration of African Americans to the South (see also page 164). Nonethnic immigration has diluted the Hispano homeland. At present, the most vigorous ethnic homelands are those of the French Canadians and Mexican Americans in south Texas.

If residents of ethnic homelands assimilate and are absorbed into the host culture, then at the very least, a geographical residue, or **ethnic substrate,** remains. The resulting culture region, though no longer ethnic, nevertheless retains some distinctiveness, whether in local cuisine, dialect, or traditions. Thus, it differs from surrounding regions in a variety of ways. In seeking to explain its distinctiveness, geographers often discover an ancient, vanished ethnicity. For example, the Italian province of Tuscany owes both its name and some of its uniqueness to the Etruscan people, who ceased to be an ethnic group 2000 years ago, when they were absorbed into the Latin-speaking Roman Empire. More recently, the massive German presence in the American heartland, now largely nonethnic, has helped shape

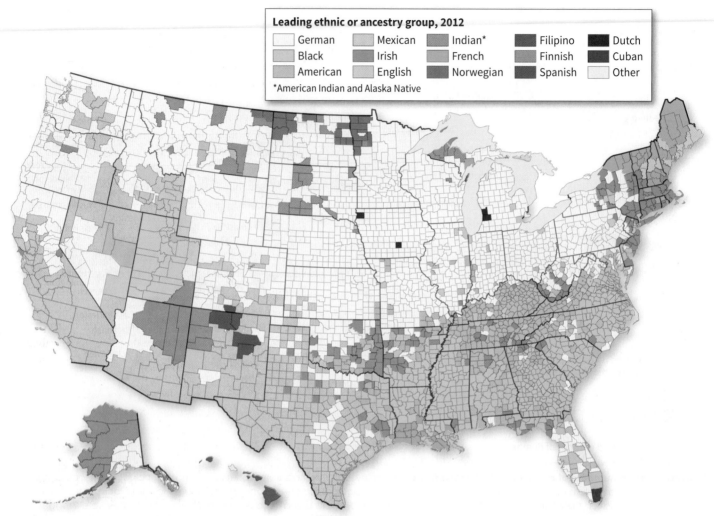

FIGURE 5.5 Ethnic and national-origin groups in the United States. Several ethnic homelands appear on this map, as do many ethnic islands. (Data from U.S. Census Bureau, 2008–2012 American Community Survey 5-Year Estimates [B02001, Race; B02006, Asian Alone by Selected Groups; B03001, Hispanic or Latino Origin by Specific Origin; B04003, Total Ancestry Reported]. Kilpinen, Jon T., 2014. "Leading Ethnic or Ancestry Group, 2012." United States Map Gallery. Map 01.01. http://scholar.valpo.edu/usmaps/1.)

the cultural character of the Midwest, which can be said to have a German ethnic substrate.

Ethnic Neighborhoods and Racialized Ghettos

Formal ethnic culture regions also occur in cities throughout the world, as minority populations initially create, or are consigned to, separate ethnic residential quarters. Two types of urban ethnic culture regions exist. An **ethnic neighborhood** is a voluntary community where people of common ethnicity reside by choice. An ethnic neighborhood has many benefits: common use of a language other than that of the majority culture, nearby kin, stores and services specially tailored to a certain group's tastes, the presence of employment that relies on an ethnically based

division of labor, and institutions important to the group— such as churches and lodges—that remain viable only when a number of people live close enough to participate frequently in their activities. Miami's orthodox Jewish population clusters in Miami Beach–area neighborhoods (**Figure 5.6**), in part because the proximity of synagogues and kosher food establishments makes religious observance far easier than it would be in a neighborhood that did not have a sizable orthodox Jewish population.

The second type of urban ethnic region is a **ghetto**. Historically, the term dates from thirteenth-century medieval Europe, when Jews lived in segregated, walled communities called ghettos (**Figure 5.7**). Ethnic residential quarters have, in fact, long been a part of urban cultural geography. In ancient times, conquerors often forced vanquished native populations to live in ghettos. Religious minorities usually received similar

FIGURE 5.6 Miami Beach, Florida, is home to a large community of Orthodox Jews, who constitute a visible ethnic presence in this neighborhood. (Courtesy of Ari Dorfsman.)

treatment. Islamic cities, for example, had Christian districts. If one abides by the origin of the term, ghettos need be neither composed of racialized minorities nor impoverished. Typically, however, the term *ghetto* is commonly used in the United States today to signal an impoverished, urban, African American neighborhood. A related term, **barrio**, refers to an impoverished, urban, Hispanic neighborhood. Ghettos and barrios are as much functional culture regions as formal ones.

Coinciding with the urbanization and industrialization of North America, ethnic neighborhoods became typical in the northern United States and in Canada around 1840. Instead of dispersing throughout the residential areas of a city, immigrant groups clustered together in a spatial expression of diaspora cultures, discussed in Chapter 2. To some degree, ethnic groups that migrated to cities came from different parts of Europe than those who settled in rural areas. Whereas Germany and Scandinavia supplied most of the rural settlers, the cities attracted those from Ireland and eastern and southern Europe. Catholic Irish, Italians, and Poles, along with Jews from eastern Europe, became the main urban ethnic groups, although lesser numbers of virtually every nationality in Europe also came to the cities of North America. These groups were later joined by French Canadians, southern blacks, Puerto Ricans, Appalachian whites, Native Americans, Asians, and other non-European groups.

Regardless of their particular history, the neighborhoods created by ethnic migrants tend to be transitory. As a rule, urban ethnic groups remain in neighborhoods while "learning the ropes," as sociologists Alejandro Portes and Rubén Rumbaut put it. As a result, their central-city ethnic neighborhoods experience a life cycle in which one group is replaced by another, later-arriving one. We can see this process in action in the

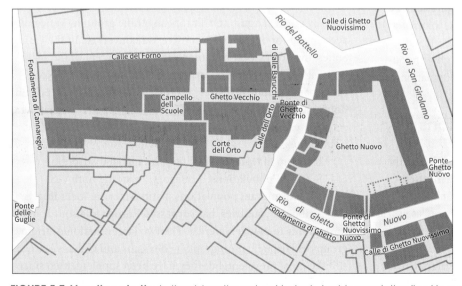

FIGURE 5.7 Venetian ghetto. In the sixteenth century, Venice's Jewish population lived in a segregated, walled neighborhood called a ghetto. On the left is a map of this early ghetto. Though most of Venice's Jews do not reside in the ghetto today (*right*), many attend religious services there, and the ghetto continues to lie at the heart of Venetian Jewish life. (Left: Adapted from the Jewish Museum of Venice; Right: Lusoltaly/Alamy)

FIGURE 5.8 Two Chinese ethnic neighborhoods. The image on the left depicts Los Angeles's traditional urban Chinatown, which is a popular tourist destination as well. The image on the right illustrates the suburban location and flavor of new ethnoburbs. This image is from the San Gabriel Valley, outside Los Angeles. (Left: Alex Millauer/Shutterstock; Right: MONICA ALMEIDA/The New York Times/Redux)

succession of groups that resided in certain neighborhoods and then moved on to more desirable areas. The list of groups that passed through one Chicago neighborhood from the nineteenth century to the present provides an almost complete history of American migratory patterns. First came the Germans and Irish, who were succeeded by the Greeks, Poles, French Canadians, Czechs, and Russian Jews, who were soon replaced by the Italians. The Italians, in turn, were replaced by Chicanos and a small group of Puerto Ricans. As this succession occurred, the established groups had often attained enough economic and cultural capital to move to new areas of the city. In many cities, established ethnic groups moved to the suburbs. Even when ethnic groups relocated from inner-city neighborhoods to the suburbs, ethnic residential clustering survived (**Figure 5.8**). For example, the San Gabriel Valley, about 20 miles (32 kilometers) from Chinatown in downtown Los Angeles, has developed as a major Chinese suburb. These suburban ethnic neighborhoods, called **ethnoburbs,** can house relatively affluent populations.

Recent Shifts in Ethnic Populations

In the United States, immigration laws have changed during the past half century, shifting in 1965 from a quota system based on national origins to one that allowed a certain number of immigrants from the Eastern and Western hemispheres, as well as giving preference to certain categories of migrants, such as family members of those already residing in the United States. These changes, and the rising levels of undocumented immigration, have led to a growing ethnic variety in North American cities.

Asia and Latin America, rather than Europe, are now the principal source of immigrants to North America

(see Figure 5.3). In 2013, Mexico alone supplied 28 percent of immigrants to the United States; China, the Philippines, India, and Vietnam were next, with a combined total of 18.4 percent. In recent years, the flow of immigrants from China and India has overtaken the numbers sent by Mexico, which are, relatively speaking, on the decline. Asian and Pacific Islanders are now the fastest-growing group in the United States, projected to be the largest immigrant group by 2055. People of Japanese ancestry form the largest national-origin group in Hawaii, and many West Coast cities, from Vancouver, Canada, to San Diego, have very sizable Asian populations. Vancouver, already 11 percent Asian in 1981, has since absorbed many more Asian immigrants, particularly from Hong Kong, which again became part of China in 1997. Canadian statistical data show that, in 2012, Chinese were by far the largest non-European ethnic group in Vancouver, accounting for one out of every five Vancouverites. In the United States, the West Coast is home to about 40 percent of the Asian population, mostly in California, whereas on the East Coast, the urban corridor that stretches from New York City to Boston houses another concentration (**Figure 5.9a**). Spatially speaking, Asians are less segregated than African American or Hispanic populations.

The map in **Figure 5.9b** shows the spatial distribution of U.S. Latino populations, a category including both U.S.-born and immigrant populations. Latin America, including the Caribbean countries, has surpassed Europe as a source of immigrants to North America. The two largest national-origin groups of Hispanic immigrants—Mexicans and Cubans—are still the most prevalent, but they have been supplemented since the 1980s by waves of Central American and, increasingly, South American arrivals. East Coast cities have absorbed large numbers of immigrants from the West Indies. The two largest

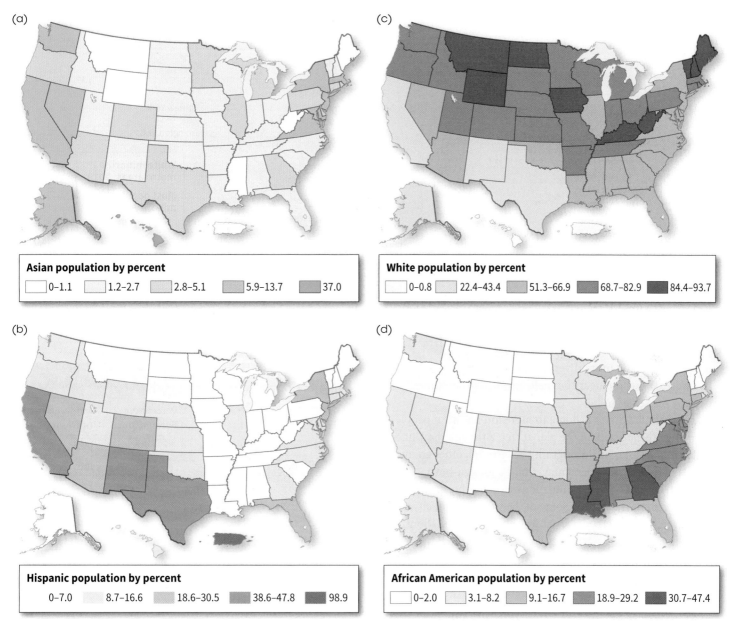

FIGURE 5.9 Ethnic and racial populations, by state. These four maps show the distribution of the four major ethnic and racial groups in the United States. Map 10a shows people indicating "Asian" alone as a percentage of the total population by state; Map 10b displays Hispanic or Latino population as a percentage of the total population by state. "Hispanic" is an ethnic category that can encompass several racial designations. Map 10c depicts those indicating "white" alone as a percentage of the total population by state. Map 10d shows people indicating "black or African American" alone as a percentage of the total population by state. (Data from U.S. Bureau of the Census, 2017.)

national-origin groups coming to New York City as early as the 1970s were from the Dominican Republic and Jamaica, displacing Italy as the leading source of immigrants. For the United States as a whole, in 2002, Latinos surpassed African Americans as the largest ethnic group after non-Hispanic whites. In some cities, Hispanics constitute majority populations (**Figure 5.10**). The Latino influx into the United States has influenced food preferences, music, fashion, and language.

As demographer William Frey has noted, ethnic populations continue to concentrate in established ports

of entry, most of them in coastal locations. There is a large swath of the United States that has remained largely non-Hispanic white (**Figure 5.9c**). **Figure 5.9d** shows that black Americans, too, remain spatially concentrated (see also Figure 2.7 in Chapter 2 for a map of Native American and Alaska Native population distribution). The 2010 census reported that 79 percent of whites live in white-majority neighborhoods, while 44 percent of African Americans live in black-majority neighborhoods and 45 percent of Hispanics reside in Hispanic-majority neighborhoods. As a nation, we

FIGURE 5.10 U.S. metro areas with minority white populations. This map shows major U.S. cities that have so-called "majority-minority" populations. In other words, in these cities, Asian, Latino, or black populations outnumber white populations. ((Source: Adapted from William F. Frey's analysis of 2010 census data.)

may be becoming more diverse—a more varied salad bowl—but we are nowhere near the melting pot we sometimes portray ourselves to be.

The United States is home to more foreign-born people than any other nation on the planet. Yet, although the United States may boast of its immigrant background, other countries are also home to high percentages of the foreign-born: 20.6 percent of Canada's population is foreign-born, as is 27.7 percent of Australia's. Some of the world's small city-states have even higher percentages of foreign-born populations, with Singapore at 43 percent and the United Arab Emirates at 71 percent foreign-born. Four U.S. urban areas—Miami, Florida; Queens, New York; and Santa Ana and Daly City, both in California—ranked in the top 10 list of the world's places that have significant proportions of foreign-born residents (**Table 5.1**). Two of the top 10 places are located in the Middle East, two in Europe, one in Canada, and one in Asia. In addition, national-origin populations prevalent in the United States are even more prevalent in other countries. Fifty million ethnic Chinese reside outside China and Taiwan. Most of these overseas Chinese live not in North America but in Southeast Asian countries. Thailand counts more than 9 million, Indonesia has nearly that number, and Malaysia more than 7 million. Parts of East Africa have long been home to relatively affluent South Asian Indian populations, whereas Lebanese populations in West Africa enjoy a similarly privileged position. Even European countries such as Germany, the United Kingdom, Italy, and Spain—long known as sources rather than destinations of migrants—today are home to millions of Africans, Turks, and Asians. Immigration-based ethnicity is far from being a phenomenon limited to North America.

TABLE 5.1 Top 10 Places by Percentage of Foreign-Born

Place	Percentage Foreign-Born	Largest Source of Immigrants
Dubai, United Arab Emirates	82	India
Brussels, Belgium	68	Italy
Luxembourg City, Luxembourg	66	France
Santa Ana, United States	53	Mexico
Daly City, United States	52	Philippines
Toronto, Canada	52	China
Miami, United States	51	Cuba
Queens, United States	48	China
Muscat, Oman	45	India
Singapore	43	Malaysia

Source: Adapted from http://en.wikipedia.org/wiki/Foreign_born.

Mobility

5.2 Explain the ways that migration and mobility shape, and are shaped by, race and ethnicity.

As we have seen, it is often through migration that a group formerly in the majority becomes, in a new place, different from the mainstream and thus labeled as ethnic or racialized. The many motives for migration can have different results in terms of where a group chooses to go, which parts of its original culture relocate and which do not, and who may become its neighbors in its new home.

Migration and Ethnicity

National Geographic's Genographic Project uses DNA samples from volunteers worldwide to substantiate the claim that all humans are descended from a group of Africans who began to migrate out of Africa about 60,000 years ago. The contemporary global pattern of ethnic populations can be traced to their specific journeys. Thus, much of the ethnic pattern in many parts of the world is the result of migration. The migration process itself often creates ethnicity, as people leave places where they belonged to a nonethnic majority and become a minority in a new home. Voluntary migrations have produced much of the ethnic diversity in the United States and Canada, and the involuntary migration of political and economic refugees has always been an important factor in ethnicity worldwide and is becoming ever more so in North America.

Forced Migration Forced migration contributes to ethnic diffusion and the formation of many ethnic culture regions. African slavery constituted the most demographically significant involuntary migration in human history (see the World Heritage Site on page 162) and has strongly shaped the geodemographic patterning of the Americas. **Refugees** from Cambodia and Vietnam created ethnic groups in North America, as did Guatemalans and Salvadorans fleeing political repression in Central America. Following forced migration and initial resettlement, the relocated group often engages in voluntary migration to concentrate in some new locality. Cuban political refugees, scattered widely throughout the United States in the 1960s, reconvened in south Florida, and Vietnamese refugees continue to gather in Southern California, Louisiana, and Texas. This is an example of stepwise migration, introduced in Chapter 1, whereby a group proceeds to its final destination via a series of intermediate migrations.

Outright warfare can also displace populations, both within the country as **internally displaced persons (IDPs)** and beyond its borders as refugees. Syria presents a recent example, with internally displaced persons far outnumbering Syrian refugees fleeing the country's borders (**Figure 5.11**).

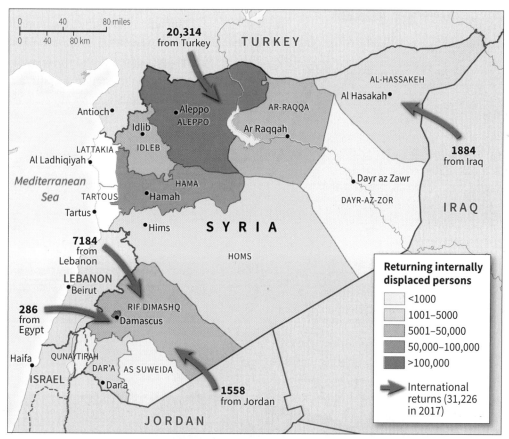

FIGURE 5.11 Syrian refugees and internally displaced persons returning home. This map shows the returns of Syrians to their homes in the first months of 2017. Starting in 2012, 1.5 million Syrians had been displaced by civil war. Some left Syria, settling as refugees in neighboring countries. There, they faced discrimination and poverty as unwanted minority populations. Far more were internally displaced persons (IDPs) who had moved within Syria, often ending up in so-called buffer zones on the borders to protect neighboring states from large flows of refugees. More are expected to return to their homes in Syria. (Source: Adapted from UNHCR, 2017.)

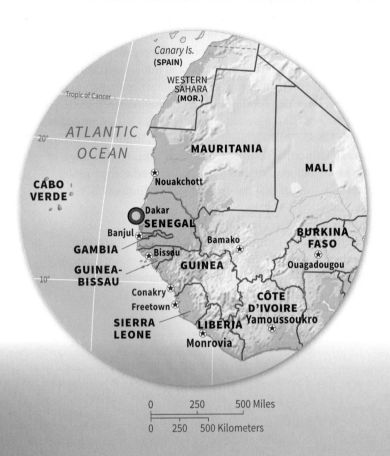

0   250   500 Miles

0   250   500 Kilometers

The small island of Gorée is located just two miles off the coast of Dakar, the capital city of Senegal. The island is positioned midway between the northern and southern tips of the West African coast, which made it centrally located for the slave trade funneling in from the West African interior. As UNESCO notes, "Gorée owes its singular destiny to the extreme centrality of its geographic positioning." The island was used as a stopover point by Europeans on the long sea voyage down the West African coast and as a warehouse for slaves. It was claimed by the Portuguese, the Dutch, the English, and the French.

- Mobility is a central dimension of the slave trade. Eleven million enslaved Africans are estimated to have crossed the ocean over the three centuries encompassing the transatlantic slave trade, from the 1500s to the 1800s, on their way to European colonies in the Americas.

- The island is a site of both painful historical memories and reconciliation. Much as with Holocaust memorials, Gorée Island is preserved and visited in order that humankind not forget the human rights atrocity that was African slavery.

- The island is home to three museums, a seventeenth-century police station, a castle, a small beach, and colonial houses.

(Barry Lewis/Alamy)

THE HOUSE OF SLAVES: Built in the 1770s, this house later belonged to a Senegalese *métis* (mixed-ancestry) woman, Anna Colas Pépin. Scholars believe that though she may have sold small numbers of slaves and kept domestic slaves herself, the actual slave transshipment site was not the house itself, but a location nearby. The house was restored and opened as a museum in 1962. Boubacar Joseph Ndiaye oversaw and curated the museum for 40 years, until his death in 2009.

(SAUL LOEB/AFP/Getty Images)

- **The Door of No Return.** The corridor ending in a gate opening onto a wharf from which slave ships departed was known as "the trip from which no one returned." Ndiaye claimed that a total of over 1 million African slaves made their final exit through the doors of this house over the four centuries of the transatlantic slave trade.

- Some scholars have contested these numbers, claiming that they were, in fact, much smaller, and that Gorée itself was not central to the Senegalese slave trade. This controversy notwithstanding, the island is hugely symbolic and therefore remains meaningful to many.

TOURISM: Gorée Island attracts so-called heritage tourists, people looking for significant traces of their own past. Indeed, UNESCO refers to it as a "memory island."

- Gorée Island is a pilgrimage site for the global African diaspora. Approximately 200,000 tourists visit the House of Slaves annually.

- Most visitors to Gorée Island, whether or not they identify as descendants of African slaves, report being extremely moved by their visit.

- The island's only real source of income is tourism, which supports the fewer than 1500 local residents.

- Because of Gorée's significance, presidents, prime ministers, and even the pope himself have visited it. The Door of No Return is a popular for photographs.

- Though debates persist about Gorée Island's historical significance in the transatlantic slave trade, it has become a symbolically important site of remembrance, particularly for African diaspora populations.

- The House of Slaves, with its Door of No Return, is a museum visited by over 200,000 tourists annually.

- Gorée Island was declared a World Heritage Site in 1978; before that, the colonial government of Senegal had declared it a National History Site in 1944.

http://whc.unesco.org/en/list/26

(Carlos Mora/Alamy)

Return Migration Return migration represents another type of ethnic mobility and involves the voluntary movement of a group back to its ancestral homeland or native country, after a long-term residency elsewhere. The large-scale return since 1975 of African Americans from the cities of the northern and western United States to the Black Belt ethnic homeland in the South is one of the most notable such movements now under way. Thanks to migration from cities like Los Angeles and Philadelphia, the South is home to 55 percent of the black population (see Chapter 3). This proportion is higher than that at any time in the preceding 50 years. Black Americans are returning to the South, particularly to cities like Atlanta, because of greater economic opportunities there than elsewhere in the country. Cultural factors play a role as well, particularly the historic attachment to place and the perception that racial prejudice has lessened in the region. Return migration has recently revitalized the once-moribund black homeland (see Figure 5.5, page 156).

Everyday Mobility and Race

Migration involves the long-term relocation of individuals, families, or entire communities. By contrast, the notion of mobility focuses on the everyday movement of people as they go about their day. When discussing mobility, it is important to note that the ability to move freely from place to place is not a birthright enjoyed equally by all. Instead, the freedom to come and go as you please is a privilege, one that typically comes with membership in dominant social categories. Men, the wealthy, and those occupying the dominant racial and ethnic strata enjoy freedom of movement, whereas women, minorities, and the poor often have their mobility curtailed, or are prohibited from moving.

The everyday mobilities of African Americans, Latinos, and Muslims are curtailed by their race and ethnicity in ways seldom experienced by white people. The greater likelihood that a black or Latino individual will be subjected to a traffic stop is a phenomenon known as "DWB" ("driving while black" or "driving while brown"), whereas the greater likelihood that a Muslim will be asked to undergo heightened security screening or denied the right to board a plane is known as "flying while Muslim." Black and Latino drivers, and Muslim air travelers, are targets of **racial profiling,** whereby a person is treated as a suspect because of his or her race, ethnicity, nationality, or religion (see **Figure 5.12**). In effect, racial profiling conveys to people of color and Muslims that they do not have the right to access or move through space without being detained and questioned. The experience of movement, rather than being empowering, is one of fear: fear of being restricted, hurt, or even incarcerated or killed. Racial profiling tells these individuals, "You do not belong." Mobility and restrictions around

FIGURE 5.12 Traffic stop. On a typical day in the United States, over 50,000 drivers are pulled over by police officers because of traffic violations. The Stanford Open Policing Project has collected data from 100 million traffic stop records from 31 states since 2015, finding significant racial bias. When pulled over for speeding, for instance, black drivers are 20 percent more likely to get a ticket (rather than a warning) than white drivers, while Hispanic drivers are more than 30 percent more likely to get a ticket. Black and Hispanic drivers are twice as likely to be searched as white drivers. (Blend Images - Hill Street Studios/Brand X Pictures/Getty Images)

movement are too often an unwelcome part of being a racialized minority (see The Video Connection).

Globalization

5.3 Interpret how globalization has deepened, transformed, or erased ethnicity, race, and racism.

The perception of differences among human groups, whether based in visible physical traits, cultural practices, language differences, or religion, is probably as old as human civilization itself. For at least a century, however, the demise of ethnic groups has been predicted. For example, we have seen that the idea of America as a melting pot has been used to describe the process by which the mixing of diverse peoples would eventually absorb everyone into one American mainstream culture. While the potent forces of globalization certainly play a role in this blending process, the fact that ethnic differences and racial strife still persist—and in some instances have worsened—points to their deeply rooted nature.

A Long View of Race and Ethnicity

In keeping with the notion that globalization is a process that has unfolded over several centuries, as opposed to only

The Video Connection

An Education in Equality
Watch at 🅜 Sapling Plus

This video profiles a young African American man, Idris Brewster, and his family over the course of his 13 years at the Dalton Academy, an elite private college-preparatory school in Manhattan. His mother is an immigrant of mixed-race Caribbean ancestry, while his father describes his past as one of economic deprivation. Both utilized education as a ticket to an upper-class lifestyle, and they want their son to use education to excel as well. Idris moves between elite and not-so-elite spaces in his daily life, and in the process he is forging his identity as a black man with a bright future in a world where race still matters.

Thinking Geographically

1. This video explores the role of education and its power to elevate an individual's class status. What sorts of expectations, hopes, and pressures do Idris's parents place on his shoulders? How does race matter in this regard?

2. Geographers and others who study race note that microaggressions are prevalent in the lives of people such as Idris Brewster. Can you find examples of racial microaggressions in the video?

3. Idris Brewster appears to lead two lives: one as a student at an elite college-prep school, and one as a young black man in Brooklyn. How do those two lives overlap, and what disconnects between them can you identify from the video?

over the past 50 years or so, it is useful to look at concepts of ethnicity and race in historical and global perspective. Shared language, religion, or ethnicity provide group members with the perception of a common history and shared destiny—that deep feeling of "we-ness"—that is the basis of the modern nation-state. Israel's identity as a Jewish nation, Turkey's common language, and Japan's distinctive cultural traditions set the stage for these groups to cohere as political entities.

Yet all three of these nation-states harbor long-standing tensions among groups that reside within their national borders but who are not considered (or do not consider themselves) part of the cultural mainstream. They may not practice the official religion of the country, as in the case of Israel's Palestinians, who are Muslims and Christians. They may not speak the official language, as with Turkey's Kurdish-speaking minority. Or they may follow cultural traditions distinct from the mainstream, as is the case with Japan's indigenous Ainu population.

Occupying a minority religious, linguistic, or ethnic position can expose a group to persecution. Violence against racial or ethnic minorities dates back to ancient times. For example, **anti-Semitism**—the persecution of Jews—began with pogroms, widespread rioting intended to persecute and massacre Jews, during the Roman Empire some two millennia ago. Across medieval Europe, violence against Jews was episodic, occurring within Spain in the eleventh century, England in the twelfth century, Germany in the fourteenth century, and Russia in the nineteenth century. Anti-Semitic hate crimes still occur today throughout Europe and the United States. Yet when Israel was founded in 1948 as the only nation-state to have Judaism as its official religion, non-Jewish minorities were, in turn, marginalized. Even minority Jewish populations, such as the Mizrahim (Arab) Jews residing in Israel, "are excluded and marginalized not only from social and political centers, but . . . from the very definition of what it is to be 'Israeli,'" claims Israeli anthropologist Pnina Motzafi-Haller. Thus, it appears that no place or time is free of tensions based on ethnic or racialized difference.

Race and European Colonization

Europe's colonization of vast territories in Africa, Asia, and Latin America was predicated on the drawing of sharp distinctions between the colonizers and the colonized. These distinctions were often depicted in racial terms. Indeed, conquered peoples were viewed as so different from Europeans that their very status as human beings was debated. The famous exchanges in the mid-sixteenth century between Caribbean-based Spanish priests and intellectuals in Spain is a case in point. Observing the abuses of indigenous populations firsthand led to an outcry by these clerics, who argued that the subjugated peoples were not animals and should not be subject to such treatment. "Are these not men? Have they not rational souls? Must you not love them as you love yourselves?" asked Antonio de Montesinos in 1511, in a sermon delivered on the island of Hispaniola (**Figure 5.13**). These same clerics, however, saw no problem with owning black slaves, simply because they did not view them as human beings.

FIGURE 5.13 La Leyenda Negra. Illustrations such as this gruesome depiction of Spanish conquistadores hanging and roasting indigenous people, while throwing their babies against a wall, were used to construct La Leyenda Negra, or "The Black Legend." La Leyenda Negra was circulated in the sixteenth century by the Dutch and the British, Spain's religious and colonial rivals. They wished to depict the Spaniards as cruel barbarians who had no moral business building empires in the New World. (AKG Images)

FIGURE 5.14 Identification card of a Rwandan Tutsi. In Rwanda, to show an ID card with "Tutsi" written on it was tantamount to a death sentence during the genocide. Whether to institute a mandatory national identification card is a hot topic in many nations faced with undocumented immigration today. Because these cards can mark some groups as undesirable, there is usually vigorous opposition to a national identification card. (Antony Njuguna/REUTERS)

European colonialism frequently drew on existing ethnic and racial cleavages in colonized societies. Rwanda provides a tragic example of this practice. Rwanda is composed of two major ethnic groups: the Hutu, who, at 85 percent of the population, constitute the majority, and the minority Tutsi. Although differences in status have long existed between the two groups, it was not until the Belgian takeover of the territory in 1918 that the distinction between Hutu and Tutsi became racialized. Tutsis were given positions of power in the colonial government, educational system, and economic structure of colonial Rwanda under Belgian rule (**Figure 5.14**). Hutus' physical differences from Tutsis were examined under a European racial lens. The shorter and darker Hutus were considered ugly, intellectually inferior, and natural slaves, whereas the taller, lighter Tutsis were extolled for their physical beauty and virtues of cultural refinement and leadership.

In 1959 the oppressed Hutu majority rebelled, leading to the deaths of some 20,000 Tutsis and the displacement of many more. Rwanda became independent from Belgium in 1961. Displaced Tutsi refugees regrouped in neighboring Uganda and in 1990 launched an invasion of Rwanda, demanding an end to racial discrimination and a reinstatement of their citizenship. Rapid deterioration of Hutu–Tutsi relations ensued, and plans to exterminate Tutsis and their Hutu sympathizers were promoted as the only way to rid Rwanda of its problems. In 1994 this **genocide** was conducted, resulting in the massacre of as many as 1 million people in the span of a few months (**Figure 5.15**).

Twentieth-Century Genocides

As these examples illustrate, ethnic and racial distinctions provided central features of European colonization and the rise of nation-states, and tensions based on race and ethnicity predate the modern era. Yet racial and ethnic

FIGURE 5.15 Rwandan genocide. These skulls are the remains of some 500,000 to 1,000,000 Tutsis, and their Hutu sympathizers, who were massacred over a period of four months in 1994 by Hutu militias. (Joe McNally/Getty Images)

TABLE 5.2 World's 10 Deadliest Genocides

Name	Place	Dates	Estimated Deaths	Percentage of Population Killed
Holocaust	Nazi-controlled Europe	1939–1945	4,200,000–11,000,000	78% of Jews in Nazi-controlled Europe
Holodomor (Ukrainian Famine-Genocide)	Ukraine and other Soviet republics	1932–1933	1,800,00–7,500,000	10% of Ukraine's population
Cambodian Genocide	Democratic Kampuchea	1975–1979	1,700,000–3,000,000	21%–33% of Cambodia's population
Belarusian Genocide	Soviet Belarus	1941–1944	1,360,000–1,970,000	25% of Belarusian population
Soviet Ethnic Cleansings	Soviet Union	1920–1951	1,000,000–1,500,000	24%–50% of total Chechen population, 30% of total Crimean Tatar population
Armenian Genocide	Ottoman Empire	1915–1922	800,000–1,500,000	75% of Armenians in Turkey
Rwandan Genocide	Rwanda	1994	500,000–1,000,000	70% of Tutsis in Rwanda
Genocide of Cathars (Albigensian Crusade)	Southern France	1209–1229	200,000–1,000,000	Cathars and Catharism extinguished
Greek Genocide	Ottoman Empire	1914–1922	289,000–750,000	13%–30% of Greeks in Anatolia
Assyrian Genocide	Ottoman Empire	1915–1923	275,000–750,000	75% of Assyrians in Turkey

conflict is far from erased from the face of today's global world (**Table 5.2**) and may, in fact, become aggravated as the bonds of the nation-state are loosened (see Subject to Debate, page 154).

Indigenous and Minority Identities in the Face of Globalization

Differences do not always cause conflict, tension, and oppression. Indeed, pride in one's ethnic or racial distinctiveness is common. Yet as revolutions in communications and transportation bring diverse peoples into heightened contact with one another the world over, distinctive markers of ethnicity come under pressure. Exposure to seductive Western cultural practices—in music, cuisine, fashion, lifestyles, and values—is thought to encourage people to shed their distinctive ways in favor of adopting a homogeneous modern, Westernized culture.

Late in the twentieth century, however, an ethnic resurgence, especially among indigenous groups, became evident in many countries around the world. In a very real way, many ethnic groups and their geographical territories have become bulwarks of resistance to globalization (see Chapter 2). Yet it is too simplistic to romanticize indigenous groups as simply the keepers of ethnic distinctiveness in a globalizing world (**Figure 5.16**). Members of indigenous groups must interpret the challenges as well as the opportunities offered by globalization and can be expected to modify their identities in complex ways.

Ethnic minority groups face similar pressures with respect to how their identities are asserted in the majority society. This

FIGURE 5.16 Native Huichol girl. Some indigenous peoples view indigenous clothing as a mark of ethnic pride and wear it daily regardless of their place of residence or level of interaction with the majority society. This Huichol girl wears a baseball cap with traditional dress, itself a blend of European and indigenous materials and styles. The Huicholes are residents of western Mexico. (Susana Valadez/Huichol Center for Cultural Survival)

is particularly true of young people. Regardless of ethnicity, this age group is typically concerned with fitting in, which often involves hiding those aspects that mark them as "different" from the mainstream in some way. These concerns are heightened for ethnic minorities and immigrants, whose differences are often markedly apparent in language, dress, religion, and cuisine.

Take the example of Brazil's Japanese-descendent population. Numbering around 1.5 million individuals, Brazil's Japanese-ancestry population is the largest outside of Japan. As we noted previously, Brazil was quite late in abolishing slavery, in 1888—in fact, it was the last country in the Western Hemisphere to do so. Yet the need for cheap agricultural labor remained, and in the first decade of the twentieth century, immigrant Japanese began to fill this need. Fleeing poverty in their home country, Japanese immigrants worked Brazil's coffee plantations. This migration reached a peak in 1931–1935, when over 70,000 Japanese immigrated to Brazil as a result of the strife in their homeland after World War I. Around this time, the Brazilian government attempted to force its immigrant populations from Germany, Italy, and Japan, as well as Jews, to "whiten" by assimilating into the mainstream through intermarriage and acculturation. They found the Japanese particularly resistant to such efforts, and by the 1970s these Japanese immigrants and their descendants, known as *Nikkeijin*, formed a distinctive and economically well-off minority group (**Figure 5.17**).

Takeyuki Tsuda's research on Brazil's *Nikkejin* youth reveals the pressures faced by ethnic minorities. For example, these youths are often stereotyped in racial ways, as being smarter, more polite, and neater than non-Japanese Brazilians. One father remarked, "When my son gets good grades in school, they say, 'Of course, it's because he's Japanese.' . . . When he keeps his desk clean in the classroom, they say, 'Of course, he is Japanese.'" To complicate matters even further, many of these Brazilian-born youths and their parents have returned to Japan in order to take advantage of the economic opportunities now available there. When they arrive, these young people face considerable pressure to speak and act Japanese. According to one young man, "The Japanese make more demands on us than on whites or other foreigners in Japan because of our Japanese faces." Tsuda found that many of these young people react by wearing Brazilian clothing, speaking Portuguese loudly in public, and in general acting "more Brazilian" than they ever did in Brazil, as ways of asserting their ethnic identity.

Thus, globalization is far from a straightforward, one-way process when it comes to race, ethnicity, or any other aspect of human geography for that matter. Rather, globalization and its associated movements of people and ideas lead to highly complex geographies of race and ethnicity.

Nature–Culture

5.4   Analyze the diverse ways in which ethnic and racial groups interact with the natural environment, and discuss how these groups may be protected or made more vulnerable.

Ethnicity is very closely linked to the nature–culture theme. The interplay between people and the physical environment is often evident in the pattern of ethnic culture regions, in ethnic migration, and in the persistence or even survival of ethnic and racialized groups and their cultures. Ethnicity and race are sometimes also linked to the environment in harmful ways, particularly when the places they inhabit are polluted.

Cultural Preadaptation

For those ethnic groups created by migration or relocation diffusion, the concept of cultural preadaptation provides an interesting approach. **Cultural preadaptation** involves a set of adaptive traits possessed by a group in advance of migration that gives them the ability to survive and a competitive advantage in colonizing the new environment. Most often, preadaptation occurs in groups migrating to a place that is environmentally similar to the one they left behind. The adaptive strategy they had pursued before migration works reasonably well in the new home.

FIGURE 5.17 Japanese Brazilians. These girls, who are of Japanese ancestry, attend a traditional Japanese day of the dead ceremony in Registro, Brazil. Registro is home to one of the oldest Japanese Brazilian communities. (YASUYOSHI CHIBA/AFP/Getty Images)

The preadaptation may be accidental, but in some cases the immigrant ethnic group deliberately chooses a destination area that physically resembles their former home. In Africa, the migration of Bantu-speaking people from Central and West Africa was probably initially driven by climate change and expansion of the Sahara. The Bantu spread south and east in search of forested lands similar to those they had previously inhabited. But their movement was finally halted because their agricultural techniques and cattle were not adapted to the drier Mediterranean climate of southern Africa.

Such ethnic niche-filling has continued to the present day. Cubans have clustered in southernmost Florida, the only part of the U.S. mainland to have a tropical savanna climate identical to that in Cuba, and many Vietnamese have settled as fishers on the Gulf of Mexico, especially in Texas and Louisiana, where they could continue their traditional livelihood. Yet historical and political patterns, as well as the factors driving chain migration, are also at work in these contemporary patterns of ethnic clustering. They act to temper the influence of the physical environment on ethnic residential selection, making it only one of many considerations.

This deliberate site selection by ethnic immigrants represents a rather accurate environmental perception of the new land. As a rule, however, immigrants tend to perceive the ecosystem of their new home as more like that of their abandoned native land than is actually the case. Their perceptions of the new country emphasize the similarities and minimize the differences. Perhaps the search for similarity results from homesickness or an unwillingness to admit that migration has brought them to a largely alien land. Perhaps growing to adulthood in a particular kind of physical environment inhibits one's ability to perceive a different ecosystem accurately. Whatever the reason, the distorted perception has occasionally caused problems for ethnic farming groups (for examples, see Chapter 8). A period of trial and error was often necessary to come to terms with the New World environment. Sometimes crops that thrived in the old homeland proved poorly suited to the new setting. In such cases, **cultural maladaptation** is said to occur.

Cultural Simplification and Isolation

When groups migrate and become ethnic in a new land, they have, in theory at least, the potential to introduce the totality of their culture by relocation diffusion. Conceivably, they could reestablish every facet of their traditional way of life in the area where they settle. However, ethnic immigrants never successfully introduce the totality of their culture. Rather, profound **cultural simplification** occurs. This happens, in part, because of chain migration: only localized fragments

of a national culture diffuse overseas, borne by groups from particular places migrating in particular eras. In other words, some simplification occurs at the point of departure.

Moreover, far more cultural traits are implanted in the new home than actually survive. Only selected traits are successfully introduced, and others undergo considerable modification before becoming established in the new homeland. In other words, absorbing barriers prevent the diffusion of many traits, and permeable barriers cause changes in many other traits, greatly simplifying the migrant cultures.

In addition, cultural choices that did not exist in the old home become available to immigrant ethnic groups. They can borrow novel ways from those they encounter in the new land, invent new techniques better suited to the adopted place, or modify existing approaches as they see fit. Most immigrant ethnic groups resort to all these devices to varying degrees.

The displacement of a group and its relocation to a new homeland can have widely differing results. The degree of isolation an ethnic group experiences in a new home helps determine whether traditional traits will be retained, modified, abandoned, or even adopted by the host culture. If the new settlement area is remote and contacts with outsiders are few, diffusion of traits from the sending area is more likely. Because contacts with groups in the receiving area are rare, little borrowing of traits can occur. Isolated ethnic groups often preserve in archaic form cultural elements that disappear from their ancestral country; that is, they may, in some respects, change less than their kinfolk back in the mother country.

Language and dialects offer some good examples of this preservation of the archaic. Germans living in ethnic islands in the Balkan region of southeastern Europe preserve archaic South German dialects better than do Germans living in Germany itself, and some medieval elements survive in the Spanish spoken in the Hispano homeland of New Mexico. The highland location of Taiwanese aboriginal peoples helped maintain their archaic Formosan language, belonging to the Austronesian family, despite attempts by Chinese conquerors to acculturate them to the Han Chinese culture and language.

Habitat and the Preservation of Difference

Certain habitats may act to shelter and protect ethnically or racially distinct populations. The high altitudes and rugged terrain of many mountainous regions make these areas relatively inaccessible and thus can provide refuge for minority populations while providing a barrier to outside influences. As we saw in the previous chapter on language, mountain dwellers sometimes speak archaic dialects or preserve their unique tongues thanks to the refuge provided by their habitat.

In more general terms, the ways of life—including language—associated with ethnic distinctiveness are often preserved by a mountainous location.

The ethnic patchwork in the Caucasus region, for example, persists thanks in no small measure to the mountainous terrain found there. As **Figure 5.18** shows, distinct groups occupy the rugged landscape of valleys, plateaus, peaks, foothills, steppes, and plains in this region. Religions and languages overlap in complex ways with ethnic identities and are made even more confounding by the geopolitical boundaries in the region. Thus, although this is one of the most ethnically diverse places on Earth, the Caucasus region has seen more than its fair share of conflict as well, much of it ethnically motivated.

Islands, too, can provide a measure of isolation and protection for ethnically or racially distinct groups. The Gullah, or Geechee, people are descendants of African slaves brought to the United States in the eighteenth and nineteenth centuries to work on plantations. Many of their nearly 10,000 descendants today inhabit the coastal islands of South Carolina, Georgia, and north Florida. Their island location has allowed much of the original African cultural roots to be preserved, so much so that the Gullah are sometimes called the most African American community in the United States. Today, as the tourist industry sets it sights on developing these coastal islands, and as Gullah youth migrate elsewhere in search of opportunity, the survival of this unique culture is in question.

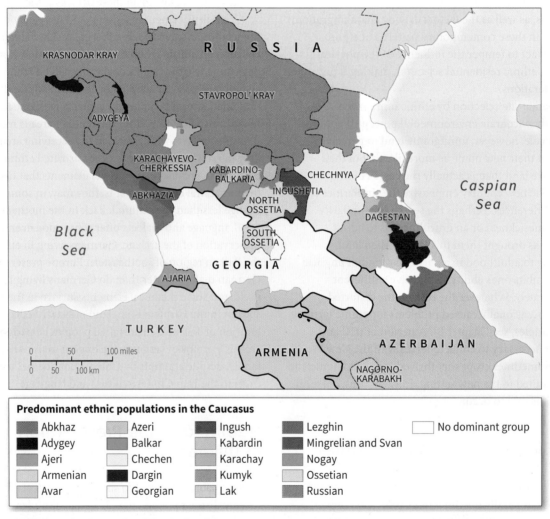

Predominant ethnic populations in the Caucasus

- Abkhaz
- Adygey
- Ajeri
- Armenian
- Avar
- Azeri
- Balkar
- Chechen
- Dargin
- Georgian
- Ingush
- Kabardin
- Karachay
- Kumyk
- Lak
- Lezghin
- Mingrelian and Svan
- Nogay
- Ossetian
- Russian
- No dominant group

FIGURE 5.18 Ethnic pluralities in the Caucasus. The Caucasus Mountains of Southwest Asia are home to one of the world's most ethnically diverse populations. Because ethnic territories often overlap, this map depicts pluralities. The populations shown make up at least 40 percent of that place's ethnic population and in most cases also constitute a majority population, although some of the more heterogeneous areas have no dominant ethnic population. The racialized term *Caucasian* is derived from this area's name, although, in fact, it has nothing to do with the people actually living in the region. (Source: Adapted from O'Loughlin et al., 2007.)

Environmental Racism

Not only people but also the places where they dwell can be labeled by society as minority. And it is a fact that, although belonging to a racial or ethnic minority group does not necessarily lead to poverty, the two often go hand in hand. In spatial terms, being poor frequently equals having the last and worst choice of where to live. In cities, that means impoverished and racialized minorities often reside in run-down, ecologically precarious, or peripheral places that no one else wants to inhabit (**Figure 5.19**). In rural areas, racialized and indigenous minorities work the smallest and least fertile lands, or—more commonly—they work the plots of others, having become dispossessed of their own lands.

Environmental racism refers to the likelihood that a racialized minority population inhabits a polluted area. As sociologist and "father of environmental justice" Robert Bullard asserts, "Whether by conscious design or institutional neglect, communities of color in urban ghettos, in rural 'poverty pockets,' or on economically impoverished Native American reservations face some of the worst environmental devastation in the nation." In the United States, Hispanics, Native Americans, blacks, and poor people of all races are more likely than wealthier whites to reside in places where toxic wastes are dumped, polluting industries are located, or environmental legislation is not enforced. Inner-city populations, which in many U.S. cities are disproportionately made up of minorities and the poor, occupy older buildings contaminated with asbestos and toxic lead-based paints. Farmworkers, who in the United States are disproportionately Hispanic, are exposed to high levels of pesticides, fertilizers, and other agricultural chemicals. Native American reservation lands are targeted as possible sites for disposing of nuclear waste.

The coexistence of nonwhite populations and industrial pollution is difficult to explain away as simply a random coincidence (**Figure 5.20**). Rather, areas such as the city of South Gate, located in Los Angeles County, have seen contaminating industries and activities purposely located in what is today a neighborhood composed of more than 90 percent Latino residents. In its early days, South Gate was inhabited mostly by white non-Hispanics. From its founding in 1923 until the 1960s, some of the largest U.S. polluters located their operations there: Firestone Tire and Rubber Company, A. R. Maas Chemical Company, Bethlehem Steel,

FIGURE 5.19 Arab slum in Delhi, India. This ethnic neighborhood is home to low-income Arabs, who are an ethnic minority population in India. Nearly 50 percent of Delhi's residents live in slums or unauthorized settlements. (Viviane Dalles/REA/Redux)

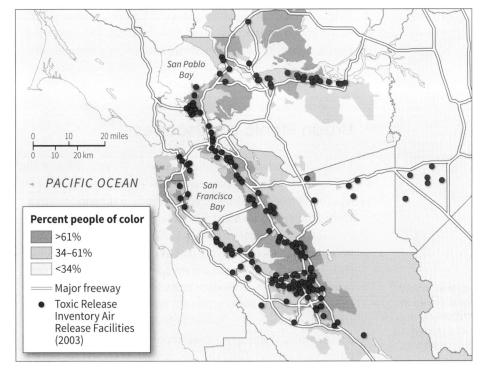

Percent people of color
- >61%
- 34–61%
- <34%

— Major freeway
- Toxic Release Inventory Air Release Facilities (2003)

PACIFIC OCEAN

San Pablo Bay

San Francisco Bay

0 10 20 miles
0 10 20 km

FIGURE 5.20 Industrial air pollution and minority neighborhoods. This figure maps the locations of TRI (Toxic Release Inventory) facilities with active air pollution releases in the San Francisco Bay Area, relative to the proportion of residences of people of color. The map clearly shows the close coincidence in space between activities harmful to human health and neighborhoods where one-third or more of the population is black, Latino, or Asian. In this example, mapping is used to promote environmental justice.

Alcoa Aluminum, and hundreds of smaller operations were allowed to set up shop in a period when few were aware of the health consequences of their activities. These industries attracted labor arriving from Mexico. During this period, the neighborhood was encircled by four major freeways and a flight path was routed overhead. Pollution resulting from transportation-related fossil fuels added to the industrial waste in this community. As the deleterious health impacts of these environmental factors became more evident, white residents began to move away, being replaced by Hispanics, some 30 percent of whom are believed to be recent undocumented immigrants. Although many of the industries are now closed, their contaminants remain in the soil and water.

Today, as well, California's cities continue to contaminate in discriminatory ways. A study conducted by Manuel Pastor and his colleagues demonstrated that racialized minorities are far more likely to live near active toxic release facilities that produce air pollution in the San Francisco Bay area (**Figure 5.21**).

Racialized minority, immigrant, and poor populations are not powerless in the face of environmental racism. Organizations such as Communities for a Better Environment work with local residents, such as those living in South Gate, to promote research, legal assistance, and activism.

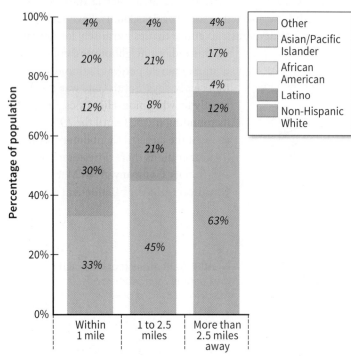

FIGURE 5.21 Air pollution and environmental racism. African American and Latino populations are far more likely to live less than 1 mile away from toxic release sites in the San Francisco Bay area. White and Asian populations, by contrast, are more likely to live 2.5 miles or more from these same sites of air pollution emissions. (Data from Pastor, Morello-Frosch, and Sadd, 2010.)

Environmental justice is the general term for the movement to redress environmental discrimination.

Environmental racism extends to the global scale. Industrial countries often dispose of their garbage, industrial waste, and other contaminants in poor countries. Industries headquartered in wealthy nations often outsource the dangerous and polluting phases of production to countries that do not have—or do not enforce—strict environmental legislation. Whether areas inhabited by poor, ethnic, or racialized minorities are deliberately targeted for unhealthy conditions, or whether they are simply the unfortunate recipients of the effects of larger detrimental processes, the connections among race, ethnicity, and the environment pose central questions for cultural geographers.

Cultural Landscape

5.5   Identify the ways that ethnicity and race leave imprints on the cultural landscape.

Ethnic and racialized landscapes often differ from mainstream landscapes in the styles of traditional architecture, in the patterns of surveying the land, in the distribution of houses and other buildings, and in the degree to which they "humanize" the land. Often the imprint is subtle, discernible only to those who pause and look closely. Sometimes it is quite striking, flaunted as an **ethnic flag**—a readily visible marker of ethnicity on the landscape that strikes even the untrained eye (**Figure 5.22**). Monuments and statues that harken to past episodes and figures can become points of contention for those who read these landscapes as supporting racism. Sometimes the distinctive markers of race and ethnicity are not visible at all but rather are audible, tactile, olfactory, or—as we explore shortly with culinary landscapes—tasty.

Urban Ethnic Landscapes

Ethnic cultural landscapes often appear in urban settings. A fine example is the brightly colored exterior mural typically found in Mexican American ethnic neighborhoods in the southwestern United States (**Figure 5.23**). These began to appear in the 1960s in Southern California, and they exhibit influences rooted in both the Spanish and the indigenous cultures of Mexico. A wide variety of wall surfaces, from apartment house and store exteriors to bridge abutments, provide the space for this ethnic expression. The subjects portrayed are also wide ranging, from religious motifs to political ideology, from statements about historical wrongs to those about urban zoning disputes. Often they are specific to the site, incorporating well-known

FIGURE 5.22 An "ethnic flag" in the cultural landscape. This maize granary, called a *cuezcomatl,* is unique to the indigenous population of Tlaxcala state, Mexico. The structure holds shelled maize. In Mexico, the cultivation of maize long remained an ethnic indigenous trait because the Spaniards preferred wheat. (Ryan Watkins/Photographer's Direct)

FIGURE 5.23 Mexican American mural located in San Francisco's Mission District, a working-class Mexican and Central American immigrant neighborhood. (Arny Raedts/Alamy)

FIGURE 5.24 House in Chinatown. The Edward Mooney House in the Chinatown neighborhood of Manhattan is painted red, a sacred color for Chinese. (Lonely Planet/Getty Images)

elements of the local landscape and thus heightening the sense of place and ethnic "turf." Inscriptions can be in either Spanish or English, but many Mexican murals do not contain a written message, relying instead on the sharpness of image and vividness of color to make an impression.

Usually, the visual ethnic expression is subtler. Color alone can connote and reveal ethnicity to the trained eye. Red, for example, is a venerated and auspicious color to the Chinese, and when they established Chinatowns in Canadian and American cities, red surfaces proliferated (**Figure 5.24**). Light blue is a Greek ethnic color, derived from the flag of their ancestral country. Not only that, but Greeks also avoid red, which is perceived as the color of their ancient enemy, the Turks. Green, an Irish Catholic color, also finds favor in Muslim ethnic neighborhoods (**Figure 5.25**) in countries as far-flung as France and China, because it is the sacred color of Islam (see Chapter 7).

FIGURE 5.25 Ethnic storefront in Coney Island, New York. The liberal use of the color green in this storefront design might indicate an Islamic presence. Indeed, this is a Pakistani ethnic grocery. (David Grossman/Alamy)

FIGURE 5.26 The planned urban landscape of Brasília. Brasília is a city of straight roadways, high-rise residential *superquadra,* and green spaces in between. The city was literally built from scratch in an area of cleared jungle in Brazil's Amazonian interior. This allowed the planners to follow a modern urban design blueprint down to the last detail. Residents of Brasília complain that its larger-than-life public spaces, high-speed vehicle traffic, and impersonal housing blocks inhibit socialization and make Brasília a difficult place in which to live. (SambaPhoto/Cassio Vasconcellos/Getty Images)

Indeed, urban ethnic and racial landscapes are visible in cities across the world. This is true even when urban planners try their best to prevent the emergence of such landscapes. When Brazil's capital was moved from Rio de Janeiro to Brasília in 1960, for example, very few pedestrian-friendly public spaces were incorporated into the urban plan of architects Lúcio Costa and Oscar Niemeyer. Rather, residences were concentrated in high-rise apartments called *superquadra;* streets were designed for high-speed motor traffic only; and no smaller plazas, cafés, or sidewalks were included—all of which discouraged informal socialization (**Figure 5.26**). As political scientist James C. Scott notes in his discussion of Brasília, the lack of human-scale public gathering places was intentional. "Brasília was to be an exemplary city, a center that would transform the lives of the Brazilians who lived there—from their personal habits and household organization to their social lives, leisure, and work. The goal of making over Brazil and Brazilians necessarily implied a disdain for what Brazil had been." But because the housing needs of those building the city and serving the government workers had not been planned, Brasília soon developed surrounding slums that did not follow an orderly layout. The desires of wealthier residents were not met by the uniform *superquadra* apartments, so they built unplanned but luxurious residences and private clubs. By 1980, 75 percent of Brasília's population lived in unanticipated settlements. And because, as previously mentioned, Brazil's social structure is, in part, based on skin pigmentation, the poorer and darker-skinned workers live in the peripheral slums, whereas the wealthier and lighter-skinned residents tend to live in isolated enclaves for the rich. Thus, a racialized landscape of wealth and poverty emerged in Brasília despite all efforts to the contrary.

The Re-Creation of Ethnic Cultural Landscapes

As mentioned previously in this chapter, migration is the principal way in which groups not ethnic in their homelands become ethnic or racialized, through relocation to a new place. As we have seen, ethnic groups may choose to relocate to places that remind them of their old homelands. Once there, they set about re-creating some—although, because of cultural simplification, not all—of their particular landscapes in their new homes.

Cuban Americans in Miami Cuban immigrants, for example, have reconstructed in Miami aspects of their prerevolutionary Cuban homeland. In the early 1960s, the first wave of Cuban refugees came to the United States on the heels of Castro's communist revolution on the island. Some went north, to places such as Union City, New Jersey, where today there is a sizable Cuban American population. Others, however, came (or eventually relocated) to Miami. Once a predominantly Jewish neighborhood called Riverside, what is today called the Little Havana neighborhood became the heart of early Cuban immigration. Shops along the main thoroughfare, Calle Ocho (or Southwest Eighth Street), reflect the Cuban origins of the residents: grocers such as Sedanos and La Roca cater to the Spanish-inspired culinary traditions of Cuba; cafés selling small cups of strong, sweet *café Cubano* exist on every block; and famed restaurants such as Versailles are social gathering points for Cuban Americans.

Because ethnicity and its expression on the landscape are fluid and ever changing, the landscape of Calle Ocho reflects the current demographic changes under way in Little Havana. Although the neighborhood is still a Latino enclave, with a Hispanic population of over 95 percent, only half identify as "Cuban." More than one-quarter of Little Havana's residents are recent arrivals from Central American countries, particularly Nicaragua and Honduras, whereas more and more Argentineans, Venezuelans, and Colombians are arriving—like the Cubans before them—on the heels of political and economic chaos in their home countries. Today, you are just as likely to see a Nicaraguan *fritanga* restaurant as a Cuban coffee shop along Calle Ocho. This has prompted some to suggest that the neighborhood's name be changed from Little Havana to the Latin Quarter (**Figure 5.27**).

Vietnamese Americans in New Orleans Geographer Christopher Airriess has studied Vietnamese refugees in the United States. These refugees were initially brought to one of four reception centers, and then further relocated to rural and urban locations throughout the country, on the theory that spatial dispersal would hasten their assimilation into mainstream U.S. society. Airriess's work reveals that a process of secondary, chain migration later led many of them to cluster in selected urban areas that offer warm weather and proximity to Vietnamese friends and relatives. Focusing on the Versailles neighborhood of New Orleans, Airriess found that the fact that the neighborhood is surrounded by swamps, canals, or bayous on three sides affords a degree of cultural isolation from other ethnic groups in the city. This has led to Versailles becoming an *ethnic neighborhood* where

FIGURE 5.27 Restaurant in Little Havana, Miami. Several different Latin and Caribbean national cultures are depicted on this restaurant façade. (Courtesy of Patricia L. Price.)

FIGURE 5.28 Lunch truck in New Orleans. *Loncheras,* or mobile lunch trucks, are a common sight in areas of New Orleans that suffered devastation during Hurricane Katrina in 2005. Mexican workers were attracted to New Orleans after 2005 because of opportunities to work in the construction sector as neighborhoods rebuilt. Lunch trucks cater to Mexican immigrant tastes and budgets. (Russell McCulley/AFP/Getty Images)

Vietnamese refugees re-create elements of their home landscape in the United States. The most prominent ethnic flag of the Vietnamese in Versailles is the vegetable gardens found on the perimeter of the neighborhood. Cultural preadaptation has led to the reproduction of Vietnamese rural landscapes here, sometimes with plants sent directly from Vietnam. It is the older Vietnamese who plant and tend the gardens. Such gardening provides a therapeutic activity, allows them to retain traditional dietary habits, enables them to produce folk medicines from plants, and reduces household food expenditures. Airriess notes that "the image of conical-hat-wearing gardeners leaning over plants within a multi-textured and moisture-soaked, vibrant green agricultural scene, coupled with frequent flyovers of military helicopters, is an eerie scene of wartime Vietnam reproduced."

After Hurricane Katrina devastated sections of New Orleans—including Versailles (today more commonly called Village de L'Est)—in 2005, the ethnic cultural landscape underwent a transformation. Many Vietnamese immigrants moved away from Versailles after the storm, whereas many Hispanics moved into the area, attracted by the many post-Katrina construction jobs. A 2006 survey found nearly 7000 Asians in New Orleans after Katrina, compared to nearly 12,000 before the storm. By contrast, Latinos have grown from about 14,000 before the hurricane to 22,000 post-Katrina—the only ethnic group, in fact, to have increased after 2005. Today, the commercial landscape reflects this transition. The first arrivals on the scene were the *loncheras,* or mobile lunch trucks (**Figure 5.28**). Then came the Hispanic-oriented restaurants, supermarkets, and services. At present, the

GEOGRAPHY @ WORK

Vivian Gonzalez

Morning Meteorologist, WSVN Channel 7, Miami, Florida

Education: BA Geography, Florida International University

BS Geoscience, Mississippi State University

Broadcast Meteorology Certificate, Mississippi State University

Q. *Why did you major in geography and decide to pursue a career in the field?*

A. Geography is a broad field of study that encompasses everything from the social, political, economic, and environmental issues applicable to the public and private sectors. It offers a wide array of opportunities that can be applied in any field. I chose the environmental sciences track because ever since I can remember, I was fascinated by the weather and natural elements. Geography helped give me a better understanding of the impact that climate change has on our society and environment. Climate change ultimately leads to weather pattern changes on a global scale, and geography gave me the basic skills and understanding needed to work in the public safety sector.

Q. *Please describe your job.*

A. I am the weekday meteorologist for the number one–rated morning show in south Florida called *Today in Florida*.

My main responsibility is to keep the public safe and informed, especially during the threat of severe weather. For example, in south Florida, during the summer months, the sea breeze each afternoon aids in the development of showers and storms, and these conditions can develop into severe weather capable of producing strong storms, tornadoes, flooding, damaging wind gusts, and hail.

Among other responsibilities, I have to receive and perform analysis on data, specialize and produce around-the-clock live severe weather and hurricane coverage, give radio weather updates, build forecast pages and high-quality graphics for on-air use, write a blog about the current weather situation on a daily basis, use multiple social media platforms such as Facebook and Twitter to stay in constant communication with the public, manage real-time Doppler radar with street-level mapper and storm tracker, perform routine updates on all weather systems to keep them running at peak efficiency. Also, it is my responsibility to stay current with all new meteorological and graphic production software.

Q. *How does your geographical background help you in your day-to-day work?*

A. Having a geographical background has helped me become a better meteorologist. It has given me the basic knowledge necessary to deliver and explain all of the factors that influence our weather from season to season and has allowed me to give the public the information they need in an "easy-to-understand" format. I joined the WSVN 7 News team in the midst of the most active hurricane season on record, in 2005. Thirty-one systems were spawned that year, with both Katrina and Wilma impacting south Florida directly. My background in geography was extremely valuable in tracking everything Mother Nature threw our way, as understanding topography played a key role in preparing the public for the intensity of the hurricanes. I feel that knowledgeable presentation and delivery of information are an art form, and to me, an exciting part of my job is coming up with different ways to make the story interesting. This is where my knowledge in geography shines, by combining what happens in the sky and how it affects what's on the ground, all through the visual medium of complex computer graphics fed by raw data.

Q. *What advice do you have for students considering a career in the geography field?*

A. The advice I would give anyone who wants to be a meteorologist or would like to attain a degree in the atmospheric science realm is to pursue a degree in geography. There are multiple career opportunities within the field of study that include job opportunities in the media and even federal, state, or local government. That being said, getting your big break or opportunity in the media industry isn't easy. There are only so many TV stations, and some small cities and rural clusters even outsource their weather programming to a regional service. Most of my colleagues tell me about their transition into the business, including some of the difficulties they encountered before getting their current opportunity. I often hear that rarely does someone start his or her career in a major market like Miami. The norm would be to start in a smaller market, gain experience, and then move your way up. Another great piece of advice is to "never stop learning." I attained my first degree in geography, but I continued to learn by earning my second degree in geosciences, a broadcast certificate in meteorology, and my AMS Seal of Approval.

remaining Vietnamese American residents are learning to adapt to the newcomers. Sara Catania recounts how Mai Thi Nguyen, a business development director in Village de L'Est, reports that her aunt, a market owner, "is learning Spanish. She's learning to say hello, how to tell customers how much something costs. It's wild. I love it. It's exciting."

History and Race in the Cultural Landscape

As you learned in Chapter 1, culture has an important symbolic dimension. Features of the built environment, such as buildings, shrines, public art, and monuments, tell a story about the history of place: who owned it, who is welcome, and what happened there. Race and ethnicity often feature prominently in these cultural stories about place (see also Doing Geography—Active Learning, Chapter 7). As we have noted above, for example, the colors used to paint buildings can mark them as the territory of particular ethnic groups.

Countries narrate their national history in part through the monuments, plaques, and statues that denote important places in that history. In the United States, these elements of the cultural landscape often refer to the Civil War era from 1861 to 1865, when the northern states fought the southern states over states' rights and the legality and ethics of slavery. Not surprisingly, monuments, plaques, and statues commemorating this defining historical period often feature individuals who were slave owners and who fought to uphold the right to own slaves (**Figure 5.29**).

These elements of the cultural landscape have become the focus of intense conflict. Many public officials, African American descendants of slaves, and progressive groups have called for their removal from public spaces, arguing that they are unwelcome reminders of a racist past that the United States would now like to move beyond. Others claim that these are simply historically accurate representations of things that happened and people who were influential in their time, but otherwise neutral in their meaning. White supremacists protested against their removal, insisting that the history of white Americans was being disrespected. Protests and violent clashes occurred across the country. As depicted in **Figure 5.30,** public works crews removing Confederate landmarks from many U.S. cities did so at night, wearing protective gear, and under police guard.

Many nations have seen racial or ethnic conflicts erupt over cultural landscapes commemorating past eras. Indeed, **iconoclasm**—the destruction of statues, monuments, plaques, and other religious or political landscape elements—dates back to biblical times. More recently, former Soviet republics have destroyed likenesses of repressive leaders such as Joseph Stalin, Venezuelan anti-colonial protesters toppled a statue of Christopher Columbus in Caracas, and the Taliban destroyed ancient statues of the Buddha in Afghanistan. These examples underscore the fact that cultural landscapes, and the elements that comprise them, are never simply passive backdrops against which life plays out. Instead, the cultural landscape is a dynamic actor in the drama of nationalism, colonialism, religion, and racism.

FIGURE 5.29 Statue of General Robert E. Lee. Statues and monuments commemorating the U.S. Civil War, such as this one in Charlottesville, Virginia, are common sights in the landscape of the U.S. South. (Joel Carillet/Getty Images)

FIGURE 5.30 Removal of Confederate Statue. Throughout southern U.S. cities, symbols of the Confederacy, such as this statue in Baltimore, were removed from the landscape. Many view these images as legitimating a racist agenda. (Denise Sanders/The Baltimore Sun via AP)

Ethnic Culinary Landscapes

"Tell me what you eat, and I'll tell you who you are." This often-quoted statement arises from the connections between identity and **foodways,** or the customary behaviors associated with food preparation and consumption that vary from place to place and from ethnic group to ethnic group (**Figure 5.31**). As geographers Barbara and James Shortridge point out, "Food is a sensitive indicator of identity and change." Immigration, intermarriage, technological innovation, and the availability of certain ingredients mean that modifications and simplifications of traditional foods are inevitable over time. An examination of culinary cultural landscapes reveals that, although landscapes are typically understood to be visual entities, they can be constructed from the other four senses as well: touch, smell, sound, and—in this case—taste.

Although Singapore occupies the southern tip of the Malaysian Peninsula in extreme Southeast Asia, its cuisine draws heavily from southern Chinese cooking. This is because three-quarters of Singapore's multiethnic population is of Chinese ancestry. Intermarriage between Chinese men, who came to Singapore as traders and settled there, and local Malay women resulted in a distinctive spicy cuisine called *nonya.* As we have already seen, there are far more ethnic Chinese in Asia than in other parts of the world, principally because of China's proximity to other Asian countries.

The corn tortilla remains a staple of central Mexican foodways and is consumed at every meal by some families. But many middle-class urban women in Mexico no longer have the time to soak, hull, and grind corn by hand to make corn tortillas from scratch. Rather, children are often sent every afternoon to the corner *tortillería* to purchase *masa,* or corn dough, or even hot stacks of the finished product. Convenience notwithstanding, old-timers complain that the uniform machine-made tortillas will never come close to the flavor and texture of a handmade tortilla. In some regions, notably in the northern part of Mexico and southern Texas, the labor-intensive corn tortilla was replaced altogether by wheat tortillas, which are much easier to make.

In the rural, mountainous Appalachian region of the eastern United States, distinct ethnic foodways persist and literally flavor this region. Geographer John Rehder observed that, during and after World War II, Appalachians maintained their cultural distinctiveness in the face of migration to northern cities in part through "care packages" of lard, dried beans, grits, cornmeal, and other foods unavailable in the North. Grocery stores in the Little Appalachia ethnic neighborhoods that formed in cities such as Detroit, Cleveland, and Chicago soon began to cater to the distinct food preferences of the population. Thus, Appalachian foodways were maintained as Appalachians moved north.

Rehder tells of sending students on a scavenger hunt during a class field trip to Pikeville, Tennessee. To one young man, he handed a card that read, "What are cat head biscuits and sawmill gravy?"

> About a half an hour later, my "biscuit man" returned with a scared look on his face and tears welling up in his eyes. I asked, "Where did you go and what happened?" He replied, "Well, . . . I . . . went to the feed store and asked an old man there, 'What are cat head biscuits and sawmill gravy?' just like you told me to do. Only the old man just growled at me and said, 'Son, if you don't know, HELL, I ain't goin' to tell you!'" And with that our lad left the feed store dejected and thoroughly upset. To calm him down, I said, "Why don't you just go on down to the café. Get a little something to drink and maybe calmly run your question by them down there." After a sufficient amount of time, I decided to go check on him. He was sitting on a red upholstered stool at the counter. As I drew closer, he looked up and, with fresh and quite different tears in his eyes, pointed to his plate: "This," he said proudly, "is cat head biscuits and sawmill gravy!" (p. 208)

Rehder explains that cat head biscuits are flour biscuits made in the exact size and shape of a cat's head and baked golden in a wood stove. Sawmill gravy is a southern-style white gravy made in a cast-iron skillet. Legend has it that the cook at the Little River Lumber Company logging camp ran out of flour for the gravy and substituted cornmeal. The loggers complained about the texture, referring to it as "sawmill gravy." For further discussion of ethnicity and food, see Doing Geography (page 182) and Seeing Geography (page 183).

FIGURE 5.31 Market in Chinatown, New York City. Grocery stores selling distinctive produce are one of the most visible landscape markers of an ethnic presence. Here, spiny fruits called *durian,* also known as "stinky fruit," hang from bags. Durian is sometimes banned from public places in its native Asia because of its pungent smell. (Kathy Willens/AP Photo)

Conclusion

The spatial expression of race and ethnicity are important to understanding the world around us. The diversity of cultural behaviors, expressed in foodways, religious observance, language, and architecture, among other things, lends richness to our landscapes and lives. How we move about in the world, whether with ease or friction, is in part shaped by our racial and ethnic identity. And though racialized tensions around migration, nationalism, settlement, and engagement with the natural world might seem to be peculiar features of twenty-first-century North America, they are in fact pervasive across human societies and historical periods. That ubiquity notwithstanding, conflict arising from intolerance of racial and ethnic difference has been an especially destructive feature of human civilizations.

Geography itself has been transformed thanks in part to our long and sometimes difficult conversations about race and other differences—sexuality, gender, and (dis) ability among them. Today, these differences are the subject of some of the most exciting research in the discipline, whereas 25 years ago there was little research if any at all on sexuality, gender, race and racism, and (dis)ability. Who geographers are has changed as well. Though the discipline of geography still has a long way to go in order to achieve a fully representative field of practitioners, there are far more nonwhite, women, queer, and differently abled geographers today than there were just a few decades ago. The scholarship and teaching of these diverse geographers has made important inroads into combating, as well as healing, the social injustices arising from intolerance.

Chapter Summary

5.1   Differentiate between race and ethnicity, and describe how ethnic groups are distributed geographically.

- "Race" and "ethnicity" are distinct notions, with race indicating a historically specific social construction, and ethnicity denoting a group identity that is often based on a common ancestry and cultural tradition.
- Minority ethnic groups adapt, to varying degrees, to the dominant ethnic culture.
- Ethnicity and race shape the settlement patterns of rural homelands and islands, as well as urban ethnic neighborhoods, ghettos, and ethnoburbs.
- Immigration is actively reshaping the ethnic makeup of many places.

5.2   Explain the ways that migration and mobility shape, and are shaped by, race and ethnicity.

- Human mobility routes—whether voluntary or involuntary, undertaken in response to persecution in a home region or opportunities beckoning beyond one's borders—shape the patterns that ethnicity and race assume in the new land.
- The everyday mobilities of driving and flying can become problematic for people of color, whose right to move freely through space can be questioned and subject to scrutiny through racial profiling.

5.3   Interpret how globalization has deepened, transformed, or erased ethnicity, race, and racism.

- Globalization appears to intensify age-old tendencies to distinguish, and discriminate, among groups.
- Europe's colonization of territories in Africa, Asia, and Latin America relied in part on creating or intensifying existing, racial and ethnic divisions in these societies.
- Genocide involves the attempted annihilation of racial or ethnic minority populations.
- Globalization can also lead to an intensification of ethnic or racial pride.

5.4   Analyze the diverse ways in which ethnic and racial groups interact with the natural environment, and discuss how these groups may be protected or made more vulnerable.

- Upon relocating through migration, ethnic minority populations replicate only portions of their cultures in their new home.
- Most migrants adopt and incorporate selected features of their host cultures, unless they are geographically isolated in their new home, in which case they tend to preserve archaic elements of their original culture.
- Similarities between home and host habitats can direct migrants to certain places rather than others.
- Certain natural habitats can protect ethnically distinct groups.
- Environmental racism illustrates how minority peoples and the places they inhabit can become the targets of undesirable practices, such as pollution.

5.5   Identify the ways that ethnicity and race leave imprints on the cultural landscape.

- Ethnic pride and the active construction of ethnically and racially distinct landscapes are evident throughout the world.
- Immigrant groups tend to reproduce the cultural landscapes of their homeland, albeit in a selective fashion.

- Elements of the cultural landscape—monuments, statues, plaques, and other symbolic depictions—can become the targets of bitter struggles over racism and history.
- Ethnic foodways are a distinctive aspect of the cultural landscape.

Key Terms

race   A classification system that is variously understood as arising from genetically significant differences among human populations, or from visible differences in human physiognomy, or as a social construction that varies across time and space (page 151).

racism   The belief that human capabilities are determined by racial classification and that some races are superior to others (page 153).

racialize   To understand people or practices through the lens of race (page 153).

ethnic group   A group of people who share a common ancestry and cultural tradition, often living as a minority group in a larger society (page 153).

acculturation   An ethnic group's adoption of enough of the ways of the host society to be able to function economically and socially (page 153).

assimilation   The blending of an ethnic group into the host society, resulting in the loss of distinctive ethnic traits (page 153).

melting pot   The belief, no longer widely held, that North America is a crucible of sorts, where ethnic differences will eventually blend together and disappear (page 153).

transculturation   The notion that people adopt elements of other cultures as well as contributing elements of their own culture, thereby transforming both cultures (page 153).

salad bowl   A contemporary approach to racial and ethnic differences in North America, which views them as mixing together, but never entirely blending and disappearing (page 153).

ethnic homelands   Sizable areas inhabited by an ethnic minority that exhibits a strong sense of attachment to the region and often exercises some measure of political and social control over it (page 153).

microaggressions   The subtle, indirect, or unintentional insults, exclusions, or small acts of violence that marginalized people encounter as they go about their daily lives (page 154).

ethnic islands   Small ethnic areas in the rural countryside; sometimes called folk islands (page 154).

ethnic substrate   The geographical residue of a formerly ethnic population that has now been absorbed into the cultural mainstream (page 155).

ethnic neighborhood   A voluntary urban community where people of like origin reside by choice (page 156).

ghetto   Traditionally, an area within a city where an ethnic group lives, either by choice or by force. Today in the United States, the term typically indicates an impoverished African American urban neighborhood (page 156).

barrio   An impoverished, urban, Hispanic neighborhood (page 157).

ethnoburb   A suburban ethnic neighborhood, sometimes home to relatively affluent immigration populations (page 158).

refugees   Persons or groups that have been forced to flee their home country. Many nations recognize as legitimate only **political refugees**—those who are at risk of harm from their own government due to the political beliefs they hold, their ethnicity or race, or because of political conflict in their home country. However, people can become **economic refugees,** driven to migrate by economic necessity; or **environmental refugees,** who are involuntarily displaced from their homeland due to natural disaster (page 161).

internally displaced persons (IDPs)   Persons or groups that have been forced to flee their homes due to conflict, natural disaster, or persecution but who remain within the borders of their home country (page 161).

racial profiling   Treating a person as a criminal suspect because of his or her race, ethnicity, nationality, or religion (page 164).

anti-Semitism   Hostility toward or persecution of Jews (page 165).

genocide   The systematic killing of members of a racial, ethnic, religious, or linguistic group (page 166).

cultural preadaptation A complex of adaptive traits and skills possessed in advance of migration by a group, giving it survival ability and competitive advantage in occupying the new environment (page 168).

cultural maladaptation Poor or inadequate adaptation that occurs when a group pursues an adaptive strategy that, in the short run, fails to provide the necessities of life or, in the long run, destroys the environment that nourishes it (page 169).

cultural simplification The process by which immigrant ethnic groups lose certain aspects of their traditional culture in the process of settling overseas, creating a new culture that is less complex than the old (page 169).

environmental racism The targeting of areas where ethnic or racial minorities, immigrants, and/or poor people live with respect to environmental contamination or failure to enforce environmental regulations (page 171).

environmental justice General term for the movement to redress environmental discrimination (page 172).

ethnic flag A readily visible marker of ethnicity on the landscape (page 172).

iconoclasm The destruction of statues, monuments, plaques, and other religious or political landscape elements (page 177).

foodways Customary behaviors associated with food preparation and consumption (page 178).

Practice at Sapling Plus

Read the interactive e-Text, review key concepts, and test your understanding.

Story Map. Explore race and ethnicity in the United States.

Web Map. Examine maps illustrating issues of race and ethnicity.

Doing Geography

ACTIVE LEARNING:
Understanding Implicit Biases

In the Subject to Debate feature, we examined the police shootings of black individuals in U.S. communities. The research that attempted to explain why some places experience more of these shootings than others relied on what are known as implicit association tests, or IATs. These tests measure biases that we are not consciously aware of, by timing the speed of our reactions to associations of individuals with concepts. For example, a person who is biased against women may take longer to associate images of women with positive words such as "joy" and "happiness." Such subtle biases, in a split-second decision such as whether to fire a weapon, may lead to the disproportionate killing of people of color by the police. In this exercise, you will take an implicit association test and discuss your experience with your classmates.

More guidance at 🌀 Sapling Plus.

EXPERIENTIAL LEARNING:
Tracing Ethnic Foodways Through Recipes

At some point, we all trace our roots back to migrants. Perhaps your ancestors walked here some 30,000 years ago. Perhaps they arrived on slave ships in the seventeenth century. Perhaps they were traders who settled and married locals. Were they part of the waves of Europeans in the eighteenth, nineteenth, and early twentieth centuries, or have you yourself only recently immigrated? More than likely, your ethnic inheritance results from a combination of different immigrant groups. As the geographer Doreen Massey wrote, "In one sense or another most places have been 'meeting places'; even their 'original inhabitants' usually came from somewhere else." In other words, if you dig into the history of any place, you'll find layers upon layers of people arriving from other places and bringing their cultural baggage—recipes and all—with them.

    For this exercise, you will analyze one of the most commonplace, yet revealing, items of ethnic geography: a recipe. Certainly, you are what you eat, but you also eat where you are, and the foodways in which you take an active part are very revealing. Even though you may not be conscious of it, the simple act of cooking a meal sets in motion all sorts of cultural geography elements: regional identity, ethnic heritage, place-specific agricultural traditions, and so on. Together, these form important components of identity and place.

Steps to Analyzing a Recipe

Step 1:   Choose a recipe to analyze—one that is used often and has been around for a while. The best candidate is a family recipe that has been passed down through the generations (see the note below). If you don't have a copy of the recipe, interview a person who does.

Step 2:   With a written recipe now in front of you, answer the following questions. You can interview someone in your family about this recipe, too.

- From where does this recipe come? With what country, or region, is it identified?

- Do any of the recipe's ingredients give clues about the origins of the recipe? Do the ingredients draw on particular animal or plant ingredients that are, or were, produced where the recipe originated?

- Has the recipe been modified to substitute ingredients that are no longer available, either because the person who used the recipe migrated or because the ingredients went out of production?

- Are there ingredients used in the recipe that are identified with particular ethnic groups and perhaps aren't consumed by others living in the same place?

- Do elements of the recipe's preparation give additional clues about the foodways of the people who developed the recipe?

- Are there special occasions, such as holidays, when this recipe is always used?

- Do any of the ingredients or methods of preparation have symbolic meanings or stories associated with them?

In sharing the results of the recipe analysis, the class can list the various places and ethnicities that together make up the foodways of the students. The class might want to organize a potluck meal where students share the dishes they have analyzed. Students may also create a cookbook and map the places from which they come.

   Note: If you are attending school in a foreign country, use a family recipe from your homeland. If you grew up in an institutional setting, where you ate food in a cafeteria or dining hall, choose a dish commonly prepared for the evening meal. Institutional foods can be very revealing of local ethnic influences.

Multigenerational cooks. Many of us learn to prepare traditional ethnic dishes by working alongside older family members. (Jacqueline Veissid/Getty Images)

SEEING GEOGRAPHY

America's Ethnic Foodscape

When does cuisine cease to be ethnic and become simply "American"? What roles do mobility and globalization play in the process?

This street scene was shot in Manhattan's Greenwich Village neighborhood, but it could have just as easily been taken in any number of major U.S. urban areas. The signs display a mixture of ethnic foods: Pakistani (Tastee Curritos), Indian (Kati Roll), and Italian (Monte's Trattoria) are the three most visible restaurants in the foreground of this photo.

As you know from Chapter 4, New York City is the most linguistically varied area on the face of the planet. This reflects the many and varied streams of immigration to the city. And with language comes food—two of the central aspects of ethnic identity. The ancestors of the owners of Monte's Trattoria may well have arrived in the United States between 1890 and the mid-1920s, the peak years for Italian immigrants. The owners of Tastee Curritos and Kati Roll probably arrived almost a century later. Changes in U.S. immigration laws in 1965 abolished the previous national quota system for immigrants, which had the effect of greatly diversifying the immigrant stream as well as increasing the total

numbers of immigrants to America. As you know from reading this chapter, the major areas sending immigrants to the United States are no longer Europe; rather, they are Asia and Latin America.

Cultural geographer Richard Pillsbury famously quipped that America has no foreign food because we have accepted every possible foreign cuisine and made it our own. Americans are for the most part very willing to consume foods from all over the globe, as this image shows. In addition, Tex-Mex, Asian fusion, and many "Chinese" dishes—among them, chop suey, General Tso's chicken, and fortune cookies—are great examples of "foreign" cuisines that have been heavily Americanized, if not invented entirely in the United States.

So what exactly is "American" food, then? Perhaps the squash, beans, and corn consumed by Native Americans are our only example of a cuisine that is truly indigenous to the United States. Or perhaps American fast food, which has itself diffused across the globe (along with the drive-in landscapes that accompany fast-food restaurants), is our only contemporary American food?

Ethnic restaurants abound in Manhattan's Greenwich Village neighborhood.
(Steve Lewis Stock/Getty Images)

Re/flecting the Border, Installation and Performance Dinner on the US/Mexico Wall, Tijuana. (Credit: Margarita Certeza Garcia in collaboration with Marcos Ramirez ERRE & Miguel Buenrostro. Photo credit: Marcos Ramirez ERRE)

chapter 6

Political Geography

A DIVIDED WORLD

Can a border be more than a dividing line?

Think about this photo and question as you read. We will revisit them in Seeing Geography on page 225.

Learning Objectives

6.1 Explain the concept of nation-states, the forces that shape political territory, and the regional characteristics of electoral politics.

6.2 Explain how mobility influences political geography, including the spatial diffusion of political authority and innovation, the displacement effects of violent political conflict, and the role of the Internet.

6.3 Explain the effects of globalization on sovereignty, supranationalism, and transnationality.

6.4 Analyze the complex interactions of politics and physical geography, including the state's role in land management, the heartland theory of world geopolitics, and the effects of rising sea levels on political territory.

6.5 Analyze the effects of national laws, international borders, and national monuments on the cultural landscape.

Re/flecting the Border, Installation and Performance Dinner on the US/Mexico Wall, Tijuana. (Credit: Margarita Certeza Garcia in collaboration with Marcos Ramirez ERRE & Miguel Buenrostro. Photo credit: Marcos Ramirez ERRE)

From the breakup of empires to regional differences in voting patterns, from the drawing of international boundaries to congressional redistricting in the United States, from the resurgence of nationalism to secession movements, human political behavior is inherently geographical. As geographer Gearóid Ó Tuathail has said, political geography "is about power, an ever-changing map revealing the struggle over borders, space, and authority." In this chapter we will examine some of the forces behind the ever-changing political map. We will explore how political phenomena diffuse in time and space, and examine some of the political forces that result in refugee crises. We will analyze the way globalization has given rise to new geographic forms of political sovereignty as well as strengthening older forms. We will learn how important nature–culture relations are to political geography, including the increasing impact of human-induced sea level rise. Finally, we will see how political authority and symbolism are inscribed in our everyday cultural landscapes.

Region

6.1 Explain the concept of nation-states, the forces that shape political territory, and the regional characteristics of electoral politics.

The theme of region is essential to the study of political geography at multiple scales of analysis. Beginning at the global scale, we find that people have divided nearly the entire surface of the Earth into bounded political units called states. These bounded territories are variously shaped, with implications for

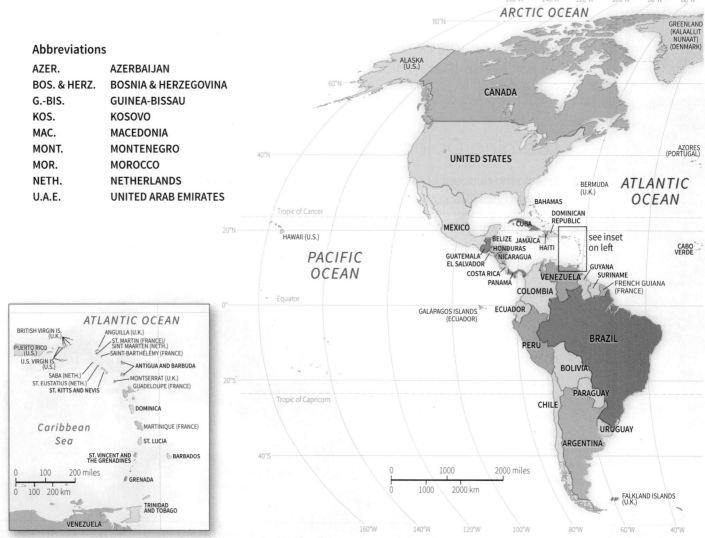

FIGURE 6.1 The independent countries of the world. In the twentieth century, the world political map was in rapid flux, and the number of states more than doubled. This process began after World War I with the breakup of the Austro-Hungarian and Ottoman empires. It then intensified after World War II when the overseas territorial empires of the British, French, Italians, Dutch, and Belgians fragmented. Most recently, the Russian-Soviet Empire disintegrated at the end of the century.

their ease of governability. They are associated with powerful political and cultural ideals, such as sovereignty, nationalism, and self-determination. States are far from fixed territories, however; they are subject to forces that hold them together and drive them apart, such as separatist movements. Even U.S. elections are influenced by geography through both the practice of gerrymandering and the system of electing presidents.

A World of Nation-States

A fundamental political-geographical fact is that the Earth is divided into roughly 200 independent countries or states (**Figure 6.1**). The **state** is a political institution that has taken a variety of forms over the centuries, ranging from Greek city-states, to Chinese dynastic states, to European feudal states.

When we talk about states today, however, we mean something geographically specific and historically recent. Modern states are independent political units with a centralized authority that claims exclusive legal, political, and economic jurisdiction over a region with defined boundaries. Such bounded regions are referred to as *state territories*.

The state claims territorial **sovereignty** within its boundaries. Sovereignty is a principle of international law that means that every state possesses the sole authority over the land and people within its boundaries, including everything from the political system to land and resource use decisions. This is sometimes referred to as *internal* sovereignty. Under international law, states recognize each other's boundaries, right to exist independently, and authority to govern their own territories. This is called *external* sovereignty. The state, in sum, is the primary form by which governance is organized spatially at the global scale.

FIGURE 6.1 (continued)

The Idea of the Nation-State Scholars generally trace the idea of the modern state to late-eighteenth-century political revolutions in North America and France. Political philosophers of the time argued that **self-determination**—the freedom of culturally distinct groups to govern themselves in their own territories—was a fundamental right of all peoples. Philosophers conceptualized the "self" in "self-determination" to be a **nation**. Nation refers to a community of people bound to a homeland and possessing a common identity based on shared cultural traits such as language, ethnicity, and religion. Political philosophers proposed the **nation-state** as the ideal political geographical unit, where the geographic boundaries of a nation (a people and its culture) would be identical to the territorial boundaries of the state (governance and authority).

Today, when we think of countries, we generally think of them in terms of the nation-state ideal. Political authorities' claims to govern in the name of all of a country's citizens, modern mass communications that link all residents, and state-based citizenship rights all work to reinforce the idea that people possess national identities. The sense of collective identity that we call **nationalism** springs from a learned and acquired attachment to a homeland, territorialized by state boundaries. Nationalism is the idea that the individual derives a significant part of his or her social identity from a sense of belonging to a nation. Geography and cultural identity are thus tightly linked in the concepts of nationalism and the nation-state. Since nationalism is a historically recent and learned emotional attachment, many scholars debate which came first, the modern territorial state or the learned feeling of belonging to a nation. They also note that the inclusive sense of belonging promoted by nationalism implies the inverse—an exclusive sense that "others" do not belong in the country. We will explore some of the consequences of this tension between inclusion and exclusion in nation-states later in this section.

The nation-state is a tangible and ubiquitous expression of **territoriality** in human society. Geographer Robert Sack defined territoriality as a cultural strategy used by individuals, groups, or organizations to claim power over an area of land and its people and resources. Territorial claims are consolidated through the establishment of precise boundaries. A **boundary** in this sense is a clearly demarcated line that marks both the limits of a territory and the division between territories. Boundaries divide the world into nation-states, but they also divide nation-states into subnational territories, such as counties, provinces, and so forth. Sack argued that the precise marking of borders is a practice that originated with European political philosophy and then spread globally, a process we will discuss in a later section. At the global scale, the term *border* is often used synonymously with *boundary*.

A **borderland** is a region straddling both sides of an international boundary where national cultures overlap and blend to varying degrees. The U.S.–Mexican border in Texas is a good example of a borderland where foods, languages, and other cultural traits mix (see Seeing Geography, page 225). "Tex-Mex" is a term frequently used to capture this distinct culture of food and music in the borderland.

Finally, nation-state territorial borders are sometimes preceded or accompanied by frontiers. A **frontier** is a region at the margins of state control and settlement, an idea that can be traced to ancient notions of the limits of civilization and the savage world beyond. Areas along international boundaries that are geographically remote from central authorities and have little or no settlement are also sometimes called frontiers.

Multinational States, Multistate Nations As it happens, the majority of independent countries do not fit the nation-state ideal. Geographers have suggested that only a few, such as Sweden, Japan, Greece, Armenia, and Finland, come close to the ideal. Instead, most countries contain multiple national, ethnic, and religious groups within their boundaries. (**Figure 6.2** illustrates this point by dividing up South Africa linguistically.) Many, though not all, multiethnic countries came into being in the twentieth century. Most of the country boundaries in Africa, South Asia, and Southeast Asia are a product of European **colonialism**, which

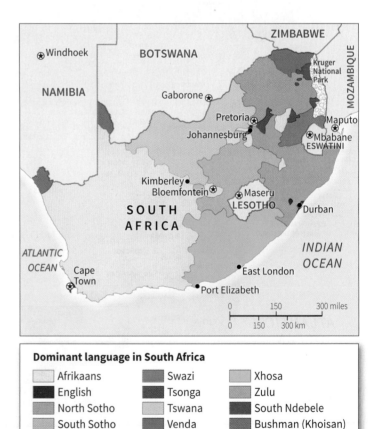

FIGURE 6.2 Languages of the Republic of South Africa, a multinational state. The mixture includes ten indigenous languages and two introduced by European settlers.

gave almost no consideration to ethnic or linguistic geographies. European colonialism, which spanned centuries, involved European countries—principally Great Britain, France, Belgium, Italy, Spain, Portugal, and the Netherlands—competing among themselves to conquer and control other peoples and their territories. The international boundaries created in this process remained in place after decolonization, when the colonies gained independence. Therefore, most modern states are ethnically and linguistically heterogeneous, and many of the world's ethnic groups are territorially divided by one or more international boundaries.

Geographic Variation in Nation-State Territories

Careful inspection of Figure 6.1 reveals that some parts of the world are fragmented into many states, whereas others exhibit much greater unity. The United States occupies about the same amount of territory as Europe, but the latter is divided into 47 independent countries. The continent of Australia is politically united in a single state, whereas South America has 12 countries and Africa has 55. Such wide variation in size is only one measure of territorial differences that may influence the governance of the nation-state; the others are shape and configuration and boundaries.

Shape and Configuration. The shape and configuration of the national territory can present challenges to a centralized authority. In the abstract, the more compact the territory, the easier the governance. Circular or hexagonal forms maximize compactness, allow for shorter communication lines, and minimize the amount of border to be defended. Of course, no country actually enjoys this abstract ideal of compactness, although some—such as France and Brazil—come close (**Figure 6.3**).

FIGURE 6.3 Differences in the distribution of national territory. The map shows wide contrasts in territorial shape. France and, to a lesser extent, Brazil approach the ideal hexagonal shape. In contrast, Chile and Russia are elongated, and Russia has an exclave in the Kaliningrad District between Poland and Lithuania. Indonesia is fragmented into a myriad of islands. The Gambia intrudes as a pene-enclave into the heart of Senegal, and South Africa has a foreign enclave, Lesotho.

Any one of several territorial distributions can make governance more difficult. Potentially most challenging are enclaves and exclaves. An **enclave** is a territory surrounded by a country but not ruled by it. Enclaves can be either self-governing (e.g., Lesotho in Figure 6.3) or an exclave of another country. In either case, its presence can pose problems for both the enclave and the surrounding country. Potentially just as disruptive is the pene-enclave, an intruding piece of territory with only the smallest of outlets. In the case of Senegal, where The Gambia intrudes as a pene-enclave, one can imagine the difficulties of Senegalese officials needing to enter and leave through two international border crossings in order to conduct business in another part of their country.

Exclaves are parts of a national territory separated from the main body of the country to which they belong (e.g., Russia's Baltic seaport, Kaliningrad, in Figure 6.3). Exclaves are particularly undesirable if a hostile power holds the intervening territory because defense of such an isolated area is difficult. Moreover, an exclave's inhabitants may feel separate and cut off from their compatriots, thereby causing internal political instability. Pakistan provides a good example of the national instability created by exclaves. Pakistan was created in 1947 as two main bodies of territory separated from each other by almost 1000 miles (1600 kilometers) of India's territory. West Pakistan held the capital and most of the territory, but East Pakistan was home to most of the people. Ethnic and religious differences between the peoples of the two sectors further complicated matters. In 1971, Pakistan broke apart and East Pakistan became the independent country of Bangladesh (see Figure 6.1).

Even when a national territory is geographically unified, difficulties can develop if the shape of the state is not compact. Narrow, elongated countries, such as Chile, The Gambia, and Norway, can be difficult to govern, as can nations fragmented into many islands (see Figures 6.1 and 6.3). In these situations, transportation and communications are difficult, causing administrative problems. Several major secession movements have challenged the multi-island country of Indonesia; one of these—in East Timor—succeeded in 2002 in creating the first new sovereign state of the twenty-first century (see Figure 6.1). Similarly, the island nations of Japan, Mauritius, Comoros, and Philippines have all faced secessionist movements in recent years.

The shape and configuration of a country can make it harder or easier to administer, but they do not determine its stability. We can think of exceptions for all of the territorial forms in Figure 6.3. For example, Alaska is an exclave of the United States, but it is unlikely to produce a secessionist movement similar to the one in Bangladesh. Conversely, the Democratic Republic of the Congo and Nigeria both have compact territories, but both have suffered through secessionist

civil wars. Geography is an important factor in national governance, but it is rarely the determining factor.

Types of Boundaries People use different criteria to establish the boundaries that create these variations in configuration. One approach is to use physical geographic features. **Natural boundaries** follow some feature of the landscape, such as a river or mountain ridge. Examples of natural boundaries are numerous; for example, the Pyrenees lie between Spain and France, and the Rio Grande serves as part of the border between Mexico and the United States. **Ethnographic boundaries** are drawn on the basis of one or more cultural traits, usually a particular ethnicity, language spoken, or religion practiced. This type of border reflects the nation-state ideal, and can be found throughout Western Europe. **Geometric boundaries** are regular, often perfectly straight lines drawn without regard for physical or cultural features of the area. The U.S.–Canada border west of the Lake of the Woods (about 93° west longitude) is a geometric boundary. Not all boundaries are easily categorized into one of these types, and some boundaries are composites of two or more types.

Territorial Coherence and Fragmentation

If there is one lesson to learn from recent world history, it is that nation-state boundaries are far from permanent. Ethnically based self-determination movements continue to fragment or threaten to fragment state territories. Some state territories have remained intact for many decades, while during the same period others have broken apart, reformed, and broken up again. In addition, there are situations where entire nations and ethnicities find themselves without a claim to any country at all. What forces drive these dynamic processes in our global system of nation-states?

Centripetal and Centrifugal Forces We begin with basic concepts. Geographers refer to a political variable that promotes national unity and coherence as a **centripetal force**. States commonly encourage centripetal forces that help fuel nationalistic sentiment. States fund and otherwise support official national languages, national folk or history museums, national parks and monuments, and sometimes even national religions. In short, state authorities actively cultivate nationalism as a means to promote and defend territorial coherence.

By contrast, a variable that disrupts internal order and furthers territorial fragmentation is called a **centrifugal force**. States that have two or more culturally distinct regions within their boundaries tend to be subject to centrifugal forces. In cases where a majority ethnic group controls the state, minority ethnic groups can be economically, politically, and culturally marginalized, producing strong centrifugal forces. Often this social marginalization is associated with and exacerbated by geographic marginalization. That is, the majority ethnic group dominates a country's geographic center, including the capital city, while minority ethnic groups are located on territorial margins near the border. Myanmar, which has suffered decades of civil war, illustrates this situation well (**Figure 6.4**).

Separatist Movements Centrifugal forces, therefore, can sometimes generate separatist movements. **Ethnic separatism** promotes the separation of an ethnically distinct group to form a politically autonomous region within an existing state or to secede and form a new nation-state. In recent decades, separatist movements have fragmented existing countries to create many new nation-states. Between 1992 and 2008, the country of Yugoslavia fragmented into seven new culturally based nation-states. As mentioned above, East Timor, which is ethnically and religiously distinct, seceded from Indonesia in 2002 after decades of political and armed struggle. More recently, the southern region of Sudan seceded after years of civil war to form the new country of South Sudan in 2011 (see Chapter 2). To this day, many separatist movements are fighting politically, and sometimes militarily, to create new nation-states. Most recently, ethnic Kurds passed a referendum to secede from Iraq to form Kurdistan. Similarly, Catalan separatists claimed victory in a referendum for independence, which the Spanish central government declared illegal (**Figure 6.5**). Both votes have created tense standoffs between separatists and the central state authorities.

Irredentist Movements Separatism is not the only way to create more ethnically homogenous national territories. Quite often ethnic groups are territorially divided by international boundaries, a situation that can generate irredentist movements. **Irredentism** is the political claim to territory in another country

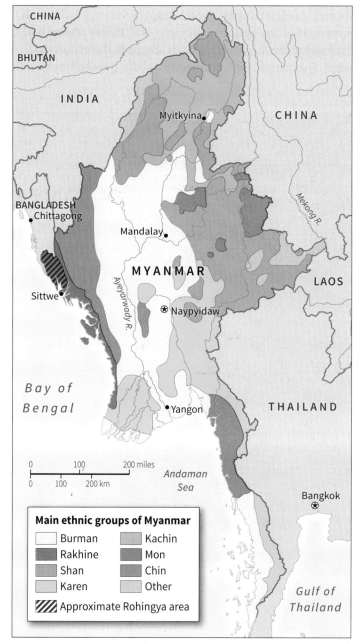

FIGURE 6.4 Ethnic divisions in Myanmar. Ethnic Burmans control the state's central territory, military, economy, and political capital, while less-powerful ethnic groups located in the hills and mountains along the country's borders are marginalized. Such situations generate strong centrifugal forces in Myanmar.

FIGURE 6.5 Protest for an independent Catalonia. Catalonian nationalists have been organizing to secede from Spain for decades. In recent years, the secessionist movement has gained public support in the region, raising the possibility of an independent Catalonia. (Marco Panzetti/NurPhoto via Getty Images)

based on ethnic affiliations and historic borders. Irredentism usually arises when an ethnic majority ruling one country seeks to reunite with people of the same ethnicity who are minorities in a neighboring country. The idea is to reclaim historically "lost" territory and thereby reunite the ethnic group within the boundaries of a single nation-state. For example, before World War II, Germany made irredentist claims on Austria and parts of then-Czechoslovakia based on the presence of German-speaking regional enclaves in those countries. Most recently, Russia annexed into its territory the Crimean Peninsula, a region of Ukraine that is ethnically Russian. In 2014, disguised Russian troops invaded the peninsula and helped Russian Crimeans forcefully take over the regional government and cede the territory to Russia.

Stateless Nations The situation of ethnic groups territorially divided by one or more international boundaries is quite common, presenting another challenge to the nation-state ideal. When ethnic groups divided this way have a sufficiently large population—that is, larger than most nation-states—they are sometimes labeled a **stateless nation**. For example, the West African Yoruba people have a population of some 40 million, but they are divided among three former colonial states: Nigeria, Benin, and Togo. Kurds, who are estimated to number 30–45 million people, are one of the most well-known stateless nations. Their historic territory, Kurdistan, is now divided among Iraq, Iran, Syria, and Turkey (**Figure 6.6**). In comparison, two countries that approach the nation-state ideal, Sweden and Finland, have respective populations of only

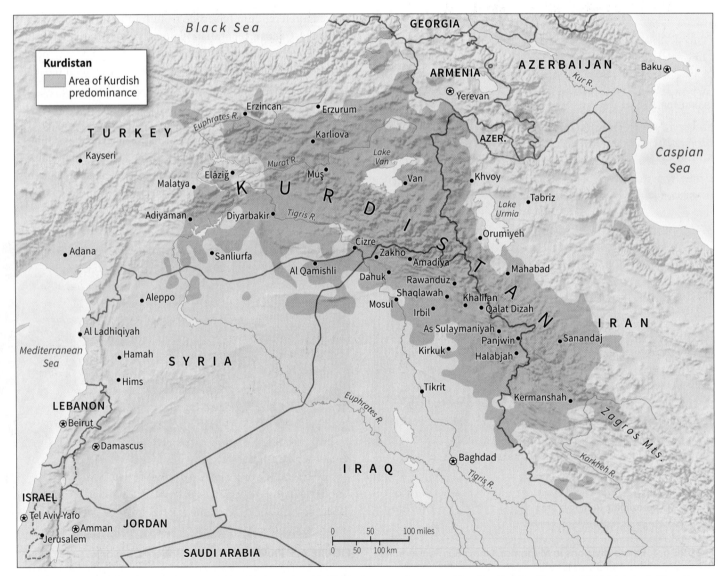

FIGURE 6.6 Kurdistan. The mountainous Kurdish homeland is divided among four countries. The Kurds, with an estimated population of 30–45 million, have lived in this region for millennia. Many of them have sought independence through political and military means, but so far it has eluded them. (Source: Office of the Geographer, U.S. Department of State.)

10 million and 5.5 million. Given the circumstances, it should not surprise us that Iraqi Kurds voted to form an independent country.

Internal Organization of Territory

States differ greatly in the way their territory is organized for administrative governance, with significant implications for territorial coherence and fragmentation. Political geographers recognize two basic types of spatial organization, the **unitary state** and the **federal state**. In unitary countries, power is concentrated centrally, with little or no authority vested in subnational political units. All major decisions come from the central government, and policies are applied uniformly throughout the national territory. France and China have unitary structures. A federal state, by contrast, disperses significant authority among subnational units such as provinces and states (in the sense of U.S. states). The United States, Canada, Germany, Australia, and Switzerland, though exhibiting varying degrees of federalism, provide examples. By allowing more latitude for subnational self-rule, the federal system can help reduce centrifugal forces and undermine public support for separatist movements.

Whether federal or unitary, a country functions through some system of political subdivisions. One increasingly common worldwide phenomenon is the establishment of territorial autonomy for indigenous peoples. For example, the Native American reservations in the United States occupy a unique and ambiguous place in the federal system of political subdivisions. These territorial enclaves have limited sovereignty and autonomy, which are derived from international treaties that the U.S. federal government signed with Native American nations during the decades of European conquest and settlement. Most reservations are small remnants of the original treaty territories. Nonetheless, Native Americans do have certain rights to self-government that differ from those of surrounding local authorities. For example, many reservations can (and do) build casinos on their land, even though gambling may be illegal in surrounding political jurisdictions. Reservations are an anomaly in the American political system of states, counties, townships, precincts, and incorporated municipalities.

Geography and the Vote

Voting is another political activity with strong regional implications. A free vote of all adult citizens can provide a clear expression of regional differences in religion, ethnicity, and cultural norms. Likewise, the regional apportionment of the collective vote can significantly influence election outcomes. **Electoral geography** is a subfield of political geography that analyzes the geographic character of political preferences and how geography can shape voting outcomes.

Gerrymandering Let us begin with the ways that geography can shape voting outcomes. Most democratic countries divide their territories into voting districts or precincts. Citizens cast their votes based on their residence within those precinct boundaries. For example, after every U.S. decennial census, the federal government conducts congressional **reapportionment**, the process by which the 435 seats in the U.S. House of Representatives are divided proportionally by population among the 50 states. In the 2010 census many northern states lost population, and so lost congressional seats through reapportionment (**Figure 6.7**). Following reapportionment, each state legislature oversees redistricting for their state. In **redistricting**, new boundaries for U.S. congressional districts are drawn to reflect the population changes since the previous census. The goal is to establish voting areas of more or less equal population and to increase or reduce the number of districts depending on the change in total population. These districts form the electoral basis for the U.S. House of Representatives. State legislatures are based on similar districts.

Geographers have historically assisted in drawing new boundaries in the redistricting process. Today, advances in GIS software make it possible for anyone with a bit of computer training to instantly draw and redraw congressional district boundaries to achieve a desired outcome. The wide availability of redistricting software has important implications for democracy. Political geographers and U.S. courts are increasingly interested in redistricting because the location of district or precinct boundaries can significantly influence election results. Boundaries can be manipulated to favor one political party, one ethnic group, or one religion over another. Historically in the United States, when a political party holds majority power in a state legislature, it will often try to redistrict to maximize and perpetuate its majority. The practice of manipulating voting district boundaries to favor a particular political party, group, or election outcome is called **gerrymandering**. Typically the resultant voting districts have odd, irregular shapes (**Figure 6.8**).

Gerrymandering can be accomplished by one of two methods (**Figure 6.9**). One is to draw district boundaries so as to concentrate all of the opposition party into one district, thereby creating an unnecessarily large majority while also ensuring that it cannot win elsewhere. In the lingo of political operators, this is known as "packing." A second method is to draw the boundaries so as to divide opposition votes into many districts. This has the effect of diluting the opposition's vote so that it does not form a simple majority in any district. This is known

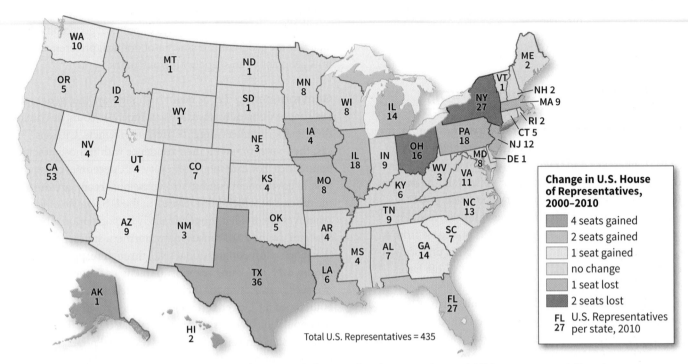

FIGURE 6.7 Reapportioning the U.S. House of Representatives seats. After every decennial U.S. census, population changes are calculated state by state to reapportion the 435 U.S. House seats. The changes from the 2010 census indicate the growing political clout of the Sunbelt states.

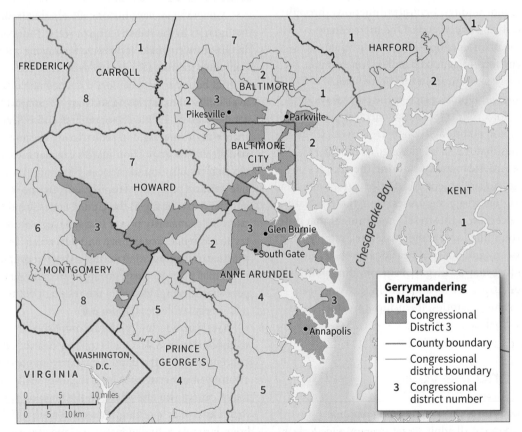

FIGURE 6.8 Gerrymandering of a congressional district in Maryland. The shape of Maryland's Third District, likened to an "amoeba convention" and "a broken-winged pterodactyl," is a case in point. While gerrymandering is difficult to prove, such shapes raise suspicions. In this case, Democrats defended the shape, while Republicans wanted the district redrawn. (Source: NationalAtlas.gov.)

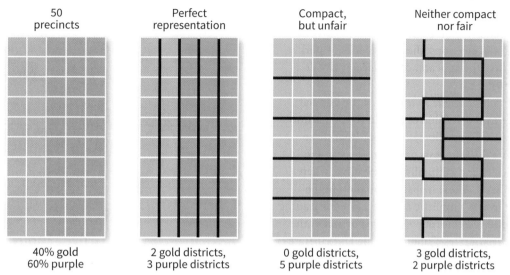

50 precincts	Perfect representation	Compact, but unfair	Neither compact nor fair
40% gold 60% purple	2 gold districts, 3 purple districts	0 gold districts, 5 purple districts	3 gold districts, 2 purple districts

FIGURE 6.9 How to gerrymander the vote. These diagrams show clearly how one can turn an electoral minority party into a de facto ruling majority by manipulating boundaries in the redistricting process. Advances in GIS software allow partisan officials to crack and pack the vote with great speed and accuracy.

as "cracking." In both cases, votes are "wasted," either because they are well beyond the number needed to win (packing) or because the boundaries have already established that there are not enough votes in the district to win (cracking). Consequently, some argue, full democratic participation in election outcomes is denied.

Gerrymandering is nothing new. It has been a recognized, if sometimes illegal, political tactic in the United States since the early 1800s. However, recent demographic shifts, historic racial and ethnic party affiliations, and advances in redistricting software have combined to make gerrymandering an ever more common and controversial practice in the United States. Texas provides a good example. The 2010 census showed the state population had grown enough to gain an additional four congressional seats. Hispanics and African Americans (historically Democratic voters in Texas) accounted for the vast majority of the population growth. The Republican Party, however, controlled the state legislature. Seventy percent of the new districts drawn by the Republican legislature had white (historically Republican supporters in Texas) majorities, even though whites' share of the statewide population dropped from 52 percent to 45 percent. Republicans consequently won an additional four districts, while Democrats gained none. Because the Texas redistricting was so blatantly discriminatory toward minority populations, it was repeatedly and successfully challenged in federal courts.

As we go to press, a case from Wisconsin is pending before the U.S. Supreme Court. Here, the issue is not racial or ethnic discrimination. Rather, this case revolves around the dilemma that the new GIS technology poses for democracy. The technology can be used to create the most equitable distribution of votes among the population or to create a distribution that all but guarantees electoral outcomes, which means that not all votes will carry equal weight. The party in power argues that it was democratically elected and therefore its redistricting reflects the will of the people. Opponents of the redistricting argue that it locks in an unfair advantage to the party in power and is inherently undemocratic. Their argument before the court is that Wisconsin's districts violate the Fourteenth Amendment of the U.S. Constitution, which in essence assures equal participation in elections through the principle of one person, one vote. Democracy and geography are tightly linked in what will likely be a landmark Supreme Court decision.

Wisconsin and Texas are two of 14 states, including most of the nation's most populous, where legislators either packed or cracked districts to assure their party would always win the majority of congressional seats. Whatever one's party affiliation, it is clear that democracy is not served when geographic boundaries are manipulated in such a way that majority preferences are not reflected in election outcomes.

Red States, Blue States We turn now to an example of how voting results may reveal regional differences. Political commentators in popular media often use maps of national election results to suggest that the United States is divided into distinct culture regions. We can trace this argument to the postelection analysis of the controversial 2000 presidential election, when, for the first time, news media adopted a universal color scheme for mapping voter preferences. States where the majority of voters favored the Democratic Party's presidential candidate were colored blue, whereas those favoring the Republican candidate were assigned red. When cartographers mapped the state-by-state results of this and subsequent presidential elections, the solid blocks of starkly contrasting colors gave the appearance of a deeply divided country (**Figure 6.10**).

The terms *red state* and *blue state* entered the popular lexicon in reference to the division of the United States into regions of "conservative" and "liberal" cultures that these maps implied. Figure 6.10 suggests, for example, that the U.S. South is politically and culturally conservative, whereas the West Coast is liberal. These maps, so common in the popular media, are overly simplistic and therefore misleading. For example, the solid red coloring of a large portion of the country in Figure 6.10 implies that most of the United States is culturally and politically conservative. However, this map gives too much visual importance to states with large areas, no matter what their population. A cartogram provides a better representation of both the true significance of states in the presidential election and the relative proportion of voters favoring Democrats or Republicans (**Figure 6.11**). When states' populations are taken into account, U.S. voters look much more equally favorable toward Democratic and Republican candidates. Indeed, the 2016 election was a rare case in which one presidential candidate, Hillary Clinton, a Democrat, received the majority of the nationwide vote, but lost the election because of the way the Electoral College apportions votes by state.

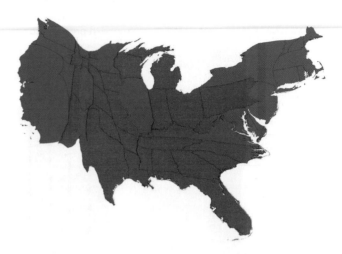

FIGURE 6.11 Cartogram of the 2016 presidential election results. This map sizes America's 50 states in proportion to their respective populations. (Courtesy of Mark Newman, University of Michigan.)

The red state/blue state designation also exaggerates the appearance of sharp geographic divisions within the country. Whether a state is designated "red" or "blue" is based on the winner-take-all system that the United States uses for presidential elections. That is, the candidate who receives a simple majority in the popular vote in a state takes all the Electoral College votes for that state. George W. Bush, for example, defeated Al Gore in Florida by only 537 votes in 2000, but he received all 25 of the state's electoral votes. This was an unusually close result, but it is common in presidential elections for only a small percentage of a state's popular vote to separate the winner and the loser. Designating a state such as Florida as "red" thus masks the narrow margin of the Republican victory and exaggerates the existence of regional polarization.

The limitations of the red state/blue state designation have led to the introduction of the term *purple state* to signify closely divided states. If we apply the purple state idea (i.e., shadings of color rather than stark contrasts) to county-level election results, we find that the simplistic division of the United States into liberal and conservative regions begins to break down (**Figure 6.12**). There is a great deal of county-by-county variation within states across the United States. New geographic patterns emerge in this map that complicate the idea of large regional divisions in the country.

Using shades of color and county-level results—rather than using contrasting colors and state-level results—provides a very different picture of U.S. regional political and cultural differences. Even this nuanced and finely grained cartographic presentation does not reveal all of the geographic complexities of U.S. presidential elections. With three-dimensional

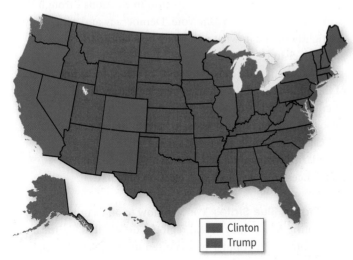

Clinton
Trump

FIGURE 6.10 2016 presidential election results. News media have adopted a standard color coding for political parties—red for Republican, blue for Democrat—in election result maps. It is now common to hear of the United States being politically and culturally divided into "red state" and "blue state" regions. Blue regions can be identified in the Northeast and the Pacific Coast. (Courtesy of Mark Newman, University of Michigan.)

FIGURE 6.12 Purple America. Rather than using stark color contrasts to represent the 2016 presidential election results, this map uses color shading. The political divisions appear less definitive and strong. (Courtesy of Mark Newman, University of Michigan.)

cartographic techniques, another pattern emerges. In the map in **Figure 6.13**, height represents voters per square mile, so that volume represents total number of voters. Here, the key geographic difference appears to exist between cities and rural areas. The general pattern we see is that rural areas are predominantly Republican and cities are predominantly Democratic.

Thus, a closer analysis of the red state/blue state phenomenon reveals some of the complexities involved in electoral geography. Choices about the scale at which data are aggregated, the cartographic techniques employed, and even the

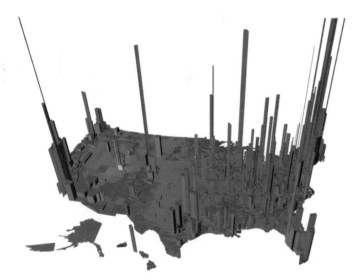

FIGURE 6.13 Presidential election results in 3D. In this map of the 2016 election, height represents voters per square mile, so urban areas stand out in contrast to suburban and rural areas. (Courtesy of Robert J. Vanderbei, Princeton University.)

map colors influence the interpretation of regional differences in election results. This is something to keep in mind in evaluating claims about a divided America.

Mobility

6.2 Explain how mobility influences political geography, including the spatial diffusion of political authority and innovation, the displacement effects of violent political conflict, and the role of the Internet.

The movement of both ideas and people affects political geography. For example, political authority and innovation diffuse in both space and time, despite barriers. Political events and developments can trigger massive human migrations, sometimes spanning continents. In recent years, violent political conflicts, including ethnic cleansing and civil wars, have produced a record number of refugees worldwide and resulted in several protracted refugee situations. Finally, the mobility of ideas over the Internet sometimes overcomes and, at other times, is stymied by political borders.

The Diffusion of Political Authority

Many independent states sprang fully formed into the world, a legacy of European colonialism. Some of the oldest states, however, expanded outward from geographic centers of power, often over the course of centuries. Historically, political power centers form in locations possessing particularly attractive resources (such as timber or fertile soils) and geographic qualities (such as natural harbors or wide river valleys). Increasing numbers of people begin to cluster there, especially if the area has some measure of natural defense against invaders. A major city emerges, with wealth enough to support a large army, which then provides the means for further diffusion of authority outward. Resources, people, and capital investment begin to flow between the center and surrounding territories, usually resulting in a further consolidation of the center's wealth and power. Ultimately, the urban power center becomes the country's capital city.

In states formed in this way, the ancient power center remains the country's single most important region, with the capital city and the cultural and economic heart of the nation. For example, France expanded from a small area around the present capital city of Paris to its current much larger

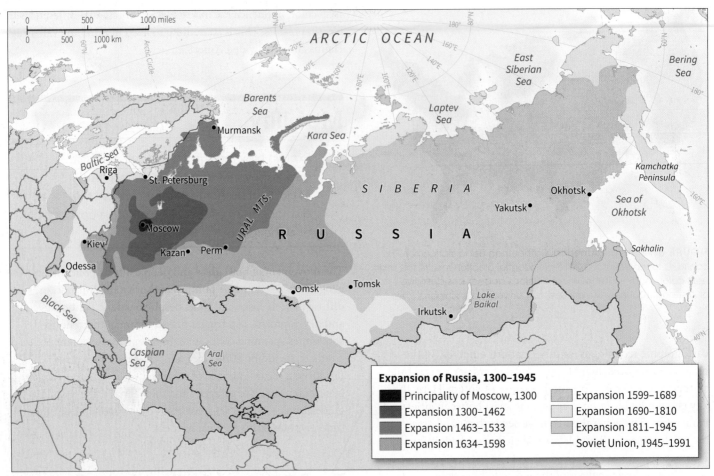

FIGURE 6.14 The diffusion of Russian political control. From a small, concentrated settlement during the Middle Ages, Moscow gradually extended its political control over an increasing amount of territory. By the nineteenth century, Russia extended from the Baltic Sea to the Pacific Ocean.

territory. China grew from a nucleus in the northeast, and Russia developed from a small region that is now Moscow (**Figure 6.14**). The United States' political authority diffused westward from an area on the Atlantic coastal plain between Massachusetts and Virginia. That region today includes the national capital, the oldest, most prestigious universities, national cultural institutions, and the nation's densest population.

Occasionally, people draw nation-state boundaries that encompass more than one geographic power center, a situation that can generate centrifugal forces. Modern Spain is illustrative. We can trace its territorial boundaries to a fifteenth-century treaty that united two monarchies (**Figure 6.15**). Present-day Barcelona, Catalonia's capital city, was part of one of those kingdoms, known as Aragon, while Madrid, the Castilian capital, belonged to the other, Castile. As the ideals of nationalism and self-determination diffused in the nineteenth century, a Catalan nationalist movement arose and agitated for independence. Barcelona and Madrid

became competing centers of power, a situation that has continued to threaten to break up Spain's territory. A similar condition can be found in the United Kingdom, between London, England, and Edinburgh, Scotland. There, too, Scottish nationalists have periodically tried to secede to form an independent Scotland.

The Diffusion of Political Innovation

One of the consequences of colonialism was the creation of new patterns of mobility among the conquered populations. One pattern in particular contained the seeds of colonialism's demise. Colonial rule in Africa depended on the establishment of a relatively small cadre of educated African civil servants. The colonial rulers selected those whom they considered the best and the brightest (and most politically cooperative) among their African subjects and sent them to the best universities in Europe. In addition to learning the skills required of colonial

FIGURE 6.15 Spain in the fifteenth century. Two powerful kingdoms, Castile (centered in Madrid) and Aragon (centered in Barcelona) were united in 1479 to form the core of modern Spain. The union, however, did not erase regional cultural and political differences, and separatist movements continue to this day.

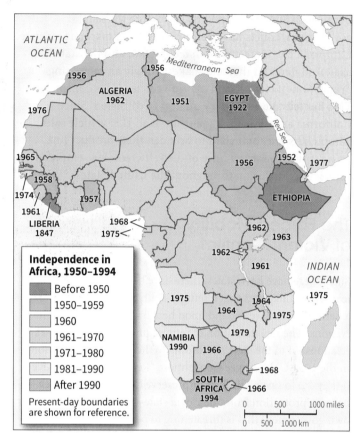

FIGURE 6.16 Independence from European colonial rule spread rapidly through Africa in the 1950s and 1960s. Prior to the 1950s, there were only two independent countries in Africa. Between 1951 and 1968, most African colonies attained self-rule. The remaining few gained freedom over the next three decades. By 1994 independence had spread from the Mediterranean to the Cape of Good Hope.

functionaries, their travels exposed them to the ideals of Western political philosophy, particularly the right of self-determination. Worldly, educated Africans returned to their countries armed with these ideas and organized new nationalist movements to oppose colonial rule.

The consequences can be seen in the spread of political independence in Africa. In 1914 only two African countries—Liberia and Ethiopia—were fully independent of European colonial or white minority rule, and even Ethiopia later fell temporarily under Italian control. Influenced by developments in India and Pakistan, the Arabs of North Africa began a movement for independence. Their movement gained momentum in the 1950s and swept southward across most of the African continent between 1960 and 1965. Many of Africa's great nationalist leaders of this period had obtained university degrees in England, Scotland, France, and other European countries (**Figure 6.16**).

Despite its rapid spread, the diffusion of African self-rule occasionally encountered hurdles. Portugal, for example, clung tenaciously to its African colonies until 1975, when a change in government in Lisbon reversed a 500-year-old policy and allowed the colonies to become independent. In colonies containing large populations of white settlers, independence came slowly and usually with bloodshed. France, for example, sought to hold onto Algeria because many French colonists had settled there. The country nonetheless achieved independence

in 1962, but only after years of violence. In Zimbabwe, a large population of white settlers refused London's orders to move toward majority rule, resulting in a bloody civil war and postponing the country's independence until 1979. White minority–ruled South Africa, which resisted democratization and self-rule the longest, delayed Africa's final decolonization until 1994.

On a quite different scale, political innovations also spread within independent countries, particularly in federal states. Federal statutes permit, to some degree, laws to be adopted by subnational territories. In the United States and Canada, for example, each state and province enjoys broad lawmaking powers, vested in the legislative bodies of these subdivisions. Sometimes diffusion results. A good example is air pollution laws, which began in California with the initiation of state legislation regulating automotive and industrial emissions. Clean air laws later spread to other states and provided the model for federal clean air legislation.

Because California enacted its law before the federal government, courts have determined that it takes precedence. California law is stricter than federal law, and the state has used it to demand that automakers supply an increasing proportion of electric and hybrid vehicles to its consumers. This has led to a secondary process of diffusion. Since California accounts for 12 percent of all U.S. vehicle sales, automakers are compelled to broaden their product line. As a result, the sale and use of fuel-efficient, low-emissions vehicles are diffusing throughout the United States.

The Forced Mobility of Violent Conflict

Since there have been states, there have been refugees. We learned in Chapter 3 that a refugee is a type of forced migrant, a definition that we expand upon here by looking to the United Nations. The UN definition of refugee pivots around people's relationship to the state. According to the extended definition in the 1951 UN Refugee Convention, "Refugees have to move if they are to save their lives or preserve their freedom. They have no protection from their own state—indeed it is often their own government that is threatening to persecute them." Often repression is driven by cultural difference—based on religion, race, or ethnicity—as much as by political disagreement. It seems that the very idea of the state contains both the promise of stability and protection and the threat of displacement and persecution. As geographers Jennifer Hyndman and Wenona Giles observe, "Displacement is the underbelly of mobility, a kind of movement that expresses the violent political relation of people to place."

Ethnic Cleansing So-called **ethnic cleansing**—the forced removal of an ethnic group by another ethnic group to create ethnically homogeneous territories—is behind many recent forced international migrations. In the 1990s, Yugoslavia's ethnic wars created refugees across Europe's Balkan Peninsula while ethnic cleansing in Rwanda sent refugees into three different countries. This phenomenon has continued, even accelerated, in the twenty-first century as dominant ethnic groups attempt to "purify" state territories. International observers accuse the military of Sudan of beginning a genocidal war in the Darfur region of that country. The people of Darfur are ethnically distinct from the dominant Arab ethnic group controlling the Sudanese state. Many thousands of Darfurians now live in refugee camps in Chad. In 2017, the Myanmar military, which is dominated by Buddhist Burmans, began an ethnic cleansing campaign against Muslim Rohingya in the country's far west (**Figure 6.17**). In an ongoing crisis, UN authorities estimated that over half the

FIGURE 6.17 Ethnic cleansing in western Myanmar. The Rohingya people, an Islamic culture group, have been driven over the border into Bangladesh. The central government, dominated by ethnic Burmans, a Buddhist culture group, does not count the Rohingya as citizens of Myanmar, though they have lived there for centuries. (Paula Bronstein/Getty Images)

Rohingya population, nearly 700,000 people, have already been forced to flee across the border to Bangladesh. Tragically, this is a small sample of the cases of ethnic cleansing in today's global political geography.

Armed Conflict Most refugees are fleeing some form of armed conflict including interstate wars, civil wars, insurgencies, and counterinsurgencies. Syria presents a tragic but all-too-common example of a twenty-first-century refugee crisis. Its citizenry is a complex assortment of religious and ethnic identities, including Arab-Sunni Muslim, Arab-Alawite (a Muslim Shi'ite sect), Kurd-Sunni, and Greek Orthodox Christian. The Baath Party controlled the government for decades, now led by President Bashir al-Assad, an Alawite. When the Arab Spring of 2011 swept through surrounding countries, overthrowing entrenched regimes, peaceful street protesters in Syria demanded a change in government. After violent government reprisals, they soon developed into armed rebel groups, with multiple militias and ethnic, political, and religious factions from Syria and nearby countries taking sides in the fight. The protests subsequently escalated into a full-blown civil war. By 2013 Syria was the source of at least 1.6 million refugees who flowed into the nearby countries of Lebanon, Turkey, Egypt, Iraq, and Jordan at a rate of 250,000 per month. Jordan hosted the largest numbers of refugees, with its al-Zaatari refugee camp vying for the dubious distinction as the world's largest.

International Response As noted in Chapter 1, the number of forced migrants worldwide has been growing in the twenty-first century. Estimates of refugees vary greatly for political reasons (national governments often impede accurate counting) as well as practical considerations (people on the move across borders are difficult to count). The United Nations High Commissioner for Refugees (UNHCR) is widely recognized as the most reliable source. In 2015, its most recent estimate, the UNHCR documented 17.2 million refugees under its mandate worldwide, plus another 5.3 million semipermanent Palestinian refugees resulting from the establishment of the state of Israel in 1948 and the 1967 Arab–Israeli war (**Figure 6.18**).

The great majority of refugees may be found in less-industrialized countries, many of which do not have adequate infrastructure or resources to manage the flow of forced migrants. This is where the UNHCR comes in. Its primary mission is to safeguard the rights and well-being of refugees. In practical terms, this means delivering shelters, clothing, and medical and personal hygiene supplies to millions of people in hundreds of refugee camps. Refugee camps are meant to be temporary settlements to house and care for displaced persons until they can be safely repatriated. Sometimes repatriation takes years, in which case the UNHCR acts as a surrogate state, securing birth certificates, facilitating schooling, and improving sanitary infrastructure.

There is thus a nearly hidden political geography of displaced persons worldwide. The UNHCR operates hundreds of camps and settlements housing millions of people in 124 countries. Although meant as temporary solutions, some camps last for decades and become major population centers that remain absent on most maps. For example, Dadaab, Kenya, is the location of one of the oldest refugee camp complexes, established in 1991 and currently housing at least 235,000 people (**Figure 6.19**). The camp complex and its population do not show up on standard world atlases. Imagine an atlas without Buffalo, New York, or Reno, Nevada, U.S. cities with approximately the same population totals. An increasing number of refugee camps fall under this designation, thus creating a geography of semipermanent settlements the size of major cities that remain invisible on the world map. The UN refers to cases such as the Dadaab camp as a **protracted refugee situation (PRS)**. People in such situations are known as **stateless people**, denied a nationality and therefore deprived of basic rights such as education, health care, employment, and freedom of movement. Millions of people are born and come into adulthood in these refugee camps with no promise of becoming citizens in a new country and little hope of returning to a homeland they never knew.

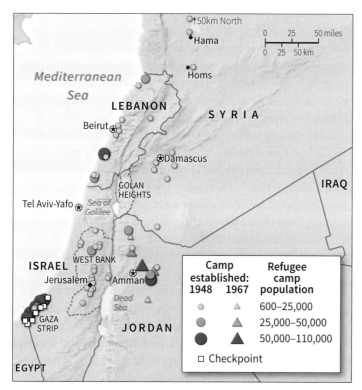

FIGURE 6.18 Refugee spaces in the Middle East. Civilians fleeing armed conflict that accompanied the establishment of the state of Israel in 1948 produced a new political geography of refugee settlements in the region. The United Nations continues to administer 58 semipermanent refugee camps in four countries. Civil and legal rights for the 1.4 million camp residents vary greatly from country to country. (Data from http://www.unrwa.org/where-we-work.)

FIGURE 6.19 City-sized refugee camp. Protracted refugee situations have led to the growth of permanent refugee camps that rival the size of major American cities. With a population of over 235,000, the camp near Dadaab, Kenya, is one of the world's largest. (Oli Scarff/Getty Images)

Geographers have taken an interest in PRSs, which present conditions of both mobility and long-term immobility. From a feminist political geography perspective, northern, developed states perceive refugees very differently if they are in a PRS or are actively on the move. Jennifer Hyndman and Wenona Giles show that refugees in PRSs are feminized based on their location and legal status. That is, refugees in PRSs are viewed as externalized, motionless, and politically passive and powerless, and thus are coded as nonthreatening/feminine. In contrast, refugees seeking asylum in northern, developed countries are viewed as internalized, on the move, and politically active and vocal, and thus are coded as threatening to both national security and the welfare state. Such feminization, Hyndman and Giles argue, can serve to reinforce the conditions that create PRSs by reducing urgency and lowering priorities for official action.

In another PRS study, geographer Adam Ramadan reveals the relative and contingent qualities of mobility/immobility stasis. Ramadan categorizes the refugee camp as a distinctive political space. PRS camps can become, in his words, "spaces of exceptional sovereignty." To illustrate his point, he examines the case of Palestinian refugee camps within Lebanon during the brief 2006 war between Israel and Hizbullah, a Shi'ite political and paramilitary organization. When Israel bombed Hizbullah positions in southern Lebanon, nearly 5000 Lebanese citizens fled to the Palestinian refugee camps of El-Bus and Rashidieh for shelter and protection. Over generations, Palestinian refugees in southern Lebanon have evolved a great degree of internal autonomy over camp political and social affairs, thus gaining a degree of sovereignty over these spaces. They exercised that sovereignty, making the decision to receive thousands of the displaced Lebanese citizens into their private homes and public buildings and care for their well-being. The study shows how the 2006 war inverted the normal relationship of host (Lebanon) and guest (Palestinian refugee). In addition, the war shifted Lebanese citizens' vision of the camps from dangerous, no-go zones to spaces of shelter and refuge. Finally, it demonstrated and reinforced Palestinian refugees' sovereignty over spaces originally conceived to be temporary and void of political status.

Transborder Mobility in Cyberspace

As the transborder movement of refugees has increased, so too has another transborder phenomenon; the Internet. What happens to political boundaries in cyberspace? The Internet and various social media have facilitated the movement of goods, money, and ideas across international borders in unprecedented new ways. These technologies make borders seem more porous, even nonexistent. We are just beginning to understand the implications for territorial sovereignty. In Germany, for example, literature espousing Nazi ideology is outlawed, yet the U.S. Constitution's First Amendment right of freedom of speech protects the dissemination of this material in the United States. What happens when such literature originates in the United States and is transmitted via social media worldwide? Should the German government prosecute the originator of the message—an American—for violating a German law?

How can countries impose their laws, defend their cultural norms, and maintain sovereignty in the Internet age? One way is to restrict citizens' access to technological hardware such as satellite dishes, computers, phone lines, and cell phones. Many countries try this, but it is becoming increasingly difficult for even the most authoritarian regimes. Another tactic deployed by states is to allow citizens' access to the web but to control the flow of information they receive. The so-called "Great Firewall of China" uses a range of computer software tools to block Chinese citizens' access to many foreign websites and to slow down Internet traffic across its borders.

One well-known example of China's effort to control its citizens' access to information is the agreement it made with Google, the company that invented the world's most popular search engine. In order to gain access to China's rapidly expanding market for Internet services, Google agreed in 2006 to create a self-censoring search engine. Google filtered out websites that violated China's censorship laws, including those promoting ideas such as free speech and freedom of worship. By 2010, Google's relationship with China had deteriorated over alleged government cyberattacks on the company. Google refused to continue censoring and threatened to leave China. By then a Google-like Chinese company had taken over the market, and Google's share of all searches within the territory dropped steeply to less than 2 percent.

As the Google case illustrates, states often rely on the cooperation of private service providers in their efforts to control Internet access. Twitter, an online social networking company, offers another example. In 2013, Twitter initiated a policy of compliance with state censorship laws. Beginning with Germany, Twitter began blocking a neo-Nazi group's tweets because the country has laws restricting the promotion of Nazi ideology. While we may agree that blocking discriminatory speech is a good thing, such censorship in the United States would violate free speech rights. The Google and Twitter cases show that states still retain significant control over global information flows across borders, sometimes forcing private service providers to choose between complying with state censorship laws or being barred from doing business.

Governments may take extreme action to halt information flow during periods of unrest. During the pro-democracy upheavals in 2011 known as the Arab Spring, several Middle Eastern and North African governments tried to limit the role of the Internet in anti-government organizing. Egypt, Libya, and Syria severed their countries' Internet connections completely, while in Tunisia and Saudi Arabia governments used Internet censorship and surveillance to control information flow (**Figure 6.20**). The fact that many authoritarian governments sought to disconnect from the Internet demonstrates the potential power of this technology to promote citizens' aspirations for political freedoms. On the other hand, many state governments now use social media and the Internet to consolidate power internally, repress opposition, and pursue cyber warfare across international borders.

These cases illustrate that political boundaries are far from irrelevant in the age of the Internet. The Internet is, to a degree, eroding political boundaries by allowing information and ideas to diffuse more rapidly and completely. As a result, political barriers to cultural diffusion have become more fragile. However, countries still have some power to control what is allowed to move across their territorial boundaries, even in cyberspace. We take an in-depth look at the political promise and peril of the Internet in the Subject to Debate feature (page 204).

FIGURE 6.20 Cyber censorship during the Arab Spring. In 2011, a wave of political protest swept across Arab countries with the aid of the Internet and various social media. Several governments responded by drastically restricting their citizens' access to the Internet. (KHALED DESOUKI/AFP/Getty Images)

Globalization

6.3 Explain the effects of globalization on sovereignty, supranationalism, and transnationality.

The effects of globalization on political geography are complex and sometimes contradictory. In some cases it has weakened national sovereignty and in other cases, strengthened it. Recently, many supranational organizations have emerged, resulting in some unresolved tensions. Finally, globalization has given rise to a new phenomenon, transnationalism, that is experienced across the social scale, from migrant workers to top executives of multinational corporations.

Globalization and Sovereignty

Many commentators have argued in recent years that globalization is eroding the sovereignty of the nation-state. Globalization, it is said, operates in networks and flows that are less and less affected by international borders. Some observers suggest that the power of states in the global order has been greatly diminished by free trade agreements, the Internet, satellite communications, and other such developments. In addition, many economists argue that "global markets" will increasingly dictate an individual state's social policies. That is, the global market will force states to reduce the cost of doing business so that they can remain competitive. From this perspective, the cumulative effect of globalizing processes is a diminution of a nation-state's control over affairs within its boundaries.

According to geographer John Agnew, the idea that globalization is diminishing the role of states is based on a misunderstanding of the history and character of state sovereignty. Agnew argues that sovereignty has been (1) mistakenly equated to territory and (2) assumed to be equally distributed among states. That is, neither the assumption that a state has total control over affairs within its territorial boundaries nor the assumption that a state's sovereignty ends at its borders is historically accurate. These mistaken assumptions lead to an all-or-nothing view of sovereignty. Agnew suggests it is more accurate to think in terms of **effective sovereignty**. That, is we should give attention to the power of states to effectively enforce or ignore sovereignty claims irrespective of territorial boundaries.

An example will help to illustrate these points. The U.S. government began holding international prisoners in a detention center on the naval base at Guantánamo Bay,

SUBJECT TO DEBATE   Whither the Political Promise of the Internet?

In the early years of the Internet, many observers celebrated its potential to liberate people from political oppression. They believed it would allow citizens to bypass the state to directly link individuals with individuals, make geography irrelevant and sovereignty obsolete, and open a new era of personal freedoms and democratic governance. Many saw the role of social media in the Arab Spring of 2010–2011, discussed in this chapter, as exemplary. In Egypt and Tunisia, pro-democracy activists used Twitter, YouTube, and Facebook to organize street protests and ultimately topple two authoritarian governments. This was the promise of the Internet realized.

Recent political uses of social media reveal that the Internet brings perils as well. The Internet can endlessly fragment and customize information to match the interests of individual consumers. As a consequence, the shared worldview that binds a country's citizens together has eroded, a situation that is increasingly exploited by antidemocratic forces. Technology experts found that organizations linked to the Russian government use social media to promote political division in an effort to destabilize established democracies in the United States and Western Europe. Hate groups in the United States use social media to promote racial and ethnic divisions among citizens. The very social media used in support of the Arab Spring are now used by violent militants in the region to organize attacks on civilians and recruit allies, with the ultimate goal of imposing their own political authority. It appears the Internet can be a tool for liberation or repression, unity or discord.

Continuing the Debate

The Internet is largely self-regulated by the corporations—Google, Facebook, Twitter, etc.—that serve as neutral platforms providing information services to consumers. Specifically, service providers are not subject to the same disclosure regulations regarding political influence as are television, newspapers, and radio. With this knowledge and given the documented uses of the Internet for political purposes, consider the following discussion questions.

- Would you characterize the Internet as promoting primarily centrifugal forces, centripetal forces, or neither?

- How might we go about changing the Internet to reduce its more antidemocratic and divisive uses?

- Do you think that governments should regulate how information is disseminated and presented on the Internet, or should individual consumers be allowed to make their own judgments on its value?

- In what ways do your own social media habits narrow or expand your access to multiple political points of view?

- When you see political information on the Internet, what methods do you use to assess its validity and reliability?

Social media in the aid of political movements. Proponents initially touted social media as democratizing force, but it can also be used by antidemocratic and hate groups such as these white supremacist demonstrators in Charlottesville, Virginia, in 2017. (Samuel Corum/Anadolu Agency/Getty Images)

Cuba, in 2002 (**Figure 6.21**). In court disputes with detention opponents, the U.S. government argued that the Guantánamo center is outside of federal court jurisdiction because it is not part of U.S. territory. At the same time, the U.S. government ignored the Cuban state's claim of sovereignty over the area and enforced sole authority within the boundaries of the naval base. In short, the United States exercised *effective* sovereignty over a space that is contained within Cuba's territorial boundaries but completely outside of the United States'. Sovereignty is not matched to nation-state territory in this case, nor is it equally distributed between Cuba and the United States. We could present many similar examples related to international migration, Internet use, global transportation, and other such globalizing processes.

How does Agnew's argument help us understand globalization's effects on state sovereignty? First, not all states will be affected by globalization in the same way, and many stronger states will have a great deal of influence over the way globalization unfolds. For example, the U.S. government's new banking regulations implemented within the country's

FIGURE 6.21 A case of effective sovereignty. Although outside of its territorial boundaries, the U.S. government exercises effective sovereignty over Guantánamo Bay, Cuba. Far from a unique case, the U.S. naval base at Guantánamo illustrates the mistake of equating state sovereignty with state territory. (Chantal Valery/AFP/Getty Images)

borders after the 2008–2009 credit crisis greatly altered the way global financial markets operate. Second, many of the challenges to sovereignty presented by globalization have always existed in some form. Globalization did not create these challenges, but has made them vastly more complicated and wide ranging. As Agnew concludes, the problem is "simply too complex for the binary thinking—globalization versus states."

Supranational Political Bodies

In recent decades, the forces of globalization have many times led independent countries to voluntarily relinquish some territorial sovereignty in exchange for greater leverage in the global sphere. **Supranationalism**, the term for this trend, exists when a collection of nation-states and their citizens relinquish some sovereign rights to a larger-scale political body that exercises authority over its member states. We call these political bodies **supranational organizations**, and they have been on the rise worldwide (**Figure 6.22**). Supranational organizations are political bodies of international integration that nation-states establish in cooperation with their neighbors for mutual political, military, economic, or cultural gain. The types of advantage sought, and the reasons for seeking them, are quite varied, leading to many different kinds of supranational organizations and outcomes.

We can trace the origins of the phenomenon to World War II and its aftermath. As the war's end drew near in 1944, 44 countries negotiated the Bretton Woods Agreement to facilitate international trade and stability in global financial markets.

It established two important supranational organizations, the World Bank and the International Monetary Fund (IMF), whose goals were to promote international economic development and regulate global financial markets, including international credit and lending policies. Both institutions today have their share of supporters and critics. In particular, the IMF has been accused of violating the sovereignty of the world's poorest and least powerful countries by dictating national economic policy.

A basic supranational organizational form is the **regional trading bloc**, a multi-country agreement that reduces or eliminates customs duties and import tariffs to promote the freer flow of goods and services across international borders. The idea is to make the border disappear for many economic activities. The North American Free Trade Agreement (NAFTA) is a good example. NAFTA is a trilateral trade agreement between Canada, the United States, and Mexico that went into full effect in 1994. In addition to reducing trade barriers, the agreement also covers a range of related issues, such as intellectual property rights, environmental protection, and international transportation policy.

Supranational organizations have grown in number and importance in recent decades, coincident with and somewhat counterbalancing the proliferation of independent countries. Some represent the vestiges of collapsed empires, such as the British Commonwealth, French Community, and Commonwealth of Independent States (CIS)—the latter a vestige of the former Soviet Union. Many supranational organizations, such as the Arab League or the Association of Southeast Asian Nations (ASEAN), possess little cohesion, as individual state interests continue to override collective interests. Such situations are common because the imbalances in economic, political, and demographic power among member states are often extreme. In other words, who gains what and how much by ceding national sovereignty varies tremendously and can inhibit commitments to collective interests.

The **European Union**, or EU, is by far the most powerful, ambitious, and successful supranational organization in the world (see Figure 6.22). The EU is a political, economic, and social union of 28 independent countries in Europe that promotes the free movement of people, goods, services, and capital among its members. It grew from a central core area of six countries that formed a regional trading bloc in the 1950s whose main goal was to lower or remove tariffs that hindered trade. From that beginning, the EU formally emerged in 1993 with a total of 12 members. Member states gradually expanded the EU's political and cultural roles as it diffused outward to encompass even more members. In 2004, it underwent its greatest expansion when it brought in most of the eastern European satellite states of the former Soviet Union. One underlying motivation behind the EU's expansion is to

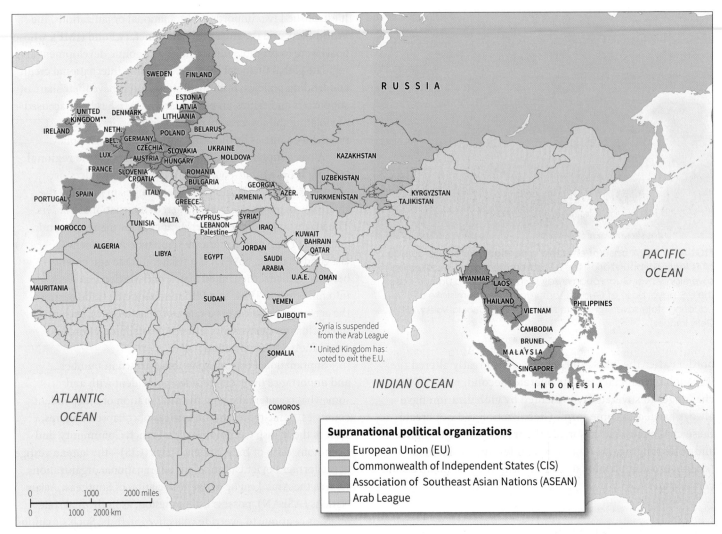

FIGURE 6.22 Some supranational political organizations in the Eastern Hemisphere. These organizations vary greatly in purpose and cohesion. ASEAN stands for the Association of Southeast Asian Nations, and its purposes are both economic and political.

weaken the power of its member countries to the point that they would never again wage war against one another—a response to the devastation of Europe in two world wars. Another was to create an economy and market on a scale large enough to compete globally with the likes of the United States and China.

The member countries have all sacrificed some of their sovereign powers to the EU administration. A single monetary currency, the euro, has been adopted by most EU members, replacing their national currencies. Most international borders within the EU were completely opened, requiring no passport checks. More importantly for citizenship and national identity, the EU has institutionalized a range of social norms related to the tolerance of religious and ethnic differences, human rights, and gender equality. In effect, the EU is pushing supranationalism to its logical conclusion. That is, at some future date nationalism as a primary identity may be obsolete

and people will think of themselves as European first and Italian, Latvian, or Polish second.

As with nation-states, supranational organizations are subject to centripetal and centrifugal forces. Some organizations have been threatened with break up. In 2016, the United Kingdom held a referendum on EU membership and its citizens voted to withdraw from the organization. The reasons are complex, but many voters were concerned that the free movement of EU citizens across member country borders had produced undesirable economic and cultural effects in the UK. Specifically, there were strong sentiments of British nationalism and nativism expressed in a vote driven by opposition to open immigration from eastern European member countries. Other EU member states face similar citizen opposition to diminished sovereignty, raising questions about the stability of the world's most notable supranational experiment. In the United States, popular opposition to

NAFTA has grown as more citizens have come to see it as a cause of job loss and declining wages. It was a major issue in the 2016 presidential election and could undergo significant restructuring, if not complete dissolution.

The Condition of Transnationality

Transnationalism is a term that we find with growing frequency in the halls of academia as well as in popular media. It refers to different ideas and phenomena, and is sometimes even used as a synonym for globalization. It also suggests that the processes of globalization are raising new questions about cultural identities, which is our concern here. We think of transnationalism as the experience of living and working in more than one country that can produce new and distinct cultural identities. Nationality and ethnicity continue to be important in an era of globalization. An increasing number of people, however, live, work, and play in a way that is rooted neither in an ethnic homeland nor in the territory of a nation-state. Rather, we see more people developing cultures of transnationalism.

Geographer Katharyne Mitchell suggests that we might think about some of the cultural implications of globalization in terms of a "condition of transnationality." She observes that the economic effects of globalization have produced a great increase in the movement of people across borders. This movement differs from other historical migrations, because it tends not to be unidirectional and permanent. With the aid of new transportation and communications technologies, people are able to maintain social networks and physically move about in ways that transcend political boundaries. The experience of cultural "in-betweenness"—of living in and being linked to multiple places around the globe, while being rooted in no single place—has become fundamental to the identity of a growing group of people.

Mitchell's study of Hong Kong Chinese immigrants in Vancouver, British Columbia, illustrates the idea of transnationalism. Immigrants entering the country under a law designed to attract business investment had to establish businesses based in Canada. The immigrants who took advantage of this law were mainly Hong Kong Chinese businesspeople leaving the colony in anticipation of its handover to the People's Republic of China.

As it happened, these immigrants maintained business and social ties in both Hong Kong and Canada, moving freely and frequently between them. In the process, a whole set of cultural conflicts were ignited between longtime Vancouver residents and the transnational Chinese. Neighborhoods were transformed as the transnationals attempted to establish themselves economically and culturally. They rapidly bought up real estate, demolished old houses, built houses in uncharacteristic architectural styles, and redesigned residential landscaping. Their mobility—a culturally defining characteristic of transnationals—made them appear transient and rootless in the eyes of longer-term residents and weakened the legitimacy of their claims and practices. This case shows us how the cultural aspects of transnationalism are bound up with the economic and political processes of globalization.

Nature-Culture

6.4 Analyze the complex interactions of politics and physical geography, including the state's role in land management, the heartland theory of world geopolitics, and the effects of rising sea levels on political territory.

How people use the land and natural resources is profoundly influenced by politics. Whether a particular habitat is conserved or degraded often has much to do with a country's land laws, tax codes, and agricultural policies. The relationship between physical geography and politics, referred to as geopolitics, is also important for explaining continental-scale political developments. Global warming and rising sea levels will become increasingly important for understanding changing nation-state boundaries and emerging international conflicts. Finally, national militaries also affect the natural environment.

State Policies and Land Management

When geographers Piers Blaikie and Harold Brookfield used the term *political ecology*, they were interested in how people made decisions about managing their lands within their specific political contexts. In particular, they wanted to know why some land managers adopt destructive practices that they know will harm the environment and therefore endanger their own futures. They observed that focusing only on proximate (i.e., direct) causes of environmental damage— for example, the farmer dumping pesticides in a river or the poor peasant cutting a patch of tropical forest—will provide a misleading explanation. They suggested that a more accurate explanation requires looking beyond proximate causes to larger-scale processes that influence land managers' decisions and actions. For example, if a farmer is in debt, she might decide to farm an area she knows is prone to soil erosion in order to expand cultivation and boost her income. She may know that this is harmful to the environment, and against her long-term interests in the land, but debt has forced her into a short-term survival strategy. We can take this analysis a step further and ask why the farmer is in debt. This might lead in any number of directions, including unfair market prices for farm produce, which make it difficult for farmers to recoup costs, forcing them to borrow. Furthermore, those unfair prices may be the result of state policies that favor cheap food for city dwellers over fair pricing for rural farmers, a fairly common political strategy. This example illustrates the importance of expanding our explanation of environmental damage beyond the immediate local circumstances to consider larger political and economic processes.

To Blaikie and Brookfield, therefore, the nation-state is key to explaining land managers' environmentally destructive practices. They demonstrate that national land laws, tax codes, credit policies, and other laws and policies all influence land-use decision making. For example, if a state assesses high taxes on land improvements, such as terracing and channeling, then its tax policies create disincentives for land managers to implement soil conservation measures. Conversely, if a state underwrites cheap loans to land managers to build such structures, its credit policies encourage soil conservation. Sometimes states provide such supports along racial and ethnic lines, favoring some groups over others. Consequently, geographic patterns of environmental damage reflect the state's differential treatment. There are many such examples of state influences on individual land-use decisions. Thus, one cannot fully explain the causes of environmental problems without analyzing the role of the state.

Geopolitics

Spatial variations in politics and the spread of political phenomena are often linked to terrain, soils, climate, natural resources, and other aspects of the physical environment. The term **geopolitics** was originally coined to describe the influence of geography and the environment on political entities. Conversely, established national political authority can be a powerful instrument of environmental modification, providing the framework for organized alteration of the landscape and for environmental protection.

Before modern air and missile warfare, a country's survival was often aided by some sort of natural protection, such as surrounding mountain ranges, deserts, or seas; bordering marshes or dense forests; or outward-facing escarpments. For example, in Egypt, desert wastelands to the east and west insulated the fertile, well-watered Nile Valley in the center.

France—centered on the plains of the Paris Basin and flanked by mountains and hills such as the Alps, Pyrenees, Ardennes, and Jura along its borders—provides another good example (**Figure 6.23**).

Expanding countries often regard coastlines as the logical limits to their territorial growth, even if those areas belong to other peoples, as the drive across the United States from the East Coast to the Pacific Ocean in the first half of the nineteenth century has made clear. U.S. expansion was justified by the doctrine of Manifest Destiny, which was based on the belief that the Pacific shoreline offered the predestined western border for the country. Underlying the doctrine was a racist ideology of Anglo-Saxon superiority, which held that Native Americans were savages blocking the progress of civilization and the productive use of the western lands. A similar doctrine led Russia to expand in the directions of the Mediterranean and Baltic seas and the Pacific and Indian oceans.

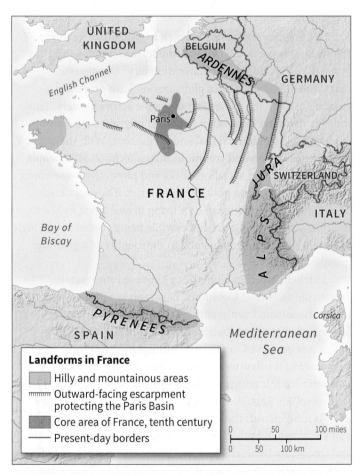

FIGURE 6.23 The distribution of landforms in France. Terrain features such as ridges, hills, and mountains offer natural protection for the country's center. Steep landforms provided physical barriers that inhibited invasions for centuries.

The Heartland Theory Beliefs about environmental influences and Manifest Destiny, combined with the history of Russian expansionism, led to the **heartland theory** of Halford Mackinder, a key figure in the history of geopolitical thought. Propounded in the early twentieth century and based on environmental determinism (see Chapter 1), the heartland theory addresses the balance of power in the world and, in particular, the possibility of world conquest based on situational advantage. It held that the Eurasian continental situation was the most likely geographic base from which to launch a successful campaign for world conquest.

In examining this huge landmass, Mackinder discerned two environmental regions: the **heartland**, which lies remote from the ice-free seas, and the **rimland**, the densely populated coastal fringes of Eurasia in the east, south, and west (**Figure 6.24**). Far from navigable seas, the heartland was invulnerable to the naval power of rimland empires. However, the rimland was vulnerable to invasion by armies from the heartland that could move overland through diverse natural gateways. Mackinder thus reasoned that a unified heartland power could conquer the maritime countries of the rimland with relative ease. He believed that the East European Plain would be the likely base for unification. As Russia had already unified this region at the time, Mackinder reasoned that the Russians were ideally situated and poised to pursue world conquest.

Following Russia's communist revolution in 1917, the leaders of rimland empires and the United States employed a policy of containment. This policy, in no small measure, found its origin in Mackinder's theory and resulted in numerous wars to contain what was then considered a Russian-inspired conspiracy of communist expansion. Overlooked all the while were the inherent fallacies of the heartland theory, particularly

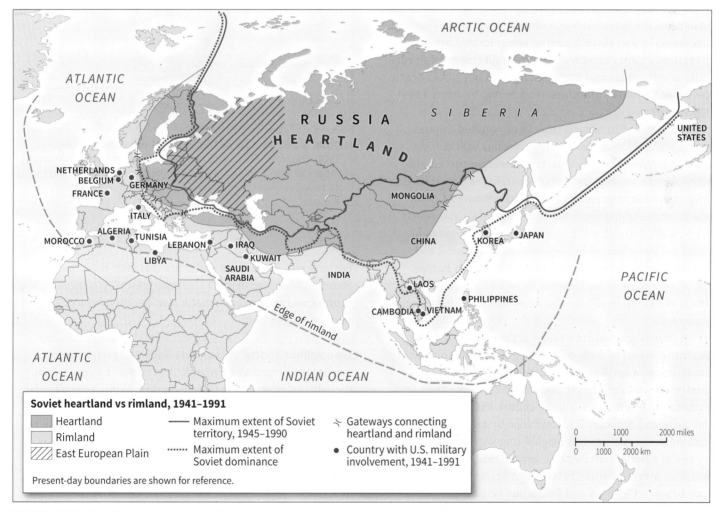

FIGURE 6.24 Heartland versus rimland in Eurasia. For most of the twentieth century, the heartland, epitomized by the Soviet Union and communism, was seen as a threat to western European countries and U.S. international power. This fear was based originally on the environmental deterministic theory of the political geographer Halford Mackinder. (Sources: Mackinder, 1904; Spykman, 1944.)

its reliance on the discredited theory of environmental determinism (see Chapter 1). Ultimately, Russia was unable to hold together its own heartland empire, much less conquer the rimland and the world. Mackinder's heartland theory is another example of the shortcomings of environmental determinism.

Buffer and Satellite States   From the very beginning of European colonialism, rival empires sought to shape global geopolitics to protect their territories. Strategic security planning historically led to the maintenance of weak states that served the geopolitical interests of more powerful states. A key strategy was to establish states that protected the boundaries of empires. These are known as **buffer states**, independent but politically and economically weak countries lying between the borders of two powerful, potentially belligerent empires or countries. A classic example is Thailand, formerly known as Siam. France and England were rival nineteenth-century colonial powers whose expanding territorial empires in Asia were a source of potential conflict. By mutual agreement, they established the boundaries of an independent state, Thailand, that would be a weakened, neutral power located between the eastern extent of England's South Asian empire of Burma, India, and Pakistan, and the western extent of France's Southeast Asian empire of Indochina (Cambodia, Vietnam, Laos) (**Figure 6.25**).

While rooted in historic rivalries of vanished empires, the geopolitical logic of buffer states remains with us today. Some observers of international relations argue that China's political and economic support of North Korea is motivated by its interest in maintaining a buffer state between China and Western-leaning South Korea and Japan. Russia's efforts to politically destabilize Ukraine are likewise thought to be motivated by its interest in keeping its neighbor weak and unable to establish strategic alliances with western Europe. Other contemporary buffer states include Mongolia between Russia and China, and Nepal between India and China (see Figure 6.1).

A related geopolitical strategy is the establishment of **satellite states**. The term is borrowed from astronomy and the situation of small celestial bodies orbiting under the gravitational pull of a larger body. A satellite state, then, is a country that is nominally independent, but is politically, militarily, and economically controlled by a more powerful state. Sometimes a stronger neighboring state assumes direct control of a buffer state, making key political and economic decisions for the satellite. Poland is an example. It was a buffer state between Germany and Russia and became a Soviet satellite state after World War II. Sometimes satellite states arise inadvertently as a result of strategic geopolitical actions.

FIGURE 6.25 Siam (now Thailand), buffer between empires, c. 1900. Thailand, which was never a European colony, remained independent and neutral. It therefore historically served as a buffer state between rival empires, the British to the west and the French to the east.

For example, some observers argue that after the United States invaded Iraq in 2003 and overthrew its government, the country was so weakened that it now functions as a de facto satellite state of Iran.

Geopolitics Today   In the post–Cold War period, geopolitics has reemerged as a dynamic field of political-geographic thought. As geographers Gerard Toal and John Agnew explain, the meaning of political geography is now reversed. Instead of focusing on the influence of geography and the environment on politics, "it now becomes the study of how geography is informed by politics." By this they mean the ways in which political goals and ideologies, based on preconceived notions of cultural identities, regional stereotypes, and regional development hierarchies, influence the ordering of the world. How does the geopolitical outlook of a state structure the world into places of crisis or stability, regions of opportunity or danger, and states of allies or enemies? Many geographers distinguish this new focus on culture by labeling their approach "critical geopolitics."

One of the important aspects of critical geopolitics is a concern with how geopolitics influences our understanding of nature–culture relations and affects the way we transform the environment. Consider, for example, the current scientific and policy debates over global climate change caused by greenhouse gas emissions. Worldwide debates must be placed in the historical geopolitical context of the Global North's political and economic domination of the Global South. From the South's perspective, according to Simon Dalby, "the North got rich by using fossil fuels for generations. Why should those in the South forgo the same possibilities just because they come on the development scene a little later?" According to advocates for the South, the North's ideas about restricting future emissions of greenhouse gases globally will have a disadvantageous effect on the South's economic development. Thus, questions of global environmental management are not restricted to the realm of science and technology; they also fall squarely in the realm of geopolitics.

Rising Seas, Changing Borders

As the world's average temperature increases, sea levels around the globe are rising. Ocean waters expand as they warm, which raises average sea level. In addition, warmer temperatures mean that water stored as ice is being released into the oceans as glaciers melt. You can expect that sea level rise (SLR) will have dramatic, worldwide effects on political geography in your lifetime (see the Video Connection).

Estimates of Sea Level Rise The type and extent of political-geographic effects will depend on how fast and how high sea levels rise. What do we know? First, we know that sea levels

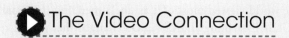

The Video Connection

Vanishing Island

Featured in this video are Edison Dardar and Chris Brunet, two of the last remaining residents of Isle de Jean Charles, an island off Louisiana's Gulf Coast. Their families have lived and worked on the island for at least three generations. As sea level has risen and soil erosion worsened, the island is becoming gradually inundated and a once-thriving community has all but disappeared. The U.S. government has no plans to build water control structures to protect the island, and it is likely to disappear within the lives of the current residents. Knowing what we do about the rate of sea level rise, we can consider Isle de Jean Charles a harbinger of the challenges facing us in the near future.

Thinking Geographically

1. What sorts of future political-geographic problems does this video suggest? Think about how those problems might vary at different levels of government (from the local to the national to the international).

Boundaries will change, of course, but also think about how political boundaries structure all of the functions managed at different government levels including taxation, elections, public education, and transportation infrastructure.

2. Early in this textbook we introduced the concept of sense of place. Do you think the residents of Isle de Charles have a strong sense of place, and if so, what are its key features? How important do you think a sense of place is in shaping the political decisions regarding the island community's future? More generally, what role do you think a sense of place might play in other coastal communities vulnerable to sea level rise? How might it influence government relocation efforts?

3. The video makes clear that the community of Isle de Charles represents a particular way of life. How do you think sea level rise will affect culture groups? Posed differently, how might sea level rise affect cultural diversity at different scales, including the global, and under what circumstances is that a political issue?

have been rising since the 1880s and are now rising at an increasing rate. The Earth's average sea level has risen about nine inches (22.86 centimeters) since then, with three inches (7.62 centimeters) of that occurring only since 1990. Forecasts of future rise are difficult, as they are hindered both by great uncertainty about humanity's responses and by incomplete scientific knowledge. We do know that no matter what human action is taken, SLR will continue because so much CO_2 has already been released into the atmosphere. Recently, scientists estimated we could see an additional 14-inch (35.56-centimeter) rise by 2050.

Recent research in the polar regions, however, suggests that that estimate is too low, and that we are facing a much faster and greater rise. Scientists found that the Arctic is warming more than twice as fast as the rest of the planet, leading to accelerated melting of polar glaciers. Without a reduction in atmospheric CO_2, scientists' mid-range forecast for SLR is another 39 inches by 2100. We are already experiencing the effects of SLR on political geography, and these forecasts suggest they will only grow more challenging.

Territorial Seas To fully understand the impacts of SLR on political geography, we first need to realize that national sovereignty does not end at the shoreline. This is a misconception reinforced by almost every world political map and globe we encounter. In actuality, territorial sovereignty extends 12 nautical miles (22 kilometers) out from the mean low-tide line of the shore, a zone referred to as **territorial seas**. Moreover, in 1982, the United Nations Convention on the Law of the Sea (UNCLOS) established the **exclusive economic zone (EEZ)**, which is the area extending 200 nautical miles (370 kilometers) from the shore. Within the EEZ, coastal states have the sole right to exploit, develop, manage, and conserve all resources. This is a huge economic benefit for coastal countries. Most of the world's undersea minerals, including rich oil deposits, and almost all ocean fisheries are found within the EEZ. UNCLOS, therefore, has altered geopolitical calculations around the globe. Most notably, formerly insignificant islands are now potential sources of vast wealth, because they are encircled by EEZs many times larger than their land areas. Consequently, countries have made competing sovereignty claims on islands, sometimes leading to tense standoffs.

With this as background, we can begin to think about the implications of SLR for political geography. Obviously, territorial boundaries will change as more and more coastal land is permanently inundated. If inundation proceeds as rapidly as forecast, international boundaries will be continually redrawn. Territorial seas will also be affected, with potential for the loss of access to important fisheries, for example. Finally, as international boundaries shift with rising waters, the potential for international disputes over sovereignty

rights will rise with them. Let's look at two cases where global warming and sea level rise are already complicating territorial sovereignty claims.

The first, Maldives, is an example of what are known as **small island developing states**, whose economies rely mostly on tourism and ocean fisheries. Maldives is a chain of more than 1000 islands in the Indian Ocean with a total land area of only 116 square miles (300 square kilometers), but with an EEZ of 331,660 square miles (859,000 square kilometers). Most of the chain is less than 3.28 feet (1 meter) above current sea level. Consequently, as sea level rises, islands will disappear. Will the Maldives retain sovereignty over its territorial seas and EEZ, the boundaries for which were established from its twentieth-century land area? This is a question of international law that has yet to be answered by the United Nations. The concept of effective sovereignty may give us a hint at the likely answer.

The second is the Arctic region itself, defined by the 66 degrees, 34 minutes north latitude line known as the **Arctic Circle** (**Figure 6.26**). In the center of the circle is the Arctic Ocean, over 5.4 million square miles (14 million square kilometers) of ice and seasonally open waters. Until 1999, the world viewed most of it as unclaimed, international waters. Global warming and UNCLOS changed that view.

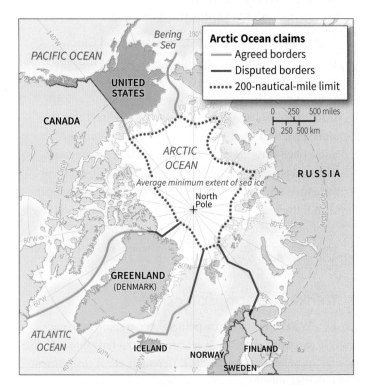

FIGURE 6.26 New race for territory in the Arctic. As the warming climate melts more ice, we are gaining easier access to a wealth of natural resources. Conflicts among the five countries that have territorial claims in the Arctic region are consequently increasing.

As temperatures warm, the extent of the region's ice is diminishing. This will likely open new shipping routes that would greatly reduce transport costs. Also, access to the region and its resources, most notably undersea oil and gas deposits, is growing, accompanied by increasing conflicts over territorial sovereignty. Five countries have territorial seas and EEZ claims to the Arctic Ocean: Canada, Denmark (through its possession of Greenland), Norway, Russia, and the United States. In 2007 Russia planted its national flag on the ocean floor near the North Pole, a provocative and legally dubious act that raised international awareness of the conflicts to come. Canada responded by expanding its military presence in the Arctic, and Denmark launched a scientific expedition to gain evidence in support of its territorial claims. In addition, Canada claims the ice-free waters of the new shipping routes are their territorial seas, a claim that other countries reject. This is just a brief introduction to the unresolved and unanticipated effects of climate change on the Arctic's political geography.

National Militaries and the Environment

The world's national armies and navies affect the environment in complex and sometimes surprising ways. Interstate warfare delivers perhaps the most spectacular and devastating effects. "Scorched earth," the systematic destruction of enemies' homes, pastures, and cultivated fields, has been a favored tactic of armies for millennia. During the Vietnam War, the U.S. military sprayed chemical defoliants on Vietnam's tropical forests and croplands in an effort to deny the enemy the cover of vegetation. Defoliants not only destroyed vegetation; they also created a secondary environmental hazard. Chemicals entered the soil and water, leading to 500,000 birth defects. In the 1991 Gulf War, the Iraqi military devised a scorched earth tactic of lighting over 600 oil wells on fire as it withdrew from Kuwait, thereby releasing tons of pollutants into the atmosphere and ground.

People fleeing wars can produce major environmental impacts elsewhere. Wars create refugees and refugees tend to concentrate in encampments, sometimes housing hundreds of thousands of people. Deforestation, soil erosion, and water pollution are common environmental problems resulting from the concentration of war refugees. For example, Kenya's Dadaab camp and its immediate surroundings have been largely deforested as the growing population scavenges for fuel, wood, and building materials.

Even military exercises and weapons tests are often environmentally destructive. American hydrogen bomb testing in the 1950s rendered uninhabitable, perhaps forever,

certain islands in the Pacific Ocean. The Hanford Nuclear Reservation in Washington State manufactured the plutonium for those bombs and many others (**Figure 6.27**). Now largely decommissioned, Hanford represents the single largest radioactive waste site in North America. Located in the upper Columbia River drainage, the site's potential to contaminate the river system poses a long-term threat to both terrestrial and marine environments over a vast territory. About a third of the underground tanks containing radioactive liquids are leaking into the soil and groundwater. It is hoped that the U.S. military and other agencies can clean up the site before contaminants enter the Columbia River itself.

Militaries and wars have less visible and dramatic effects on the environment, some of which may actually be ecologically beneficial. During Nicaragua's civil war, stretches of tropical forests became "no-person's lands," as people did not enter out of fear of both the army and insurgents who used the forests as hideouts. As a result, some wildlife populations recovered under the decreased hunting pressure. U.S. military bases have made vast areas off-limits to civilians and thereby inadvertently protected habitats for hundreds of endangered plant and animal species from commercial exploitation. When these bases are decommissioned (permanently closed), they are sometimes converted to conservation areas so that protection continues. For instance, part of the decommissioned Fort McClellan in Alabama became the 9000-acre (3642-hectare) Mountain Longleaf National Wildlife Refuge in 2004, thereby continuing protection of one of the last remaining longleaf pine forests. Part of the Mare Island Naval Shipyard in San Francisco Bay was similarly

FIGURE 6.27 Nuclear waste storage at the Hanford Nuclear Reservation. The Hanford facility, located along the Columbia River, produced plutonium for nuclear weapons for the U.S. military. Many fear that the site's stored nuclear waste may soon leak into the groundwater and spread radioactive contamination over a large region. (Jeff T. Green/Getty Images)

FIGURE 6.28 Wetlands of the former Mare Island Naval Shipyard. Because military bases prohibit commercial development, they sometimes inadvertently protect vast areas of valued natural habitats. When the Mare Island Naval Shipyard in Vallejo, California, closed, some of its land holdings were added to the San Pablo Bay National Wildlife Refuge, which protects scarce wetlands. (Courtesy of Roderick Neumann.)

passed to the San Pablo Bay National Wildlife Refuge when it was decommissioned (**Figure 6.28**).

Clearly, warfare—especially modern, high-tech warfare—can be environmentally catastrophic. Yet ecological destruction is not always the only result of war or preparation for war. As with many nature–culture phenomena, the interactions are complex and have unpredictable and unintended outcomes.

Cultural Landscape

6.5 Analyze the effects of national laws, international borders, and national monuments on the cultural landscape.

The world over, national politics are written on the landscape. For example, state laws, such as national cadastral surveys, have profound visible effects on the cultural landscape. International borders can create distinct landscapes that meld cultural elements from both sides. Finally, elements of the cultural landscape, such as monuments, are important to cultivating and maintaining nationalist sentiments.

Legal Codes and Landscape

Perhaps the best example of how political philosophy and the legal code become visible in the landscape are cadastral surveys (see Chapter 1). The U.S. government, to cite a familiar example, used a single method of surveying across much of the U.S. beginning right after national independence from England. The U.S. survey system was the brainchild of Thomas Jefferson, an early U.S. president and principal author of the Declaration of Independence. He also chaired a national committee on land surveying that resulted in the U.S. Land Ordinance of 1785. Jefferson's ideas for surveying, distributing, and settling the western frontier—as it was violently cleared of Native Americans—were based on a political philosophy of "agrarian democracy." Jefferson believed that political democracy had to be founded on economic democracy, which in turn required a national pattern of equitable land-ownership by small-scale independent farmers. In order to achieve this agrarian democracy, the western lands would need to be surveyed into parcels that could then be sold at prices within reach of family farmers of modest means.

Jefferson's solution was the **township and range system**, which established a grid of square-shaped "townships" with 6-mile (9.6-kilometer) sides across the Midwest and West. Each of these was then divided into 36 sections of 1 square mile (2.6 square kilometers), which were in turn divided into quarter-sections, and so on. Sections were to be the basic landholding unit for a class of independent farmers. Townships were to provide the structure for self-governing communities responsible for public schools, policing, and tax collection. With the exception of the 13 original colonies and a few other states or portions of states, the federal government imposed a grid-like land-division pattern on the entire country. This was the result of Jefferson's political philosophy, made visible in the landscape by the cadastral survey system (**Figure 6.29**).

FIGURE 6.29 Imposing order on the landscape. This aerial view of Canyon County, Idaho, farmland reveals a landscape grid pattern. This pattern was imposed on the landscape across much of the United States following the passage of the Land Ordinance of 1785. (David Fraizer/The Image Works)

FIGURE 6.30 Legal height restrictions, or their absence, can greatly influence urban landscapes. Singapore (*left*), the city-state at the tip of the Malay Peninsula, lacks such restrictions, and its skyline is punctuated by spectacular skyscrapers. In Rio de Janeiro (*right*), by contrast, height restrictions allow the natural environment to provide the "high-rises." (Left: David Ball/Alamy; Right: SIME/eStock Photo)

The imprint of legal codes can also be seen in the cultural landscape of urban areas. In Rio de Janeiro, height restrictions on buildings have been enforced for a long time. The result is a waterfront lined with buildings of uniform height (**Figure 6.30**). By contrast, most American cities have no height restrictions, allowing skyscrapers to dominate the central city. The consequence is high, jagged-tooth skylines in cities such as San Francisco and New York. Many other cities around the world lack legal height restrictions, such as Malaysia's Kuala Lumpur, which has the world's tallest skyscrapers.

Border Landscapes

Demarcated political boundaries can also be strikingly visible, forming border landscapes. Political borders are usually most visible where restrictions limit the movement of people between neighboring countries. Sometimes such boundaries are even lined with cleared strips, barriers, pillboxes, tank traps, and other obvious defensive installations. At the opposite end of the spectrum are international borders, such as that between Tanzania and Kenya in East Africa, that are unfortified, thinly policed, and for many miles invisible in the landscape. Even so, undefended borders of this type are usually marked by regularly spaced boundary pillars or cairns, customhouses, and guardhouses at crossing points (**Figure 6.31**).

Occasionally, present-day political boundaries reflect the durable landscape imprint of long-vanished cultures. Some of the best-known examples come from the ancient Roman Empire. In England, for example, many of today's parish and township boundaries surrounding the city of Bath follow the property markers of ancient Roman villas. Probably the best known of the Roman landscape's influence on present-day boundaries is Hadrian's Wall in northern England. Named after a Roman emperor, the wall was constructed as a fortification to defend against the unconquered peoples to the north. Today, Hadrian's Wall parallels the modern border between England and Scotland (**Figure 6.32**).

FIGURE 6.31 Even peaceful, unpoliced international borders often appear vividly in the landscape. The International Peace Garden park near Bottineau, North Dakota, and Boissevain, Manitoba, celebrates international cooperation between the United States and Canada—ironically, by emphasizing an otherwise discreet international border. (Terrance Klassen/Alamy)

World Heritage Site: Tiwanaku: Spiritual and Political Center of the Tiwanaku Culture

Tiwanaku is an ancient example of how cultural landscapes can reflect and reinforce centralized political authority. Inscribed by UNESCO in 2000, Tiwanaku served as the political and religious center of an empire that dominated the Altiplano region of the Andes Mountains between 400 and 900 C.E. It retains political-geographic significance today as a symbol of Bolivian national identity.

- Tiwanaku was a sprawling planned city of 70,000 to 125,000 residents located near the southern shores of Lake Titicaca in Bolivia. Its ruins provide an exceptional example of pre-Incan civic architecture. The architecture and city layout symbolically reflect the site's former role as an imperial center.

- Surrounding the imperial center were the towns and villages colonized by the Tiwanaku rulers. Even further out were settlements that the center dominated, but did not colonize or conquer militarily. All of the areas were knitted together economically and politically by a network of alpaca caravan trade routes that stretched across the Altiplano and beyond. The central authority at Tiwanaku controlled the caravans.

- The Tiwanaku rulers developed an ideological and religious iconography (visual images and symbols used in art) that was embedded in the center's architecture and then reproduced in the material life of the far-flung regions under its influence. For example, iconography originating from the center appeared on ceramic vessels and was reproduced on stone markers placed in peripheral settlements. Conversely, the rulers incorporated the iconography of the periphery into public architecture of the central city, thereby encouraging a variety of distant ethnic communities to identify themselves as part of a larger state. Thus, the symbolism embedded in civic architecture reinforces the idea of a periphery politically, culturally, and economically linked together through the center.

ANCIENT CIVIC ARCHITECTURE: Much of the ancient city has been erased by modern development, but the ruins of the monumental stone buildings in the ceremonial center remain protected. The architecture of the center, which is oriented toward the cardinal points, reflects both the complexity of the Tiwanaku Empire's political structure as well as its religious nature.

- Tiwanaku culture perfected stonecutting, carving, and polishing and used those skills to create a monumental architecture as a projection of its power. A series of architectural structures—Kalasasaya's Temple, Akapana's Pyramid, and Pumapumku's Pyramid—define the space of the center.

- Akapana's Pyramid, originally rising over 59 feet (18 meters), is the most commanding of the structures. Though since destroyed, a temple stood at the top, as is also common in Mesoamerican architecture.

(John Elk/Getty Images)

- Kalasasaya's Temple is a rectangular structure that likely functioned as an observatory. In its interior are two carved monoliths as well as the Gate of the Sun, a monumental sculpture standing 9.8 feet (3 meters) tall, 13 feet (4 meters) wide, and carved from a single stone. Multiple human and anthropomorphic animal carvings decorate the Gate of the Sun, with what is likely a major deity appearing at the top of the monument, at its center. Some archaeologists suggest that the carvings on the gate constituted an agricultural calendar.

- Other architectural features in the ancient city include the Palace of Putuni and Kantatillita, which stands in tribute to the political and administrative authority of imperial rule.

TOURISM: Located only 43.5 miles (70 kilometers) west of La Paz, Bolivia, and served by public transportation, Tiwanaku is relatively accessible.

- Popular literature in the United States and Europe speculated (wrongly) that the site provides evidence that an advanced race of extraterrestrials once visited Earth. This notoriety has been a major draw for international tourists, particularly during astronomical events.

- As the site is sometimes referred to as the "American Stonehenge," thousands of visitors gather during the Southern Hemisphere's winter solstice for a dawn ritual.

- Visitation has sparked a local tourist industry that has become the most important source of income in surrounding communities.

- Tiwanaku is located near the southern shores of Lake Titicaca in Bolivia.
- Tiwanaku craftspeople perfected stonecutting, carving, and polishing, creating a monumental architecture as a projection of the center's political power.
- Kalasasaya's Temple, Akapana's Pyramid, and Pumapumku's Pyramid define the space of the center.

http://whc.unesco.org/en/list/567

(AIZAR RALDES/AFP/Getty Images)

FIGURE 6.32 Hadrian's Wall, U.K. Now a World Heritage Site, this ancient Roman defensive fortification parallels the nearby modern boundary between England and Scotland. (Peter Mulligan/Getty Images)

National Iconography in the Landscape

The cultural landscape is rich in symbolism and visual metaphor, and political messages are often conveyed through such means. Statues of national heroes or heroines and of symbolic figures such as the goddess Liberty or Mother Russia form important parts of the political landscape, as do assorted monuments. Monuments such as the Statue of Liberty in New York Harbor and the statue of the naval hero Lord Nelson in London's Trafalgar Square can evoke deep nationalist sentiments (**Figures 6.33** and **6.34**). The sites of heroic (if often futile) resistance against invaders, as at Masada in Israel, prompt similar feelings of nationalism.

Some geographers theorize that the political iconography of landscape derives from an elite, dominant group in a country's population and that its purpose is to legitimize or justify its power and control over an area. The dominant group seeks both to rally emotional support and to arouse fear in potential or real enemies. As a result, the iconographic political landscape is often controversial or contested, representing only one side of an issue. Look again at Figure 6.34. The area in which Mount Rushmore stands, the Black Hills, is sacred to the Native Americans who controlled the land before whites seized it. How might these Native Americans, the Lakota Sioux, perceive this monument? Are any other political biases contained in it? Cultural landscapes are always complicated and subject to differing interpretations and meanings, including political ones (see the World Heritage Site feature).

The landscapes of capital cities are important in shaping a sense of national identity and belonging. Geographer Diana Ter-Ghazaryan's study of Armenia's national capital, Yerevan, shows how central and contentious urban landscape design can be to national identity. After gaining independence from the Soviet

FIGURE 6.33 The Statue of Liberty and Lord Nelson in Trafalgar Square. These are widely recognized national symbols of the United States and England, respectively. (Left: Cameron Davidson/Getty Images; Right: lachris77/iStock/Getty Images)

landscape: Opera Square, Northern Avenue, and Republic Square (**Figure 6.35**). The post-Soviet development of each of these sparked street protests over what Yerevan will become and what it means to be Armenian. National identity, especially in times of major political transition such as Armenia has experienced, can be a very contentious issue that is expressed through struggles over the physical landscape of a nation's capital.

Geographer Gail Hollander has demonstrated the important symbolic role that the landscape and environment of the Florida Everglades have played in U.S. presidential elections (**Figure 6.36**). The symbolic role of the Everglades has changed over time, from the 1928 presidential campaign to the present. In 1928 it was presented as worthless swampland. As such, it played a key role in the election of President Herbert Hoover, who promised to drain it for agricultural development. By the 1970s, it was seen as an endangered wetland in need of protection and ecological restoration. Thus, in virtually every recent presidential campaign, the candidates' positions on the Everglades are seen as indicators of their commitment to the environment. Photo opportunities in the park give both Democratic and Republican candidates alike a chance to symbolically link their political campaigns to the ecological restoration of what has come to be viewed as a treasured national landscape.

FIGURE 6.34 **Mount Rushmore,** in the Black Hills of South Dakota, presents a highly visible expression of American nationalism, an element of the political landscape. (PAUL DAMIEN/National Geographic Creative)

Union, the Armenian political elite began redesigning the capital city as part of a process of symbolically constructing a national identity for a modern, democratic Armenia. Ter-Ghazaryan identified three key sites that symbolically anchored the Yerevan

FIGURE 6.35 **Opera Square in Yerevan, Armenia's capital.** Opera Square is an iconic place in the city's landscape and serves as a touchstone for Armenian national identity. As such, it has served as a locus for contentious debates over what it means to be Armenian in the post-Soviet era. (Luis Dafos/Alamy)

FIGURE 6.36 **Everglades National Park in Florida has long played a role in U.S. presidential politics.** Visits to iconic national parks and protected landscapes are common in U.S. political campaigns and are meant to symbolize a candidate's commitment to protecting national heritage. Here, Vice President Joe Biden speaks during an Everglades tour while U.S. Senator Bill Nelson and Representative Alcee Hastings look on. (Taimy Alvarez/Sun Sentinel/MCT via Getty Images)

Conclusion

Political spatial variations—from local voting patterns to the spatial arrangement of international power blocs—are of key interest in human geography. The nation-state organizes the relations between regions of the globe and provides the basis for one of the strongest of collective human sentiments—nationalism. We learned that globalization has had contradictory effects on state sovereignty and nationalism, in some times and places strengthening them and in others weakening them. We have also seen how important the nation-state is to the physical environment, from the structure of basic state laws and policies to the practices of its militaries.

Nature–culture relations are important to political geography, and the tools of political ecology help us understand the links between systems of power and the physical environment. Countries do not exist in an environmental vacuum. The spatial patterns of landforms often find reflection in boundaries, core areas, and geopolitical strategies. Likewise, political culture very much influences our ideas and judgments about landscape and environment. Finally, politics leaves diverse imprints on the cultural landscape, and landscapes often provide the symbolism and visual metaphors to support or refute political ideologies.

Chapter Summary

6.1 Explain the concept of nation-states, the forces that shape political territory, and the regional characteristics of electoral politics.

- A world divided into bounded political territories, along with the associated concepts of national identity, sovereignty, and self-determination, are defining characteristics of the modern era.
- Different political systems—unitary versus federal—are founded on different internal territorial divisions.
- The political boundaries that divide the world are in nearly constant flux under the pressures of centripetal and centrifugal forces that cause national territories to cohere or fragment.
- Election outcomes have strong regional associations, as shown by gerrymandering and the red state/blue state dichotomy in U.S. presidential elections.

6.2 Explain how mobility influences political geography, including the spatial diffusion of political authority and innovation, the displacement effects of violent political conflict, and the role of the Internet.

- Historically, political authority has often diffused from a geographic center of political power to incorporate more and more territory into the nation-state.
- Political innovation, such as the principle of self-determination, has diffused around the globe and is often an important justification for major geographic and territorial changes.
- Violent political conflicts, often related to efforts to create more culturally homogenous national territories, have produced a growing refugee crisis in the world.
- The Internet and social media have accelerated the geographic movement of political ideals and created new opportunities for political action across borders, with varying outcomes for democratic governance.

6.3 Explain the effects of globalization on sovereignty, supranationalism, and transnationality.

- The concept of effective sovereignty helps us understand how, depending on the context, globalization may either weaken or strengthen state sovereignty.
- Supranationalism—in a variety of institutional forms, such regional trading blocs and global financial institutions—both propels and is propelled by globalization.
- Transnationalism and transnational identities are new phenomena enabled by globalization.

6.4 Analyze the complex interactions of politics and physical geography, including the state's role in land management, the heartland theory of world geopolitics, and the effects of rising sea levels on political territory.

- In searching for explanations of land managers' environmentally damaging actions, it is important to examine state laws and policies influencing managers' decision making.

- Geopolitical factors such as relative location and physical geographic characteristics are important to the formation and success of many modern states.
- Climate change and accompanying sea level rise will have continuing and increasingly complex effects on the sovereignty of individual countries.
- National militaries have a wide range of effects on countries' environments, from protection to devastation.

6.5 Analyze the effects of national laws, international borders, and national monuments on the cultural landscape.

- National legal codes and policies, particularly those associated with land survey systems, have a visible effect on landscape.
- The regions straddling international borders often produce distinctive cultural landscapes.
- Landscape elements are often viewed as symbolic expressions of national identity and are therefore often highly contested.

Key Terms

state Independent political units with a centralized authority that makes claims to sole legal, political, and economic jurisdiction over a bounded territory. Often used synonymously with "country" (page 187).

sovereignty The right of individual states to control political and economic affairs within their territorial boundaries without external interference (page 187).

self-determination The freedom of culturally distinct groups to govern themselves in their own territories (page 188).

nation A community of people bound to a homeland and possessing a common identity based on a shared set of cultural traits such as language, ethnicity, and religion (page 188).

nation-state An independent country dominated by a relatively homogeneous cultural group (page 188).

nationalism The sense of belonging to and self-identifying with a national culture (page 188).

territoriality A learned cultural response, rooted in European history, that produced the external bounding and internal territorial organization characteristic of modern states (page 188).

boundary A precisely demarcated line that marks both the limits of a territory and the division between territories (page 188).

borderland A region straddling both sides of an international boundary where national cultures overlap and blend to varying degrees (page 188).

frontier A region at the margins of state control and settlement, an idea that can be traced to ancient notions of the limits of civilization and the savage world beyond (page 188).

colonialism Centuries-long process of European countries competing among themselves to conquer and control other peoples and their territories across the world (page 188).

enclave A piece of territory surrounded by, but not part of, a country (page 190).

exclave A part of national territory separated from the main body of a country by the territory of another country (page 190).

natural boundary A political border that follows some feature of the natural environment, such as a river or mountain ridge (page 190).

ethnographic boundary A political boundary that follows some cultural border, such as a linguistic or religious border (page 190).

geometric boundary A political border drawn in a regular, geometric manner, often a straight line, without regard for environmental or cultural patterns (page 190).

centripetal force Any factor that supports the internal unity of a country (page 190).

centrifugal force Any factor that disrupts the internal unity of a country (page 190).

ethnic separatism Separation of an ethnically distinct group to form a politically autonomous region within an existing state, or to secede and form a new nation-state (page 191).

irredentism The political claim to territory in another country based on ethnic affiliations and historic borders (page 191).

stateless nation A large ethnic group whose population is divided by one or more international boundaries (page 192).

unitary state An independent state that concentrates power in the central government and grants little or no authority to subnational political units (page 193).

federal state An independent country that disperses significant authority among subnational units (page 193).

electoral geography A subfield of political geography that analyzes the geographic character of political preferences and how geography can shape voting outcomes (page 193).

reapportionment The process by which the 435 seats in the U.S. House of Representatives are divided proportionately by population among the 50 states following every U.S. census (page 193).

redistricting The process of drawing new boundaries for U.S. congressional districts to reflect the population changes since the previous U.S. census (page 193).

gerrymandering The practice of manipulating voting district boundaries to favor a particular political party, group, or election outcome (page 193).

ethnic cleansing The forced removal of an ethnic group by another ethnic group to create ethnically homogeneous territories (page 200).

protracted refugee situation (PRS) Results when people are born and mature into adulthood in refugee camps with no promise of becoming citizens in a new country and little hope of returning to their home country (page 201).

stateless people Displaced people, often long-term refugees from armed conflict, who are denied a nationality and therefore deprived of basic rights such as education, health care, employment, and freedom of movement (page 201).

effective sovereignty The idea that states' power to effectively enforce or ignore sovereignty claims irrespective of territorial boundaries varies in time and from country to country (page 203).

supranationalism When a collection of nation-states and their citizens relinquish some sovereign rights to a larger-scale political body that exercises authority over its member states (page 205).

supranational organization Political bodies of international integration that nation-states establish in cooperation with their neighbors for mutual political, military, economic, or cultural gain (page 205).

regional trading blocs Multi-country agreements that reduce or eliminate customs duties and import tariffs to promote the freer flow of goods and services across international borders (page 205).

European Union (EU) A political, economic, and social union of 28 independent countries in Europe that promotes the free movement of people, goods, services, and capital among its members (page 205).

transnationalism The experience of living and working in more than one country that can produce new and distinct cultural identities. Sometimes used synonymously with globalization to refer to economic, political, or cultural processes operating across international borders (page 207).

geopolitics A term originally coined to describe the influence of geography and the environment on political entities (page 208).

heartland theory A 1904 theory of Harold Mackinder that the key to world conquest lay in control of the continental interior of Eurasia (page 209).

heartland The interior of a sizable landmass, removed from maritime connections; in particular, the interior of the Eurasian continent (page 209).

rimland The maritime fringe of a country or continent; in particular, the western, southern, and eastern edges of the Eurasian continent (page 209).

buffer state Independent but politically and economically weak country lying between the borders of two more powerful, potentially belligerent empires or countries (page 210).

satellite state A nominally independent country, but politically, militarily, and economically directly controlled by a more powerful state (page 210).

territorial seas The zone of territorial sovereignty extending 12 nautical miles (22 kilometers) out from the mean low-tide line of a country's coastline (page 212).

exclusive economic zone (EEZ) The zone of resource control established by the UN Convention on the Law of the Sea extending 200 nautical miles (370 kilometers) out from the mean low-tide line of a country's coastline (page 212).

small island developing states Small island countries whose economies rely mostly on tourism and ocean fisheries (page 212).

Arctic Circle The latitude line of 66 degrees, 34 minutes north (page 212).

township and range system The land survey system created by the U.S. Land Ordinance of 1785, which divides most of the country's territory into a grid of square-shaped "townships" with 6-mile (9.6-kilometer) sides (page 214).

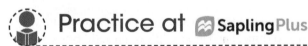

Practice at Sapling Plus

Read the interactive e-Text, review key concepts, and test your understanding.

Story Maps. Explore various types of borders around the world.

Web Maps. Examine issues in political geography.

Doing Geography

ACTIVE LEARNING:
The Role of Twitter Bots in Democratic Elections

Throughout this chapter we have drawn attention to the importance of social media to political geography. For example, it is now widely recognized that the use of online "bots" allows interested groups to easily spread misinformation—"fake news"—to influence elections. Bots, computer software applications programmed to run automated tasks, are a common component of the Internet. Increasingly, bots are used to create thousands of phony social media accounts to sway public opinion. Bot accounts allow one individual to appear to be a large group representing a seeming tide of public sentiment. Russian operatives used bots posing as actual living Americans with Twitter and Facebook accounts to spread fake news during the 2016 election. The problem for citizens is that it is difficult to distinguish a bot from a real person. For this exercise, you will explore the political effectiveness of bots and evaluate whether the source of a political tweet is a bot or a real person.

More guidance at SaplingPlus

EXPERIENTIAL LEARNING:
The Complex Geography of Congressional Redistricting

Congressional redistricting normally happens every 10 years in the United States, following each national census. In some cases, such as Texas in 2002, redistricting can occur between censuses. Although soon proved wrong, newspaper accounts at the time predicted that Texas's mid-census redistricting would protect the Republican Party's majority in the U.S. Congress for the foreseeable future. Does this mid-census redistricting fall under the category of Republican gerrymandering, as some claim, or is it, as the Texas Republican Party argues, a case of necessary adjustments in response to population shifts? Either way, the case of Texas demonstrates how critically important the drawing of congressional district boundaries is to democratic governance.

This exercise requires you to identify cases of possible gerrymandering in your home state or an adjacent state. To do so, follow the steps on the next page.

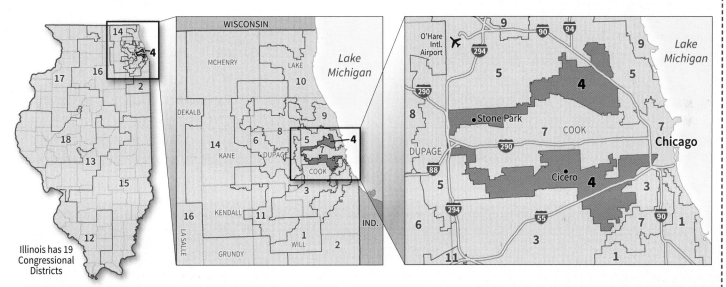

Congressional District 4 in Illinois

14 Other congressional districts   —— Congressional district boundary   COOK County   ---- County boundary   —— State boundary

The complex, convoluted shapes of some voting district boundaries, such as U.S. Congressional District 4 in Illinois, are evidence of gerrymandering. (Source: National Atlas.gov.)

Doing Geography (continued)

Steps to Identifying Gerrymandering

Step 1: Obtain a map of congressional district boundaries in your chosen state (http://nationalmap.gov is one possible source). Once you have done so, see if you can visually identify districts that may have been gerrymandered. Figure 6.8 (page 194) and the discussion on pages 193–197 should be helpful to you in identifying such districts.

Step 2: Having identified your candidate(s) for gerrymandering, address the following questions:

- What was it about the configuration of the boundaries that made you think the district(s) may have been gerrymandered?

- When were the boundaries drawn?
- Can you identify which of the major political parties was in power when the boundaries were drawn?
- Which party do these boundaries favor and why? That is, what are the racial, economic, religious, and ethnic characteristics of the district(s) that may suggest a particular party affiliation?

Step 3: Look at the proportion of major party registration in nearby districts to see if you can determine whether the boundary lines were drawn in order to dilute or to concentrate opposition votes.

SEEING GEOGRAPHY

Art in the Borderland

Can a border be more than a dividing line?

Are borders lines that divide cultures or sites for cultural blending? As the mobility of people has increased worldwide, there has been a corresponding reaction to restrict movement across borders. The reaction is founded on the belief that international borders should be clear lines of separation, where one nationality or culture ends and another begins. Historically, however, international borders function as definite dividing lines mostly during periods of open hostility between countries. More typical is peaceful coexistence and the consequent formation of borderland cultures, where a distinctive sense of place is created through the daily interactions and movements across borders. Borderland cultures consequently have produced distinctive visual and performance art forms. Nowhere is this more evident than in the U.S.–Mexico borderland.

The U.S.–Mexico borderland is currently a politically contentious region. Millions of U.S. citizens support the construction of an imposing, impenetrable wall. Artists on both sides of the border are actively producing art that both challenges the wall's desirability and reflects a distinctive borderland culture

that has been generations in the making. For example, a group of artists from the cross-border cities of Ciudad Juárez, Mexico, and El Paso, Texas, refer to themselves as *transborderistas* and proclaim that their art comes from the borderland, not separate countries or cultures. Their art, sampled here, celebrates borderland culture. Artists also organize borderland community events that attract people from both sides. Their activities work to both assert and help create a cultural identity that is transborder, a blending of both American and Mexican sensibilities and values. In some cases, their work recalls the recent past when routine back-and-forth movement across the border was common. One artist featured an abandoned trolley line that had once linked Ciudad Juárez and El Paso, when people moved daily across the border to shop, work, and recreate. This is a tiny slice of borderland life that we find all along the 2000-mile (3220-kilometer) international boundary. Music, painting, dance, and theater in the borderland are thriving, despite, or perhaps because of, efforts to build a wall that threatens vibrant transborder communities.

Re/flecting the Border, Installation and Performance Dinner on the US/Mexico Wall, Tijuana. (Credit: Margarita Certeza Garcia in collaboration with Marcos Ramirez ERRE & Miguel Buenrostro. Photo credit: Marcos Ramirez ERRE)

Parking lot shrine to the Virgin of Guadalupe, Self-Help Graphics and Art, East Los Angeles. (Courtesy of Patricia L. Price.)

chapter 7

The Geography of Religion

SPACES AND PLACES OF SACREDNESS

How can an ordinary landscape, such as a parking lot, become a sacred space?

Think about this photo and question as you read. We will revisit them on page 265.

Learning Objectives

7.1 Define the differences among religions as well as the basic tenets of the world's principal religions, and locate their primary regional expressions on a world map.

7.2 Describe how and why the world's religions became distributed as they are today.

7.3 Analyze how globalization has affected the practice of religion.

7.4 Explain specific ways in which the natural environment shapes, and is shaped by, religious beliefs and practices.

7.5 Evaluate how religions have left their particular mark on the cultural landscape.

Religion is a core component of culture.

For many, religion is the most profoundly felt dimension of their identities. For this reason, it is important to clearly state what is meant by the term and provide a sense of the many ways in which religion can be manifest in people's lives. Religion can be defined as a more or less structured set of beliefs and practices through which people seek mental and physical harmony with the powers of the universe. The rituals of religion mark the events in our lives—birth, puberty, marriage, having children, and death—that are observed and celebrated. Religions often attempt not only to accommodate but also to influence the awesome forces of nature, life, and death. In literal terms, the word *religion*—derived from the Latin *religare*—means "to fasten loose parts into a coherent whole." Religions help people make sense of their place in the world.

The importance of religion to the study of human geography will be explored through our five themes. First, religious beliefs differ from one place to another, producing spatial variations that can be mapped as culture regions. Second, a high degree of mobility is characteristic of religions as they have spread historically through conquest, trade, and missionary activity. Violent conquests, diasporas, and the geographic expansion of religions through conversion have all played a role in shaping the contemporary world map of religious belief. Pilgrimages, too, illustrate how people can literally be set in motion through religion. Third, the global spread of religious beliefs has led to the reach of some religions, such as Hinduism, beyond their traditional hearths, whereas other religions, such as Judaism, have become diluted in part through migration-related diffusion. Religions, like all elements of cultural geography, must either adapt, or not, to the globalizing world. But religions influence globalization as well; here, we will consider the faith-based dimensions of the Internet. Fourth, the natural environment frequently plays an important part in faith-based belief structures, either because the forces of nature are viewed as potentially negotiable or because features of the natural landscape, such as rivers and mountains, are thought to exert a powerful influence over human destinies. Thus, the nature–culture dimension of religion is a key aspect of its cultural geography. Fifth, and finally, religious beliefs are often visible on the cultural landscape. For example, religious architecture, such as mosques, temples, and shrines, literally marks the landscape with the imprint of particular religious beliefs. How the spiritual shaping of some spaces as sacred comes to be is an important statement about what, and who, matters, culturally speaking.

Region

7.1 Define the differences among religions as well as the basic tenets of the world's principal religions, and locate their primary regional expressions on a world map.

Most religions incorporate a sense of the supernatural that can be manifest in the concept of a God or gods that play a role in shaping human existence, in the notion of an afterlife that may involve a place of rest (or torment) for those who have died, or in ideas of a soul that exists apart from our physical bodies and that may be released, or even born again, once we have died. This sense of the otherworldly is often spatially demarcated through the designation of sacred spaces, such as cemeteries, religious buildings, and sites of encounters with the supernatural.

In addition, religion is often at the heart of how people with very different worldviews can come to understand one another. So, on the one hand, the conquest of the Americas by the Iberians (people of modern-day Spain and Portugal) was often a violent affair, in which the religious conversion of indigenous peoples to Christianity accompanied the political takeover. Temples were destroyed and coercion was often used to convert the natives to Christianity. On the other hand, the Virgin of Guadalupe, said to have appeared to the indigenous Mexican convert Juan Diego in 1531—only 10 years after Cortes's conquest of Mexico—is believed to be one of the most powerfully healing figures in the Americas to this day. Her portrait is a mixture of European and indigenous American symbols (**Figure 7.1**). In her kind manner of speaking to Juan Diego in Nahuatl (an indigenous language spoken by central Mexicans) and her resemblance to the Earth goddess Tonantzín, she made sense to native Mexicans. For the Spaniards, dark-skinned virgins had long been part of their religious symbolism. In the midst of the violence of conquest, then, the Virgin of Guadalupe provided a mother figure that was readily acceptable to native Mexicans, was familiar to Spaniards, and acted as a bridge by which people from these two very different cultures could understand one another.

Classifying Religions

Each of the world's major religions is organized according to more or less standardized practices and beliefs, and each is practiced in a similar fashion by millions, even billions, of adherents worldwide. Yet many people also express their

FIGURE 7.1 The Virgin of Guadalupe. This is the image believed to have appeared to Juan Diego. As he opened his cape in the presence of Bishop Zumárraga of Mexico City, roses of Castile (a powerful symbol to Spaniards) fell onto the floor and this image was left behind on the fabric of the cape. The Virgin's downcast gaze and dark features spoke to Mexican Indians, as did the belt about her waist, which indicates that she is pregnant. (Leemage/Corbis via Getty Images)

religious faith in individual ways. Rituals and prayers can be adapted to fit particular circumstances or performed at home alone, or over the Internet. Some religions, including the Taoic religions of East Asia, as well as Hinduism and Buddhism, are largely individual or family-oriented practices. Some people do not observe a widely recognized religion at all. They may be secular, holding no religious beliefs, or express skepticism—even hostility—toward organized religion. They may consider themselves to be faithful but not follow an organized expression of their beliefs. Or they may practice an unconventional belief system, or **cult**. The term *cult* is often used in a pejorative sense because it conjures images of mind control, mass suicide, and extreme veneration of a human leader. However, the term is also sometimes used to denote a belief system that is emerging but has yet to become mainstream.

Proselytic and Ethnic Religions Different types of religion exist in the world. One way to classify them is to distinguish between proselytic and ethnic faiths. **Proselytic religions**, such as Christianity and Islam, actively seek new members and aim to convert all humankind. For this reason, they are sometimes also referred to as **universalizing religions**. They instruct their faithful to spread the Word to all the Earth using persuasion and sometimes violence to convert nonbelievers. The colonization of peoples and their lands is sometimes a result of the desire to convert them to the conqueror's religion. By contrast, each **ethnic religion** is identified with a particular ethnic or tribal group and does not seek converts. Judaism provides an example. In the most basic sense, a Jew is anyone born of a Jewish mother. Though a person can convert to Judaism, it is a complex process that has traditionally been discouraged. Proselytic religions sometimes grow out of ethnic religions—the evolution of Christianity from its parent Judaism is a good example.

Monotheistic and Polytheistic Religions Another distinction among religions is the number of gods worshipped. **Monotheistic religions**, such as Islam and Christianity, believe in only one God and may expressly forbid the worship of other gods or spirits. **Polytheistic religions** believe there are many gods. For example, Vodun (also spelled Voudou in Haiti or Voodoo in the southern United States) is a West African religious tradition with adaptations in the Americas wherever the enslavement of Africans was once practiced. Although, as with most major religions, there is one supreme God, it is the hundreds of spirits, or *iwa*, that Vodun adherents turn to in times of need. Some of the better-known spirits in the Haitian Voudou tradition are Danbala Wedo, the peaceful snake-god who brings rain and fertility; Legba, the keeper of crossroads and doorways, who is invoked at the beginning of all rituals; and Ezili Danto, the protective mother figure portrayed as a dark-skinned country woman.

Syncretic and Orthodox Religions Finally, the distinction between syncretism and orthodoxy is important. **Syncretic religions** combine elements of two or more different belief systems. Umbanda, a religion practiced in parts of Brazil, blends elements of Catholicism with a reverence for the souls of Indians, wise men, and historical Brazilian figures, along with a dash of nineteenth-century European spiritism, which was a set of beliefs about contacting spirits through mediums. Caribbean and Latin American religious practices often combine elements of European, African, and indigenous American religions. Sometimes, in order to continue practicing their religions, people in this region would hide statues of Afrocentric deities within images of Catholic saints. Or they would determine which Catholic figures were most like their own deities and focus their worship on those parallel Catholic saints. Note in **Figure 7.2** the comparison between Danbala, the snake-god of Haitian Voudou, with the Catholic Saint Patrick, who is also associated with snakes.

Orthodox religions, by contrast, emphasize purity of faith and are generally not open to blending with elements of other belief systems. The word *orthodoxy* comes from Greek and literally means "right" (*ortho*) "teaching" (*doxy*). Many religions, including Christianity, Judaism, Hinduism, and Islam, have orthodox strains. So, for instance, although some orthodox Jews closely follow a strict interpretation of the Oral Torah (a specific version of the Jewish holy book), moderate but committed Jews may observe only some or perhaps none of the dietary, marriage, and worship proscriptions observed by orthodox Jews. Intolerance of other religions, or of those fellow believers not seeming to follow the "proper" ways, is associated with **fundamentalism** rather than orthodoxy. Many who consider themselves orthodox are, in fact, quite tolerant of other beliefs (see Subject to Debate on page 231). **Religious extremism**, by comparison, is an intolerant expression of religious fundamentalism, resulting in violence.

Religious Culture Regions

Because religion has a strong territorial association, religious culture regions abound. The most basic kind of formal religious culture region depicts the spatial distribution of organized religions (**Figure 7.3**). Some religions, such as Hinduism and Buddhism, are strongly associated with one world region, while others, such as Islam and Christianity, are dispersed across many regions. Some parts of the world exhibit an exceedingly complicated pattern of religious adherence, and the boundaries of formal religious culture regions, like most cultural borders, are rarely sharp. While roughly three-quarters of the world's

FIGURE 7.2 Danbala and Saint Patrick. Danbala in Haitian Voudou is parallel to the Catholic Saint Patrick; both are associated with snakes. (Left: Roudy Azor/Indigo Arts Gallery; Right: Popperfoto/Getty Images)

SUBJECT TO DEBATE Religious Fundamentalism

Throughout history, religions have been one of the main ways in which people have attempted to make sense of the changing world around them. Although we may associate globalization with the fast pace and seemingly shrinking world of the past half century or so, people from diverse cultures have, in fact, been in contact across the globe for thousands of years. So how have religions helped people to cope with the changes brought about by new ideas, new ways of doing things, and new belief systems?

One response is acceptance. The Muslim conquest of Iberia in the eighth century brought a centuries-long flourishing of culture to modern-day Spain and Portugal. Islamic rule here was noted for its humane and enlightened nature, whereas the rest of western Europe languished in the Dark Ages. The Muslims who ruled Iberia were renowned for their religious tolerance, and they included Jews and Christians as valued members of their governmental, scientific, and artistic communities.

Another response is intolerance. When Buddhism branched off from Hinduism, its parent religion, not all Hindus were pleased at the way Buddhism replaced Hinduism in some areas. Angkor Wat, a temple complex in Cambodia, is replete with bas-relief carvings of Buddha figures with their faces either entirely chipped off or recarved to resemble Hindu deities. Hindus intolerant of the Khmer King Jayavarman VII's Buddhist beliefs were responsible for defacing these sacred images on the king's death in 1220 C.E.

Yet another response is syncretism, where elements of preexisting faiths are blended with new ones in order to make them more palatable. An example is the scheduling of the Christmas holidays during the older Roman holiday period celebrating the Roman deity Saturn. This occurred in the third and fourth centuries C.E., as Rome was beginning to Christianize, as a way to encourage Romans to adopt Christianity. Feasting, gift giving, singing in the streets (caroling), and consuming human-shaped cookies (gingerbread men) are some of the traditions from the Saturnalia festivities that have found their way into Christmas celebrations.

Today many religions, including Christianity, Judaism, Hinduism, and Islam, are experiencing intense fundamentalist movements. Fundamentalism means a return to the founding principles of a religion, which may include a literal interpretation of sacred texts and an attempt to follow the ways of a religious founder as closely as possible. Fundamentalists draw a sharp distinction between themselves and other practitioners of their religion, whom they do not believe to be following the proper religious principles, and between themselves and adherents of other faiths. Fundamentalism can be seen as an attempt to purify religious belief and practice in the face of modern influences thought to debase the religion. These tendencies have led to fundamentalists being regarded as anti-modern and intolerant, although this view is strongly disputed by fundamentalists themselves.

Fundamentalism is an emotionally charged term because it is often used in a derogatory manner to portray its followers as radical extremists. The tendency of the U.S. media to use the term *Islamic fundamentalists* as a synonym for *terrorists* is an unfortunate example of this. Yet there are connections between politics and religious fundamentalism. The political agenda of the U.S. government on matters of abortion, adoption, marriage, foreign policy, domestic security, gay rights, and the curriculum in public schools has been notably influenced by politically conservative religious groups. Although not all of these groups are entirely fundamentalist in nature, many embrace fundamentalist Christian beliefs that espouse creationism and the sinfulness of homosexuality, as well as questioning the separation of church and state. Islamism, a political ideology based in conservative Muslim fundamentalism, holds that Islam provides the political basis for running the state. Similar to the influence of Christian fundamentalism, Islamist influences in several Muslim-majority countries have set a conservative social agenda and have strongly questioned the separation of church and state.

Continuing the Debate

We often hear about religion in connection with violence. The next time you watch a news video or read a newspaper, keep these questions in mind:

- Do all religions have a dark, violent side to them, even those that profess a peaceful worldview?
- Is violence a necessary corollary to religious fundamentalism?
- What role does the media play in creating the perception—or inciting the practice—of religious antagonism?

Religious fundamentalists protested outside the Supreme Court building in Washington, DC, where arguments on same-sex marriage were being heard. (B Christopher/Alamy)

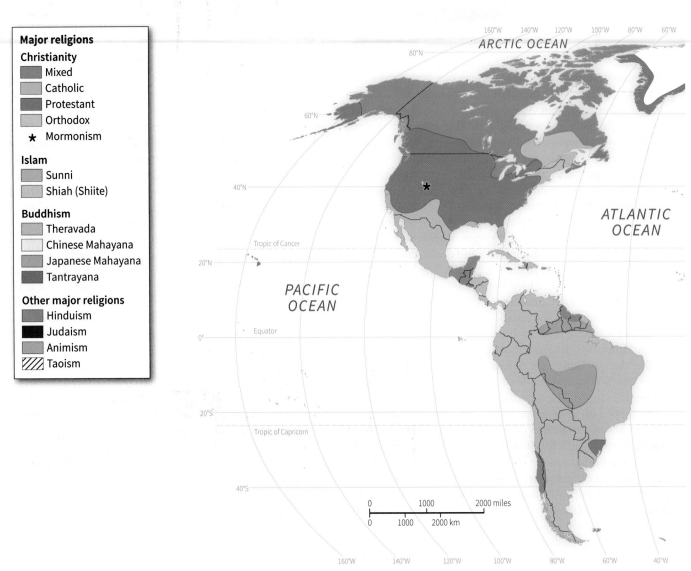

FIGURE 7.3 The world distribution of major religions. Much overlap exists that cannot be shown on a map of this scale. This map shows which faith is dominant in a region. "Animism" includes a wide array of diverse belief systems. Taoistic areas are cross-hatched to show the overlap with Buddhism in these areas. "Mixed" Christianity means that none of the three major branches of that faith has a majority.

inhabitants live in countries where their religion is in the majority, persons of different faiths often live in the same province or town (**Figure 7.4**).

Judaism Judaism is a 4000-year-old religion and the first major monotheistic faith to arise in southwestern Asia. It is the parent religion of Christianity and is also closely related to Islam. Jews believe in one God who created humankind for the purpose of bestowing kindness on us. As with Islam and Christianity, people are rewarded for their faith, punished for violating God's commandments, and can atone for their sins. The Jewish holy book, or Torah, consists of the first five books of the Hebrew Bible. In contrast to other monotheistic faiths, Judaism is not proselytic and has remained an ethnic religion through most of its existence.

Judaism has split into a variety of subgroups, partly as a result of the **Diaspora**, a term that refers to the forced dispersal of the Jews from Palestine in Roman times and the subsequent loss of contact among the various colonies. Jews, scattered to many parts of the Roman Empire, became a minority group wherever they went. In later times, they spread throughout much of Europe, North Africa, and Arabia. Those Jews who lived in Germany and France before migrating to central and eastern Europe are known as the *Ashkenazim*, those who never left the Middle East and North Africa are called *Mizrachim*, and those from Spain and Portugal are known as *Sephardim*. Spain expelled its Sephardic Jews in 1492, the same year that Christopher Columbus set sail for the New World. It was not until the quincentennial of both events, in 1992, that the Spanish government issued an official apology for the expulsion.

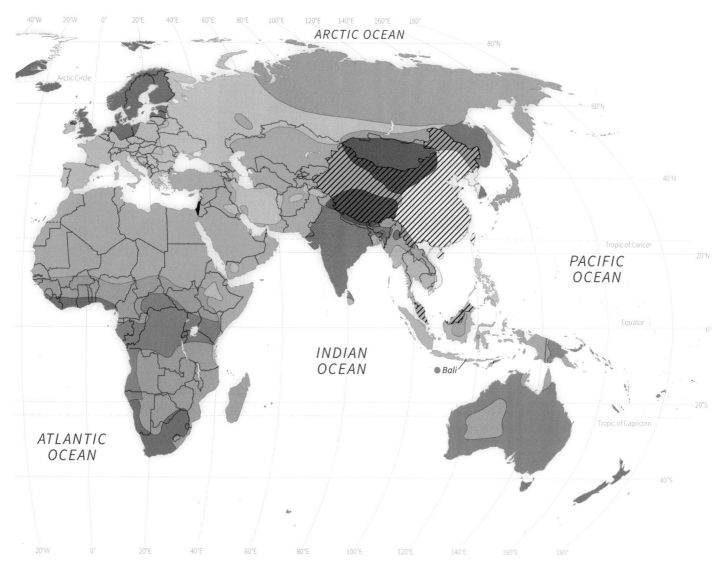

FIGURE 7.3 (continued)

The late nineteenth and early twentieth centuries witnessed large-scale Ashkenazic migration from Europe to America. The Holocaust that decimated European Judaism during the Nazi years involved the systematic murder of perhaps a third of the world's Jewish population, mainly Ashkenazim. Because of the Holocaust, Europe ceased to be the primary homeland of Judaism, and many of the survivors fled overseas, mainly to the Americas and later to the newly created state of Israel. Today, Judaism has about 15 million adherents throughout the world. At present, a roughly equal number—6 million—of the world's Jews reside in North America and Israel; Europe and Latin American are home to most of the rest.

Christianity **Christianity**, a proselytic faith, is the world's largest religion, both in area covered and in number of adherents, claiming 2.3 billion people, or about a third of the global population (**Figure 7.5**). Christians are monotheistic, believing that God is a Trinity consisting of three persons: the Father,

the Son, and the Holy Spirit. Jesus Christ is believed to be the incarnate Son of God who was given to humankind for the sake of redemption some 2000 years ago. Through Jesus's death and resurrection, all of humanity is offered redemption from sin and provided eternal life in heaven and a relationship with God.

Christianity, Islam, and Judaism are the world's three great monotheisms and share a common culture hearth in southwestern Asia (see Figure 7.14). All three religions venerate the patriarch Abraham and thus are often called Abrahamic religions. Because Judaism is the parent religion of Christianity, the two faiths share many elements, including the Torah, the first five books of what Christians call the Old Testament (which forms part of the Christian Bible or holy book); the practice of prayer; and a clergy. These common elements are the basis for the term *Judeo-Christian*, which is used to describe beliefs and practices shared by the two faiths. Because Jesus was born into the Jewish faith, many Christians still accord Jews a special status as a chosen people, and see Christianity as the natural continuation or fulfillment of Judaism.

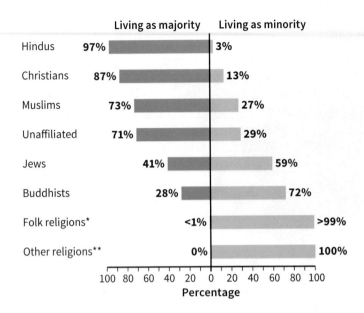

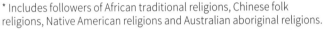

* Includes followers of African traditional religions, Chinese folk religions, Native American religions and Australian aboriginal religions.
** Includes Baha'i's, Jains, Sikhs, Shintoists, Taoists, followers of Tenrikyo, Wiccans, Zoroastrians, and many other faiths.

FIGURE 7.4 Majority or minority? Most people live in a country where their religion is the majority faith. Others, however, are religious minorities in their country of residence. This chart shows the percentage of majority and minority adherents among the world's major faiths. (Data from Pew Research Center's Forum on Religion and Public Life, Global Religious Landscape, 2012.)

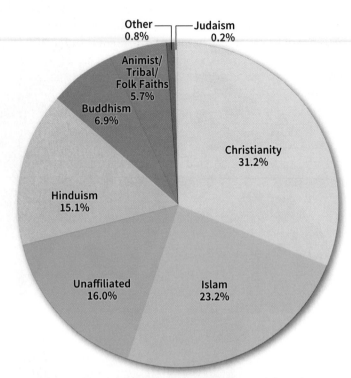

FIGURE 7.5 Major religions of the world. The world's major religions, by numbers of adherents expressed as a percentage. The third largest religious group in the world is, in fact, composed of those who are not affiliated with any established faith. Note that although Judaism is a prominent religion in the United States, it does not rank among the major world religions in terms of total number of adherents. (Data from Pew Research Center Demographic Projections: The Changing Global Religious Landscape, 2015.)

Christianity has long been fragmented into separate branches (see Figure 7.3). The major division is threefold, made up of Roman Catholics, Protestants, and Eastern Christians. Western Christianity, which now includes Catholics and Protestants, was initially identified with Rome and the Latin-speaking areas that are largely congruent with modern western Europe, whereas the Eastern Church dominated the Greek world from Constantinople (now the city of Istanbul, Turkey). Belonging to the Eastern group are the Armenian Church, reputedly the oldest in the Christian faith and today centered among a people of the Caucasus region; the Coptic Church, originally the religion of Christian Egyptians and still today a minority faith there, as well as being the dominant church among the highland people of Ethiopia; the Maronites, Semitic descendants of seventh-century heretics who disagreed with some of early Christianity's beliefs and retreated to a mountain refuge in Lebanon; the Nestorians, who live in the mountains of the Middle East and in India's Kerala state; and Eastern Orthodoxy, originally centered in Greek-speaking areas. Having converted many Slavic groups, Orthodox Christianity is today made up of a variety of national churches, such as Russian, Greek, Ukrainian, and Serbian Orthodoxy, with a collective membership of some 260 million (**Figure 7.6**).

FIGURE 7.6 Greek Orthodox monastery. The bells of St. John's Monastery in Patmos, Greece. (Courtesy of Patricia L. Price.)

Western Christianity splintered, most notably with the emergence of Protestantism in the Reformation of the 1400s and 1500s. As with all religious reformation movements, Protestantism sought to overcome what were viewed to be the wrong practices of Roman Catholicism, such as the need for priestly or saintly mediation between humans and God, while still staying within the general framework of Christianity. Since then, the Roman Catholic Church, which alone includes 1.2 billion people, or more than one-sixth of humanity, has remained unified; but Protestantism, from its beginnings, tended to divide into a rich array of **sects**, which together and including Anglican and independent Christians have a total membership today of about 900 million worldwide.

In the United States and Canada, the denominational map vividly reflects the fragmentation of Western Christianity and the resulting complex pattern of religious culture regions (**Figure 7.7**). Numerous denominations imported from Europe were later augmented by Christian denominations that originated in America. The American frontier was a breeding ground for new religious groups, as individualistic pioneer sentiment found expression in splinter Protestant denominations. Today, in many parts of the country, even a relatively small community may contain the churches of half a dozen religious groups. As a result, we can find about 2000 different religious denominations and cults in the United States alone. In a broad Bible Belt across the South, Baptist and other

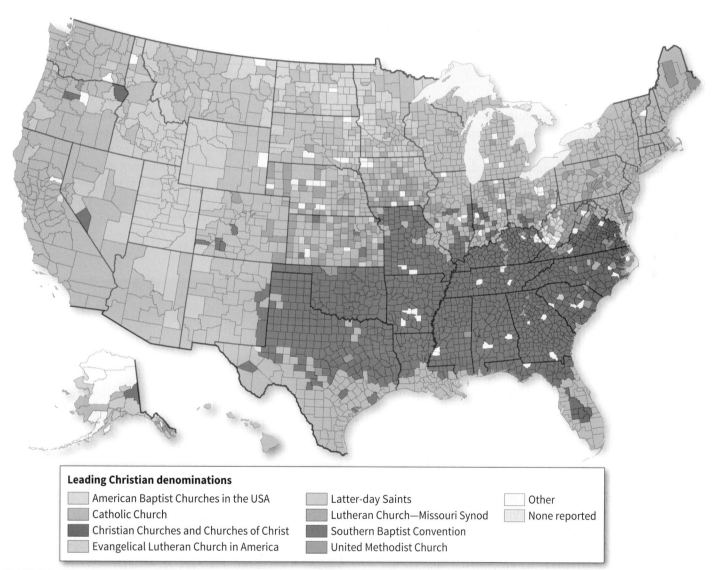

Leading Christian denominations

- American Baptist Churches in the USA
- Catholic Church
- Christian Churches and Churches of Christ
- Evangelical Lutheran Church in America
- Latter-day Saints
- Lutheran Church—Missouri Synod
- Southern Baptist Convention
- United Methodist Church
- Other
- None reported

FIGURE 7.7 Leading Christian denominations in the United States, shown by counties. In the majority counties, the church or denomination indicated claims 50 percent or more of the total church membership. The most striking features of the map are Baptist dominance throughout the South, a Lutheran zone in the upper Midwest, Mormon (Latter-Day Saints) dominance in the interior West, and the zone of mixing in the American heartland.

 The Video Connection

Bible Belt Atheist
Watch at Sapling Plus

Jerry Dewitt, formerly a pastor in DeQuincy, Louisiana, became an atheist. He founded an atheist congregation in a Lake Charles, Louisiana, coffee shop. Today, he preaches the virtues of living in the here-and-now, as opposed to preparing for the next life.

Thinking Geographically

1. Using Figure 7.7, identify where the Bible belt is located. What does the man who is briefly featured mean when he says that the Louisiana community portrayed in the video is located in the "buckle on the Bible belt"?

2. Dewitt and his atheist congregation are, for all intents and purposes, a religious minority in the heavily Christian southern United States. What similarities and differences do they share with religious minorities in other places?

3. Dewitt claims that his atheism is not a choice, but a realization that came about over time. Do other faiths work as choices, as gradual realizations, or in other ways? Describe how.

---

conservative fundamentalist denominations dominate, and Utah is at the core of the Mormon realm. A Lutheran belt stretches from Wisconsin westward through Minnesota and the Dakotas, and Roman Catholicism is dominant in southern Louisiana, the southwestern borderland, and heavily industrialized areas of the Northeast. The Midwest is a thoroughly mixed zone, although Methodism is the largest single denomination.

Today, Christianity is geographically widespread and highly diverse in its local interpretation. Although it is not as fast growing as Islam, intense missionary efforts strive to increase the number of adherents. Because of this proselytic work, Africa and Asia are the fastest-growing regions for Christianity.

Islam Islam, another great proselytic faith, claims 1.8 billion followers, largely in the desert belt of Asia and northern Africa and in the humid tropics as far east as Indonesia and the southern Philippines (see Figures 7.3 and 7.5). Adherents of Islam, known as Muslims (literally, "those who submit to the will of God"), are monotheists and worship one absolute God known as Allah. Islam was founded by Muhammad, considered to be the last and most important in a long line of prophets. The word of Allah is believed to have been revealed to Muhammad by the angel Gabriel (Jibrail) beginning in 610 C.E. of the Christian calendar in the Arabian city of Mecca (**Figure 7.8**). The Qur'an, Islam's holy book, is the text of these revelations and also serves as the basis of Islamic law, or Sharia. Most Muslims consider both Jews and Christians to be, like themselves, "people of the Book," and all three religions share beliefs in heaven, hell, and the resurrection of the dead. Many biblical figures familiar to Jews and Christians, such as Moses, Abraham, Mary, and Jesus, are also venerated as prophets in Islam. Adherents to Islam are expected to profess belief in Allah, the one God whose prophet was Muhammad; pray five times daily at established times; give alms, or *zakat*, to the poor; fast from dawn to sunset during the holy month of Ramadan; and make at least one pilgrimage, if possible, to the sacred city of Mecca in Saudi Arabia. These duties are known as the Five Pillars of the faith.

FIGURE 7.8 Muslims at prayer in Mecca, Saudi Arabia. (ayazad/Shutterstock)

Although not as severely fragmented as Christianity, Islam, too, has split into separate groups. Two major divisions prevail. Shi'ite (or Shi'a) Muslims, some 10–13 percent of the Islamic total in diverse subgroups, form the majority in Iran and Iraq. Shi'ites believe that Ali, who was Muhammad's son-in-law, should have succeeded Muhammad. Sunni Muslims, who represent the Islamic orthodoxy (the word *Sunni* comes from *sunnah*, meaning "tradition"), form the large majority worldwide (see Figure 7.3). Islam has not undergone a reformation parallel to that undergone by Christianity with Protestantism. However, reformation is the goal of liberal movements within Sunni and Shi'ite Islam.

Islam's strength is greatest in the Arabic-speaking lands in Southwest Asia and North Africa, although the world's numerically largest Islamic population is found in Indonesia. Other large clusters live in Pakistan, India, Bangladesh, and western China. Because of successful conversion efforts in non-Muslim areas and high birthrates in predominantly Muslim areas, Islam is the fastest-growing world religion. In the United States, more people convert to Islam than to any other religion, and Islam is predicted to surpass Judaism to become America's second-largest faith around 2035.

Hinduism **Hinduism**, a religion closely tied to India and its ancient culture, claims about 1.1 billion adherents (see Figures 7.3 and 7.5). Hinduism is a decidedly polytheistic religion. Although Hindus recognize one supreme god, Brahman, it is his many manifestations that are worshipped directly. Some principal Hindu deities include Vishnu, Shiva, and the mother goddess Devi. Ganesh, the elephant-headed god depicted in **Figure 7.9**, is often revered by Hindu university students because Ganesh is the god of wisdom, intelligence, and education. Believing that no one faith has a monopoly on the truth, most Hindus are notably tolerant of other religions (but see Subject to Debate, page 231).

Hindus strive to locate the harmonious and eternal truth, *dharma*, which is within each human being. Social divisions, or castes, separate Hindu society into four major categories, or *varna*, based on occupational categories: priests (Brahmins), warriors (Kshatriyas), merchants and artisans (Vaishyas), and workers (Shudras). Castes are related to dharma inasmuch as dharma implies a set of rules for each varna that regulate their behaviors with regard to eating, marriage, and use of space. All Hindus also share a belief in reincarnation, the idea that, although the physical body may die, the soul lives on and is reborn in another body. Related to this is the notion of *karma*. Karma can be viewed almost as a causal law, which holds that what an individual experiences in this life is a direct result of that individual's thoughts and deeds in a past life. Likewise, all thoughts and deeds, both good and bad, affect an individual's future lives. Ultimately, *moksha*, or liberation of the soul from

FIGURE 7.9 Ganesh. This image of Ganesh, the elephant-headed god of wisdom, intelligence, and education, is being immersed in the Indian Ocean at Pattinapakkam Beach in Chennai, at the culmination of his annual birthday celebration. Across India, tens of thousands of Ganesh idols are submerged in a ritual send-off of the deity to his sacred mountain home of Kailash. (ARUN SANKAR/AFP/Getty Images)

the cycle of death and rebirth, will occur when the worldly bonds of the material self fall away and one's pure essence is freed. *Ahimsa*, or the principle of nonviolence, involves veneration of all forms of life. This implies a principle of non-injury to all sentient creatures, which is why many Hindus are vegetarians.

Hinduism has splintered into diverse groups, some of which are so distinctive as to be regarded as separate religions. Jainism, for instance, is an ancient outgrowth of Hinduism, claiming perhaps 7 million adherents, almost all of whom live in India, and traces its roots back more than 25 centuries. Although they reject Hindu scriptures, rituals, and priesthood, the Jains share the Hindu belief in *ahimsa* and reincarnation. Jains adhere to a strict asceticism, a practice involving self-denial and austerity. For example, they practice veganism, a form of vegetarianism that prohibits the consumption of all animal-based products, including milk and eggs. Sikhism, by contrast, arose much later, in the 1500s, in an attempt to unify Hinduism and Islam. Centered in the Punjab state of northwestern India, where the Golden Temple at Amritsar serves as the principal shrine, Sikhism has about 24 million followers. Sikhs are monotheistic and have their own holy book, the Adi Granth.

No standard set of beliefs prevails in Hinduism, and the faith takes many local forms. Hinduism includes very diverse peoples. This is partly a result of its former status as a proselytic religion. A Hindu majority on the Indonesian island of Bali suggests the religion's former missionary activity (**Figure 7.10**). Today, most Hindus consider their religion an ethnic one, whereby one is acculturated into the Hindu community by birth; however, conversion to Hinduism is also allowed. Hindus form a majority in only three countries: Nepal, India, and Mauritius. Yet 97 percent of the world's Hindus live in these three countries; thus, Hindus are more likely than any other faith to constitute a majority in the countries where they reside (see Figure 7.4).

Buddhism Hinduism is the parent religion of **Buddhism**, which began 25 centuries ago as a reform movement based on the teachings of Prince Siddhartha Gautama, "the awakened one" (**Figure 7.11**). He promoted the four "noble truths": life is full of suffering, desire is the cause of this suffering, cessation of suffering comes with the quelling of desire, and an Eightfold Path of proper personal conduct and meditation permits the individual to overcome desire. The resultant state of enlightenment is known as *nirvana*. Those few individuals who achieve nirvana are known as Buddhas. The status of Buddha is open to anyone regardless of social status, gender, or age. Because Buddhism derives from Hinduism, the two religions share many beliefs, such as dharma, reincarnation, and *ahimsa*. Buddhism and its parent religion, Hinduism, as well as the related faiths Jainism and Sikhism, are known as dharmic religions.

Today, Buddhism is the most widespread religion in South and East Asia, dominating a culture region stretching from Sri Lanka to Japan and from Mongolia to Vietnam. In the process of its proselytic spread, particularly in China and Japan, Buddhism fused with native ethnic religions, such as Confucianism,

FIGURE 7.11 Buddhism is one of the religious faiths of South Korea. Here, an image of the Buddha is carved from a rocky bluff to create sacred space and a local pilgrimage shrine. (Courtesy of Terry G. Jordan-Bychkov.)

FIGURE 7.10 Hindu temple in Bali, Indonesia. Besakih, known as the Mother Temple of Bali, is the largest Hindu temple complex on the island, occupying the slopes of Mount Agung. (MuYeeTing/Getty Images)

Taoism, and Shintoism, to form syncretic faiths that fall into the Mahayana division of Buddhism. Southern, or Theravada, Buddhism, dominant in Sri Lanka and mainland Southeast Asia, retains the greatest similarity to the religion's original form, whereas a variation known as Tantrayana, or Lamaism, prevails in Tibet and Mongolia (see Figure 7.3). Buddhism's tendency to merge with native religions, particularly in China, makes it difficult to determine the exact number of its adherents. Estimates center on 500 million people (see Figure 7.5). Although Buddhism in China has mingled with local faiths to become part of a composite ethnic religion, elsewhere it remains one of the three great proselytic religions in the world, along with Christianity and Islam.

Taoic Religions Confucianism, Shinto, and Taoism together make up the faiths that center on Tao, the force that balances and orders the universe. Derived from the teachings of the philosopher K'ung Fu-tzu (551–479 B.C.E.), **Confucianism** was later promoted by China's Han dynasty (206 B.C.E.–220 C.E.) as the official state philosophy. Thus, it has bureaucratic, ethical, and hierarchical overtones that have led some to question whether it is more a way of life than a religion in the proper sense. In China, Confucianism's formalism is balanced by Taoism's romanticism. **Taoism**, which is both an established religion and a philosophy, emphasizes the dynamic balance depicted by the Chinese yin-yang symbol. The "three jewels of Tao" are humility, compassion, and moderation. **Shinto**, once the state religion of Japan, has long been blended with Buddhism, as well as infused with a Confucian-derived legal system, but at its core Shinto is an animistic religion. Because Tao is found in nature, there is some overlap with the animistic faiths discussed next. In addition, because of their fluid nature, Taoic religions tend toward syncretism and have blended with other faiths, particularly Buddhism. People may simultaneously practice elements of all these faiths. For these reasons, it is difficult to provide an exact number of adherents, although estimates hold that there are about half a billion. Taoic religions center on East Asia (see Figure 7.3).

Animism Peoples in diverse parts of the world often retain indigenous **animistic religions** (**Figure 7.12**). Animistic faiths subscribe to the idea that souls or spirits exist not only in humans, but also in animals, plants, rocks, natural phenomena such as thunder, geographic features such as mountains or rivers, or other entities of the natural environment. For the most part, animists do not form organized or recognized religious groups, but instead practice as ethnic religions common to clans or tribes. In addition, most animists follow oral, rather than written, traditions and thus do not have holy books. A tribal religious figure, called a *shaman*, may serve as an intermediary

between the people and the spirits, usually by going into a trance. Such belief systems are referred to as **shamanism**.

Currently numbering perhaps 240 million, animists believe that nonhuman beings and inanimate objects possess spirits or souls. These spirits are believed to live in rocks and rivers, on mountain peaks and celestial bodies, in forests and swamps, and even in everyday objects. As such, objects are considered to be alive, and depending on local beliefs, objects and animals may assume human forms or engage in human activities such as speech. For some animists, the objects in question do not actually possess spirits but rather are valued because they have a particular potency to serve as a link between a person and the omnipresent god. Followers of Wicca, a contemporary neo-pagan religion derived from pre-Christian European practices of reverence for the mother goddess and the horned god, worship the god and goddess who are thought to inhabit everything. Ritual tools used by Wiccans include the *athame* (dagger), *boline* (knife), and *besom* (broom), which are used in ceremonies to direct energy as practitioners communicate with the god and goddess.

Animistic elements can also pervade established religions. Adherents of Shintoism, a Japanese religion related to Buddhism, worship ancestors and *kami*, or spirits, which inhabit natural objects, such as waterfalls and mountains. We should not classify such systems of belief as primitive or simple, because they can be extraordinarily complex. Even in places such as the United States, where the majority of the inhabitants would not see themselves as animists, such beliefs are pervasive. For example, many people in the United States believe that their pets possess souls that ascend to heaven upon death (**Figure 7.13**).

FIGURE 7.12 Druids from the Mistletoe Foundation bless mistletoe in Worcestershire, England. Celtic Druids have their roots in the ancient, pre-Christian reverence for natural elements, such as streams, hills, and plants. Mistletoe is sacred to the Druidic faith. (Andrew Fox – Corbis/Getty Images)

FIGURE 7.13 Pet grave marker. Pumpkin the cat was apparently baptized into the Christian faith and now lies at rest in Pet Heaven Cemetery in Miami, Florida. (Courtesy of Patricia L. Price.)

Mobility

7.2   Describe how and why the world's religions became distributed as they are today.

Religions, according to religious studies scholar Thomas Tweed, can be thought of as "flows that intensify joy and confront suffering by drawing on human and suprahuman forces to make homes and cross boundaries." In other words, religions have always been on the move, crossing oceans and borders to establish new dwelling places. To a remarkable degree, the origin of the major religions was concentrated spatially in three principal culture hearth areas (**Figure 7.14**). A **culture hearth** is a focused geographic area where important innovations are born and from which they spread. Many religions mandate periodic return of the faithful to these culture hearths in order to confirm or renew their faith.

The Semitic Religious Hearth

All three of the great monotheistic faiths—Judaism, Christianity, and Islam—arose among Semitic peoples who lived in or on the margins of the deserts of southwestern Asia, in the Middle East (see Figure 7.14). Judaism, the oldest of the three, originated some 4000 years ago. Only gradually did

its followers acquire dominion over the lands between the Mediterranean and the Jordan River—the territorial core of modern Israel. Christianity, whose parent religion is Judaism, originated here about 2000 years ago. Seven centuries later, the Semitic culture hearth once again gave birth to a major faith when Islam arose in western Arabia, partly from Jewish and Christian roots.

Religions spread by both relocation and expansion diffusion. Recall from Figure 1.26 (page 20) that expansion diffusion can be divided into hierarchical and contagious subtypes. In hierarchical diffusion, ideas become implanted at the top of a society, leapfrogging across the map to take root in cities and bypassing smaller villages and rural areas. Because their main objective is to convert nonbelievers, proselytic faiths are more likely to diffuse than ethnic religions, and it is not surprising that the spread of monotheism was accomplished largely by Christianity and Islam, rather than Judaism. From Semitic southwestern Asia, both of the proselytic monotheistic faiths diffused widely, as shown in Figure 7.14.

Christians, observing the admonition of Jesus in the Gospel of Matthew—"Go ye therefore and teach all nations, baptizing them in the name of the Father, and of the Son, and of the Holy Ghost, teaching them to observe all things whatsoever I have commanded you"—initially spread through the Roman Empire, using the splendid system of imperial roads to extend the faith. In its early centuries of expansion,

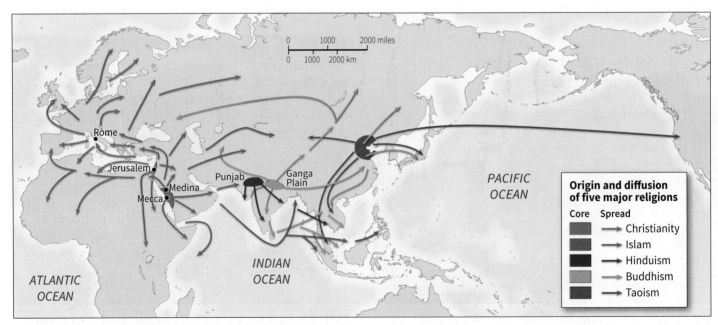

FIGURE 7.14 The origin and diffusion of five major world religions. Christianity and Islam, the two great proselytic monotheistic faiths, arose in Semitic southwestern Asia and spread widely through the Old World. Hinduism and Buddhism both originated in the northern reaches of the Indian subcontinent and spread throughout southeastern Eurasia. Taoic religions originated in East Asia, relocating with Chinese and Japanese migration regionally and, more recently, to North and South America.

Christianity displayed a spatial distribution that clearly reflected hierarchical diffusion. The early congregations were established in cities and towns, temporarily producing a pattern of Christianized urban centers and pagan rural areas. Indeed, traces of this process remain in our language. The Latin word *pagus*, "countryside," is the root of both *pagan* and *peasant*, suggesting the ancient connection between non-Christians and the countryside. Today, the term **pagan** is used to denote those who follow an unconventional religion, such as the Wiccans and Druids mentioned in the previous section (see Figure 7.12).

The scattered urban clusters of early Christianity were created by missionaries such as the apostle Paul, one of Jesus's disciples, who moved from town to town bearing the news of the emerging faith. In later centuries, Christian missionaries often used the strategy of converting kings or tribal leaders, setting in motion additional hierarchical diffusion. The Russians and Poles were converted in this manner. Some Christian expansion was militaristic, as in the reconquest of Iberia from the Muslims and the invasion of Latin America. Once implanted in this manner, Christianity spread farther by means of contagious diffusion. When applied to religion, this method of spread is called **contact conversion** and is the result of everyday associations between believers and nonbelievers.

The Islamic faith spread from its Semitic hearth area in a predominately militaristic manner. Obeying the command in the Qur'an that they "do battle against them until there be no more seduction from the truth and the only worship be that of Allah," the Arabs expanded westward across North Africa in a wave of religious conquest. The Turks, once converted by the Arabs, carried out similar Islamic conquests. In a different sort of diffusion, Muslim missionaries followed trade routes eastward to implant Islam hierarchically in the Philippines, Indonesia, and the interior of China. Sub-Saharan Africa is the current major region of Islamic expansion, an effort that has produced competition with Christians for the conversion of local animists. As a result of missionary successes in sub-Saharan Africa and high birthrates in its older sphere of dominance, Islam has become the world's fastest-growing religion in terms of the number of new adherents.

The Indus-Ganges Religious Hearth

The second great religious hearth area lay in the plains fringing the northern edge of the Indian subcontinent. This lowland, drained by the Ganges and Indus rivers, gave birth to Hinduism and Buddhism. Hinduism, which is at least 4000 years old, was the earliest faith to arise in this hearth. Its origin apparently took place in Punjab, from where it diffused to dominate the subcontinent, although some historians believe that the earliest

form of Hinduism was introduced from Iran by emigrating Indo-European tribes about 1500 B.C.E. Missionaries later carried the faith, in its proselytic phase, to overseas areas, but most of these converted regions were subsequently lost to other religions.

Branching off from Hinduism, Buddhism began in the foothills bordering the Ganges Plain about 500 B.C.E. (see Figure 7.14). For several centuries, it remained confined to the Indian subcontinent, but missionaries later carried the religion to China (100 B.C.E. to 200 C.E.), Korea and Japan (300 to 500 C.E.), Southeast Asia (400 to 600 C.E.), Tibet (700 C.E.), and Mongolia (1500 C.E.). Buddhism developed many regional forms throughout Asia, except in India, where it was ultimately accommodated within Hinduism.

The diffusion of Buddhism, like that of Christianity and Islam, continues to the present day. As many as 4 million Buddhists live in the United States. Mostly, their presence is the result of relocation diffusion by Asian immigrants to the United States, where immigrant Buddhists outnumber Buddhist converts by three to one.

The East Asian Religious Hearth

K'ung Fu-tzu and Lao Tzu, the respective founders of Confucianism and Taoism, were contemporaries who reputedly once met with each other. Both religions were adopted widely throughout China only when the ruling elite promoted them. In the case of Confucianism, this was several centuries after the master's death. During his life, K'ung Fu-tzu wandered about with a small band of disciples trying to convince rulers to put his ideas on good governance into practice. But he was shunned even by lowly peasants, who criticized him as "a man who knows he cannot succeed but keeps trying." Thus, early attempts at contagious diffusion were unsuccessful, whereas hierarchical diffusion from politicians and schools eventually spread Confucianism from the top down. Taoism, as well, did not gain wide acceptance until it was promoted by the ruling Chinese elite.

After 1949, China's communist government officially repressed organized religious expression, dismissing it as a relic of the past. In other words, the government attempted to erect an absorbing barrier that would not only halt the spread of religion, but would erase it from Chinese public and private life. As noted in Chapter 1, absorbing barriers are rarely completely successful, and this example is no exception. Although temples were converted to secular uses, and even looted and burned, religion was driven underground rather than eradicated. After the end of the Cultural Revolution (1966–1976), which aimed to purify Chinese society of bourgeois excesses such as religion, tolerance for religious expression grew. However, the Chinese

FIGURE 7.15 Shinto shrine in the United States. With the migration of Japanese people in the nineteenth and twentieth centuries, particularly to the West Coast of the United States, Shinto traditions and religious structures followed. This image depicts the Tsubaki Shrine in Granite Falls, Washington State. (© Alexander Marten Zhang)

Communist Party's official stance still holds that religious and party affiliations are incompatible; thus, some party officials are reluctant to divulge their religious status. This, combined with the fact that many Chinese practice elements of several Taoic faiths simultaneously, makes the precise enumeration of adherents difficult.

Taoism and Confucianism have spread with the Chinese people through trade and military conquest. Thus, people in Taiwan, Malaysia, Singapore, Korea, Japan, and Vietnam, along with mainland China, practice these beliefs or at least have been influenced by them. Today, Chinese and Japanese migrants alike have relocated their belief systems across the globe, as the image of the Tsubaki Shinto Shrine in Washington State shown in **Figure 7.15** confirms.

FIGURE 7.16 Religious pilgrims. Visitors number in the millions annually and can provide a holy site's main source of revenue. *Top:* Spiritual pilgrims visit Machu Picchu, an ancient Incan ceremonial center in the Peruvian Andes. *Middle:* Catholic pilgrims greet Pope Francis in Saint Peter's Square, The Vatican, Rome. *Bottom:* Buddhist pilgrims meditate underneath the Bodhi Tree, a descendent of the tree where the Buddha attained enlightenment, at the Mahabodhi Temple in Bodhgaya, India. (Top: Bernard Roussel/Getty Images; Center: powerofforever/iStock Unreleased/Getty Images; Bottom: Mauricio Abreu/ Getty Images)

Religious Pilgrimage

For many religious groups, journeys to sacred places, or **pilgrimages**, are an important aspect of faith-based mobility (**Figure 7.16**). Pilgrimages are typical of both ethnic and proselytic religions. They are particularly significant for followers of Islam, Hinduism, Shintoism, and Roman Catholicism. Along with missionaries, pilgrims constitute one of the largest groups of people voluntarily on the move for religious reasons: religious pilgrimage is the number one driver of tourism. Pilgrims do not aim to convert people to their faith through their journeys. Rather, they enact in their travels a connection with the sacred spaces of their faith. Indeed, some religions mandate pilgrimage, as is the case with Islam, where pilgrimage to Mecca is one of the Five Pillars of the faith.

The sacred places visited by pilgrims vary in character (**Figure 7.17**). Some have been the setting for miracles; a few are regions where religions originated or areas where the founders of the faith lived and worked; others contain sacred physical features, such as rivers, caves, springs, and mountain peaks; and still others are believed to house gods or are religious administrative centers where leaders of the religion reside. Examples include the Arabian cities of Mecca and Medina in Islam, which are cities where Muhammad resided and thus form

Islam's culture hearth; Rome, the home of the Roman Catholic pope; the French town of Lourdes, where the Virgin Mary is said to have appeared to a girl in 1858; the Indian city of temples on the holy Ganges River, Varanasi, a destination for Hindu, Buddhist, and Jain pilgrims; and Ise, a shrine complex located in the culture hearth of Shintoism in Japan. Places of pilgrimage might be regarded as places of spatial convergence, or nodes, of functional culture regions.

Religion provides the stimulus for pilgrimage by offering those who participate in the purification of their souls or the attainment of some desired objective in their lives, by mandating pilgrimage as part of devotion, or by allowing the faithful to connect with important historic sites of their faith. For this reason, pilgrims often journey great distances to visit major shrines. Other sites of lesser significance draw pilgrims only from local districts or provinces. Pilgrimages can have tremendous economic impact because the influx of pilgrims is a form of tourism.

In some localities, the pilgrim trade provides the only significant source of revenue for the community. Lourdes, a town of about 16,000, attracts between 4 million and 5 million pilgrims each year, with many seeking miraculous cures at the famous grotto where the Virgin Mary reportedly appeared. Not surprisingly, among French cities, Lourdes ranks second

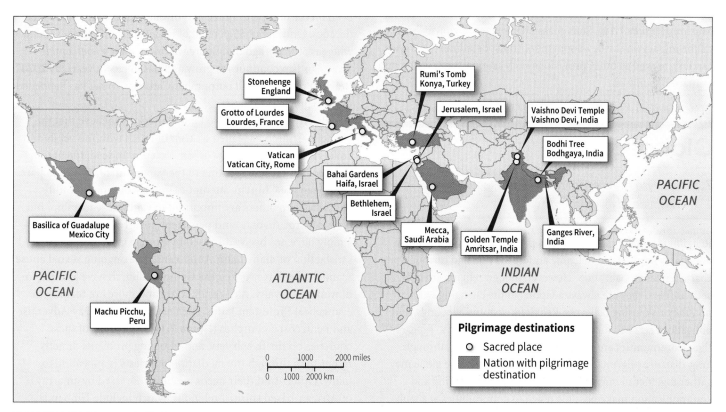

FIGURE 7.17 Places of pilgrimage. This map displays the most popular sites of religious or spiritual pilgrimage. Taken together, these 14 sites attract 100 million visitors annually.

FIGURE 7.18 Religious segregation. Access to Mecca is restricted, allowing Muslims only. The Saudi Arabian government believes that permitting non-Muslim tourists to visit Mecca and Medina would disturb the sanctity of these sacred places. (vario images GmbH & Co. KG/Alamy)

only to Paris in its number of hotels, although most of these are small. Mecca, a small city, annually attracts millions of Muslim pilgrims from every corner of the Islamic culture region as they perform the **Hajj**, or Fifth Pillar of Islam, which all able-bodied Muslims who can afford the journey are encouraged to do at least once in their lifetime. By land, by sea, and (mainly) by air, the faithful come to this hearth of Islam, a city closed to all non-Muslims (**Figure 7.18**). Such mass pilgrimages obviously have a major impact on the development of transportation routes and carriers, as well as other provisions such as inns, food, water, and sanitation facilities.

Globalization

7.3   Analyze how globalization has affected the practice of religion.

Religion is seen by many people as providing a stable anchor, one that is particularly welcome amidst the flux and turmoil that characterizes globalization. However, though religions may be seen as fixed, they are always adapting to the changing world; if not, they risk losing believers. Religious customs, meetings, and obligations serve to form a strong basis of community. Typically, these communities are experienced locally; however, through long-distance pilgrimage, the faithful from across the globe may gather together in one place. Today's increased mobility and communication make possible religious communities at a truly global scale. The rise of the Internet has reshaped how people worship and interact with others, allowing for online communities

of faith to arise. In some places organized religion appears to be losing ground, while in others the numbers of the faithful are on the rise. All in all, religion is an enduring feature of human existence that will likely transcend the forces of globalization.

The Rise of Evangelical Protestantism in Latin America

As economic, social, and political changes occur in places, cultural forces such as religion must adapt. For instance, in Chapter 4, we saw how languages change over time in response to immigration, urbanization, and trade-based exchanges. At other times, cultural forces drive changes in politics, economics, and societies. In Chapter 5, we saw how Hispanic immigration to the United States has resulted in major shifts in consumption, political behavior, and social attitudes.

A good example of how the cultural, economic, social, and political arenas work together is provided by the contemporary religious landscape of Latin America and the Caribbean. In this region, Roman Catholicism has dominated the religious landscape since the time of the Iberian conquerors in the late 1400s. More Catholics live in this region than in any other on Earth, with three out of every five Catholics residing here. Brazil is the largest Catholic country in the world in terms of population, with more than 120 million adherents. However, the Catholic Church has been on the decline in Brazil and throughout the region in recent decades. Only about 65 percent of Brazilians are Catholics today, whereas 50 years ago that figure was more than 95 percent. If Protestantism continues to grow as it has in recent years, by 2020 Brazil's Protestants will outnumber Catholics. What has happened to the region's Catholics?

Briefly, more and more Latin Americans believe that the Catholic Church has failed to keep in touch with the needs and concerns of contemporary urban societies. Indeed, the naming of Pope Francis of Argentina to the papacy in 2013 may, in part, have been the Church's attempt to underscore that humility and service to others are important to Catholic leadership. But birth control, divorce, and persistent poverty are issues that simply have not been addressed by Roman Catholicism to the satisfaction of many Latin Americans. The ongoing sexual abuse scandals that have rocked the Catholic Church have also led to diminished loyalty. As a result, more and more are turning to evangelical Protestant faiths, such as the Seventh-Day Adventist and Pentecostal churches (**Figure 7.19**). The focus of these churches on thrift, sobriety, and resolving problems directly rather than through the mediation of priests is appealing to many. Others find their needs are best addressed by an array of African-based spiritist religions, which include Umbanda, Candomblé, and Santería. Whether Catholicism will ultimately reform from within to become more current, or whether it will

FIGURE 7.19 Evangelical Protestants on the march. Over a million evangelical Christians participated in the annual March for Jesus in São Paulo, Brazil. (Nelson Antoine/AP Photo)

continue to lose out to rival faiths, is a key question in regard to the Latin American religious landscape.

Religion on the Internet

In the mid-fifteenth century, Gutenberg's printing press made the Bible available to a mass audience. The Internet, proponents argue, represents a similar technological revolution, one that will inevitably attract new adherents to religious faiths. Gathering together regularly in a holy place—a mosque, temple, church, or shrine—is at the heart of most organized religions. Traditionally, people have come together to receive sacraments, sing, pray, and celebrate major life events in houses of worship. With the tens of thousands of religious web sites now in existence, it is no longer necessary to physically go to a place of worship. Rather, you can sit in front of a computer, at any time of the day or night, and read a holy book, submit and read online prayers, engage in theological debate, or watch broadcasts of religious services (**Figure 7.20**).

For cultural geographers, the most interesting dimension of online worship concerns what it does to the role of place in a global society. Clearly, the Internet has made it possible to practice religion in a way that is not linked to a specific place of worship—but is this a good thing? Supporters of practicing religion online contend that it allows people to become members of virtual communities that would not otherwise have been available to them in times of spiritual need. Now illness, invalidism, or the pressures of a busy life need not present a barrier to worship. Detractors argue that solitary worship online erodes the place-based communities that are at the heart of many religions. Recall from Chapter 1 that debates center on whether globalization dilutes the role

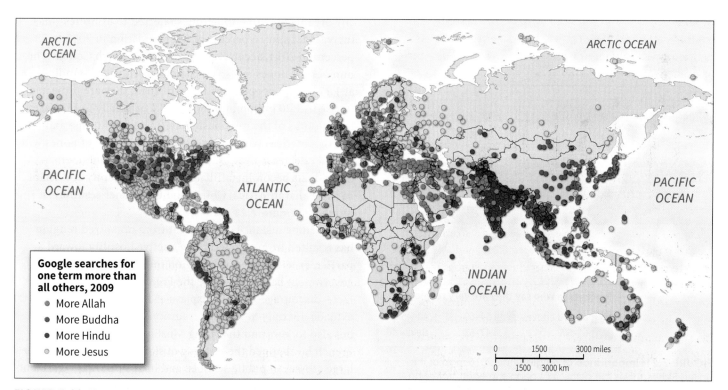

Google searches for one term more than all others, 2009
- More Allah
- More Buddha
- More Hindu
- More Jesus

FIGURE 7.20 The geography of religion, according to Google. This map of Google searches for one religious figure above all others depicts an online geography that mirrors regional religious patterns on the ground. (Source: http://www.floatingsheep.org/2010/01/googles-geographies-of-religion.html with permission from Matthew Zook.)

of place and of place-based differences. As people pick and choose elements from the different faiths available online, will religions converge into a watered-down, homogeneous "McFaith"?

Religion's Relevance in a Global World

A 2016 Gallup poll revealed that 53 percent of Americans consider religion to be very important in their daily lives, with an additional 22 percent claiming religion to be fairly important, making the United States one of the more religious nations on Earth. Across the United States, the importance of religion varies, as depicted in **Figure 7.21**.

Newer religious influences, too, are making an entrance onto the American religious stage. An important dynamic of religious change in North America has long been immigration. Waves of Catholics, Protestants, Jews, Muslims, and others have arrived on the shores of North America and continue to do so, shaping and reshaping the religious landscape. Another source of religious variety is the cultural popularity of selected elements of religions that are not dominant in North America. For example, if you have ever practiced yoga or meditation, you are engaging in Hindu and Buddhist practices (**Figure 7.22**). Many Americans do yoga or meditate to enhance physical well-being, to promote spiritual growth, to relieve

FIGURE 7.22 Yoga class in session. Many people in the United States have incorporated Hindu and Buddhist spiritual practices into their lives. (Ryan McVay/Getty Images)

stress, or even to keep up with the latest trends, rather than as part of a religious practice. Yet they are among the growing number of Americans who have found a blend of Eastern and Western religious practices to be compatible with their beliefs and lifestyles.

The percentage of Americans who do not belong to any organized faith—sometimes referred to as **nones**—has increased, from 16 percent of the population in 2007 to 23 percent in 2014, according to the Pew Research Center. The numbers of nones is also on the rise worldwide. Standing at 1.17 billion people worldwide in 2015, they are projected to rise to 1.20 billion by 2060, thanks primarily to increases in the ranks of the religiously unaffiliated in Europe and North America. Worldwide, however, the share of nones will actually decrease over time, because of rising numbers of individuals becoming affiliated with Christianity and Islam in sub-Saharan Africa, China, and the former communist republics (**Figure 7.23**).

In some instances, the retreat from organized religion has resulted from a government's active hostility toward a particular faith or toward religion in general. The French government has long—since the French Revolution, in fact—discouraged public displays of religious faith. This extends not only to France's traditional Roman Catholicism but also to Jews and the rising Muslim population. In 2004 French law banned the wearing of skullcaps, headscarves, or large crosses in public schools, and in 2011 burqas—garments worn by some Muslim women—were outlawed in all public areas.

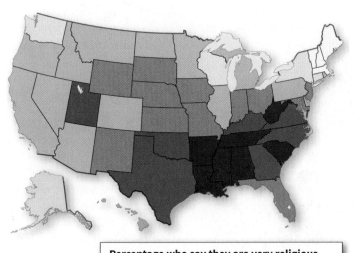

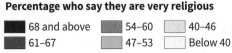

Percentage who say they are very religious

| ■ 68 and above | ■ 54–60 | 40–46 |
| ■ 61–67 | ■ 47–53 | Below 40 |

FIGURE 7.21 Importance of religion. This 2016 map of the United States shows the percent of people in each state who indicated that religion was very important in their lives. (Source: www.pewforum.org.)

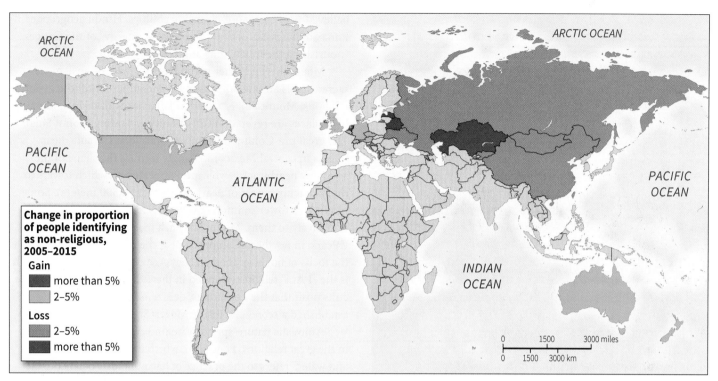

FIGURE 7.23 Change in religiously unaffiliated population, 2005–2015. In some world regions, the nones are on the rise; while in others, there is a proportional increase in adherents of Christianity and Islam that outstrips the rise of the religiously unaffiliated. Rising numbers of Christians and Muslims are due both to births and to conversions into these two faiths. (Source: Adapted from World Religion Forum.)

Far from being a contemporary phenomenon, nonbelief has always been a feature of the religious landscape. In medieval Europe, according to historian Molly Worthen, "Ordinary people often skipped church and had a feeble grasp of basic Christian dogma. Many priests barely understood the Latin they chanted—and many parishes lacked any priest at all. Bishops complained about towns that used their cathedrals mainly as indoor markets or granaries." Such patterns once again reveal the inherent spatial variety of humankind.

Nature–Culture

7.4 Explain specific ways in which the natural environment shapes, and is shaped by, religious beliefs and practices.

Our religious beliefs shape how we engage with the natural world. Some religions posit that humans transcend nature and are given license to dominate and exploit the natural world. Others see humans as merely co-dwellers in the natural world, alongside and interconnected with other inhabitants of nature.

Thus, religions can inspire environmental stewardship when nature is viewed as a sacred part of creation, or it can encourage depletion and destruction of the natural world in the service of humankind. Finally, there is a spatial expression of religious practices, teachings, and prohibitions around food and drink that is reflected in the distribution and spread of certain animals and plants.

Appeasing the Forces of Nature

One of the main functions of many religions is the maintenance of harmony between a people and their physical environment. Thus, religion is perceived by its adherents to be part of an adaptive strategy; for that reason, physical environmental factors, particularly natural hazards and disasters, exert a powerful influence on the development of religions.

Environmental influence is most readily apparent in animistic faiths. In fact, an animistic religion's principal goal is to mediate between its people and the spirit-filled forces of nature. Animistic ceremonies are often intended to bring rain, quiet earthquakes, end plagues, or in some other way manipulate environmental forces by placating the spirits believed responsible for these events.

FIGURE 7.24 Holy water from the Jordan River. This water is taken from the Jordan River, at the site where Jesus is believed to have been baptized by John the Baptist. It can be purchased and used in rituals. (Courtesy of Patricia L. Price.)

Rivers, mountains, trees, forests, and rocks often achieve the status of sacred space, even in the great religions. The Ganges River and certain lesser streams such as the Bagmati in Nepal are holy to the Hindus, and the Jordan River has special meaning for Christians, who often transport its waters in containers to other continents for use in baptism (**Figure 7.24**). Most holy rivers are

believed to possess soul-cleansing abilities. Hindu geographer Rana Singh speaks of the "liquid divine energy" of the Ganges "nourishing the inhabitants and purifying them."

Mountains and other high places likewise often achieve sacred status among both animists and adherents of the great religions. Mount Fuji is sacred in Japanese Shintoism, and many high places are revered in Christianity, including Mount Sinai. The great pre-Columbian temple pyramid at Cholula, near Puebla in central Mexico, strikingly mimics the shape of the awesome nearby active volcano Popocatépetl, which towers to the menacing height of nearly 18,000 feet (5500 meters). Some mountains tower so impressively as to inspire cults devoted exclusively to them. Mount Shasta, a massive snowcapped volcano in northern California, near the Oregon border, serves as the focus of no fewer than 30 New Age cults, the largest of which is the "I Am" religion, founded in the 1930s (**Figure 7.25**). These cults posit that the Lemurians, denizens of a lost continent, established a secret city inside Mount Shasta.

Animistic nature-spirits lie behind certain practices found in the great religions. **Feng shui**, which literally means "wind and water," refers to the practice of harmoniously balancing the opposing forces of nature in the built environment. A feature of Asian religions that emphasizes Tao, the dynamic balance found in nature, feng shui involves choosing environmentally auspicious sites for locating houses, villages, temples, and graves. The homes of the living and the resting places of the dead must be aligned with the cosmic forces of the world in order to ensure good luck, health, and prosperity. Although the practice of feng shui dates back some 7000 years, contemporary people practice its principles. **Figure 7.26** depicts a high-rise condominium in Hong Kong's Repulse Bay neighborhood that incorporates feng shui principles in its design. The square opening in the building's center is said to provide passage to the bay for the

FIGURE 7.25 Two high places that have evolved into sacred space. *Left:* The reddish sandstone Uluru, or Ayers Rock, in central Australia is sacred in Aboriginal animism. *Right:* Snowy Mount Shasta in California is venerated by some 30 New Age cults. (Left: Hans-Peter Merten/Getty Images; Right: tntemerson/iStock/Getty Images)

FIGURE 7.26 Condominium in Hong Kong. The square opening in this building is supposed to allow for the passage of the dragon that resides in the hill behind. (Courtesy of Ari Dorfsman.)

FIGURE 7.27 Pray for rain. A church in Clovis, New Mexico, does its part to end a drought. (Joe Raedle/Getty Images)

dragon that dwells in the hill behind the building, allowing the dragon to drink from the waters of the bay and return to its abode unencumbered. Some Westerners have also adopted the principles of feng shui. Office spaces as well as homes are arranged according to its basic ideas. For example, artificial plants, broken articles, and paintings depicting war are thought to bring negative energy into living spaces and so should be avoided.

Although the physical environment's influence on the major Western religions is less pronounced, it is still evident. Some contemporary adherents to the Judeo-Christian tradition believe that God uses plagues to punish sinners, as in the biblical account of the 10 plagues inflicted on Egypt, which allowed the Israelites to go into the wilderness. Modern-day droughts, earthquakes, and hurricanes are interpreted by some as God's punishment for wrongdoing, whereas others argue that this is not so. Environmental stress can, however, evoke a religious response not so different from that of animistic faiths. Local ministers and priests often attempt to alter unfavorable weather conditions with special services (**Figure 7.27**).

The Impacts of Belief Systems on Plants, Animals, and Food

Every known religion expresses itself in food choices, to one degree or another. In some faiths, certain plants and livestock, as well as the products derived from them, are in great demand because of their roles in religious ceremonies and traditions. When this is the case, the plants or animals tend to spread or relocate with the faith.

For example, in some Christian denominations in Europe and the United States, celebrants drink from a cup of wine that they believe is the blood of Christ during the sacrament of Holy Communion. The demand for wine created by this ritual aided the diffusion of grape growing from the sunny lands of the Mediterranean to newly Christianized districts beyond the Alps in late Roman and early medieval times. The vineyards of the German Rhine were the creation of monks who arrived from the south between the sixth and ninth centuries. For the same reason, Catholic missionaries introduced the cultivated grape to California, in an example of relocation diffusion. In fact, wine was associated with religious worship even before Christianity arose. Vineyard keeping and winemaking spread westward across the Mediterranean lands in ancient times in association with worship of the god Dionysus.

Religious taboos can even function as absorbing barriers, preventing diffusion of foods, drinks, and practices that violate the taboo. Mormons, who are encouraged to avoid caffeine, have not taken part in the American fascination with coffee. Sometimes these barriers are permeable. Certain Pennsylvania Dutch churches, for example, prohibit cigarette smoking but do not object to member farmers raising tobacco for sale in the commercial market.

Some religions set forth very specific guidelines for food preparation and consumption. Jews, for instance, are forbidden

GEOGRAPHY @ WORK

Robert Szypko

First-Grade Teacher, Success Academy, Bedford-Stuyvesant, New York

Education: BA Geography, Dartmouth College

Courtesy of Robert Szypko.

Q. *Why did you major in geography and decide to pursue a career related to the field?*

A. I am fascinated by the ways that individuals negotiate cultural, social, and ethnic differences across space both in global and local (particularly urban) contexts. I initially chose geography because of the field's strong tradition in development studies, but grew to love the larger analytical frameworks and technological skills emphasized in the discipline. I appreciate nuance, and the cultural geography courses I took brought my understanding of the nuances of culture, landscapes, and other topics to a new level. The more geography courses I took, the more I realized the breadth of geography's implications. Suddenly, my interests in music, education, graffiti, and journalism gained a greater richness as I saw them through a geographer's lens.

Q. *Please describe your job.*

A. I lead a classroom of 25 first-grade students who come from a variety of neighborhoods across Brooklyn. My responsibilities include planning and executing lessons, analyzing and responding to classroom achievement data, and maintaining a positive and hardworking classroom culture. I work closely with parents to leverage them in the education of their children, and work to better understand the backgrounds and home lives of each student in order to best serve them in the classroom.

Q. *How does your geographical background help you in your day-to-day work?*

A. My geographical background has helped me more fully understand and account for my students' varied backgrounds and the geographical diversity they represent (both in terms of where their family is from and where they live now, in different neighborhoods throughout Brooklyn). Geography's emphasis on complexity when dealing with cultures and communities at any scale has given me the critical eye to maintain an open mind about parents, students, the neighborhood of my school, and so on. It has also helped me as I try to lead students toward a better understanding of the world around them. Children's literature contains many different geographic representations, and I firmly believe that geographic nuance is important even at such a young age, from breaking down the homogenous identity of "Africa," to developing a deep understanding of the different landscapes and cultures of the arctic zone.

Q. *What advice do you have for students considering a career in the geography field?*

A. Keep an open mind about what geography can do for you in your career. The critical thinking skills and technological skills (e.g., GIS) you acquire as a geographer can serve you well in a variety of industries and jobs.

---

from consuming animals deemed "unclean," including shellfish, pork, and insects. Meat and milk must not be stored, prepared, or consumed together, which leads observant Jewish households to have two sets of dishes, two sinks, and even two ovens and refrigerators. *Kosher* food—food that is fit for consumption by observant Jews—allows only meat slaughtered by a trained individual using a sharp knife, drained of blood, and deemed disease-free. Muslims are commanded to consume only *halal*, or permissible, foods. Pork and alcohol are prohibited, while halal livestock should be fed with "clean" food that is free of animal

by-products. Many non-Jews and non-Muslims view the kosher and halal certifications as an indication of wholesomeness, and seek out such foods for health rather than religious reasons. Indeed, the top three reasons given for purchasing kosher products were food quality, safety, and healthfulness; religious reasons ranked sixth (Rashtogi, 2010). But in the Netherlands, animal rights activists have deemed halal practices unacceptable because they do not stun an animal prior to slaughter, and the Dutch have banned halal meat, to the consternation of some who see this as a backlash against Muslim immigrants to the country.

The case of India's sacred cows provides an intriguing example of how religious beliefs shape the role of animals in society, in this instance avoiding the use of cows as food while emphasizing the animal's sacred as well as practical functions. Although only one out of every three of India's Hindus practices vegetarianism, almost none of India's Hindu population will eat beef, and many will not use leather. Why, in a populous country such as India, where many people are poor and where food shortages have plagued regions of the country in previous decades, do people refuse to consume beef? There are several quite legitimate reasons. First, the dairy products provided by cow's milk and its by-products, such as yogurt and *ghee* (clarified butter), are central to many regional Indian cuisines. If the cow is slaughtered, it will no longer be able to provide milk. Second, in areas of the world that rely heavily on local agriculture for food production, such as India, cows provide valuable agricultural labor in tilling fields. Cows also provide free fertilizer in the form of dung. Finally, the value of the cow has been incorporated into Indian Hindu beliefs and practices over many centuries and has become a part of culture. Krishna, an incarnation of the major Hindu deity Vishnu, is said to be both the herder and the protector of cows. Nandi, who is the deity Shiva's attendant, is represented as a bull.

Think for a moment about what you consider appropriate to eat. Perhaps beef is part of your diet, but would you eat horse meat? What is so different about a cow and a horse? How about dog or cat meat? What makes certain animals pets in some cultures and dinner in others? Through this sort of questioning, you may come to realize that practices that seem second nature—such as what you will and will not eat—are, in fact, the result of long histories that are very different from place to place.

Ecotheology

Ecotheology is the name given to a rich and abundant body of literature studying the role of religion in habitat modification. More exactly, ecotheologians ask how religious teachings and worldviews are related to our attitudes about modifying the physical environment. In some faiths, human power over natural forces is assumed. In others, nature is a sacred creation that should be respected and protected by humans as an ethical component of the faith's practice.

Judaism, Christianity, and Nature The Judeo-Christian tradition teaches that humans have dominion over nature, but it goes further, promoting the view that the Earth was created especially for human beings, who are separate from and superior to the natural world. This view is implicit in God's message to Noah after the Great Flood, promising "every moving thing that lives shall be food for you, and as I gave you the green plants, I give you everything." The same theme is repeated in the Book of Psalms, where Jews and Christians are told "the heavens are the Lord's heavens, but the Earth he has given to the sons of men." Humans are not part of nature, but separate, forming one aspect of a God–nature–human trinity.

Believing that the Earth was given to humans for their use, early Christian thinkers adopted the view that humans were God's helpers in finishing the task of creation, that human modifications of the environment were therefore God's work. Small wonder, then, that the medieval period in Europe witnessed an unprecedented expansion of agricultural acreage, involving the large-scale destruction of woodlands and the draining of marshes. Nor is it surprising that Christian monastic orders, such as the Cistercian and Benedictine monks, supervised many of these projects, directing the clearing of forests and the establishment of new agricultural colonies.

Subsequent scientific advances permitted the Judeo-Christian West to modify the environment at an unprecedented rate and on a massive scale. This marriage of technology and theology is one cause of our modern ecological crisis. The Judeo-Christian religious heritage, in short, has for millennia promoted an instrumentalist view of nature that is potentially far more damaging to the ecology than an organic view of nature in which humans and nature exist in balance. Yet there is considerable evidence to the contrary. For example, the Orthodox Church in Russia is working to create wildlife preserves on monastery lands. The Patriarch of Constantinople, leader of Eastern Orthodox Christianity, has made the fight against pollution a church policy, declaring that damaging the natural habitat constitutes a sin against God. The Church of England has declared the abuse of nature blasphemous, and throughout the monotheistic religions, the green teachings of long-dead saints, such as Christianity's St. Francis of Assisi, who treasured birds and other wildlife, now receive heightened attention. In 1986 the Assisi Declarations united representatives of five of the world's major religions—Buddhism, Christianity, Hinduism, Islam, and Judaism—in Assisi, Italy, to discuss how faiths could work together to honor the sacredness of nature and stop its destruction.

Some fundamentalist Protestant sects herald ecological crisis and environmental deterioration as a sign of the coming Apocalypse, Christ's return, and the end of the present age. Thus, they welcome ecological collapse and, obviously, are unlikely to be interested in addressing the problem of environmental degradation. Other conservative Protestants, however, have adopted conservationist views, inspired by biblical admonitions such as the Great Flood story from the Old Testament, in which Noah saves animals by bringing them onto the ark. This story is viewed by some fundamentalists as a call to protect endangered species. An ecotheological focus underlies the multidenominational National Religious Partnership for the Environment, which includes many evangelical Protestant members. The hope is to mobilize the Christian Right against wanton environmental destruction in the same way they oppose abortion.

Buddhism and Hinduism and Cremation Real-world practices do not always reflect the stated ideals of religions. Both Buddhism and Hinduism protect temple trees but demand huge quantities of wood for cremations (**Figure 7.28**). Traditional Hindu cremations, for example, place the corpse on a pile of wood, cover it with more wood, and burn it during an open-air rite that can last up to six hours. The construction of funeral pyres is estimated to strip some 50 million trees from India's countryside annually. In addition, the ashes are later swept into rivers, and the burning itself releases carbon dioxide into the atmosphere, contributing to the pollution of waterways and the atmosphere. The use of wood, and its placement on the ground, provide an important symbolic connection between the body and the Earth. It does not, however, lead to efficient burning; on average, 880 pounds (400 kilograms) of wood are required to cremate a single corpse. A "green cremation system" has recently been developed by a New Delhi–based nonprofit organization. By placing the first layer of wood on a grate, and placing a chimney over the pyre, wood use can be reduced by 75 percent, and the burning time shortened from three days for traditional cremation to two hours. Although traditional Hindus balk at

FIGURE 7.28 Wood gathered for Hindu cremations at Pashupatinath, on the sacred river Bagmati in Nepal. These cremations contribute significantly to the ongoing deforestation of Nepal and reveal the underlying internal contradiction in Hinduism between conservation as reflected in the doctrine of *ahimsa* and sanctioned ecologically destructive practices. (Cormac McCreesh/Getty Images)

the notion of breaking with conventional practice, severe wood shortages and the escalating cost of wood will likely make green cremations an increasingly popular option in the future.

Ecofeminism Another interpretation of the relationship between humans and nature comes from **ecofeminism**. This movement arose in the 1970s and identified a parallel between how women are discriminated against by a masculinist society, and how nature—often portrayed as a feminine entity—is similarly degraded, exploited, and oppressed. Ecofeminists point out that the rise of the all-powerful male sky-deity of Semitic monotheism came at the expense of Earth goddesses of fertility and sustainability. In their view, because the Judeo-Christian tradition elevated a sky-god remote from the Earth, the harmonious relationship between people and the habitat was disrupted. The ancient holiness of ecosystems perished, endangering huge ecoregions. The **Gaia hypothesis** arose from an ecofeminist spirit, in which the Earth is seen as a mother figure who reacts to humankind's environmental depredations through a variety of self-regulating mechanisms.

Cultural Landscape

7.5 Evaluate how religions have left their particular mark on the cultural landscape.

Religion is a vital aspect of culture; its visible presence can be quite striking, reflecting the role played by religious motives in the human transformation of the landscape. In some places, the religious aspect is the dominant visible evidence of culture, producing sacred landscapes. At the opposite extreme are areas almost purely secular in appearance. Religions differ greatly in visibility, but even those least apparent to the eye usually leave some subtle mark on a place.

Religious Structures

The most obvious religious contributions to the landscape are the buildings erected to house divinities or to shelter worshippers. These structures vary greatly in size, function, architectural style, construction material, and degree of ornateness (**Figure 7.29**). To Roman Catholics, for example, the church building is literally the house of God, and the altar is the focus of key rituals. Partly for these reasons, Catholic churches are typically large, elaborately decorated, and visually imposing. In many towns and villages, the Catholic house of worship is the focal point of the settlement, exceeding all other structures in size and grandeur. In medieval European towns, Christian

FIGURE 7.29 Traditional religious architecture takes varied forms. St. Basil's Cathedral on Red Square in Moscow (*above*) reflects a highly ornate Russian landscape presence, whereas the plain brown chapel in South Dakota (*top right*) demonstrates the opposite tendency of favoring visual simplicity by British-derived Protestants. The ornate Hindu temple in New Delhi, India (*bottom right*), offers still another sacred structure. (Above: vladimir zakharov/Getty Images; Top right: Medioimages/Photodisc/Getty Images; Bottom right: b n khazanchi/Getty Images)

cathedrals were the tallest buildings, representing the supremacy of religion over all other aspects of life.

For many Protestants—particularly the traditional Calvinistic chapel-goers of British background, including Presbyterians and Baptists—the church building is, by contrast, simply a place to assemble for worship. The result is a small, simple structure that appeals less to the senses and more to personal faith. For this reason, traditional Protestant houses of worship typically are not designed for comfort, beauty, or high visibility but instead appear deliberately humble (see Figure 7.29, middle). Similarly, the religious landscape of the Amish and Mennonites in rural North America, the "plain folk," is very subdued because they reject ostentation in any form. Some of their adherents meet in houses or barns, and the churches that do exist are very modest in appearance, much like those of the southern Calvinists.

So-called **megachurches** present a recent twist on the Protestant house of worship, one that shuns humbleness and plainness for size and grandeur (**Figure 7.30**). Located in suburban areas, but also on the rise in Protestant areas worldwide, megachurches count from 2000 to 10,000 members and rely on a business-oriented approach to designing spaces and services in line with their congregation's wishes. These include plenty of parking and bathrooms, but also recreational facilities, high-tech audiovisual sermon delivery, and casual dress codes. Megachurches also foster a sense of connectedness and community in an increasingly secular world. Geographers Barney Warf and Morton Winsberg, who have studied megachurches, claim that they are "the new face of American religion, a threat that many traditional denominations have understandably viewed with alarm."

Islamic mosques are usually the most imposing structures in the landscape, whereas the visibility of Jewish synagogues varies greatly. Hinduism has produced large numbers of visually striking temples for its multiplicity of gods, but much worship is practiced in private households with their own personal shrines (**Figure 7.31**).

Nature religions such as animism generally place only a subtle mark on the landscape. Nature itself is sacred, and imposing features already present in the landscape—mountains, waterfalls, and grottos, for example—mean that few human-made shrines are needed.

FIGURE 7.30 Megachurch. Yoido Full Gospel Church, in Seoul, South Korea, is the largest megachurch in the world, with approximately 830,000 members. (Pascal Deloche/GODONG/Getty Images)

FIGURE 7.32 Billboard in Three Forks, Montana. Protestant billboards often use American-style advertising slogans to promote church attendance. (Courtesy of Nancy H. Belcher.)

FIGURE 7.31 Temples dedicated to ancestors in Bali, Indonesia. Bali's population practices a blend of Hinduism, animism, and ancestor worship. These temples to ancestors are a feature of every Balinese home, and offerings of incense, food, and flowers are made three times per day to show reverence. (Courtesy of Ari Dorfsman.)

Paralleling this contrast in church styles are attitudes toward roadside shrines and similar manifestations of faith. Catholic culture regions typically abound with shrines, crucifixes, and other visual reminders of religion, as do some Eastern Orthodox Christian areas. Protestant areas, by contrast, are bare of such symbolism. Their landscapes instead display such features as signboards advising the traveler to "get right with God," a common sight in the southern United States (**Figure 7.32**).

Faithful Details

Not only buildings but also other architectural choices, such as color and landscaping elements, can convey spiritual meaning through the landscape. Consider the color green. The image of a Muslim mosque in northern Nigeria in **Figure 7.33** depicts how green—sacred in Islam—appears often on mosque domes and minarets, the tall circular towers from which Muslims are called to prayer. The prophet Muhammad declared his favorite color to be green, and his cloak and turban were said to be green. In Christianity, green also has positive connotations, symbolizing fertility, freedom, hope, and renewal. It is used in decorations, such as banners, and for clergy vestments, especially those associated with the Trinity and with the season of Epiphany.

Water, often used as a decorative element in fountains, baths, and pools, has religious meaning. In fact, all three of the great desert monotheisms—Islam, Christianity, and Judaism—consider water to have a special status in religious rituals. In general, followers of all three religions believe water has the ability to purify and cleanse the body as well as to purify and cleanse the soul of sins. Muslims use water in ablutions, or washing, of the hands, face, or ears before daily prayers and other rituals. Footbaths, or ablution fountains, are important features located outside of mosques, and decorating with water, particularly pools and fountains, is a common practice in Islamic architecture. Central to Christianity is baptism, in which the newborn infant or adult convert is sprinkled with or fully immersed in holy water, symbolizing the cleansing of original sin for infants and repentance or cleansing from sin for adults.

FIGURE 7.33 Muslim mosque in northern Nigeria. Areas with Muslim populations enjoy a landscape where mosque minarets and domes are frequently green in color. (Laurie Strachan/Alamy)

Jews may engage in a ritual bath, known as a *mikveh*, before important events, such as weddings (for women), or on Fridays, the evening of the Jewish Sabbath (for men). The story of the Great Flood, found in the Book of Genesis (common to both Jews and Christians), depicts the washing away of the sins of the world so that it could be born anew.

Landscapes of the Dead

Religions differ greatly in the type of tribute each awards to its dead. These variations appear in the cultural landscape. The few remaining Zoroastrians, called Parsees, who preserve a once-widespread Middle Eastern faith now confined to parts of India, have traditionally left their dead exposed to be devoured by vultures. Thus, the Parsee dead leave no permanent mark on the landscape. Hindus cremate their dead. Having no cemeteries,

their dead, as with the Parsees, leave no obvious mark on the land. Yet on the island of Bali in Indonesia, Hinduism has blended with animism, such that temples to family ancestors occupy prominent places outside the houses of the Balinese, a landscape feature that does not exist in India itself (see Figure 7.31).

In Egypt, spectacular pyramids and other tombs were built to house dead leaders. These monuments, as well as the modern graves and tombs of the rural Islamic folk of Egypt, lie on desert land not suitable for farming (**Figure 7.34**). Muslim cemeteries are usually modest in appearance, but spectacular tombs are sometimes erected for aristocratic persons, giving us such sacred structures as the Taj Mahal in India (**Figure 7.35**), one of the architectural wonders of the world.

Taoic Chinese typically bury their dead, setting aside land for that purpose and erecting monuments to their deceased kin. In parts of China—reflecting its pre-communist history—cemeteries and ancestral shrines take up as much as 10 percent of the land in some districts, greatly reducing the acreage available for agriculture.

Christians also typically bury their dead in sacred places set aside for that purpose. These sacred places vary significantly from one Christian denomination to another. Some graveyards, particularly those of Mennonites and southern Protestants, are very modest in appearance, reflecting the reluctance of these groups to use any symbolism that might be construed as idolatrous. Among certain other Christian groups, such as Roman Catholics and members of the various Orthodox

FIGURE 7.34 Farmer with Great Pyramids in the background. The foreground of this image, shot near Cairo along the fertile banks of the Nile River in Egypt, shows that this land is heavily utilized for agriculture. In the background are the Great Pyramids of Giza, originally built to house the dead, situated in the adjacent arid desert lands. (Universal Images Group North America LLC/DeAgostini/Alamy)

FIGURE 7.35 The Taj Mahal in Agra, India. Built as a Muslim tomb, the Taj Mahal is perhaps the most impressive religious structure in the world. (Pallava Bagla/Getty Images)

Churches, cemeteries are places of color and elaborate decoration (**Figure 7.36**).

Cemeteries often preserve truly ancient cultural traits because people as a rule are reluctant to change their practices relating to the dead. The traditional rural cemetery of the southern United States provides a case in point. Freshwater mussel shells are placed atop many of the elongated grave mounds, and rose bushes and cedars are planted throughout the cemetery. Recent research suggests that the use of roses may derive from the worship of an ancient, pre-Christian mother goddess of the Mediterranean lands. The rose was a symbol of this great goddess, who could restore life to the dead. Similarly, the cedar evergreen is an age-old pagan symbol of death and eternal life, and the use of shell decoration derives from an animistic custom in West Africa, the geographic origin of slaves in the American South. Although the present Christian population of the South may be unaware of the origins of their cemetery symbolism, it seems likely that their landscape of the dead contains animistic elements thousands of years old, revealing truly ancient beliefs and cultural diffusions.

Sacred Spaces

Sacred spaces consist of natural or human-made sites that possess special religious meaning that is recognized as worthy of devotion, loyalty, fear, or esteem. By virtue of their sacredness, these special places might be avoided by the faithful, sought out by pilgrims, or barred to members of other religions (see Figure 7.18). Often, sacred space includes the site of supposed supernatural events or is viewed as the abode of gods. Cemeteries, too, are also generally regarded as a type of sacred space. So is Mount Sinai, described by geographer Joseph Hobbs as endowed "with special grace," where God instructed the

FIGURE 7.36 Two different Christian landscapes of the dead. In the Yucatán Peninsula of Mexico, the dead rest in colorful, aboveground crypts, whereas in Amana, Iowa, communalistic Germans prefer tidiness, order, and equality. (Left: Macduff Everton/Getty Images; Right: Alan Solomon/TNS/ZUMAPRESS.com)

Hebrews to "mark out the limits of the mountain and declare it sacred." Conflict can result if two religions venerate the same space. In Jerusalem, for example, the Muslim Dome of the Rock, on the site where Muhammad is believed to have ascended to heaven, stands above the Western Wall, the remnant of the ancient Jewish temple (**Figure 7.37**; see the World Heritage Site). These two religions literally claim the same space as their holy site, which has led to conflict.

Sacred space is receiving increased attention in the world. In the mid-1990s, the internationally funded Sacred Land Project began to identify and protect such sites, 5000 of which have been cataloged in the United Kingdom alone. Included are such places as ancient stone circles, pilgrim routes, holy springs, and sites that convey mystery or great natural beauty. This last type falls into the category of mystical places—locations unconnected with established religion where, for whatever reason, some people believe that extraordinary, supernatural things can happen. The Bermuda Triangle in the western Atlantic, where airplanes and ships supposedly disappear, is a mystical place and, in effect, a vernacular culture region. Sometimes the sacred space of vanished ancient religions never loses—or later regains—the functional status of mystical place, which is what has happened at Stonehenge in England.

In his book *The Sacred and the Profane*, the renowned religious historian Mircea Eliade suggested that all societies, past and present, have sacred spaces. According to Eliade, sacred spaces are so important because they establish a geographic center on which society can be anchored. Through what he termed *hierophany*, a sacred space emerges from the profane, ordinary spaces surrounding it. A contemporary example of hierophany is provided by the appearance of an image of the Virgin of Guadalupe on a bank building in Clearwater, Florida, transforming the profane space of finance into a Catholic pilgrimage site (**Figure 7.38**).

FIGURE 7.37 Sacred spaces. The three so-called Abrahamic faiths—Judaism, Christianity, and Islam—arose in the same place and thus their adherents revere some of the same holy sites. Mount Sinai, in North Africa, is where Moses is said to have received the Ten Commandments as the Israelites continued their journey out of Egypt; it is also mentioned several times in the Qur'an. (Eddie Gerald/Alamy Stock Photo)

FIGURE 7.38 Our Lady of Clearwater. This 60-foot (18-meter) image appeared in December 1996 in the window of a bank building in Clearwater, Florida. Because of its resemblance to Our Lady of Guadalupe, Catholics—including a sizable Mexican immigrant population as well as Catholic pilgrims from around the region—gather in the parking lot every day for the celebration of Mass. (Courtesy of Patricia L. Price.)

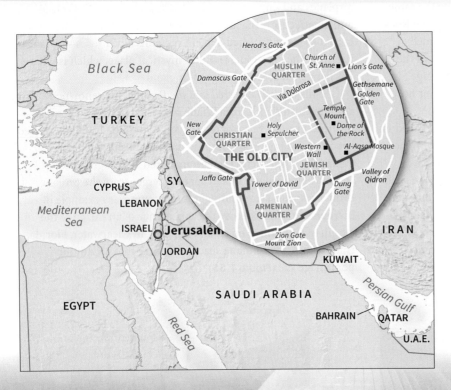

The Old City of Jerusalem and its walls were added to the list of World Heritage Sites in 1981. The area is less than one-half of a square mile, set within modern-day Jerusalem. The Old City has been occupied since the fourth millennium B.C.E., thus making Jerusalem one of the world's oldest urban settlements (see Chapter 10). Divided into four quarters—Armenian, Christian, Jewish, and Muslim—these different "neighborhoods" were formed based on major ethnic and religious divisions dating back centuries.

- Greater Jerusalem holds tremendous significance for the major world religions, as it is the place where Jews built and rebuilt the temple that served as a center of worship, Christ was crucified, and Mohammed ascended to heaven.

- The Old City includes the city's holiest sites: the Temple Mount and the portion of the wall of the Jewish Second Temple known as the Western Wall, the Church of the Holy Sepulcher, the al-Aqsa Mosque, and the Dome of the Rock. Access to these sites, some of which are off-limits to Jews, is hotly disputed today.

- The Dome of the Rock, pictured here, is considered to be the "most contested piece of real estate on Earth" due to its religious importance. For thousands of years, Jews, Christians, and Muslims have shared Jerusalem's city spaces—sometimes cooperatively and at other times in conflict.

(Jon Arnold Images Ltd/Alamy)

(MARCO LONGARI/AFP/Getty Images)

WAILING WALL: Pictured below is a portion of the Jewish Second Temple that was built in 515 B.C.E. after the first temple, constructed in 957 by King Solomon, was attacked by the Egyptian pharaoh and then destroyed by the Babylonians in 586. The Second Temple, itself threatened by Greek and Roman invaders, was ultimately destroyed by the latter in 70 C.E. A remnant of the Western Wall encloses the Second Temple courtyard and is one of Judaism's holiest sites.

- It is often referred to as the "Wailing Wall" because of the Jewish custom of mourning the destruction of the Second Temple at this spot. Jews pray here and place slips of paper containing written prayers inside the crevices of the wall. More than 1 million notes are placed within the wall's crevices each year, and some organizations allow the faithful to e-mail notes that are later placed within the wall.

- As with holy sites from many faiths, the Wailing Wall is segregated by gender. Men visit one side, while women and children are restricted to a much smaller and more crowded portion of the wall on the right-hand side.

- Some women argue that ultra-Orthodox restrictions against women and men sharing the same space in places of worship illegally restrict women's right to space, and that bans against women wearing prayer garments and reading aloud from the Torah at the Wall are unconstitutional under Israeli law.

TOURISM: Jews, Christians, and Muslims have for millennia shared Jerusalem's city spaces in ways that involve cooperation as well as conflict. Its streets, apartments, and public spaces bustle with nearly 1 million inhabitants of diverse faiths and ethnicities going about their daily lives. It is fascinating to experience the mixture of language, dress, and customs interacting in Jerusalem.

- In addition, 3.5 million tourists visit every year—roughly four times the city's resident population. Given the number of religious sites in Jerusalem, pilgrimage alone accounts for many tourist visits. The country of Israel, where Jerusalem is located, is also home to many sites of historical, archaeological, cultural, and religious significance, as well as being near Mediterranean beaches and the greater Middle Eastern region.

- In 1982, Jerusalem's Old City and Walls were placed on the list of World Heritage Sites in danger, due in part to the deterioration and damage stemming from tourism and a lack of adequate conservation efforts. In other words, Jerusalem's Old City is literally in danger of being loved to death.

- Approximately 3 million tourists each year
- Divided into four quarters: Armenian, Christian, Jewish, and Muslim
- Home to important holy sites: the Temple Mount and Western Wall, Church of the Holy Sepulcher, the al-Aqsa Mosque, and the Dome of the Rock
- Sometimes a site of conflict due to shared religious spaces

http://whc.unesco.org/en/list/148

(Lior Mizrahi/Getty Images)

Conclusion

Religion is firmly interwoven in the fabric of culture. Religions vary greatly in their content, as well as their spatial distribution from one region to another. Religious differences are at the heart of conflicts the world over. Yet religions tend to share some common elements, the most basic being that they provide a structure of faith that anchors communities. For this reason, religion forms one of the common threads of human geography, and understanding them is therefore vitally important.

Chapter Summary

7.1 Define the differences among religions as well as the basic tenets of the world's principal religions, and locate their primary regional locations on a world map.

- The world's religions can be classified in several ways. Proselytic religions are those that actively seek new members through conversion, while ethnic religions are identified with specific groups and do not seek converts.
- Monotheistic religions worship one god, while polytheistic religions believe there are many gods.
- Syncretic religions combine elements of two or more belief systems, while orthodox religions emphasize purity of faith and are not open to accepting elements of other religions.
- Fundamentalism involves a return to the founding principles of a religion, whereas religious extremism can involve violent expressions of religious intolerance.
- The world's major faiths share common characteristics depending on where they arose: Judaism, Christianity, and Islam are related faiths, as are Hinduism and Buddhism, and Confucianism, Taoism, and Shinto.
- Animistic religions believe that inanimate objects and natural phenomena have souls or spirits.

7.2 Describe how and why the world's religions became distributed as they are today.

- The world's religions have spread from their respective culture hearths through relocation and expansion diffusion, through migration of the faithful, imperial conquest, and militaristic imposition.
- Religious pilgrimage to visit holy sites is the primary driver of tourism.

7.3 Analyze how globalization has affected the practice of religion.

- If religions do not adapt to a changing world, they risk losing followers, as is the case with Catholicism in Latin America.
- The Internet has reshaped how people worship and interact with communities of the faithful.
- The proportion of the world's population that does not belong to any organized faith is on the increase in North America and Europe, while these "nones" are outstripped by growing numbers of new followers of Christianity and Islam in other parts of the world.

7.4 Explain specific ways in which the natural environment shapes, and is shaped by, religious beliefs and practices.

- Religions attempt to appease the forces of nature, and natural features often achieve the status of sacred spaces.
- Feng shui is a feature of Asian religions that emphasizes Tao, the dynamic balance of nature, through the arrangement of elements in a place.
- Religious beliefs, practices, and taboos are often expressed through food and drink, therefore shaping the distribution and spread of certain animals and plants.
- Ecotheology examines how religious worldviews shape attitudes toward nature.

7.5 Evaluate how religions have left their particular mark on the cultural landscape.

- Religious architecture exhibits a great range of sizes and styles, from very grandiose and large megachurches, mosques, and temples, to humble Protestant churches and Balinese household shrines.
- Decorative elements, such as color and water, symbolize aspects of faith.

- Religions memorialize their dead in diverse ways, ranging from the spectacular tombs of ancient Egypt and the Taj Mahal, to modest cemeteries with locally meaningful features.
- Sacred spaces arise through their association with gods, events, or religious figures, and can become the focus of conflict if claimed by more than one religion.

Key Terms

religion   A social system involving a set of beliefs and practices through which people seek harmony with the universe and attempt to influence the forces of nature, life, and death (page 228).

cult   An unconventional belief system, often focusing on a person (page 229).

proselytic or universalizing religions   Religions that actively seek new members and aim to convert all humankind (page 229).

ethnic religion   A religion identified with a particular ethnic or tribal group; does not seek converts (page 229).

monotheistic religion   The worship of only one God (page 229).

polytheistic religion   The worship of many gods (page 229).

syncretic religions   Religions, or strands within religions, that combine elements of two or more belief systems (page 229).

orthodox religions   Strands within most major religions that emphasize purity of faith and are not open to blending with other religions (page 230).

fundamentalism   A movement to return to the founding principles of a religion, which can include literal interpretation of sacred texts, or the attempt to follow the ways of a religious founder as closely as possible (page 230).

religious extremism   An intolerant expression of religious fundamentalism, resulting in violence (page 230).

Judaism   Monotheistic ethnic faith of the Jews, and the parent religion of Christianity, whose principles are rooted in the Torah (page 232).

Diaspora   When written with a capital "D," *Diaspora* refers to the forced dispersal of the Jews from Palestine in Roman times and the resulting wide spatial reach of contemporary Jewish populations across the globe. When written with a lowercase "d," *diaspora* refers to the dispersal across places of people who have a common religious, ethnic, or national background (page 232).

Christianity   Monotheistic proselytic faith of Christians, related to Judaism and to Islam, whose principles are rooted in the teachings of Jesus as recorded in the Old and New Testaments of the Bible (page 233).

Sect   A subgroup holding somewhat different religious beliefs from the main religion of which it is a part (page 235).

Islam   Monotheistic proselytic faith of the Muslims, whose principles are established in the word of Allah as conveyed by the Prophet Mohammed and recorded in the Qur'an (or Koran) and practiced in the Five Pillars of the faith (page 236).

Hinduism   Polytheistic ethnic faith and the parent religion of Buddhism, closely tied to India and its ancient culture (page 237).

Buddhism   Monotheistic religion based on the teachings of Prince Siddhartha Gautama (also known as Buddha or "the awakened one"), which posits that the root of suffering is desire (page 238).

Confucianism   Derived from the teachings of Chinese philosopher K'ung Fu-tzu, this has bureaucratic, ethical, and hierarchical overtones and is more of a way of life than a religion (page 238).

Taoism   Both an established religion and a philosophy, Taoism emphasizes the dynamic balance depicted by the Chinese yin-yang symbol, and is centered on the "three jewels of Tao"—humility, compassion, and moderation (page 239).

Shinto   An animistic faith that was once the state religion of Japan and has long been blended with Buddhism and Confucianism; Shinto adherents worship ancestors and *kami*, or spirits, that inhabit natural objects such as waterfalls and mountains (page 239).

animistic religion   The belief that souls or spirits exist not only in humans but also in animals, plants, rocks, natural phenomena such as thunder, geographic features such as mountains or rivers, and other entities of the natural environment (page 239).

shamanism   An animistic belief system wherein a shaman serves as an intermediary between people and spirits (page 239).

culture hearth   A focused geographic area where important innovations are born and from which they spread (page 240).

paganism   A term used to refer to rural, polytheistic non-Christians during the Roman empire; today *paganism* denotes an unconventional religion (page 241).

contact conversion   The spread of religious beliefs by personal contact (page 241).

pilgrimages   Journeys to sacred places or places of religious importance (page 243).

Hajj   The so-called Fifth Pillar of Islam is an annual religious pilgrimage to the city of Mecca. Able-bodied Muslims who can afford the journey are encouraged to perform the Hajj at least once in their lifetime (page 244).

nones   Refers to those who do not belong to any organized religious faith (page 246).

feng shui   The practice of harmoniously balancing the opposing forces of nature in the built environment by choosing environmentally auspicious sites for locating houses, villages, temples, and graves (page 248).

ecotheology The study of the influence of religious belief on habitat modification (page 251).

ecofeminism A movement arising in the 1970s that identifies a parallel between how women are discriminated against by a masculinist society, and how nature—often portrayed as a feminine entity—is similarly exploited, degraded, and oppressed (page 252).

Gaia hypothesis The theory that there is one interacting planetary ecosystem, Gaia, that includes all living things and the land, waters, and atmosphere in which they live;

further, that Gaia functions almost as a living organism, acting to control deviations in climate and to correct chemical imbalances, so as to preserve Earth as a living planet (page 252).

megachurch Large Protestant church structures, which have large congregations (2000–10,000 members) and utilize business models to tailor their spaces and services to their congregation's needs (page 253).

sacred spaces Areas recognized by a religious group as worthy of devotion, loyalty, esteem, or fear to the extent that they become sought out, avoided, inaccessible to nonbelievers, or removed from economic use (page 256).

Practice at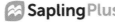

Read the interactive e-Text, review key concepts, and test your understanding.

 Story Map. Explore sacred spaces in some of the world's urban areas.

 Web Map. Examine the geography of religious and spiritual places.

Doing Geography

ACTIVE LEARNING:
Navigating the Religious Landscape of the United States

The Pew Research Center is one of the most important sources of data about culture, trends, and attitudes in the United States. A major research focus area for the Pew Foundation is religion. In this exercise, you will explore data collected by the Pew Foundation on the religious landscape of the United States, and how it varies by state, as well as by region of the country.

More guidance at **Sapling** Plus

EXPERIENTIAL LEARNING:
The Making of Sacred Spaces

For this exercise, you will use Professor Kenneth Foote's ideas about how spaces become sacred. In his book *Shadowed Ground: America's Landscapes of Violence and Tragedy*, Foote argues that what a society chooses to forget about its past is at least as important as what it remembers in shaping an image of itself that can be presented to the world. Foote discusses how certain places in the United States have, or have not, become sacred sites. He notes, for example, that the battlefields of the Revolutionary and Civil wars are well marked and visited by thousands every year. Yet the 1692 execution site of the supposed witches of Salem, Massachusetts, is unmarked. How could the exact site of such an infamous event in U.S. history be impossible to locate? Foote realized that this scenario was not limited to the United States, writing, "Indeed, soon after my first trip to Salem, I found myself in Berlin before the reunification of Germany. There I came across similar places—Nazi sites such as the Gestapo headquarters and Reich chancellery—that have lain vacant since just after World War II and seem to be scarred permanently by shame."

Professor Foote suggests that sites of important events can be sanctified, obliterated, purposely ignored, or fall somewhere in between. Sanctified sites are set apart from other sites, are carefully maintained, are often publicly owned, and attract annual visitors and ceremonies. At the other end of the spectrum, the sites of particularly shameful, stigmatized, or violent events may simply be obliterated from the landscape in an attempt to forget about them.

Steps to Understanding Sacred Spaces

Step 1: Think about sites near where you now live and where a violent or tragic event occurred, for example, a murder, a freak accident, a natural disaster, an assassination, an act of terrorism, or a heinous crime. Virtually all places experience some such event. If nothing readily comes

to mind, inquire, and inevitably you will discover an example for this exercise, even if you live in a rural area or a college town.

Step 2: Once you have identified a site, consider its current status. Is it sanctified, is it obliterated, or does it fall somewhere in between? Elements of the landscape, such as signs or monuments (or the lack of them), will indicate the status of the site.

Step 3: Take note of any debates surrounding what to do with this site. Has the status of the site changed over time? How?

When people are unsure of how an event fits into their history, the site where the event occurred may inhabit a sort of limbo, remaining unmarked until the society in question comes to terms with the event. This is the case with many sites of racially motivated, violent, or shameful acts in the United States, such as the Memphis, Tennessee, motel where Martin Luther King, Jr. was assassinated in 1968, or the Manzanar, California, internment camp where Japanese Americans were confined during World War II. Only recently have the two sites become prominent on the landscape after decades of neglect. Consider the following questions:

- Are any newsworthy events happening now that may mark a place as a site of tragedy?
- What sites, whether nearby or far away, do you expect to become more sacred over time?
- Can you think of an event that is so horrific that the place where it occurred will be forever forgotten and erased from the map?

Cemetery at Manzanar National Historic Site. The Japanese inscription reads "Soul Consoling Tower." Fifteen of the 146 Japanese Americans who died at this Relocation Center are buried here. A total of 110,000 Japanese Americans were interned during World War II in camps like Manzanar. After the 1941 Japanese attack on Pearl Harbor, Hawaii, Japanese Americans were deemed "enemy aliens." It was not until 1992—nearly 50 years after the end of the war—that Manzanar was declared a National Historic Site. (George Ostertag/AGE Fotostock)

SEEING GEOGRAPHY

Parking Lot Shrine

How can an ordinary landscape, such as a parking lot, become a sacred space?

This large statue of the Virgin of Guadalupe inhabits the corner of the parking lot of Self-Help Graphics and Art, an artists' cooperative in East Los Angeles. According to Michael Amezcua, a metalwork artist at Self-Help, neighborhood residents actively use this shrine, leaving offerings and petitions to the Virgin in the outdoor grotto. Every December 12, neighborhood residents gather here to celebrate the feast day of the Virgin of Guadalupe.

We could interpret this unique parking lot shrine to be a variation on the yard shrine, a common feature of the residential landscape of the southwestern United States. Throughout the Mexican American homeland of the Southwest, it is common to see a small shrine to the Virgin of Guadalupe in people's front yards. This, in turn, is a variation on the Mexican custom of maintaining elaborate shrines and altars to important religious figures. These are sometimes located inside the house in a space reserved especially for altars (called a *nicho*). Or they can be located outside the house, near the front door, because Mexican homes usually do not have yard space in front of the house. On December 12, altars all over Mexico and Mexican American areas in the United States are elaborately decorated with candles, ribbons, and flowers. Devotees of the Virgin of Guadalupe celebrate with family and neighbors throughout the night. Since turquoise blue is considered to be the favorite color of the Virgin of Guadalupe, this color abounds. Because the Virgin of Guadalupe is also the official patron saint of Mexico, the red, white, and green of the Mexican flag are prevalent in altar decorations.

In Miami, another Latino population center in the United States, yard shrines are built to La Virgen de la Caridad de Cobre. Like the Virgin of Guadalupe, La Virgen de la Caridad is an American manifestation of the Virgin Mary. This Virgin hails from the copper mining village of Cobre, in Cuba. She is believed to protect seafarers and has been adopted by Cuban rafters as their patron saint. Yemayá, who is an *orisha*, or spirit, in the Afrocentric Cuban religious tradition of Santería, is often equated with La Virgen de la Caridad de Cobre, as both are associated with the sea.

What does the fact that this particular shrine is built in a parking lot rather than a yard say about what, and how, spaces become sacred? Can you think of a more mundane place than a parking lot for such a hallowed figure as the Virgin of Guadalupe? Perhaps the drive-through architecture of strip malls, highways, and parking lots that is so closely associated with the urban landscape of Los Angeles has something to do with it. Where else, other than in Los Angeles, would you find a sacred parking lot?

Parking lot shrine to the Virgin of Guadalupe, Self-Help Graphics and Art, East Los Angeles. (Courtesy of Patricia L. Price.)

Two types of contemporary agricultural landscapes.

(Top: Airphoto/Jim Wark; Bottom: chris warren/CW Images/Alamy Stock Photo)

chapter 8

The Geography of Agriculture and Food

SHAPING THE LAND, FEEDING THE WORLD

What differences can you "read" in these landscapes? Can you determine their locations?

Think about these photos and questions as you read. We will revisit them in Seeing Geography on page 309.

Learning Objectives

8.1 Identify the major agricultural production types, their regions, and the social and environmental factors shaping them.

8.2 Describe the ancient and modern patterns of domesticated plant and animal diffusion and the importance of labor mobility to industrialized agriculture.

8.3 Explain how the processes of globalization have restructured the geography of agricultural production and food consumption and the sorts of problems that have arisen from that restructuring.

8.4 Explain the complex interactions between industrial agriculture and the environment, especially with regard to issues of agrichemical use and climate change.

8.5 Analyze the influence of agriculture in shaping the cultural landscape and how that influence varies geographically.

(Airphoto/Jim Wark)

Food is a basic human need. Where and how to get food has been a primary enterprise of humankind since the origin of our species. Food acquisition has also been, until recently, the primary way that most humans interacted with the natural environment. Consequently, geographers have long made food and agriculture a major focus of their investigations. In this chapter, the five themes will guide our examination of some of this work as well as our exploration of the major trends shaping food and agriculture today. We will learn that agriculture has a strong regional expression, that mobilities have been important to agriculture from its very beginnings to the present, and that globalization has fundamentally altered the way people source the foods they need. Furthermore, agriculture is both highly dependent on nature and highly transformative of the natural landscape, more so than any human activity prior to the Industrial Revolution. Finally, agricultural landscapes are vital for understanding our cultural roots and collective histories.

Region

8.1 Identify the major agricultural production types, their regions, and the social and environmental factors shaping them.

In this section we introduce a classification system to organize the regional variations in farming. There are also other food production practices that do not fit neatly into our farming categories, but that nevertheless have strong regional expression. Lastly, we show how the von Thünen model helps explain the spatial patterning of farming systems at a regional scale.

Classifying Agricultural Practices

Agriculture, the cultivation of crops and rearing of livestock to produce food, animal feed, drink, and fiber, has a strong regional expression. Climate is one of the main variables determining what kind of agriculture is possible. Therefore, regional agricultural patterns tend to reflect regional climate patterns. From these patterns we can identify regional farming systems. **Farming systems** are populations of farms that share a similar resource base, including climate, water availability, slope, topography, and soil quality, and a similar set of practices including crop and livestock types, property ownership, farm size, temporal and spatial patterns of cultivation and animal rearing, and the use of machinery or irrigation. Stated succinctly, a farming system is defined by its distinctive mix of three fundamental variables: land, labor, and capital.

We can classify farming systems broadly into paired types. First, farming systems can be either extensive or intensive. **Extensive agriculture** refers to the practices of crop cultivation and livestock rearing that require low levels of labor or capital input relative to the areal extent of land under production. Thus, extensive systems rely mostly on natural soil fertility and prevailing climate conditions. **Intensive agriculture**, in contrast, uses high levels of labor or capital relative to the size of the landholding. Intensive agriculture can overcome local environmental constraints and boost productivity through the use of irrigation, machinery, synthetic fertilizers, and manufactured chemicals to control pests and weeds.

Second, systems can be either primarily subsistence or commercial. **Subsistence agriculture** is food production mainly for consumption by the farming family and local community, rather than for sale in the market. **Commercial agriculture** is oriented toward the production of agricultural commodities for sale in the market.

Finally, we can classify farming systems based on what they produce. Some systems are primarily based on cultivating plant crops (fruits, vegetables, and cereals). Other farming systems primarily rear animals (livestock). Bear in mind that all of these are ideal types and that in actuality matters of emphasis distinguish farming systems. We turn first to crop-based farming systems.

Crop-Based Systems

There is a wide variety of crop-based farming systems. Among them are slash-and-burn agriculture, paddy rice farming, cereal–root crop farming, plantations, market gardening and truck farming, and grain farming.

Slash-and-Burn Agriculture Many of the peoples of tropical lowlands and hills in the Americas, Africa, and Southeast Asia practice a land-rotation farming system known as slash-and-burn agriculture. **Slash-and-burn agriculture** is a cultivation method that involves cutting small plots in forests or woodlands, burning the cuttings to clear the ground and release nutrients, and planting in the ash of the cleared plot. It is also known as *swidden cultivation*. *Swidden* is a term derived from an Old English term meaning "burned clearing," reflecting the ancient and widespread use of this method.

Working with digging sticks or hoes, the farmers plant a variety of crops in the ash-covered clearings, varying from the maize (corn), beans, bananas, and manioc of the Americas to the yams and dryland rice grown in Southeast Asia (**Figure 8.1**). Different crops typically share the same clearing, a practice called **intercropping**. This technique allows taller, stronger crops to shelter lower, more fragile ones; reduces the chance of total crop

FIGURE 8.1 A farmer in Tojo, central Sulawesi, Indonesia, surveys the results of clearing and burning a patch of forest for planting. When slash-and-burn agriculture is practiced sustainably, what looks like destruction is actually part of a cycle that begins and ends in forest. (YUSUF AHMAD/REUTERS/Newscom)

losses from disease or pests; and provides the farmer with a varied diet. The complexity of many intercropping systems reveals the depth of knowledge acquired by swidden cultivators over many centuries. Relatively little tending of the plants is necessary until harvest time, and no fertilizer is applied to the fields because the ashes from the fire are a sufficient source of nutrients.

The planting and harvesting cycle is repeated in the same clearings for perhaps three to five years, until soil fertility begins to decline as nutrients are taken up by crops and not replaced. Subsequently, crop yields decline. These fields then are temporarily left fallow, and new clearings are prepared to replace them. Because the farmers periodically shift their cultivation plots, another commonly used term for the system is *shifting cultivation*. The abandoned cropland lies fallow for 10 to 20 years before farmers return to clear it and start the cycle again. Swidden cultivation is an extensive system and primarily, though not exclusively, a form of subsistence agriculture.

Although the technology of swidden may be simple, it has proved to be an efficient and adaptive strategy. Swidden plots can return more calories of food for the calories spent on cultivation than does modern mechanized agriculture. When population density is low, slash and burn can leave most of the forest intact over centuries of continuous use. Nonetheless, swidden cultivation can be environmentally destructive under certain conditions. In poor tropical countries with large landless populations, one often finds a front of pioneer swidden farmers advancing on the forests, causing the spread of deforestation. Swidden cultivation, still widely practiced throughout the

tropics, is thus a highly variable system, occurring in both sustainable and unsustainable forms.

Paddy Rice Farming Peasant farmers in the humid tropical and subtropical parts of southern and eastern Asia practice a highly distinctive type of agriculture called paddy rice farming. **Paddy rice farming** is a system of wet rice cultivation on small, level fields bordered by impermeable dikes that are flooded with 4–6 inches (10–15 centimeters) of water for about three-quarters of the growing season. Like swidden cultivation, it is a very ancient practice. Unlike swidden cultivation, paddy farming is considered a form of intensive agriculture. Rice, the dominant paddy crop, forms the basis of civilizations in which almost all the caloric intake is of plant origin. From the monsoon coasts of India through the hills of southeastern China and on to the warmer parts of Korea and Japan stretches a broad region of diked, flooded rice fields, called paddies. Throughout the region, paddies are perched on terraced hillsides, creating a striking cultural landscape (**Figure 8.2**; see also World Heritage Site, page 302).

Paddy rice farming is a form of subsistence agriculture, usually a plot of only 3 acres (1.2 hectares). Yet it can feed a family because wet rice provides a very large output of food per unit of land. Still, paddy farmers must till their small patches intensively to harvest enough food. A system of irrigation that can deliver water when and where it is needed is often key to success. In addition, large amounts of fertilizer, typically in the form of manure, must be applied to the land. Paddy farmers often plant and harvest the same parcel of land twice per year, a practice known as **double-cropping**. These systems are extremely productive, yielding more food per acre than many forms of industrialized agriculture in the United States.

FIGURE 8.2 Cultivation of rice on the island of Bali, Indonesia. Paddy rice farming traditionally entails enormous amounts of human labor and yields very high productivity per unit of land. (Denis Waugh/Getty Images)

Cereal–Root Crop Mixed Farming In colder, drier farming regions that are climatically unsuited to wet rice and other crops, farmers practice a diversified system of agriculture based on cereal grains, root crops (e.g., potatoes, yams), and herd livestock, referred to as **cereal–root crop farming**. **Cereal grains** are the seeds that come from a wide variety of grasses cultivated around the world (e.g., wheat, maize, rice, millet, sorghum, rye, barley, and oats). This system can be found in the river valleys of the Middle East, parts of Europe, much of central and southern Africa, and the mountain highlands of Latin America and New Guinea (**Figure 8.3**). Many geographers refer to these farmers as peasants, recognizing that they often represent a distinctive folk culture and socioeconomic class strongly rooted in the land. **Peasants** are small-scale farmers who own their fields, rely chiefly on family labor, and produce both for their own subsistence and for sale in the market. The dominant grain crops vary by continent and include wheat and barley (e.g., the Middle East), sorghum and millet (e.g., Africa), and oats and maize (e.g., the Americas). In most cases, root crops such as cassava, yams, and potatoes provide the greatest caloric output and are used primarily for family subsistence. Common cash crops—some of them raised for export—include cotton, flax, hemp, coffee, and tobacco (**Figure 8.4**).

These farmers also raise herds of cattle, pigs, sheep, and, in South America, llamas and alpacas. The livestock pull the plow; provide milk, meat, and wool; serve as beasts of burden; and produce manure for the fields. They also consume a portion of the grain harvest. In some areas, such as the Middle Eastern

FIGURE 8.4 Peasant producers in global trade. Miguel, a peasant farmer in Mexico, joined a fair trade cooperative in an effort to get better prices for his coffee crop in the global market. (BONNIER ERICK/SIPA/Sipa Press/Newscom)

river valleys, irrigation helps support this peasant system. In general, however, most modern agricultural technologies are beyond the financial reach of most peasants. Due to the complex variety of forms that this mixed farming system can take, it is difficult to classify. In relative terms, it tends to be more intensive than extensive, more subsistence than commercially oriented, and involves more crop cultivation than livestock rearing.

Plantation Agriculture In certain tropical and subtropical areas, Europeans and Americans introduced a commercial farming system called plantation agriculture. A **plantation** is a large landholding devoted to capital-intensive, specialized production of a single tropical or subtropical crop for the global marketplace. Plantations maximize the production of crops such as sugarcane, bananas, coffee, coconuts, spices, tea, cacao, pineapples, rubber, and tobacco for European and American markets (**Figure 8.5**). Similarly, textile factories require cotton, sisal, jute, hemp, and other fiber crops grown on plantations. Workers usually live right on the plantation, where a rigid social and economic segregation of labor and management produces a two-class plantation society of wealthy landowners and poor workers.

Plantations provided the base for European and American colonial expansion into tropical Asia, Africa, and Latin America. The plantation system originated in the 1400s on

FIGURE 8.3 Cereal-root crop farming in Africa. Peasant farmers in the Democratic Republic of the Congo prepare their field for planting. The photo illustrates the important role of women's labor in African food production. (Brent Stirton/Getty Images for WWF-Canon)

large amounts of manual labor, initially in the form of slave labor and later as wage labor. Most of the capital comes from investors in distant countries, and therefore much of the profit from plantations is exported, along with the crops themselves, to Europe and North America. As a result of the concentration of ownership and production, a handful of multinational corporations, such as Chiquita and Dole, control the largest share of plantations globally.

Market Gardening and Truck Farming The growth of urban markets in the last few centuries also gave rise to other commercial forms of agriculture, notably market gardening. **Market gardening** is a small-scale farming system of one to a few acres, usually farmer-owned and managed, that produces a mixture of vegetables and fruits, mostly for local and regional markets. Truck farming is a term often used synonymously with market gardening, but it differs in important ways. **Truck farming** is a scaled-up version of market gardening, with larger acreages, less crop diversity, and more distant markets. *Truck* does not refer to a vehicle, but rather is derived from a French word for "barter" or "exchange." Both systems are classified as intensive agriculture, but neither system is involved in livestock rearing.

In the United States, a broken belt of market gardens extends from California eastward through the Gulf and Atlantic coast states, with scattered districts in other parts of the country. New Jersey, for example, is called the "Garden State" because of the historic importance of its market gardens in feeding Philadelphia and New York City. The lands around the Mediterranean Sea are dominated by market gardens. In

FIGURE 8.5 Banana plantation in Puerto Viejo, Costa Rica. (John Coletti/Getty Images)

Portuguese-owned sugarcane-producing islands off the coast of tropical West Africa—São Tomé and Principe—but the greatest concentrations now exist in the American tropics, Southeast Asia, and tropical South Asia (**Figure 8.6**). Historically, most plantations have been located near seacoasts, close to the shipping lanes that carry their produce to consumers in Europe, the United States, and Japan. Many plantation regions are defined by a single crop—referred to as monocropping—such as bananas or palm oil. Plantation agriculture long relied on

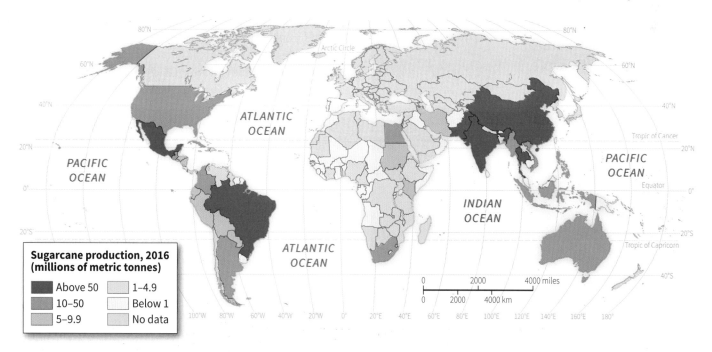

Sugarcane production, 2016 (millions of metric tonnes)

- Above 50
- 10–50
- 5–9.9
- 1–4.9
- Below 1
- No data

FIGURE 8.6 **Sugarcane producers of the twenty-first century.** Sugarcane plantations have expanded into nearly every tropical and subtropical region in the world. Brazil's production now dwarfs that of the once-predominant producers of the Caribbean. (Source: UNFAO.)

regions with mild climates, such as California, Florida, and the Mediterranean region, winter vegetables are a common truck farm crop raised for sale in colder regions of the higher latitudes. Many regions concentrate on a single product, such as wine grapes, olives, oranges, apples, lettuce, or tomatoes (**Figure 8.7**). Many truck farmers participate in cooperative marketing arrangements, whereby they pool their resources to get their produce to market more cheaply than if it were done by individual farmers.

Grain Farming **Grain farming** is an intensive, highly mechanized commercial farming system that specializes in the production of cereal grains, the most globally important of which are rice, maize (corn), and wheat. Grain farming is practiced on very large parcels of land, often hundreds of acres each. Widespread use of machinery, synthetic fertilizers, pesticides, and genetically engineered seed varieties enables grain farmers to operate on this large scale. The planting and harvesting of grain are more completely mechanized than any other form of agriculture. Commercial rice farmers employ such techniques as sowing grain from airplanes, and wheat growers use lasers to guide tractors with 60-foot (18.3-meter)-wide plows. Harvesting is often done by hired migratory crews operating corporation-owned machines (**Figure 8.8**).

While irrigation can expand the range of grain farming, the type of cereal grain grown is still greatly dependent on climatic factors, especially the length of the growing season. Therefore,

FIGURE 8.8 Mechanized wheat harvest on the Great Plains of the United States. An aerial view of combines harvesting wheat in Colorado. North American grain farmers operate in a capital-intensive manner, investing in machines, chemical fertilizers, and pesticides. (Glowimages/Getty Images)

rather than a single grain farming region, we have multiple grain regions, often referred to as "belts." Wheat belts, for example, stretch through Australia, the Great Plains of interior North America, the steppes of Russia and Ukraine, and the pampas of Argentina. Farms in these areas generally are very large, ranging from family-run wheat farms to giant corporate operations (see Figure 1.2). The United States, Canada, Australia, the European Union (EU), and Argentina together account for more than 85 percent of all wheat exports, while the United States accounts for about 70 percent of corn exports (**Figure 8.9**).

Animal-Rearing Systems

Animal-rearing farm systems include pastoralism, livestock raising and fattening, and dairy farming.

Pastoralism Sometimes called pastoral nomadism or nomadic herding, **pastoralism** is a system of breeding and rearing domesticated herd animals by moving them over expansive areas of open pasturelands. While systems vary by type of livestock, most commonly cattle, sheep, and goats, a common

FIGURE 8.7 Market gardening in the Pacific Northwest of the United States. Some market gardeners specialize in one kind of crop. A common market garden crop in the Pacific Northwest is hops, grown mostly for beer brewing in the region. (Gary J. Weathers/Getty Images)

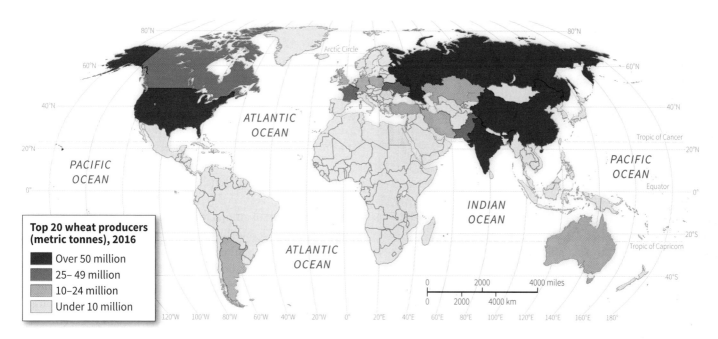

FIGURE 8.9 World's top wheat producers. Mechanized grain farming has spread rapidly over the past few decades, introducing new leading players into the global spotlight. (Source: UNFAO.)

Top 20 wheat producers (metric tonnes), 2016
- Over 50 million
- 25– 49 million
- 10–24 million
- Under 10 million

characteristic of all nomadic herding is mobility. In a practice known as **transhumance**, herders move with their livestock in search of the best forage for the animals as seasons and pasture conditions change. Some herders move their animals in a vertical pattern from lowlands in winter to mountain highlands in summer (**Figure 8.10**). Others shift horizontally, following the seasonal movement of rainfall in order to herd animals to the areas with the most plentiful pastures.

In general, pastoralism occurs in areas that are climatically unsuited to crop cultivation. Nomadic herders can be found in the more arid or colder zones of the Eastern Hemisphere, particularly in the deserts, prairies, and savannas of Africa, the Arabian Desert, and the interior of Eurasia. Pastoral nomadism is very closely associated with ethnic and indigenous tribal culture groups, and the type of livestock and manner of herding are important in creating distinctive cultural identities. For example, some nomads herd while mounted on horses, such as the Mongol people of East Asia, or on camels, such as the Bedouin people of the Arabian Peninsula. Others, such as the Rendille tribe of East Africa, herd cattle, goats, and sheep on foot. North of the tree line in Eurasia, the cold tundra forms a zone of nomadic herders who raise reindeer, such as the Sámi people.

Nomadic cultures' pasture and livestock management strategies are rational responses to climatically variable and unpredictable environments. In particular, mobility is a strategy that allows herders to take fullest advantage of the resulting variations in pasture productivity. Their need for mobility dictates that the nomads' few material possessions be portable, including the tents used for housing (**Figure 8.11**). Their mobile lifestyle also affects how wealth is measured. Typically, in nomadic cultures, wealth is based on the size of livestock holdings rather than on the accumulation of property and personal possessions. Usually, the nomads obtain nearly all of life's necessities, including food, clothing, and shelter, from livestock products or by bartering with the farmers of adjacent river valleys and oases. While some herders raise livestock for the market, pastoralism is an extensive system oriented mostly toward subsistence.

FIGURE 8.10 Transhumance in France (Régis Domergue)

FIGURE 8.11 Nomadic pastoralists in West Africa. This household in Chad is moving their livestock to new pastures. Mobility is key to their successful use of the variable and unpredictable environment of this region of Africa. They are taking with them all of their possessions, including their shelter—a hut, which they have packed on top of their cows. (Michael Nichols/National Geographic/Getty Images)

Livestock Ranching **Livestock ranching** is the practice of using extensive tracts of land to raise herds of livestock to market as meat, hides, or wool. Ranching might seem similar to nomadic herding. It is, however, a fundamentally different livestock-raising system with very different geographic and historical origins. Both nomadic herders and livestock ranchers specialize in animal husbandry to the exclusion of crop raising, and both live in the more semiarid regions. Livestock ranchers, however, have fixed places of residence, called ranches, with dwellings, livestock barns, and fenced corrals. Ranchers operate as individuals or individual families rather than within a communal or tribal organization, and there is no close association between ranching and ethnic identities. In addition, ranchers raise livestock on a large scale for commercial purposes, not for their own subsistence.

Livestock ranching can be traced to herding practices in the Iberian Peninsula in the late medieval period, but it took its full form in the Americas after European colonizers introduced horses, cattle, and sheep. Originally, workers known as cowboys in North America, *gauchos* in South America, and *vaqueros* in Mexico herded cattle and sheep on horseback over private ranchlands or public pastures or a combination of the two. Horses are still used in herding, but increasingly are supplemented by trucks and off-road vehicles.

Ranching is found in the temperate zone worldwide in areas where environmental conditions make crop farming difficult.

Ranchers raise only two kinds of animals in large numbers: cattle and sheep. Ranchers in the United States, Canada, tropical and subtropical Latin America, and parts of Australia specialize in cattle raising. Midlatitude ranchers in cooler and wetter climates specialize in sheep. Sheep ranching is geographically concentrated to such a degree that only four countries—Australia, China, New Zealand, and the United Kingdom—account for 62 percent of the world's wool exports.

Livestock Fattening Ranching is closely associated in some countries with livestock fattening. **Livestock fattening** is an intensive system of animal feeding, utilizing fenced enclosures to finish livestock, mostly cattle and hogs, for slaughter and processing for the market. The logic of this system is to have the animals gain as much body mass in as short a time as possible by confining immature animals to a fenced **feedlot** (to limit movement and associated weight loss) and feeding them large amounts of grain. In the case of cattle, animals are bought from ranchers when they are about one year old and are then fed for approximately 200 days before being slaughtered. This intensive, commercial animal rearing system was invented in the United States in the late nineteenth century, but did not become prevalent until the 1960s. One of the most highly developed fattening areas is the famous Corn Belt of the U.S. Midwest, where farmers raise corn and soybeans exclusively to feed to cattle and hogs (**Figure 8.12**). Typically, slaughterhouses and meatpacking plants are located near feedlots, creating new meat-producing regions, which are often dependent on mobile populations of cheap immigrant labor for butchering and packing. A similar system prevails over much of western and central Europe, though the feed crops there more commonly are oats and

FIGURE 8.12 Cattle feedlot for beef production. This feedlot, in Colorado, is reputedly the world's largest. (Glowimages/Getty Images)

potatoes. Other zones of commercial livestock fattening appear in overseas European settlement zones such as southern Brazil and South Africa.

One of the central characteristics of livestock fattening is the combination of feed crops and animal husbandry. Many feedlots purchase all of their grain from farmers in the region, although in the case of hogs many farmers feed animals with the corn cultivated on their own farms. Today, livestock fatteners are specializing their activities: some concentrate on breeding animals, others on finishing them for market. Increases in the amount of land and crop harvest dedicated to beef production have accompanied the growth in feedlot size and number. In the United States, across Europe, and in European settlement zones around the globe, the majority of the cereal harvest goes to livestock fattening. The fact that more grain is raised to feed animals rather than humans highlights a key drawback of this system: it is a highly inefficient way to supply human nutrition.

The livestock fattening and slaughtering industry has become increasingly concentrated, on both the national and global scales. For example, in 1980 in the United States, the top four beef producers accounted for 41 percent of all production. Today, the top four companies account for over 80 percent of beef production. One company alone accounts for 25 percent. Corporate conglomerates such as ConAgra and Cargill control much of the beef supply through their domination of the grain market and ownership of feedlots and slaughterhouses. The concentration of the industry has extended north and south across the border, primarily spurred by the 1994 North American Free Trade Agreement (NAFTA). Cargill owns beef operations in more than 60 countries.

Dairying **Dairying** is a farming system that specializes in the breeding, rearing, and utilization of livestock (primarily cows) to produce milk and its various by-products, such as yogurt, butter, and cheese. As with many farming systems, there is a great range in farm size and degree of industrialization. Most dairy farms in the United States are small-scale family operations of 100 or fewer cows, but over half of all dairy cows belong to only 20 percent of the dairy farms. Those larger farms are highly mechanized and may have up to 5000 dairy cows or more.

In the large dairy belts of the northern United States from New England to the upper Midwest, western and northern Europe, southeastern Australia, and northern New Zealand, the keeping of dairy cows depends on the large-scale use of pastures. Dairy products vary from region to region, depending, in part, on how close the farmers are to their markets. Dairy belts near large urban centers usually produce milk, which is more perishable, while those farther away specialize in butter, cheese, or processed milk.

As with livestock fattening, increasing numbers of dairy farmers have adopted the feedlot system and now raise their cattle on feed purchased from other sources. Often situated on the suburban fringes of large cities for quick access to market, the dairy feedlots operate like factories. Like industrial factory owners, feedlot dairy owners rely on hired laborers to help maintain their herds. By easing trade barriers, globalization compels U.S. dairy farmers to compete with producers in other parts of the world. Huge feedlots, a factory-style organization of production, automation, the concentration of ownership, and the increasing size of dairy farms are responses to this intense competition.

Other Food Production Systems

In addition to the basic crop-based and animal-based farm systems, there are farms in cities, fish farms, and nonagricultural systems of food production.

Urban Agriculture The United Nations (UN) calculated that in 2008 the human species passed a milestone. For the first time in history, more people live in cities than in the countryside. As this global-scale rural-to-urban migration gained momentum, a distinct form of agriculture rose in significance. **Urban agriculture** is the practice of growing fruits and vegetables on small private plots or shared community gardens within the confines of a city. Millions of city dwellers, in developed and less-developed countries alike, now produce enough vegetables, fruit, meat, and milk from tiny urban or suburban plots to provide most of their food, often with a surplus to sell (**Figure 8.13**). Thus, it is an intensive farming system that can be oriented toward subsistence or commercial production.

In China, urban agriculture now provides 90 percent or more of all the vegetables consumed in the cities. In the African metropolis of Kampala, Uganda, 70 percent of the poultry consumed by residents comes from urban lands. In Dar es Salaam, Tanzania, 90 percent of the leafy vegetables consumed by residents comes from urban gardens. In 2010, the UN Food and Agriculture Organization convened the first symposium on urban agriculture in Africa in recognition of the important role it plays in feeding the continent's twenty-first-century urban residents.

Geographer Susanne Freidberg has conducted research demonstrating the importance of urban agriculture to family income and food security in West Africa. **Food security** is officially defined by the UN as when all people, at all times, have physical and economic access to sufficient safe and nutritious food to meet their dietary needs and food preferences for an active and healthy life. Focusing on the city of Bobo-Dioulasso in Burkina Faso, Freidberg found that urban farming could provide food security for families and substantial incomes from vegetable sales in both the domestic and export markets.

FIGURE 8.13 Urban agriculture. Farmers work a rice field near Antananarivo, Madagascar. Urban agriculture is a critical food source for many African city dwellers. (Martin Harvey/Getty Images)

In the developed world, urban agriculture is becoming more prevalent despite, or perhaps in reaction to, the fact that food sourcing is increasingly globalized. The reasons for the boom in urban agriculture in cities such as Vancouver, Canada, and Oakland, California, are complex. However, almost everywhere it is seen as a partial solution to the problems of food insecurity and lack of access to nutritious foods in cities. The work of geographer Nik Heynen has shown that solving food insecurity is integral to improving urban social justice in the United States. People who lack access to a healthy diet tend to be poorer residents, often members of marginalized minority groups. Similarly, the World Health Organization (WHO) has identified the poor availability of and inequitable access to nutritious vegetables and fruits in European cities as a major twenty-first-century problem for the region. According to the WHO, the promotion of urban food production in European cities will help reduce urban poverty and inequalities.

Aquaculture Most of us don't think of the ocean when discussing agriculture. In fact, however, every year more and more of our animal protein is produced through **aquaculture**—the cultivation and harvesting of aquatic organisms under controlled conditions. Aquaculture includes **mariculture**, which refers to the farming of saltwater species such as shrimp, oysters, marine fish, and more. In the heart of Brooklyn, New York, urban aquaculture thrives in a laboratory, where fish bound for New York City restaurants are harvested from indoor tanks. Aquaculture has its own distinctive cultural landscapes of containment ponds, rafts, nets, tanks, and buoys (**Figure 8.14**). As with terrestrial agriculture, there is a marked regional character to aquacultural production.

Aquaculture is an ancient practice dating back at least 4500 years. Older local practices, sometimes referred to as traditional aquaculture, involve simple techniques such as constructing retention ponds to trap fish or "seeding" flooded rice fields with shrimp. Commercial aquaculture, the industrialized, large-scale protein factories driving today's production growth, is a contemporary phenomenon. The greatest leaps in technology and production have occurred mostly since the 1970s.

At the turn of the twenty-first century, aquaculture was experiencing phenomenal growth worldwide, growing nearly four times faster than all terrestrial animal food-producing sectors combined. Aquaculture is a key reason that per capita protein availability has more than kept pace with population growth in recent decades. For example, China's population grew 63 percent from 1970 to 2010, while its annual per capita protein supply derived from aquaculture went from 2.2 pounds (1 kilogram) to over 79.3 pounds (36 kilograms). In 2014, global aquaculture production surpassed wild-caught seafood production for the

FIGURE 8.14 Emerging mariculture landscapes. As mariculture expands worldwide, many coastal landscapes and ecosystems are being transformed, as in the case of marine fish farming off Langkawi Island, Malaysia. (age fotostock/Superstock)

first time. Aquaculture's share of production, which was only 4 percent in 1970, is projected to continue growing as demand for seafood increases and wild fish stocks decline.

The extraordinary expansion of food production by aquaculture has come with high costs to the environment and human health. As with industrialized agriculture, most commercial aquaculture relies on large energy and chemical inputs, including antibiotics and artificial feeds made from the wastes of poultry and hog processing. Such production practices tend to concentrate toxins in farmed fish, creating a potential health threat to consumers. The discharge from a fish farm, which is equivalent to the sewage from a small city, can pollute nearby natural aquatic ecosystems. In the tropics, especially in Asia, the expansion of commercial shrimp farms is contributing to the loss of highly biodiverse coastal mangrove forests.

Although aquaculture can take place just about anywhere that water is found, strong regional patterns do exist. Mariculture is prevalent along tropical coasts, particularly in the mangrove forest zone. Marine coastal zones in general, especially in protected gulfs and estuaries, have high concentrations of mariculture. On a global scale, China dwarfs all other regions, producing over two-thirds of the world's farmed seafood (**Figure 8.15**). Asia and the Pacific regions combined produce over 90 percent of the world's total.

If you order shrimp, trout, or salmon for dinner at any of the myriad restaurant chains, you will almost certainly be eating farmed protein. As the populations of wild species decline and the price of captured fish increases, farmed seafood will increasingly move into the breach. If trends continue, we may be among the last generations to enjoy affordable wild-caught seafood.

Nonagricultural Food Production

Areas of extreme climate, particularly deserts and subarctic forests, do not support any form of agriculture. Such lands are found predominantly in much of Canada and Siberia. Often these areas are inhabited by indigenous **hunter-gatherers**, such as the Inuit of arctic Greenland, Alaska, and Canada, who gain a livelihood by hunting game, fishing where possible, and gathering edible and medicinal wild plants. Twelve thousand years ago, all humans lived as hunter-gatherers.

Today, less than 1 percent of humans are hunter-gatherers. For the most part, present-day hunter-gatherers are confined to regions climatically unsuited to cultivation. In most hunting-and-gathering societies, a division of labor by gender occurs. As a broad, but not universal, generalization, males do most of the hunting and fishing, whereas females carry out the equally

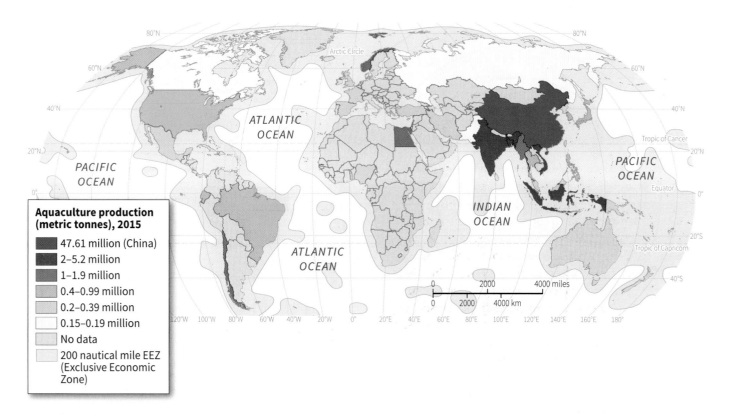

FIGURE 8.15 Global aquaculture and fisheries. Aquaculture is expanding rapidly around the globe, led by China, the world's top producer. Virtually all mariculture takes place within countries' 200-mile (322-kilometer) exclusive economic zone (EEZ) (see Chapter 6). The EEZ is also the site of most commercial fishing, which has greatly depleted wild fish stocks and increased the demand for mariculture production. (Source: UNFAO.)

important task of harvesting wild plants. Hunter-gatherers generally rely on a great variety of animals and plants for their food, with plants providing the larger share of nourishment.

The Von Thünen Model

Geographers and others have long tried to understand the distribution and intensity of agriculture in relation to transportation costs to market. The nineteenth-century German scholar-farmer Johann Heinrich von Thünen developed a spatial model to address this issue. For his model, von Thünen imagined an "isolated state" that has (1) no trade connections with the outside world, (2) only one, centrally located market, (3) uniform soil and climate, and (4) level terrain throughout. He further assumed that all farmers located the same distance from the market had equal access to it and that all farmers sought to maximize their profits and produced solely for market. Von Thünen created this model to analyze the influence of distance from market and transport costs on the type and intensity of agriculture.

Figure 8.16 presents a modified version of von Thünen's isolated-state model, which reflects the effects of improvements in transportation since the 1820s, when von Thünen proposed his

theory. The model's fundamental feature is a series of concentric zones, each occupied by a different type of agriculture, located at progressively greater distances from the central market.

For any given crop, the intensity of cultivation declines with increasing distance from the market. Farmers near the market have minimal transportation costs and can invest most of their resources in labor, equipment, and supplies to augment production. Indeed, because their land is more valuable and subject to higher taxes, they have to farm intensively to make a profit. With increasing distance from the market, farmers invest progressively less in production per unit of land because they have to spend progressively more on transporting produce to market. The effect of distance means that highly perishable products such as milk, fresh fruits, and garden vegetables need to be produced near the market, whereas more durable products, such as cheese or dried fruit, can be produced farther away.

The concentric-zone model describes a situation in which highly capital-intensive forms of commercial agriculture, such as market gardening and feedlots, lie nearest to market. The increasingly distant, successive concentric belts are occupied by progressively less intensive types of agriculture, represented by dairying, livestock fattening, grain farming, and ranching. How well does this modified model describe reality? As we would

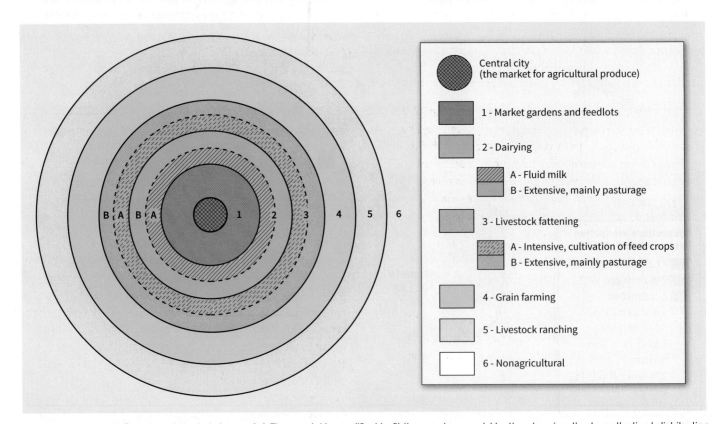

FIGURE 8.16 Von Thünen's isolated-state model. The model is modified to fit the modern world better, showing the hypothetical distribution of types of commercial agriculture. Other causal factors are held constant to illustrate the effect of transportation costs and differing distances from the market. The more intensive forms of agriculture, such as market gardening, are located nearest the market, whereas the least intensive form (livestock ranching) is most remote.

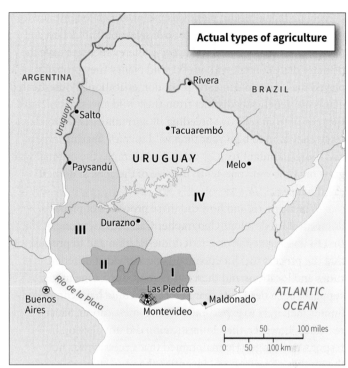

Predicted and actual agriculture in Uruguay

I ▮ Market gardening   II ▮ Dairying   III ▮ Grain farming   IV ▢ Livestock ranching   ⊠ Central city

FIGURE 8.17 Ideal and actual distribution of types of agriculture in Uruguay. This South American country possesses some attributes of von Thünen's isolated state in that it is largely a plains area dominated by one city.

expect, the real world is far more complicated. For example, the emergence of refrigeration and mechanized transport has collapsed distance. Still, on a world scale, we can see that intensive commercial types of agriculture tend to occur near the huge urban markets of northwestern Europe and the eastern United States. An even closer match can be observed in smaller areas, such as in the South American nation of Uruguay (**Figure 8.17**).

Mobility

8.2 Describe the ancient and modern patterns of domesticated plant and animal diffusion and the importance of labor mobility to industrialized agriculture.

Mobility, as we shall see, is important for understanding the origins and functioning of modern agriculture. Diffusion has played a great role in agriculture, from its ancient origins to modern-day technology transfers, such as the green revolution.

In addition, the natural cycles of agriculture have made migratory labor critical to industrialized farming systems.

Ancient Origins and Diffusion of Domestication

Agriculture began with the process of domestication. **Domestication** is the multigenerational process through which humans selectively breed, protect, and care for individuals taken from populations of wild plant and animal species to create genetically distinct species, known as domesticates. It should be emphasized that domestication is an ongoing process of trial and error, not a singular event such as a discovery. Domestication began as the gradual culmination of thousands of years of close association between hunter-gatherers and the natural environment.

Plant Domestication Archaeological evidence tells us that humans domesticated plants before livestock. A **domesticated plant** is one that is deliberately planted, protected, cared for, and used by humans. The first step in plant domestication was perceiving that a certain plant was useful, which led initially to its protection and, eventually, to seed selection and deliberate planting. Because women in hunter-gatherer societies have the primary daily contact with useful wild plants, they probably played the larger role in early plant domestication. Domesticated plants are genetically distinct from their wild ancestors because they result from selective breeding by agriculturists. Accordingly, they tend to be bigger than wild species, bearing larger and more abundant fruit or grain. For example, the original wild grew on a cob only one-tenth the size of the cobs of domesticated maize.

When, where, and how did these processes of plant domestication develop? Geographer Carl Sauer was one of the first to investigate the origin of domestication and to propose that the process was independently developed at different times and locations and then diffused outward. One might suppose that a desperate search for food could have compelled hunter-gatherers to experiment with domestication. Sauer reasoned, however, that domestication did not develop in response to hunger. He maintained that necessity was not the mother of agricultural invention, because starving people must spend every waking hour searching for food and have no time to devote to the leisurely experimentation required to domesticate plants. Instead, he suggested that domestication was accomplished by peoples who had enough food to remain settled in one place and devote considerable time to plant care. Sauer therefore reasoned that the first farmers were probably sedentary folk rather than migratory hunter-gatherers. He further postulated that hearth areas of domestication must have

been in regions of great biodiversity where many different kinds of wild plants grew, thus providing abundant vegetative raw material for experimentation and crossbreeding. For the most part, Sauer's early suppositions have proved correct.

Geographers, archaeologists, and, increasingly, genetic scientists continue to investigate the geographic origins of domestication. Because the conditions conducive to domestication are relatively rare, researchers agree that agriculture arose independently in at most nine regions (**Figure 8.18**). All of these ancient regions have made significant contributions to the modern diet. For example, the Fertile Crescent in the Middle East is the origin of the great cereal grains of wheat, barley, rye, and oats that are so key to our modern diets. This region is also home to the first domesticated grapes, apples, and olives. China and New Guinea provided rice, bananas, and sugarcane, while the African centers gave us peanuts, yams, and coffee. Native Americans in Mesoamerica created another important center of domestication, from which came crops such as maize (corn), tomatoes, and beans. Farmers in the Andes domesticated the potato. While crop diffusions out of these nine regions have occurred over the millennia, the forces of globalization have now made even the rarest of local domesticates available around the world.

The dates of earliest domestication are continually being updated by new research findings. Until recently, archaeological evidence suggested that the oldest center was the Fertile Crescent, where crops were first domesticated roughly 10,000 years ago. However, domestication dates for other regions are constantly being pushed back by new discoveries. Most

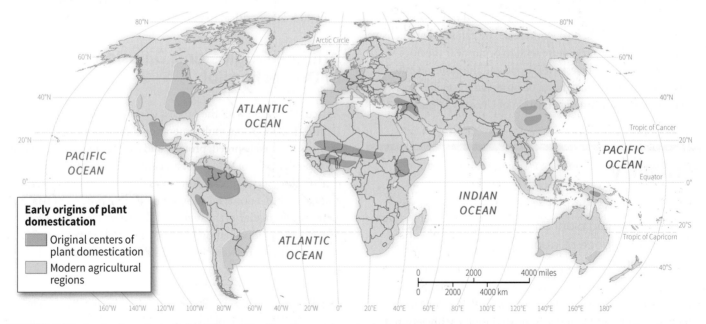

FIGURE 8.18 Ancient centers of plant domestication. New archaeological discoveries and new technologies such as genetic science are changing our understanding of the geography and history of domestication. This map represents a synthesis of the latest findings. (Source: Diamond, 2002.)

dramatically, in the Peruvian Andes archaeologists recently excavated domesticated seeds of squash and other crops that they dated to 9240 years before the present. These seeds were associated with permanent dwellings, irrigation canals, and storage structures, suggesting that farming societies were established in the Americas 10,000 years ago, similar to the dating of the Fertile Crescent.

Animal Domestication A **domesticated animal** is one that depends on people for food and shelter and that differs from wild species genetically, physically, and behaviorally as a result of human-controlled breeding and frequent human contact. Animal domestication apparently occurred later in prehistory than did the first planting of crops—with the probable exception of the dog, whose companionship with humans appears to be much more ancient and related to hunting rather than agriculture. The most recent theories of animal domestication suggest mutually beneficial multispecies interactions in which nonhuman species actively sought contact, rather than a process driven solely by human interests and agency. The early farmers of Eurasia deserve credit for the first great animal domestications, most notably that of herd animals. The wild ancestors of major herd animals—such as cattle, pigs, horses, sheep, and goats—lived primarily in a belt running from Syria and southeastern Turkey eastward across Iraq and Iran to the Indus Valley of South Asia.

Over several thousand years, animal domesticates diffused into Africa and across the Eurasian continent. For example, cattle were domesticated in at least two centers, the Fertile Crescent and the Indus Valley. Genetic analysis reveals that farmers migrating out of those regions brought domesticated cattle with them and so spread the innovation far and wide in highly complex patterns of diffusion. To add to the complexity, there is evidence, still not scientifically definitive, that Africa was the site of a third center of cattle domestication some 10,000 years ago and that this African stock mixed with the stock later brought from India. The diffusion of domesticates is an ongoing puzzle that continues to draw attention from a range of scientific fields.

Many other animal domestications and innovations can be traced to antiquity. Humans independently domesticated other wild animal species in other regions. In the Americas, for example, people domesticated the llama, alpaca, guinea pig, Muscovy duck, and turkey. Farmers in Eurasia were the first to combine domesticated plants and animals in an integrated system, the antecedent of the peasant grain, root, and livestock farming described earlier. Current evidence suggests that roughly 5000 years ago, people of the Indus Valley civilization began using oxen to pull the plow, a revolutionary invention that greatly increased the acreage under cultivation and thus greatly increased food supplies.

Modern Diffusions

Over the past 500 years, European exploration and colonialism were instrumental in redistributing a wide variety of crops on a global scale: maize and potatoes from North America went to Eurasia and Africa; wheat and grapes from the Fertile Crescent went to the Americas, and West African rice went to the United States and Brazil.

The diffusion of specific crops continues, extending the process begun many millennia ago. The introduction of the olive, grape, and date palm by Spanish missionaries in eighteenth-century California, where no cultivation previously existed, is a recent example of relocation diffusion (**Figure 8.19**). This exchange also was part of a larger process of multidirectional diffusion. Eastern Hemisphere crops were introduced to the Americas, Australia, New Zealand, and South Africa through the mass emigrations from Europe over the past 500 years. Crops from the Americas diffused in the opposite direction. For example, chili peppers and maize, carried by the Portuguese to their colonies in South Asia, became staples of diets all across that region.

In human geography, our understanding of agricultural diffusion focuses on more than just the crops; it also includes an analysis of the cultures and indigenous technical knowledge systems in which they are embedded. For example, geographer

FIGURE 8.19 U-pick olive grove in California. When the Spanish colonized California, they introduced European agriculture and crops, especially staples such as olives. Currently there is something of a resurgence in olive cultivation in California, reflecting an interest in locally sourced foods. (Courtesy of Roderick Neumann)

Judith Carney's study of the diffusion of African rice (*Oryza glaberrima*), which was domesticated independently in the inland delta area of West Africa's Niger River, shows the importance of indigenous knowledge. European planters and slave owners carried more than seeds across the Atlantic from Africa to cultivate in the Americas. The Africans taken into slavery, particularly women from the Gambia River region, had the knowledge and skill to cultivate rice. Slave owners actively sought slaves from specific ethnic groups and geographic locations in the West African rice-producing zone, suggesting that they knew about and needed Africans' skills and knowledge.

One of the more recent cases of agricultural diffusion is the known as the green revolution of the 1960s and 1970s. The **green revolution** refers to the U.S.-supported development of high-yielding varieties (HYVs) of cereal crops and accompanying agricultural technologies for transfer to economically less developed countries. It was built upon post–World War II scientific advances in plant breeding, synthetic fertilizers, chemical pesticides, and irrigation (discussed in detail in a later section). From the perspective of the U.S. government, it was intended as a technological fix to political issues of uneven land distribution and rural poverty in the world's less developed regions. Scientists and their funders directed the earliest efforts toward HYVs of wheat in Mexico and later expended the program to rice in the Philippines and other parts of Asia. While there is some dispute over how much the green revolution contributed, there were remarkable increases in cereal production in Mexico, Central America, and many parts of Asia in the latter half of the twentieth century.

The HYVs of the green revolution should be viewed as the establishment of bundled industrial technologies rather than simply as improved crop seeds. First, HYVs tend to be less resistant to insects and diseases, necessitating the accompanying use of newly developed pesticides. In order to meet their yield potential, they also require significant amounts of synthetic fertilizers and irrigation water. The green revolution, then, promises larger harvests but ties the farmer to greatly increased expenditures for a bundle of expensive technologies including seed breeding, irrigation infrastructure, and synthetic fertilizers and pesticides. It thereby enmeshes farmers in the global corporate economy, creating, for better or worse, relations of interdependency over which they have little control. This constitutes a fundamental shift in a farming economy and culture that had persisted for thousands of years.

The green revolution illustrates how cultural and economic factors influence patterns of hierarchical diffusion. In India, for example, new HYV rice and wheat seeds first appeared in 1966. These crops required chemical fertilizers and protection by pesticides, but with the new HYVs, India's 1970 grain production output was double its 1950 level. However, poorer farmers—the great majority of India's agriculturists—could not afford the capital expenditures for chemical fertilizer and pesticides, and the gap between rich and poor farmers widened. Many of the poor became displaced from the land and flocked to the overcrowded cities of India, aggravating urban problems. To make matters worse, the improper use of agrichemicals on the land heightened environmental damage and health problems.

The widespread adoption of HYV seeds has created another problem: the loss of plant diversity and genetic variety. Before HYV seeds diffused around the world, each farm community developed and exchanged its own distinctive seed types through the annual harvest-time practice of saving seeds from the better plants for the next season's sowing. Enormous genetic diversity vanished almost instantly when farmers began purchasing HYV seeds rather than saving seed from the last harvest. "Seed banks" have belatedly been set up to preserve what remains of domesticated plant variety, not just in the areas affected by the green revolution but also in the American Corn Belt and many other agricultural regions where HYVs are now dominant. In sum, the green revolution has been a mixed blessing.

Agricultural Labor Mobility

Agriculture, more than any other modern economic endeavor, is constrained by the rhythms of nature. Biological cycles associated with planting and harvesting are reflected in cycles of labor demand. Seeds must be planted at the right moment to take advantage of seasonal conditions. When crops ripen, they must be harvested in a matter of a few days or weeks. In between these peak periods, labor demand is minimal, so farmers face a dilemma. They need to mobilize a large labor force for periods of peak need, but paying a year-round work force would raise farm production costs. One of farmers' primary solutions has been to employ seasonal migratory labor.

In the United States, the use of migrant workers has been central to the growth and profitability of farming. In his study of labor and landscape in California, geographer Don Mitchell found that the mechanization and intensification of farming created ever-greater extremes in seasonal labor demand, lessening the demand for labor during nonharvest periods while still requiring manual labor for the resultant increased harvests. California thus developed an agricultural industry based on migratory labor, often employing cultural and racial stereotypes to depress farm wages and tighten employers' control over farmworkers. Mitchell noted that through much of the twentieth century, California growers surmised that "Hispanic, black, or Asian workers . . . were 'naturally' better suited to agricultural tasks" than white workers. Prevailing racist attitudes allowed them to "pay nonwhite workers a lower wage than white workers." Ultimately, the federal government provided the legal mechanism, called the Bracero Program, by which contract workers were brought from Mexico to California during periods of peak labor demand. Migrant workers lived in substandard housing, were paid less than a living wage, and could be deported back to Mexico if they complained too much.

Low wages, job insecurity, and all-around poor working conditions led migrant farm workers to try to organize in order to bargain collectively with farmers. Ultimately, they formed the United Farm Workers of America (UFW), a labor union created specifically for the conditions in California agriculture. Through boycotts and other efforts, the UFW was able to win collectively bargained concessions for workers, including medical insurance, improved housing, and living wages.

The case of California's migrant workers is not unusual. Geographer Gail Hollander has documented the use of migratory labor in the development of Florida's sugarcane region. Florida produces 20 percent of the U.S. sugar supply, which until the 1990s was harvested entirely by hand by migrant workers imported seasonally from "former slave plantation economies of the Caribbean." Like their California

counterparts, growers relied on racial stereotypes to argue that only blacks were suitable for cutting cane in Florida. The federal government also established a special federal immigration program similar to Bracero to import Afro-Caribbean migrant workers for the Florida harvest and repatriate them afterward.

Migrant farmworkers remain ubiquitous today in other regions, too, such as the European Union. A recent agricultural boom in Mediterranean coastal Spain relies heavily on North African migrant workers, many of them undocumented immigrants, who are subject to persistent racial prejudices (**Figure 8.20**). U.K. farmers face a potential labor crisis because Brexit is likely to end the stream of farm workers migrating from the EU states of eastern Europe, such as Poland.

Globalization

8.3 Explain how the processes of globalization have restructured the geography of agricultural production and food consumption and the sorts of problems that have arisen from that restructuring.

It is hard to overstate the importance of globalization for agriculture. As we have learned, agricultural production is very much place-based, dependent on local conditions including climate, soils, and terrain. Increasingly, however, many locally produced crops and food products are distributed and consumed globally. Hence, we are experiencing an intensifying two-way interdependence between local production and global consumption. This is altering worldwide one of humanity's distinguishing cultural characteristics: the foods we eat. In this section we discuss the key elements of our global system of food production, distribution, and consumption, including the importance of industrialization and corporatization. We also highlight the key dilemmas facing the global system, such as foodborne diseases and persistent malnutrition, and review alternatives, including the slow-food movement and fair trade.

FIGURE 8.20 Migrant farmworkers: a global phenomenon.
International migrant farmworkers, such as these African immigrants harvesting lettuce in southern Spain, are critical to agricultural production in the European Union. (Mark Eveleigh/Alamy)

The Global Food System

The term *global food system* is a convenient shorthand for discussing the globalization of food and agriculture. The **global food system** is the agro-industrial complex, organized at the global scale, that encompasses all elements of growing, harvesting, processing, transporting, marketing, consuming, and disposing of food for people. A very small number of multinational corporations now dominate this system. Let's begin our close examination of this system with a look at its historical evolution.

Historical Perspective The global trade in agricultural products is inseparable from the European exploration and colonization of the world's tropical and subtropical regions. As European maritime explorers brought far-flung cultures into contact, myriad crops were diffused around the globe. The processes of exploration and colonization created new regional cuisines (imagine "traditional" Italian cuisine without tomatoes from the Americas) and at the same time homogenized diverse diets by establishing a disproportionate global reliance on only three cereal grains: wheat, rice, and maize.

The expansion of European empires in the seventeenth and eighteenth centuries was closely associated with the expansion of tropical plantation agriculture. Plantations in warm climates produced what were then luxury foods for markets in the Global North, such as sugar, tea, coffee, bananas, and other tropical crops. The expansion of plantation agriculture had profound effects on local ecology, all but obliterating, for example, the forests of the Caribbean and the tropical coasts of the Americas. The trend in plantation expansion continues today with new food crops, such as palm oils, being added to the list of plantation crops. (**Figure 8.21**).

In sum, we have witnessed the development over centuries of a global food system that, among other changes, has freed affluent consumers in the Global North from the constraints of local growing conditions. Fresh strawberries, bananas, pears, avocados, pineapples, and many, many other types of temperate, subtropical, and tropical produce are now available in our urban supermarkets any day, any time of the year. On the other hand, the emphasis on a relatively small number of staple crops desired by northern consumers can mean the abandonment of local crop varieties and a decline in the associated biological diversity.

Agricultural Industrialization and Technological Change

The emergence of the global food system is a story of the increasing industrialization of agricultural production and the transportation of those crops to consumers. There are three key elements of the industrialization process: use of industrial inputs, mechanization, and bioengineering.

First, there has been an increasing reliance on the use of synthetic fertilizers, pesticides, herbicides, and other industrial inputs. A **synthetic fertilizer** is a nutrient needed for plant growth, most commonly nitrogen, phosphorus, and potassium, that is industrially manufactured using petroleum by-products. **Pesticides** are industrial chemicals that kill or repel animals, insects, or plants that can damage, destroy, or inhibit the growth of crops. **Herbicides** are a common type of pesticide designed to kill or inhibit the growth of unwanted plants (weeds) that compete with crop plants. These chemicals are produced within the petrochemical industry, which developed them in the twentieth century as by-products of oil refining. Since then, petrochemical use has become pervasive in conventional agriculture. Collectively, pesticides and synthetic fertilizers are known as agrichemicals. Artificial fertilizer application stimulates crop growth and pesticides reduce crop losses. Their application has been responsible for major increases in yields per acre of many crops. (The negative environmental and health effects are discussed in the next section.)

The livestock sector has seen a similar rise in the use of industrial inputs. Productivity increases there largely come from shortening the time from birth to slaughter. Advances in our scientific understanding of animal nutrition and growth led to the introduction of concentrated feeds and synthesized vitamins and hormones. Hormones stimulate the animals' growth rates and boost the efficiency with which they convert feed into meat. These industrial inputs reduced the need for extensive pasture areas and allowed farmers to move livestock production into confinement facilities. The technical term for this animal rearing system is **concentrated animal feeding operation (CAFO)**, which confines livestock, including cattle, sheep, turkeys, chickens, and hogs, at high densities in cages only large enough to allow the animal body to grow and to accommodate equipment for feeding and refuse removal (**Figure 8.22**). The close confinement of animals has required the additional industrial inputs of vaccines and antibiotics in order to combat the spread of diseases that reduce productivity. As with plant crops, industrialization of the livestock sector has led to large productivity increases.

FIGURE 8.21 An oil palm plantation in Malaysia. Plantation agriculture continues to expand in many Third World countries. While plantation-grown export commodities can be important to national economies, the accompanying destruction of tropical rain forests is a high ecological price to pay. (Stuart Franklin/Magnum Photos)

FIGURE 8.22 CAFO facility. Most hogs are now raised in concentrated animal feeding operations like this one in Indiana. (John Gress/Corbis via Getty Images)

Second, nearly every part of the production process, from field preparation to harvesting, is increasingly mechanized. Industrialized farms are dependent on fossil fuels to propel the cultivators and harvesters, pump the irrigation water, heat buildings, and so much more. Machinery is growing ever more sophisticated (and expensive), incorporating computer systems, lasers (for guiding plows and field leveling), and GPS. Greater mechanization for both crop-based and animal-based farming has greatly increased both labor productivity and, indirectly, the size of farms. This means fewer workers are needed, and vast portions of rural areas around the world are experiencing depopulation even as they become more productive.

One of the most important aspects of agricultural mechanization is the development of so-called cool chains. **Cool chain** refers to the system that uses new refrigeration and food-freezing technologies to keep farm produce fresh in climate-controlled environments at every stage of transport from field to retail grocers and restaurants. It is an integrated system of climate-controlled warehouses, ships, planes, trucks, and railcars. Cool chains are the backbone of a food system that allows gourmet ice cream produced from northern California cows to be sold in an upscale grocery in Atlanta, Georgia, and green beans grown in Kenya to be delivered fresh to stores in Manchester, England. Using cool chains, low-weight, high-value agricultural products can be flown to any major city in the world and sold fresh to shoppers in less than 48 hours after harvest.

The third and final element of industrialized agriculture is the emergence of biotechnology and the introduction of genetic engineering into farming systems. **Genetic engineering** is a technology that artificially modifies, manipulates, or recombines an organism's DNA. Pieces of DNA can be recombined with the DNA of other organisms to produce new properties, such as pesticide tolerance or disease resistance. DNA can be transferred not only between plant species but also between plant and animal species, which makes this technology truly revolutionary and unlike any other development since the beginning of domestication. The new varieties that are produced through genetic engineering are known as **genetically modified organisms (GMO)**, and the resulting products are called GMO foods.

Genetic engineering is transforming agriculture rapidly in ways that are just beginning to become evident. Commercial production of GMO foods began first in the United States only in 1996. The technology has now diffused around the globe to 26 countries, but the United States still dominates, accounting for 40 percent of the world's acreage (**Figure 8.23**). Today, over 90 percent of all soybeans, corn, sugar beets, and canola produced in the United States are genetically modified varieties. Their genetically engineered resistance to disease and drought are an important reason for the spread of GMOs, but that's only part of the story. For example, all of the GMO soybean seeds in the United States are engineered to tolerate greater doses of synthetic herbicides produced by petrochemical companies. Farmers are therefore not buying just GMO seeds, but an entire integrated product line of supporting industrial inputs. Productivity increases from the use of GMOs have been recorded, mostly as a result of reduced losses to diseases and insects. GMOs are not yet integral to animal rearing systems. In 2017, however, the first animal GMO consumed as food—a genetically engineered, farm-raised salmon called AquaAdvantage—became available to consumers (**Figure 8.24**).

To summarize, the global food system is characterized by a decades-long trend of industrialization and the introduction of new technologies. This industrial model of food production and distribution, developed largely in the United States, is now commonplace worldwide. Especially in the case of major cereal grains, the farming practices of South America, North America, Europe, and Australia are nearly indistinguishable. In many instances they are even conducted by the same multinational corporation, a topic we will take up next.

The Rise of Multinational Corporations in Agriculture The global food system is dominated by a handful of multinational corporations, collectively engaged in agribusiness. An **agribusiness** is a firm engaged in all parts of farming operations, from cultivation to storage, transport, processing, and marketing. Agribusiness can also refer to the entire system—entirely commercial, industrial, large-scale, and mechanized. Most agribusiness farming uses **monoculture**, the cultivation of a single commercial crop on expansive tracts of land. One of the distinguishing features of agribusiness is its vertically integrated structure. **Vertical integration** is when a single firm is engaged

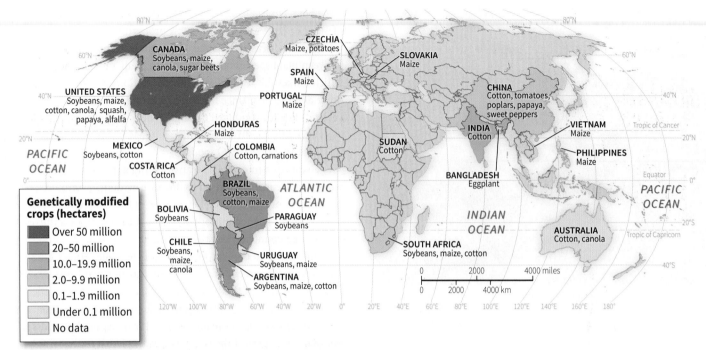

FIGURE 8.23 Worldwide use of genetically altered crop plants, especially maize and soybeans. This diffusion has occurred rapidly despite the concerns of health scientists, environmentalists, and consumers. (Source: ISAAA, 2016.)

FIGURE 8.24 Fast-growing GMO salmon. The larger fish, which was genetically modified to grow faster and larger, is the same age as the naturally occuring Atlantic salmon in the foreground. (AquaBounty Technologies, Inc.)

at multiple stages of the production process. Vertical integration takes a variety of forms depending on the nature of production processes and markets. The case of chicken provides a useful illustration of vertically integrated agribusiness as well as the interconnections of changing cultural values and global food production.

As background, we begin by noting the shift from beef to poultry as American consumer preferences underwent fundamental changes following World War II. From 1945 to 2017, annual per capita consumption of chicken in the United States rose from 5 to 90 pounds (2.25 to 40.8 kilograms), and in the 1990s surpassed that of beef. This was a startling development in American culture, where the "steak and potatoes" dinner and the myth of cowboys riding the open range had been central to an imagined national identity.

Changes in consumer preferences have been accompanied by increasing production efficiencies in the poultry industry, including a shift toward CAFOs. Advances in breeding, nutrition, housing, and processing largely account for increases in production efficiency (**Figure 8.25**). During the 1950s, a few companies began to vertically integrate all of these activities within a single multinational corporate structure. Such integrated companies own (1) the mills that convert raw materials to poultry feed, (2) the hatcheries where eggs are incubated, (3) the slaughtering plants where the birds are butchered, (4) the secondary processing facilities where the meat is turned into products (think chicken nuggets), and (5) the product brands that we buy.

Where are the farmers in this scheme? They are contractor farmers between steps 2 and 3 of our list. **Contract farming** is an arrangement between an independent farmer and an agribusiness company in which the farmer receives all the inputs from the company needed to produce a crop under conditions specified by the company in exchange for a guaranteed buyer and price for the crop. It is a growing and widespread practice in the production of many different crops worldwide. Farmers who produce a single commodity, such as chicken, must produce far more than the local market can consume in order to be profitable. They must therefore sell in both national and

FIGURE 8.25 Industrialized poultry. Poultry production and processing has become increasingly industrialized, while consumer preferences have shifted from whole chickens to parts. Here, workers process chicken carcasses in a factory's disassembly line in Bangkok, Thailand. (Robert Nickelsberg/The LIFE Images Collection/Getty Images)

global markets, access to which requires a dependence on multinational agribusinesses.

In the case of chickens, the farmer owns only the land and the poultry sheds. The multinational company provides all the hatchlings, feed, drugs, technical advice, and transport to and from the farm and promises to buy the finished chickens based on weight gained. Proponents argue that such an agreement gives the farmer a stable and predictable income without bearing the cost of inputs. Critics argue that the farmer bears all the risks in the event of crop failure and all the costs of farm maintenance. In any event, about 90 percent of all U.S. chickens are now grown under contract, with farmers acting as way

stations in agribusiness's vertically integrated production and marketing chain.

The taste for chicken has been spreading worldwide and new producers are now challenging the United States' dominance of the expanding global market. In the last decades of the twentieth century, world chicken trade grew nearly 500 percent, while the U.S. share of that trade doubled (**Figure 8.26**). China had been the hottest import market, because rising affluence there led to increasing per capita consumption. Now China, along with Brazil, Turkey, and Japan, is increasing its production and its exports of poultry. Thus, the U.S. share of the global market is shrinking in the twenty-first century even as per capita poultry consumption continues to rise worldwide.

The trend toward vertically integrated agribusiness has driven and is driven by the twin processes of consolidation and concentration. **Corporate consolidation** is when a business grows by amalgamating many companies into one large company through mergers and acquisitions. Consolidation has been the dominant trend in agribusiness, leading to growing industry concentration. **Industry concentration** refers to a process whereby an increasing proportion of production is controlled by decreasing numbers of companies. In agribusiness, this means, for example, that only six companies—the so-called "big six" of Germany's BASF and Bayer, Switzerland's Syngenta, and the United States' Dow Chemical, DuPont, and Monsanto—control about 70 percent of the global seed market and 100 percent of the market in genetically engineered seed stock. The big six will likely soon become the big four, further concentrating ownership of the world's commercial crop seeds.

We will look at the case of Monsanto Corporation to illustrate these agribusiness trends. Monsanto, based in St. Louis,

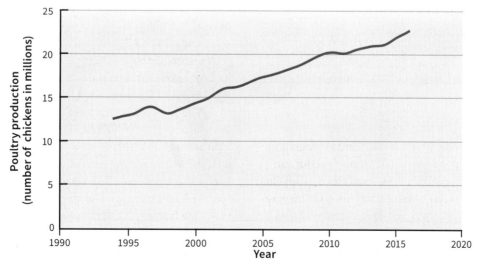

FIGURE 8.26 The global taste for chicken. Poultry production has skyrocketed worldwide, propelled by new industrialized systems of meat production and changing food preferences.

Missouri, started in 1901 as a chemical company specializing in food additives and pharmaceuticals. In the middle decades of the twentieth century, it became a leader in the emerging agrichemical industry by acquiring chemical and pharmaceutical companies. It subsequently produced some of the biggest-selling pesticides in agriculture, including the herbicide Roundup. In the 1980s, it began acquiring commercial seed companies and biotechnology companies to become a pioneer in GMO foods. It genetically engineered and patented a number of varieties of maize, wheat, soybeans, and vegetables, including the first crop variety engineered to produce its own pesticides. After decades of acquisitions and diversification, Monsanto became the world's largest single producer of commercial seeds, controlling over one-quarter of the global market by 2017.

Problems and Alternatives

The global food system has unquestionably upended Thomas Malthus's prediction that as the world's population grows, famine and starvation are inevitable (see Chapter 3). It feeds billions more people than Malthus could ever have imagined inhabiting the Earth. The rise in agricultural productivity, however, has come with costs. Now a growing number of detractors suggest the costs may be too high.

The Persistence of Food Deprivation In the words of development specialist Raj Patel, the global food system has produced a world divided into the "stuffed" and the "starved." He means that in the world's affluent regions, people suffer from the overconsumption of food—with resultant high rates of heart disease, obesity, and other health problems—while in the poorer regions, people suffer from chronic malnutrition and even starvation. Yet food production has grown more rapidly than the world population over the past 50 or 60 years. Per capita, more food is available today than in 1950, when fewer than half as many people lived on Earth. And production continues to increase. Global projections show continuing annual production increases of 1.5 percent through 2030. At the same time, today nearly 1 billion people worldwide are malnourished, some to the point of starvation (**Figure 8.27**). Thus, we face a paradox. On a global scale, there is enough food produced to feed everyone, but famine and malnutrition prevail for one out of six of the world's population.

What explains this paradox of dearth amidst plenty? As geographer Thomas Bassett and economist Alex Winter-Nelson show in their *Atlas of World Hunger*, one must ask both where and why people are hungry. We see in Figure 8.27 where people are going hungry on a global scale. Why they are hungry is complex and varies geographically and historically. For example, Bassett and Winter-Nelson explain that the crisis of HIV/AIDS in southern Africa affects mostly 15- to 49-year-olds, the most productive segment of the population. The epidemic thus negatively affects food production and raises the level of food insecurity, especially in rural areas.

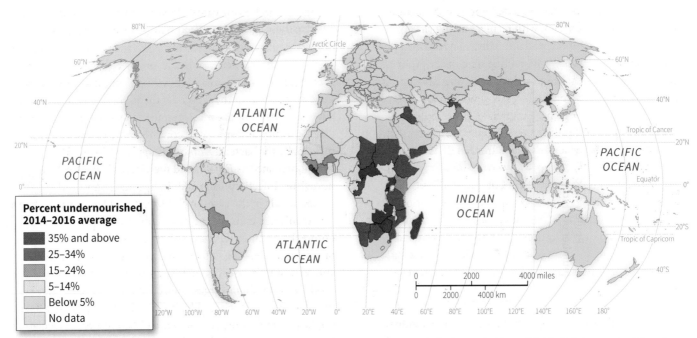

FIGURE 8.27 Mapping hunger worldwide. While food supply has outpaced population growth on a global scale, many world regions continue to suffer from malnutrition. (Source: UNFAO, 2017.)

To a great extent, international political economics, not global food shortages, causes hunger and starvation. International trade favors the farmers in wealthier countries through systems of government subsidies that keep the prices of their agricultural exports artificially low. Third World farmers find it difficult to compete. Many Third World countries do not grow enough food to feed their populations, and they cannot afford to purchase enough imported food to make up the difference. As a result, famines can occur even when plenty of food is available. For example, Bangladesh suffered a major famine in 1974, a year of record agricultural surpluses in the world.

Internal government policies are also an important cause of famine. The roots of the largest famine of the twentieth century may be traced to the agricultural policies of the Chinese government's 1958–1961 "Great Leap Forward" initiative. At this time, the Chinese government required peasants to abandon their individual fields and work collectively on large, state-run farms. Thirty million rural Chinese died of starvation during the Great Leap Forward. Misguided government policies triggered one of the first famines of the twenty-first century as well. In the early 2000s, Zimbabwe's President Robert Mugabe clung to power by demonizing white commercial farmers, threatening them with imprisonment. Other government policies discouraged planting and cultivation, thus resulting in another human-caused famine. In the case of Somalia during the famine of 2011–2012, warring factions prevented food aid from reaching starving populations. Over one-quarter of a million people died needlessly during that period of drought.

Finally, the global food system suffers from persistent and massive problems of wastage. The UN Food and Agriculture Organization estimates that one-third of all the world's postharvest production is wasted annually. The reasons are complex, found in both the production and consumption ends of the system, and are different in developed and developing countries. In developed countries losses tend to occur at the retail end, where consumer preferences overemphasize appearance. Good food is thrown away by grocery stores and restaurants simply because consumers think it does not "look right." The amount of food wasted by consumers in affluent regions is roughly equivalent to the net food production of Africa, the region most beset by famine and malnutrition. In developing countries, postharvest wastage tends to occur further down the chain at the production end. Bad roads, unreliable energy sources, and inadequate storage facilities are systemic problems, which translate to good food rotting before it can even get to market.

Food Fears A globalized food system is particularly vulnerable to food contaminations that can cause illness and even death. That is because even the simplest processed foods, such as

breakfast cereals or snack bars, may be comprised of ingredients that are grown in several different countries, shipped around the world, processed, packaged, and then sent to retailers. Each link in the chain "from farm to fork" is another opportunity for contamination with unwanted viruses and bacteria. National-level recalls of contaminated foods seem to be in the news almost daily. Data collected from hospitals and clinics indicate that every year there are 48 million cases of illnesses resulting from food contamination in the United States.

News of disease outbreaks, some life-threatening, has raised anxiety levels among consumers about eating food transported from distant places. Responses to such crises in the global food system vary culturally and individually and may be based more on the perception of risk than on the probability of illness or death. The way in which national governments and consumers respond to a particular food safety problem can fundamentally reshape geographic patterns of agricultural production on a global scale. For example, Japan banned all imports of U.S. beef when a single cow in Washington was found to have bovine spongiform encephalopathy (BSE), commonly known as "mad cow disease." The ban meant that the United States lost its largest beef export market. Conversely, the idea that a culturally symbolic food may be tainted and life-threatening can shake the strongest of cultural identities. For example, the 1990s cases of mad cow disease in Great Britain—where beef and dairy production and consumption have long been associated with British cultural identity—challenged culturally defined notions of a proper British meal.

As a sign of the growing prevalence of food contamination, the U.S. Centers for Disease Control (CDC) initiated a program to monitor and investigate **foodborne outbreaks**—illnesses that occur when two or more people consume the same contaminated food or drink. The CDC focuses its investigations on tracing the outbreak source, a complex geographical puzzle that requires investigators to move methodically back down the cool chain until they discover the point of contamination. CDC records show the most frequent contaminants are various genetic strains and species of common viruses and bacteria, principally the bacteria *Salmonella* spp. and *Escherichia coli* (abbreviated as *E. coli*), and the viruses hepatitis A and norovirus. Infections can result in hospitalization and even death, mostly in the cases of children, the elderly, and the infirm.

Ironically, many of the most common foodborne outbreaks occur in what nutritionists consider the basics of a healthy diet—fresh fruits and vegetables, nuts, seeds, and lean protein sources such as chicken and fish. Fresh produce is the single largest source of foodborne outbreaks. In particular, prepackaged leafy greens have repeatedly been implicated in outbreaks of *E. coli* and *Salmonella*. As we write, two people are reported dead and 64 ill in Canada and the United States after consuming leafy greens contaminated with

FIGURE 8.28 Vulnerabilities in the global food system. While leafy greens like this lettuce are considered key to a healthy diet, large-scale production and processing have created dangerous bacterial contamination problems. (GomezDavid/Getty Images)

E. coli. Through CDC monitoring, contamination is often detected before an outbreak occurs. In 2013, Taylor Farms, a multinational company with farms and processing facilities in the United States and Mexico, recalled from 39 states its baby green spinach, which was potentially contaminated with *E. coli* (**Figure 8.28**). Such preemptive recalls are becoming commonplace, and large agribusinesses may face several recalls each year.

The nature of the global food system has made foodborne outbreaks both more common and more difficult to control. Take the case of an outbreak of hepatitis A in 10 U.S. states that afflicted 161 people coast to coast, of whom 70 were hospitalized. The product common to all cases was Townsend Farms Organic Antioxidant Blend. The strain of hepatitis A was identified as common in the Middle East and North Africa. Using this genetic clue and information about the geographic origins of the product's component ingredients, investigators narrowed the likely source to pomegranate seeds from a global exporter in Turkey. The location of the seeds' production has not been identified, but the CDC noted that a similar outbreak in Europe was traced to seeds grown in Egypt. This case illustrates how the growing complexity of the global food system increases consumer choice while simultaneously exposing consumers to risks that transcend national borders.

Alternatives to the Global Food System We turn now to three problems of the global food system and the alternative responses to each.

First, as we described earlier, more and more control over the food system is concentrated into fewer corporate hands. In the case of seeds, the big six own the majority of the world's patented hybrid seeds and all GMO seeds. Whoever controls seeds controls access to the next crop harvest, because patents do not allow seeds to be replanted, requiring farmers to buy seeds every year.

Some farmers and consumer activists consider this to be an undesirable situation for political, economic, and ecological reasons. Consequently, alternatives have emerged. One is the food sovereignty movement. **Food sovereignty**, a term coined in the 1990s, is the principle that all people have a right to sufficient, healthy, and culturally appropriate food and that control over food production should remain in local farmers' hands rather than those of multinational corporations. Food sovereignty has become a rallying cry, particularly among Third World peasant farmers and indigenous peoples worldwide.

Another response to the corporate control of hybrid and GMO seeds is the movement to save and distribute heirloom seeds. **Heirloom seeds**, also called heritage seeds, are locally bred seeds of crop varieties that for marketing reasons are not part of the global food system and are open-pollinated rather than produced through hybridization. We might think of heirloom seeds as those produced by folk cultures. The patented hybrid seeds of the global food system are often created for marketing purposes (e.g., long shelf life, uniform color) rather than for nutrition and taste. Heirloom seeds may be ill suited to mass consumer markets, but they have the advantage of promoting varied, healthy diets. Moreover, the seeds are not patented and farmers can replant and exchange them without monetary cost. There are now numerous heirloom seed exchanges available to farmers and home gardeners, mostly in developed countries.

Second, the global food system emphasizes the production of highly processed foods with low nutritional value and questionable taste. An opposition movement began in Italy in the 1980s, partly in reaction to the spread of American fast-food chains such as McDonalds, but more generally to create an alternative to the globalization of food. Known as the **slow-food** movement, it is now an international organization promoting local food production and traditional cuisines and food ways in over 100 countries. People who dedicate themselves to slow-food diets and to obtaining as much of their nutrition as possible from local farmers are known as **locavores**. As we know, the meaning of "local" is highly subjective, and so there is no agreed-upon definition. The U.S. Department of Agriculture defines local as either produced within 400 miles (644 kilometers) of the consumer or within the same state.

Third, because rising agricultural productivity reduces labor needs and because small farms have difficulty competing with multinational corporations, many rural areas are losing population and smaller farmers cannot stay in business. Consumers interested in local, nutritious food sources are sympathetic to the plight of small farmers, and a number of marketing alternatives have arisen in recent decades. **Community-supported agriculture (CSA)** is a direct-to-consumer marketing arrangement whereby the farmers are guaranteed buyers for their produce at guaranteed prices and consumers receive fresh food directly from the producers. A CSA is a prepaid subscription service that consumers sign up for with a local farmers' organization. Another direct-to-consumer alternative is local farmers' markets. A **farmers' market** provides a setting (ranging from a few stalls in the street to covered enclosures extending a few city blocks) for farmers to sell their produce directly to consumers. The benefit to farmers is that more of the profit from crop sales is retained on the farm rather than going to wholesalers and food processors. Consumers get fresher food and a direct relationship with the person producing what they eat, thereby lessening fears of foodborne diseases.

Not all food, such as coffee, tea, and bananas, can be obtained locally. On the global scale, consumers can choose the alternative of fair trade–labeled products. **Fair trade** is a third-party certification program that supports good crop prices for Third World farmers and environmentally sound farming practices.

These alternatives to the global food system are clearly interlinked and interdependent. Slow food advocates, for example, depend upon the preservation of heirloom seeds and locavores depend on farmers' markets and CSAs. They have all arisen and grown in parallel with and in opposition to the increasing globalization of agriculture. Despite the growth and ubiquity of such alternatives, their impact remains minuscule compared to that of the global food system. Consequently, they have yet to significantly affect the problem we will address in our next section—the environmental costs.

Nature-Culture

8.4 Explain the complex interactions between industrial agriculture and the environment, especially with regard to issues of agrichemical use and climate change.

In this section we explore the interactions between the natural environment and agriculture. First, there are consequences to the rising dependence on irrigation and agrichemicals. Second, agriculture influences global climate change on agriculture

and vice versa. Third, there are efforts to reduce agriculture's negative environmental effects, including organic agriculture and sustainable agriculture.

The Environmental Costs of Industrial Agriculture

As we learned in the previous section, we can attribute the modern advances in agricultural productivity to the rising use of agrichemicals, mechanization, and other industrial technologies. We begin by looking at the environmental costs associated with the industrial water control technologies that allowed the tremendous expansion of land under cultivation.

The Limits of Industrial Water Control **Land reclamation** is the term used in reference to efforts to drain land inundated with either fresh or salt water. The Netherlands pioneered this technology in the fifteenth century, and today much of its agriculture relies on a system of dikes and canals that hold back the North Sea. Mechanization and industrialization have more recently accelerated the drainage of freshwater wetlands worldwide. In the United States, for example, European settlers drained for cultivation more than half of the wetlands once found in the lower 48 states. California sacrificed an astonishing 91 percent of its wetlands to attain its position as one of the most productive agricultural regions in the world. The ecological losses, especially as measured in terms of biodiversity, are incalculable. Wetlands provide habitat for wide range of species and are among the Earth's most biologically productive and diverse ecological communities. Wetlands species are widely threatened with extinction worldwide. Most countries now have regulations inhibiting further wetland drainage, and 169 countries have signed the International (Ramsar) Convention on Wetlands, which promotes the global protection of remaining wetlands.

If drainage technologies seek to remove water where it's not wanted by agriculture, then irrigation is its counterpart, seeking to add water where it's needed. **Irrigated agriculture** is farming that relies on the controlled application of water to cultivated fields. In the past century, land under irrigation has expanded with the development of new technologies that facilitate the capture and delivery of water. Two capture technologies in particular should be highlighted. One is surface impoundment through large dam construction. The dams of recent decades are industrial and engineering marvels, among the largest human-made structures on Earth. During the 1930s and 1940s, the U.S. government launched a dam-building initiative across the arid West, allowing millions of desert hectares to be irrigated and transformed into fertile farms. Since then, large dam construction and irrigated agriculture have accelerated globally. Many of the world's biggest dams are now found in Egypt,

Pakistan, and China. The other technology is deep groundwater extraction. While less dramatic than dam construction, deep-well drilling technology and the development of powerful industrial pumping systems have been no less important to the expansion of irrigated agriculture since the 1930s. Deep drilling has provided access to huge aquifers, subterranean water deposited hundreds of thousands of years ago. Today, 3280-foot (1000-meter)-deep wells are common. Well-and-pump technology now irrigates over 12.3 million acres (5 million hectares) of the semiarid Great Plains, much of it used to grow feed corn for the region's livestock fattening sector.

Recent irrigation efforts have allowed "deserts to bloom" and greatly increased the amount of land under cultivation, but the environmental costs are significant and are yet to be paid. Beginning with dams, the immediate effects are easily visible. Vast areas of natural habitat and associated biodiversity are lost when they are submerged by water impounded behind dams. Downstream aquatic ecology is also negatively affected. Dams block fish from migrating to spawn and reproduce and alter water quality, thus creating unlivable conditions for many native aquatic species. Deep-well critics point out that it is a form of water mining in that it is depleting a nonrenewable resource. As water is pumped out, the shallower wells run dry. In addition, pumping can reduce stream flows and sometimes cause them to dry up completely. Scientists have also noted that land subsidence is another common effect with undesirable environmental consequences.

We illustrate the potential costs of large-scale irrigation with a case from the former Soviet Union. The location is the arid borderland between Kazakhstan and Uzbekistan in central Asia, where irrigation has had severe ecological consequences. The once-huge Aral Sea has become so diminished by the impoundment and diversion of water from the rivers flowing into it that large areas of lakebed now lie dry and exposed (**Figure 8.29**). Not only was the local fishing industry destroyed, but also noxious, chemical-laden dust storms now blow from the barren lakebed onto nearby settlements, causing assorted health problems. Irrigation water diverted to arid lands destroyed an aquatic ecosystem and, ironically, produced a toxic desert.

What makes such environmental costs particularly troublesome is the fact that irrigated agriculture based on large dams and deep-water mining has a limited life span. All modern dams have a planned life expectancy, based partly on an estimation of how fast sediment washed down from upstream will fill up the reservoir. The large dams of the modern era are gradually filling up with sediment, reducing water storage capacity and available irrigation water. The U.S. government estimates that by 2020, 65 percent of the country's dams will have reached their projected life spans. As for water mining, groundwater is nearly everywhere being pumped to the surface faster than it can be replenished. For example, well-and-pump irrigation has drastically depleted the great Ogallala aquifer, which stretches under the Great Plains from South Dakota to Texas. Government scientists estimate that very soon the Ogallala will support only half of current irrigated acreage. Drilling deeper, while technologically feasible, is becoming economically prohibitive as costs exceed revenue from crop sales. As the bill comes due, the question of how we will sustain the great productivity advances of modern irrigated agriculture has yet to be answered.

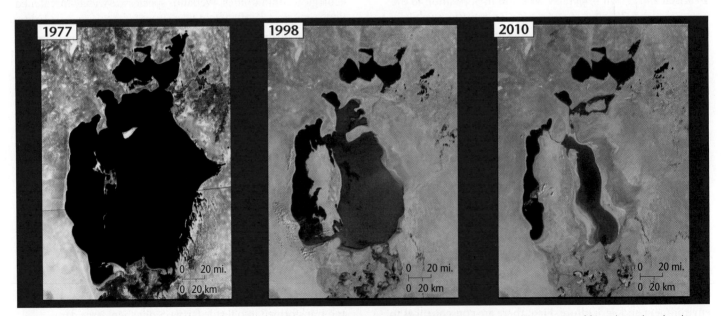

FIGURE 8.29 The incredible shrinking sea. The Aral Sea in central Asia has nearly disappeared over the course of four decades due to the diversion of water for irrigated agriculture. While agricultural productivity increased, the price paid was environmental devastation. (U.S. Geological Survey Department of the Interior/USGS, U.S. Geological Survey/photo by Jane Doe)

Other ecological problems result from applying large amounts of water to cultivated soils. Salinization, the process of salt accumulation in the top layers of soil, is a major problem for irrigated agriculture around the globe. It is particularly a problem in the world's arid regions, such as the U.S. West. There irrigation water evaporates quickly, leaving behind dissolved mineral salts. Eventually, the soil becomes too saline for crops to grow. More ecological problems arise from the runoff of polluted irrigation water into the surrounding environment. These are extensive and complex enough to warrant separate treatment in the next section.

Agrichemicals: Faustian Bargain? Industrialization has greatly increased the productivity of farms in recent decades. Those benefits must be weighed against the demonstrable environmental and human health costs of agrichemical use. Some observers believe industrialized agriculture is a Faustian bargain, promising a bounty of food, but with a high price tag only later revealed. Let us take a close look at this interpretation, starting with the effects of pesticide use.

Farmers commonly apply manufactured pesticides in aerosol form through spraying, either from the air or on the ground. Depending on the method and on weather conditions, as much as 95 percent of the pesticide spray may not reach its pest target, but is instead dispersed into the surrounding environment. A number of negative environmental effects can result. First, some pesticides kill insects indiscriminately. Consequently, insect-feeding wildlife, especially birds, will decline. Some pesticides also kill bees, which are necessary for the pollination of many agricultural crops. Second, pesticides can enter the bodies of living animals and inhibit reproduction, along with a number of other negative health effects that result in declining wildlife populations. Third, many pesticides are known to have adverse health effects on humans, including hormone disorders, skin and neurological diseases, and cancers.

One of the most vulnerable human populations in the U.S. are the field workers, mostly migrant and resident Latino farm workers, who come into direct contact with pesticides on a daily basis. Geographer Laura Pulido has shown that in the United States, farmworkers' exposure to pesticides is one manifestation of Latinos' position as an oppressed minority. Because they are a marginalized group within the dominant society, their exposure to harmful agrichemicals had not been a primary concern of mostly white environmentalists campaigning to limit pesticide use. Environmentalists, alarmed by the negative effects on wildlife and birds, saw pesticides as an environmental problem. Pulido argues that for farmworkers, pesticides are an environmental justice issue. Through her careful examination of the UFW's labor contract negotiations of the 1970s, she shows that farmworkers wanted safeguards in place on the farms and in the fields to protect them from exposure. In other words, we can see pesticide use as an environmental justice issue that reflects the racialized structure of farm labor that we discussed in the Mobility section of this chapter.

Pesticides also enter the environment as runoff from farms, as residue on food crops, and in the tissues of livestock. In this way, pesticides are spread far beyond the crop fields. After decades of pesticide use, the chemicals are found in every ecosystem worldwide. These chemicals ultimately reduce both aquatic and terrestrial biodiversity and cause increases in the rates of cancer and birth defects in humans and animals far from the cultivated fields. For example, a recent study by the U.S. Geological Survey found that every stream and 90 percent of the wells sampled contained pesticide residue. Humans, we may conclude, ingest pesticide residues daily, but especially when eating the food produced by industrialized agriculture. It is important to note that the most dangerous pesticides have been banned in most countries, that recent improvements in application procedures have limited the spread of pesticides, and that newer pesticides have been developed that are less harmful. On a global scale, however, environmental and health problems from pesticide use are increasing due to poor regulation and enforcement in developing countries.

Synthetic fertilizers are another form of agrichemical with well-known negative environmental effects. Even the most careful application will not guarantee that all of the fertilizer will be taken up by the target crops. As with pesticides, residual fertilizer ends up in the wider environment. Unlike pesticides, however, synthetic fertilizers are simply manufactured forms of naturally occurring plant nutrients, so their environmental effects are different. In the case of CAFOs, the concentration of large numbers of animals in confined spaces produces high volumes of manure and other wastes that can escape the site.

With both crops and CAFOs, problems occur when residual nutrients enter runoff from the fields and find their way to surface waters. When excess synthetic fertilizer or animal wastes enter aquatic systems they produce a sudden flush of nutrients that can trigger rapid plant growth, most notably algae. Nutrients in runoff result in "algae blooms," sudden spurts of algae growth that turn bodies of water green. The growth is so rapid and extensive that it consumes most of the dissolved oxygen in the water. Aquatic life is essentially deprived of the oxygen needed to survive, resulting in massive fish die-offs in both freshwater and nearshore marine environments. Scientists have recently identified so-called "dead zones"—sections of a water body where there is very little aquatic life—appearing seasonally at almost all U.S. river mouths. Maryland's Chesapeake Bay and the Mississippi River mouth in the Gulf of Mexico each have summer dead zones covering over 7700

FIGURE 8.30 Gulf of Mexico dead zone visible from space. Nutrient pollution from agriculture in the Mississippi River watershed causes annual summer algae blooms that deplete oxygen in the water, resulting in massive marine life die-offs. The summer 2017 dead zone was the largest ever recorded, 8776 square miles (22,730 square kilometers). (USGS)

square miles (20,000 square kilometers) (**Figure 8.30**), an area greater than the state of Connecticut. Another facet of the Faustian bargain is revealed: higher yields on land come at the price of damaged ocean fisheries.

Agriculture and Global Climate Change

Since climate is arguably the most important factor influencing agriculture, any changes in the global climate will have direct consequences for the global food system. On the other hand, modern agriculture influences global climate change in a number of ways.

The Unfolding Effects of Global Change The Intergovernmental Panel on Climate Change (IPCC) is the scientific body that the United Nations designated to track global climate change. Evaluating the research of hundreds of scientists from around the world, its most recent report concludes that the evidence for unprecedented global warming is "unequivocal." Without reductions in greenhouse gases (GHG), such as CO_2, climate conditions will soon be significantly different from those under which humans invented agriculture. That is, the climate conditions that prevailed for millennia and to which our global food system is adapted are beginning to shift, perhaps radically and irreversibly.

One of the IPCC's predictions for global warming's effects is an increase in extreme weather, such as droughts and floods, that can reduce crop yields. Also, regional climates are expected to shift geographically, which means that some areas currently cultivated will no longer be suitable for agriculture. As it happens, global warming will first be felt in those regions where poorer farmers operate without significant capital investment at the margins of suitable agricultural environments. Sub-Saharan Africa is a case in point (**Figure 8.31**). As we have seen, sub-Saharan Africa faces relatively greater challenges than other regions, including a rapidly growing population, high poverty rates, and high rates of malnutrition. Unfortunately, recent studies indicate declining productivity under most climate change models. In particular, under global warming scenarios, productivity projections for African cereal grains are negative throughout the region, with a few local exceptions.

Great concern exists about the near-term prospects for agriculture that relies on large-scale irrigation. Climate data suggest that the U.S. Southwest has likely entered a much drier period than ever before experienced by European settlers. The Colorado River Basin provides a disturbing example. Its dams and reservoirs irrigate 5.5 million acres (2.2 million hectares) of farmland. Lake Mead, the giant reservoir in Arizona and Nevada that supplies water to California's farms, has not reached full capacity since 1983 and is unlikely ever to be full again (**Figure 8.32**). The year 2013 marked 14 years of drought in the region, the worst in 100 years. Subsequently, the lake's water level hit a record low in 2016. The heavy snowfall and consequent rise in water level in 2017 is likely a temporary reprieve. The U.S. Southwest, the most agriculturally productive

FIGURE 8.31 Regions at risk in a warming future. People in Africa's Sahel region, which marks the transition from the Sahara Desert to moister conditions to the south, will likely suffer disproportionately from global climate change. This region is characterized by low income levels, high birth rates, and low agricultural productivity, which means the predicted drier conditions will probably result in regional food shortages. (Ariadne Van Zandbergen/Alamy)

FIGURE 8.32 Drought and the limits of irrigated agriculture. In the mid-twentieth century, irrigation water from Lake Mead, the reservoir behind the Hoover Dam, enabled the conversion of millions of acres of arid land to agriculture in the Southwest. The reservoir has not reached capacity for four decades, however (drought marks are clearly shown in this photo), and likely never will again, putting the future of some irrigated lands into question. (mtcurado/iStock/Getty Images)

FIGURE 8.33 Forest clearing for soybeans. Brazil has experienced rapid growth in soybean production at the expense of its tropical forests. (Ricardo Beliel/Brazil Photos/LightRocket via Getty Images)

are in the world, will face increasingly difficult challenges to the viability of current farming and land-use practices.

Almost every region in the world faces similar threats to future agricultural productivity from climate change. Consequently, global climate change presents one of the greatest challenges to food security. Declining yields translate to declining earnings for farmers, thereby threatening food security in the more impoverished rural regions of the world. There are some regions that are expected to see rising levels of crop productivity from increased temperatures and levels of atmospheric CO_2, but globally the negative effects on food security far outweigh the positive.

The Effects of Agriculture on Climate Modern agriculture contributes to global warming in several ways. The clearing and burning of forests for cultivation both increases atmospheric CO_2 (from forest combustion) and reduces the amount of CO_2 removed from the atmosphere. (Forests are considered carbon sinks because in growing they take up atmospheric CO_2 and store it for decades in living biomass.) For centuries, increased food production has been achieved by converting forests to pastures and plowed fields. In extensive parts of Eurasia and North America, temperate forests have virtually vanished. This pattern continues today with the clearance of tropical forests in developing countries (**Figure 8.33**). Because agriculture's impact on climate change is so tied to land-use changes such as forest clearance, the United Nations Food and Agricultural Organization

(UNFAO)—the international agency charged with producing, coordinating, and disseminating information on food and agriculture—bundles agriculture, forestry, and land-use change when assessing climate–food interactions. According to the UNFAO, these bundled activities account for one-fifth of human-caused global warming, second only to the energy sector.

Globally, agriculture contributes 14 percent of the world's total of anthropogenic CO_2 and 42 percent of methane. It is the largest source of another GHG, nitrous oxide, accounting for 75 percent of the world's emissions. There is, however, significant regional variation in agriculture's GHG contributions. Most of the methane, which has the greatest impact on warming, comes from livestock (in gases released in the digestion process) in Europe and North America. CO_2 emissions come from fossil fuel use in industrial agriculture in the Global North and from burning for forest clearance in the South. In the past two decades, the positive contribution of CO_2-absorbing forests has declined across Asia, but increased across Europe. Such regional variability suggests that efforts to reduce agriculture's contribution to global warming will require localized strategies.

Sometimes land converted to pasture and cultivation is not well managed, is just not suitable for long-term agricultural production, or both. In such cases a process of land degradation takes place whereby the condition of the pastures and fields becomes so poor that livestock and crops can no longer be raised. In extreme cases, such degradation over an extensive area results in desertification. **Desertification** is the process whereby once-fertile land is converted to desert either through climate variation, human activities, or a combination of the two. This occurs most often in arid and semiarid regions of the world.

Desertification contributes to global warming primarily by reducing grasslands' absorption of CO_2 from the atmosphere.

The contribution of these agricultural activities to global warming, significant as they are, do not include all aspects of the global food system. If we consider emissions generated from the manufacture of agrichemicals, the use of machinery in production, and postproduction emissions from transportation, processing, and retailing, the total contribution to global GHGs is even greater. Most of these types of emissions are currently attributed to the energy sector, thereby masking their origin in the global food system. If we consider all points of the food production–consumption pathway, our current agricultural system turns out to be one of the most significant contributors to global warming.

Reducing Agriculture's Environmental and Human Health Damage

The environmental problems associated with our current global food system have compelled farmers, scientists, social activists, and political leaders to consider alternatives. In this section we examine a few of the most prominent.

Sustainable Agriculture Given the sorts of problems we have identified in this chapter, many critical observers of the current industrialized agro-food system argue that it favors short-term gains over long-term viability. In a word, many argue that the dominant practices are unsustainable. In the 1980s, the idea of an alternative, sustainable agriculture took hold and is now endorsed worldwide, at least in word if not in deed. At the most basic level, **sustainable agriculture** is the production of food and fiber in a manner that does not diminish the future capacity of the land, harm the environment, or damage human health. Such a definition is anchored in an ecological approach to agriculture, sometimes called agro-ecology. Sustainable agriculture has taken on an ever-widening meaning, incorporating ideas of economic viability, social justice, and the ethical treatment of animals. As its meaning expanded, consumer choice in support of sustainable practices became a central focus. One consequence is the often confusing product labeling we all face at the grocery store touting food that is "all natural," "antibiotic free," "hormone-free," "cage-free," and so on.

As cultural geography studies have shown, folk and indigenous knowledge about ecological conditions can be a foundation for sustainable agriculture. Many indigenous land-use systems have been in place for decades, even centuries, without significant land degradation. An example of sustainable indigenous agriculture is the paddy rice farming that occurs near the margins of the Asian wet-rice region, where unreliable rainfall causes harvests to vary greatly from one year to the next. Farmers have developed complex cultivation strategies to avert periodic famine, including growing many varieties of rice. Many farmers, including those in parts of Thailand, rejected the green revolution, viewing the advice given to them by agricultural experts as inappropriate for their marginal lands. Based on generations of experimentation, the local farmers knew that their traditional adaptive strategy of diversification was superior.

In West Africa, peasant grain, root, and livestock farmers have also developed adaptations to local environmental influences. They raise a multiplicity of crops on the more humid lands near the coast. Moving inland toward the drier interior, farmers plant fewer kinds of crops but grow more drought-resistant varieties. Having observed many cases like these in which local practices have proved effective and sustainable, most geographers now agree that agricultural experts need to consider indigenous knowledge when devising sustainable agriculture projects.

Don't Panic, It's Organic Alarmed by the ecological and health hazards of chemical-dependent, industrialized agriculture, a small counterculture movement emerged in the United States and Europe in the 1960s and 1970s. Geographer Julie Guthman labels this the organic farming movement, which we can consider as a branch of the sustainable agriculture movement. **Organic agriculture** is the production of crops and livestock using ecological processes, natural biodiversity, and renewable resources rather than relying on industrial practices and synthetic inputs. According to Guthman, the organic farming movement in the United States saw in organic agriculture a solution to a range of social, cultural, political, and environmental ills. These included the loss of small family-owned and -operated farms, environmental pollution from industrial agriculture, corporate control of the food system, and the nutritional deficiencies of highly processed foods.

When the organic food movement was in its infancy, there was no way to differentiate organically produced animals and crops from the products of what has come to be called conventional agriculture. **Conventional agriculture** is really another term for farming that depends on manufactured synthetic inputs, GMO seeds, and other such industrial practices that we have discussed in this chapter. Movement advocates, many of them based in California, invented new certification systems in the 1970s that focused on the technical aspects of organic agriculture, particularly the absence of agrichemicals. In 1979, California passed the first law on organics in the United States. At the federal level, opposition from corporate agribusiness interests delayed regulatory legislation on organic farming until 1990 and full implementation for another decade.

By legislating regulatory standards for organic agriculture, the state and federal governments provided organic farmers the basis for differentiating their products in the marketplace. This,

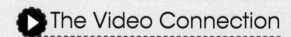 The Video Connection

The Marijuana Divide
 Watch at Sapling Plus

Recreational marijuana use is now legal in Colorado and eight other U.S. states, creating an economic boom in many agricultural regions. This video examines the different approaches to this newly legal crop in two Colorado towns. Gunnison, an agricultural community, is contrasted with Crested Butte, a resort community. Gunnison residents voted to ban the cultivation and sale of marijuana in their town, while Crested Butte residents voted in favor. Marijuana cultivation in Colorado now generates tens of millions of dollars in sales and tax revenue every month, influencing Gunnison residents to reconsider.

Thinking Geographically

1. We have noted that many agriculture communities face declining incomes and population loss. How does Gunnison residents' ambivalence toward marijuana cultivation reflect concerns over rural economic decline? Based on this chapter's discussion of organic agriculture, what effect do you think marijuana legalization might have on Gunnison's family farms?

2. In the video, several residents express concerns that marijuana is changing or will change cultural norms in rural communities. How do these concerns relate to the general differences between folk and popular cultures? How do residents in the communities express a sense of place in relation to their respective approaches to marijuana farming?

---

in turn, allowed producers and retailers to charge consumers a premium. The U.S. grocery chain Whole Foods, for example, labels their produce "conventional" or "organic," with the latter having a higher price. The organic food market is now the most rapidly growing (and most profitable) agricultural sector worldwide. Sales of organics in the United States, while only a bit more than 5 percent of total food and beverage sales, are growing at the phenomenal pace of 11 to 15 percent annually. The lure of premium pricing and growing consumer demand has encouraged many producers to switch to organic agriculture.

Land-use practices reflect these market shifts. In the United States, land under organic production increased 27 percent from 2011 to 2016. European consumers are likewise seeking more access to organic food, and farmers have responded. In the EU, the area of land under organic cultivation grew nearly 19 percent between 2012 and 2016. With only 0.7 percent of its agricultural land in organic production, the U.S. does not come close to the scale of these shifts in the EU. In several European countries, including Austria, Estonia, and Sweden, over 18 percent of all agricultural land is used in organic production. Globally, Australia has roughly six times the total amount of land under organic production as the United States, and Argentina has nearly double.

In Guthman's assessment, the U.S. organic farming movement, because it focused solely on the technical aspects of organic production, fell short of addressing many of the social and political ills it sought to correct. Her study of California organic agriculture demonstrated that organic farming is readily incorporated into large-scale agribusiness enterprises. Indeed, some existing agribusinesses merely purchased existing organic farms as a means of diversifying their operations and tapping the profits from premium pricing. While organic agriculture has produced environmental and health benefits (for farmworkers and consumers), its effect on the historic decline of family farms and their rural communities has been minimal.

Green Fuels from Agriculture Early in the twentieth century, Henry Ford campaigned for fueling cars with alcohol from crop wastes and Rudolf Diesel ran his first diesel engines on fuel made from peanut oil. Today we call these energy sources **biofuels**, fuels derived from organic wastes or plant material. Auto pioneers, however, quickly abandoned biofuels for nonrenewable fossil fuels, and the rest is history. The modern global economy is dependent on fossil fuels to produce everything that we eat, wear, listen to, read, and live in. By definition, nonrenewable fuel sources will eventually run out, causing both prices and political uncertainties to increase. In addition, fossil-fuel combustion is the largest source of atmospheric GHGs, which contribute to global warming. Therefore, an urgent search is under way to find alternative, renewable fuel supplies, and agriculture has become one of the main sources.

How can agriculture be a source of energy for industry? By tapping the simple process of fermentation, many plant materials can be converted to a combustible alcohol, ethanol.

FIGURE 8.34 Corn Belt ethanol plant. Plants such as this one in Colorado have been springing up in cornfields across the midwestern United States as demand for alternative fuel sources increases. Corn is the main source of ethanol as a gasoline additive in the United States. (RICK WILKING/REUTERS)

In the United States, corn is the main source of ethanol, and new ethanol plants dot the landscape of the midwestern Corn Belt (**Figure 8.34**). Responding to federal legislation, U.S. ethanol production expanded more than 600 percent during the first decade of the twenty-first century.

This shift in corn production from food to fuel affects nearly every aspect of the field crop sector, as well as livestock production, habitat protection, food retailing, and the global grain trade. For example, total planted corn acreage in 2013 was the highest in 75 years, and the corn harvest of 2016–2017 was an all-time record of 15.1 billion bushels. Farmers have expanded corn acreage by reducing fallow periods and plowing land set aside for conservation. Additionally, ethanol for fuel creates a new source of consumer demand that competes with existing demand for food. As a result, the price of corn on the global market rose, with anticipated negative consequences for the poor and undernourished.

Brazil, which is the closest international rival of the United States in ethanol production, grows and ferments sugarcane, which yields twice as many gallons of ethanol per acre as does corn. The policies of the Brazilian government have encouraged the creation of a delivery infrastructure (plants, tanks, and pumps) and the manufacture and sale of "flex" cars that can run on either gasoline or ethanol. The country has freed itself from

a dependence on imported fossil oil, an achievement that many countries can only dream of emulating.

Will biofuels from agriculture produce a sustainable, environmentally friendly alternative to fossil fuels? Biofuels appear to offer tremendous promise, yet they are not without drawbacks. For example, switching farms from food to fuel production can threaten food security. The environmental benefits of biofuels also may be reduced by some of the associated costs of increasing ethanol and biodiesel production. Biodiesel, fuel made from vegetable oils, takes less energy to produce than crop-based ethanol, but it is typically more expensive than petroleum-derived diesel. These issues are explored further in Subject to Debate.

Cultural Landscape

8.5 Analyze the influence of agriculture in shaping the cultural landscape and how that influence varies geographically.

Our transformation of nature to produce food was the most fundamental cultural landscape development in human history prior to the Industrial Revolution. Today we find that many of our most striking cultural landscapes are agricultural: the vineyards of Mediterranean Europe, the ranchlands of the U.S. West, the terraced rice paddies of East and Southeast Asia. In this section we investigate the historical interactions of farming, culture, and regional settlement patterns. We begin by examining how agricultural landscapes reflect patterns of property ownership, then highlight field borders as one example of agriculture's distinctive landscape features. We close with an analysis of the importance of agricultural landscapes for cultural identities.

Cadastral Surveys and Field Patterns

The agricultural landscapes of countries that use cadastral surveys (see Chapter 1) look very different from those that do not, because the property lines and associated field patterns have different origins. Major regional contrasts exist in property lines and field patterns, such as unit-block versus fragmented landholdings and geometric government survey lines versus irregular customary property boundaries. Let us look at these differences in detail.

Fragmented farms are found throughout Eurasia and Africa and in many indigenous cultural regions worldwide. In one common settlement pattern, farmers live in nucleated settlements of small towns and villages. Their landholdings are fragmented into many separate fields situated at varying distances and lying in various directions from the settlement.

SUBJECT TO DEBATE Can Biofuels Save the Planet?

In response to diminishing oil reserves and global warming, the U.S. government passed the 2007 Energy Independence and Security Act. It mandated that ethanol be added to gasoline. Governments around the world have implemented similar initiatives to increase renewable fuel use. Globally, the most promising environmental outcome of increased biofuel use is a decrease in greenhouse gases. Growing plants consume atmospheric carbon dioxide. Using them for fuel thus recycles an important greenhouse gas, in contrast with fossil fuels, which release stored carbon into the atmosphere when combusted.

Biofuel demand is transforming agriculture around the world, but the energy and environmental benefits are disputed. Because U.S. corn cultivation is so thoroughly industrialized, ethanol production consumes nearly as much fossil fuel as it replaces. By some estimates, corn ethanol production uses more energy than it supplies. Brazil's sugarcane ethanol industry has a far better energy balance of 1 unit of fossil fuel input to 8 units of biofuel output. These energy gains may be offset by other environmental costs. Sugarcane cultivation has created a vast and expanding monoculture. Many fear continued expansion will contribute to deforestation. Likewise, in the United States, portions of 35 million acres (14.2 million hectares) of land set aside for soil and wildlife conservation have been plowed to grow corn for ethanol. Due to industrial agriculture's GHG contributions, it is not at all clear that biofuel production and use will help reduce global warming.

Continuing the Debate

Contemplate the future role biofuels will play in addressing the linked crises of energy supply and global warming and consider these questions:

- What do you think can be done to make biofuels more promising environmentally?
- How can food security for the poor be assured as biofuel use increases?
- Who do you think will benefit the most from the expansion of biofuel use? Small farmers or agribusiness? High-income countries or low-income countries? Consumers or corporations?

The biofuel debate. Many U.S. politicians and farm lobbyists promote ethanol from corn as an alternative fuel source for cars. Many environmental scientists and economists argue otherwise. (Pat LaCroix/Getty Images)

The individual landholdings generally have an irregular shape, as the boundaries were established by customary use rather than formal surveys. In Eurasia, many fragmented landholdings date back to an early period of peasant communalism. By custom, each farm household in the community had access to land of varying soil composition and terrain. Similarly, in the East African highlands farmers may have multiple fields at different elevations and with different exposures to allow for diversification of crops and extended harvest periods. Fragmented holdings remain a prominent feature shaping the world's cultural landscapes.

Unit-block farms, by contrast, are those in which all of the farmer's property is contained in a single, contiguous piece of land. Such farms are found mainly in the European settler colonies of the Americas, Australia, New Zealand, and South Africa, where much of the land was surveyed prior to and in preparation for settlement. In the United States, early settlement was surveyed in a metes and bounds system, which used natural features such as trees, boulders, and streams and therefore produced irregularly shaped property boundaries (**Figure 8.35**). Most often, however, settler colonies reveal a regular, geometric land survey. The township and range system, discussed in Chapter 6, is a good example. That survey imposed a grid pattern that resulted in today's checkerboard character of the American agricultural landscape (**Figure 8.36**). Regions with these survey systems tend to have patterns of dispersed

FIGURE 8.35 Original land-survey patterns in the United States and southern Canada. The cadastral patterns still retain the imprint of the various original survey types. The map is necessarily generalized, and many local exceptions exist.

FIGURE 8.36 American township and range survey creates a checkerboard, illustrated well by irrigated agriculture in the desert of California's Imperial Valley. (Glowimages/Getty Images)

settlement, with individual family farmsteads constructed on the unit blocks. Unlike fragmented farms, therefore, farmers live and farm on the same parcel of land, producing a contrasting agricultural landscape. The Seeing Geography feature of this chapter explores the differences in these landscape types.

Another unit-block type is the long-lot farm, where the landholding consists of a long, narrow unit block stretching back from a road, river, or canal (**Figure 8.37**). Long-lot surveys tend to organize farmers' dwellings in a linear settlement pattern. Long lots occur widely in the hills and marshes of central and western Europe, in parts of Brazil and Argentina, along the rivers of French-settled Québec and southern Louisiana, and in parts of Texas and northern New Mexico. These unit-block farms are elongated because such a layout provides each farmer with fertile valley land, water, access to transportation facilities, or some combination of these. This property system also allows for the greatest number of farmers, because it divides access to a scarce resource equally rather than concentrating it in the hands of a few. In French America, long lots appear in rows along streams, because waterways provided the chief means of transport in colonial times. In the hill lands of central Europe, a road along the valley floor provides the focus, and long lots reach back from the road to the adjacent ridgetops. Hispanic settlers introduced the long lot system to New Mexico's Rio Grande Valley in the 1700s so that every farmstead would have direct access to water, either in canals or in natural watercourses. Water access was essential because the land is too arid to farm without irrigation.

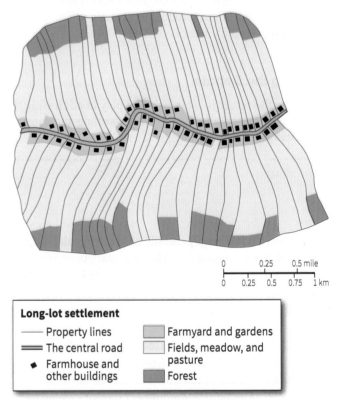

Long-lot settlement

— Property lines

═ The central road

◆ Farmhouse and other buildings

▨ Farmyard and gardens

▫ Fields, meadow, and pasture

▨ Forest

FIGURE 8.37 A long-lot settlement in the hills of central Germany. Each property consists of an elongated unit block of land stretching back from the road in the valley to an adjacent ridgetop, part of which remains wooded.

Fencing and Hedging

Property and field borders are often marked by fences or hedges, heightening the visibility of these lines in the agricultural landscape. Fences and hedges add a distinctive touch to the cultural landscape. Because different cultures have their own methods and ways of enclosing land, types of fences and hedges can be linked to particular groups. In New England, western Ireland, or the Yucatán Peninsula, we see mile upon mile of stone fencing that typifies those landscapes. Barbed-wire fences swept across the American countryside a century ago, but remnants of older styles can still be seen. Like most visible features of culture, fence types can serve as indicators of cultural diffusion.

The hedge is a living fence. In Brittany and Normandy in France and in large areas of the United Kingdom, hedgerows are a major aspect of the rural landscape (**Figure 8.38**). To walk or drive the paths and roads of hedgerow country is to experience a unique feeling of confinement as one moves between living, green walls. In recent decades, farmers have consolidated smaller fields into larger, contiguous landholdings, eliminating hedgerows in the process. The removal of hedgerows means not only the loss of a defining feature of the rural landscape but also a decline in habitat for many rare plants, mammals, and birds. In response, the U.K. government passed regulations to protect hedgerows in England and Wales.

Across much of the western United States, another type of field border appears: the windbreak. A windbreak (or

FIGURE 8.39 Windbreak on the Great Plains. First introduced on a mass scale following the 1930s Dust Bowl, these iconic landscape features have deteriorated as trees matured and died. The U.S. government recently created a funding initiative to revitalize them and maintain the soil protection they provide. (Oklahoma Forestry Services)

shelterbelt) is a row of trees or shrubs planted on the edges of cultivated fields to slow the wind and reduce soil erosion. We find them across the U.S. West, primarily in arid and semiarid regions of the Great Plains and in major valleys, such as those of central California. Such locations combine two characteristics that can cause rapid soil loss: strong, persistent winds and long dry seasons when exposed soils can dry out and become susceptible to erosion. Windbreaks became ubiquitous in the West after the environmental disaster of the 1930s Dust Bowl. As part of disaster recovery efforts, the U.S. government initiated the Great Plains Shelterbelt project, providing money and expertise for planting windbreaks from Canada to Texas. Windbreak trees are now part of the cultural landscape of the Great Plains (**Figure 8.39**). Like the U.K. hedgerows, they also provide for greater biodiversity in an otherwise ecologically impoverished monocrop farming system.

Agricultural Landscape Protection

As we noted in Chapter 6, U.S. national identity and the U.S. democratic system of government are historically rooted in the idea of a nation of independent farmers and ranchers. We have since grown farther from that vision with every passing decade. The U.S. Census Bureau reports that in 1910, one in two people (54.4 percent) lived in rural areas. By 2010, only one in five did (19.3 percent). Despite the sentimental references to the

FIGURE 8.38 Hedgerows in England. Archetypical hedgerow landscapes, such as this one in Devon, England, historically common throughout western Europe, are now disappearing as land-ownership has become increasingly concentrated. (Tony Eveling/Alamy)

World Heritage Site: Honghe Hani Rice Terraces

The cultural landscape of the Honghe Hani Rice Terraces in Yunnan, China, is one of the most recent additions to the World Heritage Site list. Inscribed in 2013, the site offers visually striking examples of humans' creative response to environmental limits and of long-term sustainable agricultural production.

- Thousands of terraced rice paddies, built over the course of 1300 years, reach down the slopes of the Ailao Mountains to the valley of the Hong River. The entire site covers 386 square miles (1000 square kilometers) and three distinct terraced areas: Bada, which is for the most part gently sloped; Duoyishu, which is steeper; and Laohuzui, which is very steep. Eighty-two villages of 50 to 100 households each are located among the terraced fields.

- The Hani people, whose ancestors migrated to the region some 2000 years ago, are the predominant ethnic group resident in it. As far back as China's Tang dynasty (seventh to tenth century C.E.), outside commentators have remarked on the Hani's skill and ingenuity in modifying the mountainsides for agriculture. The Hani migrants initially encountered a forested ridge cut by numerous surface springs fed by underground seepage. The Hani left forest pockets intact to protect the spring sources, built rock terraces to create level ground, and then channeled the water from the springs to irrigate the fields.

- Hani religion is rooted in a belief in the sacredness of nature. This spirituality supports ecologically beneficial land-use practices, such as the maintenance of sacred forests and trees in and around Hani villages. The Hani's efforts over the centuries have produced an integrated nature–culture system that unites forest, dwelling, water, and land to sustainably produce both abundant food and environmental benefits.

ON THE TERRACES: Each household has access to one or two segments of the irrigated terrace (as well as to a house site, fuel and building material from the forest, and pasture). The terraced fields are maintained and cultivated through a complex system of communal obligations and individual entitlements. A fundamental task of this system is the regulation of the movement of water onto and off of tens of thousands of individual fields.

(Chen Haining/Xinhua News Agency/Newscom)

- Roughly 276.5 miles (445 kilometers) of water channels, including 4 trunk canals and 392 branch ditches, crisscross the mountainside. Residents employ bars of wood or stone to direct water through these channels. Each bar is carved with a different-sized outlet so that sufficient water flows to the lower ditches on the slope and each ditch receives the correct amount of water. Ditch leaders are appointed from the villages to maintain the ditches, allocate water, and settle disputes. The villagers, in turn, pay the ditch leader in rice for the service of managing the ditches.

- The water itself is nutrient rich, having passed through the forest litter and picked up key minerals. In late winter, this nutrient-laden water is diverted to the fields to fertilize the soil. In addition, residents have established manure ponds where livestock manure is collected during the winter to create a source of organic fertilizer. At the start of spring planting, nutrient-rich water from the ponds is diverted onto the terraces to fertilize the crop.

- Along with multiple varieties of rice, residents harvest from the terraces many aquatic herbs and vegetables. Eels, bivalves, snails, fish, shrimp, and other aquatic animals are raised in the water of the flooded fields. Ducks are also part of the system and feed on aquatic animals and vegetation. Soybeans and livestock, including pigs, cattle, and buffalo, are raised on the land between flooded paddies.

TOURISM: The region has no history of mass tourism, though its scenic beauty and dramatic irrigation structures have long attracted photographers from all over the world.

- Recent developments, however, including its WHS designation and local transportation infrastructure upgrades, are likely to initiate a wave of touristic activity. The Chinese government has formulated a plan for an ecotourism industry, founded on the region's reputation as an example of sustainable agriculture within an intact traditional culture.

- Future projects include the designation of scenic tourist routes, an information center focused on local culture, the maintenance of traditional architecture, and the promotion of place-based agricultural products.

- Beginning 1300 years ago, the Hani people of China's Ailao Mountains constructed an ingenious system of terraced agriculture.

- By protecting forests, cycling organic nutrients, and carefully controlling the distribution of water, Hani farming is both highly productive and environmentally sustainable.

- Long appreciated by photographers for its scenic beauty, the landscape of the Honghe Hani rice terraces will soon be developed for tourism.

http://whc.unesco.org/en/list/1111

(Yang Zongyou/Xinhua News Agency/Newscom)

family farm expressed in popular media, the United States is predominantly an urban culture, with only tenuous emotional attachments to highly industrialized agricultural landscapes. This might help explain why so few Americans spend getaway weekends in farming regions, the California wine country excepted. It might also help to explain why there is not one "living" agricultural landscape found among the many World Heritage Sites in the United States.

The situation elsewhere is quite different. Asian countries have proposed many living agricultural landscapes for WHS designation, such as the terraced rice paddies of China featured in this chapter. In western Europe, there are many government and nongovernment efforts to protect agricultural landscapes. France alone has at least four World Heritage Sites designed to protect agricultural landscapes, including the Causses and the Cévennes, a huge 747,046-acre (302,319-hectare) site where one of the last active transhumance systems in Europe is practiced. The United States also has no equivalent to the United Kingdom's Campaign to Protect Rural England, a nonprofit organization that advocates for the preservation of "a beautiful and living countryside." The U.K. governments in Scotland,

England, and Wales have many local zoning initiatives that protect agricultural landscapes from industrial development and urban sprawl.

At the EU level, agricultural landscape protection is a major policy initiative that touches upon international trade, economic development, environmental conservation, and other policy sectors. EU policies, for example, link cultural landscape protection with biodiversity conservation, noting that biodiversity is generated from traditional agricultural practices. Therefore, some EU conservation funding is directed toward keeping traditional agricultural landscapes intact. These landscapes also represent the cultural heritage of nations and regions and are therefore attractive to tourists, both domestic and international. The EU offers subsidies to farmers who use the traditional farming practices that work to maintain heritage landscapes (thereby claiming a cultural exception to international trade agreements that bar agricultural subsidies). The contrast of Europe with the United States is another example of how the values associated with cultural landscapes are rooted in the social histories of specific places.

Conclusion

A griculture is one of the oldest and most enduring areas of human geographic study. We have seen that this ancient human endeavor varies markedly from region to region, resulting in formal agricultural regions identifiable by their particular farming systems. We learned how central mobilities are to the geography of food and agriculture, from the diffusion of the first domesticates to the present dependence on migratory labor. The processes of globalization have resulted in the emergence of a global food system that links place-based agricultural producers to consumers worldwide, creating both advantages and vulnerabilities. We explored the complex, multidirectional interactions between industrialized agricultural practices and the wider environment. Finally, we noted how agricultural production leaves its mark on the landscape and how important agricultural landscapes are for cultural heritage.

The first agricultural revolution fundamentally changed humankind. In equally dramatic fashion, the Industrial Revolution sparked further changes. We turn to this topic in the next chapter.

Chapter Summary

8.1 Identify the major agricultural production types, their regions, and the social and environmental factors shaping them.

- Farming systems can be classified as extensive or intensive, subsistence or commercial, and crop-based or animal-based.
- There are six basic kinds of crop-based systems, their geographic distribution determined largely by regional climates.
- There are four basic types of animal-rearing systems, strongly influenced by climate, but also by cultural and economic factors.
- In addition to the formal regions of based on farming system types, there are also urban- and aquatic-based agricultural practices.
- In some of the world's most extreme climates, a small number of people continue to procure food primarily through hunting and gathering rather than agriculture.

- The von Thünen model helps us understand the spatial patterning of farming systems surrounding major urban markets.

8.2 Describe the ancient and modern patterns of domesticated plant and animal diffusion and the importance of labor mobility to industrialized agriculture.

- Agriculture began with the process of domestication in a few ancient centers of origin around the world, and then diffused outward from those centers.
- Modern agricultural diffusions are linked to the movement of crops and livestock through European exploration and colonization and through technology transfers, exemplified by the green revolution.
- Labor demands vary seasonally in agricultural production, a problem solved in the modern era through the use of migratory labor.

8.3 Explain how the processes of globalization have restructured the geography of agricultural production and food consumption and the sorts of problems that have arisen from that restructuring.

- Most people now procure at least a portion of their sustenance through a global food system that is dominated by a few vertically integrated multinational corporations and based in industrial agricultural practices such as the use of agrichemicals and GMO seeds.
- Some of the problems associated with the global food system, such as unequal access to nutrition and foodborne illnesses, have prompted farmers and consumers to develop alternatives such as food sovereignty, community-supported agriculture, and slow food.

8.4 Explain the complex interactions between industrial agriculture and the environment, especially with regard to issues of agrichemical use and climate change.

- Much industrial agriculture is based on the unsustainable use of water and the use of agrichemicals that have significant human health and environmental impacts such as increased cancers and water pollution.
- Global climate change will continue to present increasingly difficult challenges to agriculture due to phenomena such as more frequent and intense droughts, while industrial agriculture exacerbates climate change by contributing greenhouse gases to the atmosphere.
- Sustainable agriculture, organic agriculture, and the farming of biofuel crops are possible ways of reducing some of industrial agriculture's most damaging environmental and human health effects.

8.5 Analyze the influence of agriculture in shaping the cultural landscape and how that influence varies geographically.

- The two basic farm field patterns, unit-block and fragmented landholdings, are expressions of historic property surveys and settlement that greatly influence the cultural landscapes of agricultural regions.
- One of the most distinctive features in agricultural landscapes are the fences, hedgerows, and windbreaks that mark the boundaries of fields.
- In some regions of the world, particularly Europe, governments fund the protection of agricultural landscapes because people perceive them as important to cultural identities and national heritage.

Practice at Sapling Plus

Read the interactive e-Text, review key concepts, and test your understanding.

 Story Map. Explore how people use technologies to farm in inhospitable places.

 Web Map. Examine issues related to the geography of agriculture.

Key Terms

agriculture The cultivation of crops and rearing of livestock to produce food, animal feed, drink, and fiber (page 268).

farming systems Populations of farms that share a similar resource base, including climate, water availability, slope, topography, and soil quality, and a similar set of practices including crop and livestock types, property ownership, farm size, temporal and spatial patterns of cultivation and animal rearing, and the use of machinery or irrigation (page 268).

extensive agriculture Crop cultivation and livestock rearing systems that require low levels of labor or capital input relative to the areal extent of land under production (page 268).

intensive agriculture Crop cultivation and livestock rearing systems that use high levels of labor or capital relative to the size of the landholding (page 268).

subsistence agriculture Food production mainly for consumption by the farming family and local community, rather than principally for sale in the market (page 268).

commercial agriculture Farming oriented exclusively toward the production of agricultural commodities for sale in the market (page 268).

slash-and-burn agriculture A cultivation method (also called swidden or shifting cultivation) that involves cutting small plots in forests or woodlands, burning the cuttings to clear the ground and release nutrients, and planting in the ash of the cleared plot (page 268).

intercropping The farming practice of planting multiple crops together in the same clearing (page 268).

paddy rice farming A system of wet rice cultivation on small, level fields bordered by impermeable dikes that are flooded with 4–6 inches (10–15 centimeters) of water for about three-quarters of the growing season (page 269).

double-cropping The farming practice of planting and harvesting the same parcel of land twice per year (page 269).

cereal–root crop farming A diversified system of agriculture based on the cultivation of cereal grains, root crops (e.g., potatoes, yams), and the rearing of herd livestock (page 270).

cereal grains Seeds that come from a wide variety of grasses cultivated around the world including wheat, maize, rice, millet, sorghum, rye, barley, and oats (page 270).

peasants Small-scale farmers who own their fields, rely chiefly on family labor, and produce both for their own subsistence and for sale in the market (page 270).

plantation Large landholding devoted to capital-intensive, specialized production of a single tropical or subtropical crop for the global marketplace (page 270).

market gardening A small-scale farming system of one to a few acres, usually farmer-owned and managed, that produces a mixture of vegetables and fruits for mostly local and regional markets (page 271).

truck farming A scaled-up version of market gardening, with larger acreages, less diversity of crops, and oriented toward more distant markets (page 271).

grain farming An intensive, highly mechanized commercial farming system that specializes in the production of cereal grains, the most globally important of which are rice, maize (corn), and wheat (page 272).

pastoralism A system of breeding and rearing herd livestock (also called pastoral nomadism or nomadic herding) by moving them over expansive areas of open pasturelands (page 272).

transhumance A herding practice in which herders move with their livestock seasonally in search of the best forage for the animals as pasture conditions change (page 273).

livestock ranching The practice of using extensive tracts of land to rear herds of livestock to market as meat, hides, or wool (page 274).

livestock fattening An intensive system of animal feeding, utilizing fenced enclosures to finish livestock, mostly cattle and hogs, for slaughter and processing for the market (page 274).

feedlot A fenced enclosure used for intensive livestock feeding that serves to limit livestock movement and associated weight loss (page 274).

dairying A farming system that specializes in the breeding, rearing, and utilization of livestock (primarily cows) to produce milk and its various by-products, such as yogurt, butter, and cheese (page 275).

urban agriculture The practice of growing fruits and vegetables in small private plots or shared community gardens within the confines of a city (page 275).

food security When all people, at all times, have physical and economic access to sufficient safe and nutritious food to meet their dietary needs and food preferences for an active and healthy life (page 275).

aquaculture The cultivation and harvesting of aquatic organisms under controlled conditions (page 276).

mariculture A form of aquaculture oriented to the farming of saltwater species such as shrimp, oysters, marine fish, and more (page 276).

hunter-gatherers Culture groups (e.g., the Inuit of arctic Greenland, Alaska, and Canada) that gain a livelihood by hunting wild game, fishing where possible, and gathering edible and medicinal wild plants (page 277).

domestication The multigenerational process through which humans selectively breed, protect, and care for individuals taken from populations of wild plant and animal species to create genetically distinct species, known as domesticates (page 279).

domesticated plant A plant that is bred, planted, protected, cared for, and used by humans (page 279).

domesticated animal Animals that depend on people for food and shelter and that differ from wild species genetically, physically, and behaviorally as a result of human-controlled breeding and frequent human contact (page 281).

green revolution The U.S.-supported development of high-yielding varieties (HYVs) of cereal crops and accompanying agricultural technologies for transfer to economically less developed countries (page 282).

global food system The agro-industrial complex, organized at the global scale, that encompasses all elements of growing, harvesting, processing, transporting, marketing, consuming, and disposing of food for people (page 283).

synthetic fertilizer A nutrient needed for plant growth, most commonly nitrogen, phosphorus, and potassium, that is industrially manufactured using petroleum by-products (page 284).

pesticide An industrially manufactured chemical that kills or repels the animals, insects, or plants that can damage, destroy, or inhibit the growth of crops (page 284).

herbicide A common type of pesticide that is designed to kill or inhibit the growth of unwanted plants (i.e., weeds) that compete with crop plants (page 284).

concentrated animal feeding operation (CAFO) The practice of confining livestock, including cattle, sheep, turkeys, chickens, and hogs, at high densities in cages only large enough to allow the animal's body to grow and to accommodate equipment for feeding and refuse removal (page 284).

cool chain The system that uses new refrigeration and food-freezing technologies to keep farm produce fresh in climate-controlled environments at every stage of transport from field to retail grocers and restaurants (page 285).

genetic engineering A technology that artificially modifies, manipulates, or recombines an organism's DNA (page 285).

genetically modified organisms (GMO) Living organisms, including crops and livestock, that are produced through genetic engineering (page 285).

agribusiness A firm engaged in all parts of farming operations, from cultivation to storage, transport, processing, and marketing. The term can also refer to the entire enterprise of commercial, industrial, large-scale, and mechanized agriculture (page 285).

monoculture The cultivation of a single commercial crop on extensive tracts of land (page 285).

vertical integration When a single firm is engaged at multiple stages of the production process (page 285).

contract farming An arrangement between an independent farmer and an agribusiness company in which the farmer receives all the inputs from the company needed to produce a crop under conditions specified by the company in exchange for a guaranteed buyer and price for the crop (page 286).

corporate consolidation When a business grows by amalgamating many companies into one large company through mergers and acquisitions (page 287).

industry concentration A process whereby an increasing proportion of production is controlled by decreasing numbers of companies (page 287).

foodborne outbreaks Illnesses that occur when two or more people consume the same contaminated food or drink (page 289).

food sovereignty The principle that all people have a right to sufficient, healthy, and culturally appropriate food and that control over food production should remain in local farmers' hands rather than those of multinational corporations (page 290).

heirloom seeds Locally bred seeds of crop varieties (also called heritage seeds) that for marketing reasons are not suitable to the global food system and are open pollinated rather than produced through hybridization (page 290).

slow food A global consumer movement and an international organization promoting local food production and traditional cuisines and foodways (page 290).

locavores People who dedicate themselves to slow-food diets and to obtaining as much of their nutrition as possible from local farmers (page 290).

community-supported agriculture (CSA) A direct-to-consumer marketing arrangement whereby the farmers are guaranteed buyers for their produce at guaranteed prices and consumers receive fresh food directly from the producers (page 291).

farmers' market A setting (ranging from a few stalls in the street to covered enclosures extending a few city blocks) for farmers to sell their produce directly to consumers (page 291).

fair trade A third-party certification program that supports good crop prices for Third World farmers and environmentally sound farming practices (page 291).

land reclamation The term used in reference to efforts to drain land inundated with either fresh or salt water (page 291).

irrigated agriculture Farming that relies on the controlled application of water to cultivated fields (page 291).

desertification The process whereby once-fertile land is converted to desert either through climate variation, human activities, or a combination of the two (page 295).

sustainable agriculture The production of food and fiber in a manner that does not diminish the future capacity of the land, harm the environment, or damage human health (page 296).

organic agriculture The production of crops and livestock using ecological processes, natural biodiversity, and renewable resources rather than relying on industrial practices and synthetic inputs (page 296).

conventional agriculture Used in contrast to organic agriculture, this is another term for farming that depends on manufactured synthetic inputs, GMO seeds, and other such industrial practices (page 296).

biofuel A fuel derived from organic wastes or plant material (page 297).

Doing Geography

ACTIVE LEARNING:
Analyzing the Geography of Foodborne Outbreaks

In this chapter we learned that the global food system poses unique challenges. One of the most daunting is foodborne outbreaks. The costs of contaminations are high in terms of public health, but also in terms of the economic costs of tracking them and recalling the products. For example, in 2018 one ice cream company recalled an entire year's worth of product distributed to 35 major grocery chains because *Listeria*—a bacterium that causes severe illness and sometimes death—was found at its factory. Such a contamination could bankrupt a company, one reason that tracing foodborne outbreaks is critical for many segments of society.

For this exercise, you will analyze the complex interlinkages of geography, agricultural production, food consumption, and pathologies by using the U.S. Centers for Disease Control's Foodborne Outbreak Online Database (FOOD) tool.

More guidance at 🔵 **Sapling** Plus

EXPERIENTIAL LEARNING:
The Global Geography of Food

Until recently, people throughout history obtained the food they needed either by growing it themselves or procuring it directly from nearby farmers. The choice and availability of food was limited and changed seasonally. Around the turn of the twentieth century, this situation began to change dramatically as the pace of urbanization and industrialization accelerated. Today, very few people in developed countries grow their own food or even know where the food they eat was produced or who produced it. Your task for this exercise is to find out.

This exercise can be organized as a group or individual project. As a group project, students can be assigned to research particular categories of food. As an individual project, you should begin with a typical day's meals and identify all their ingredients (don't forget the seasonings and cooking oils used in preparing them).

Steps to Tracing the Global Geography of Food

Step 1: The project starts at the food markets where you usually shop. For much of the information you will need, you can refer to the labels on the food items. For some items, such as fish, poultry, and meat, you may need to speak to the butcher or store manager. Find out, as specifically as possible, where the food item was produced. Find out the name of the company that marketed the product and, if available, the name of the parent company.

Step 2: After collecting this basic information, you will need to do further Internet research. Organize a list of companies and the food products they market; then locate the geographic origins of each food product.

Step 3: Now look for patterns.

- Which and how many companies are involved and what proportion of the food supply does each control?

- What proportion and which kinds of food are produced in other countries? Do certain kinds of foods tend to be produced closer to the market than others?

- Do certain kinds of food tend to be marketed by large corporations more than other kinds of food?

- Can you think of explanations for the patterns you identify?

A bountiful produce display, common in U.S. supermarkets. Where does the bounty come from, who grows it, and how is it made available for our tables no matter what the season? (Geri Lavrov/Getty Images)

SEEING GEOGRAPHY

Reading Agricultural Landscapes

What differences can you "read" in these landscapes? Can you determine their locations?

Take a careful look at the two photos shown here and systematically identify the differences between them, beginning with the one on the top. The most striking aspect of this aerial landscape shot is the abrupt division between the cultivated land at the top and the uncultivated land at the bottom. Looking closely, we see that an irrigation channel forms the boundary between the two. A second prominent feature of the landscape is the checkerboard pattern of the fields and the straight roads forming their boundaries. Other details emerge as you look more closely. For example, the settlement pattern consists of isolated, sparsely arranged farmsteads separated by large expanses of cultivated fields. You might also note that the uncultivated land is brown and treeless and that trees in the cultivated portion are found only along the watercourses.

The landscape features in the photo on the bottom are nearly the opposite of those in the image on the top. Settlement is clustered in a densely populated village centered on a church and town square. The fields are of irregular sizes and shapes and form a band of cultivated land around the concentrations of houses, some of which are built of stone. There is no evidence of irrigation. Trees and shrubs are concentrated in the outermost band but also occur throughout the landscape, which overall appears verdant.

Putting all these visual clues together leads us to conclude that the landscape on the top must be somewhere in the western United States. We know this region was surveyed and settled under the township and range system, which explains the isolated farmsteads and checkerboard pattern. We also know that much of the western United States is arid or semiarid, which explains the need for irrigation and the general lack of trees and green vegetation in the bottom half of the photo. In fact, this is a photo of Mack, Colorado, where irrigation meets the desert. The landscape on the bottom is located in Europe. The large church in the center and the dense, clustered housing suggest the settlement pattern of a historical, western European market town. The irregular fields and their close proximity to the town are explained by deep historical patterns of land ownership and the reliance on foot travel in preindustrial agriculture. The verdant landscape and absence of irrigation suggest the temperate climate characteristic of western Europe. In fact, you are looking at the vineyard region of Saône-et-Loire, France.

Two types of contemporary agricultural landscapes. (Top: Airphoto/Jim Wark; Bottom: CW Images/Alamy Stock Photo)

Clockwise from top left: China's Mongolia mining region; Foxconn workers assembling smartphones; using the latest iPhone; children sifting through electronic waste in Manila, Philippines.

chapter 9

Development Geography

TRANSFORMING LANDSCAPES OF WELL-BEING

Have you ever wondered where your smartphone started its life, and where it will go when it dies?

Think about this question as you read. We will revisit these photos in Seeing Geography on page 349.

Learning Objectives

9.1 Discuss the history of development, the diverse ways to measure it, and how development has shaped, and reshaped, regions.

9.2 Analyze the relationships between transportation and development.

9.3 Explain how formerly poor places are today significant agents of development.

9.4 Recognize the role that nature has played over time as concepts and practices of development have changed.

9.5 Describe how development has shaped landscapes.

VVK/EPA/REX/Shutterstock

You may have had the experience of traveling to places where the everyday conditions of life seem so much more difficult than the conditions you are accustomed to at home. If you haven't traveled to such places, you certainly have seen enough television, movies, and Internet content to understand that the everyday lives of people in different parts of the world, including their access to housing, food, and health care, may differ greatly from yours. You may also have noticed differences in the quality of life within your own neighborhood, town, or city.

Being "developed" as opposed to "underdeveloped" is typically understood as primarily an economic issue. Wealthy places are considered to be developed, whereas poor places are labeled as underdeveloped. Indeed, as we will soon explore, the key terms relating to development geography have traditionally been economically based. Yet development is a much more comprehensive notion, encompassing such things as happiness, well-being, and autonomy. The question of how to achieve, measure, and sustain this broader understanding of development is every bit as important today as it was several centuries ago.

In this chapter, we will explore the history of development as a concept and a practice. Different understandings of what development means have led to different strategies for accomplishing it. Regardless of the approach taken, there are some basic divisions of economic activities that guide our thinking about economies as they grow and change. Transportation technologies have also changed over the centuries, and are central drivers of how people, ideas, and goods move from place to place within economic systems. Transportation is but one example of how some places and people have access to technologies that place them in powerful positions, while others do not. However, the rise of the so-called Global South is reshaping power dynamics around who and where innovation and economic growth arise. The extraction and transformation of the Earth's resources has been central to past economic development, while the wise use of resources holds the key to the future health and well-being of human populations. Finally, development efforts past and present have dramatically shaped the world around us, resulting in the ravaged landscapes of mineral extraction to the factories, shopping malls, and Internet commerce landscapes that shape today's development panorama.

Region

9.1   Understand the history of thinking about development, the diverse ways to measure it, and how development has shaped, and reshaped, regions.

We've already touched on many development-related topics throughout this book, which tells you that development is a concept that reaches into many areas of human geography.

Development includes but exceeds mere wealth as defined in economic terms. It is important to understand exactly what is meant by development and how this idea has evolved, because the term is a powerful one that has literally shaped and reshaped the world in which we live.

The idea of development has a remarkably influential but relatively short history. Various theories of development, dating from the period immediately after World War II, arose in an attempt to explain why it is that some regions of the world enjoy relative prosperity, while others do not. These theories prioritized some factors over others as key to explaining these differences, and as a result, some understandings of development focus on measuring economic growth, while others place relatively more emphasis on quality of life. Yet despite these different approaches, it is true that as economic systems grow and evolve, distinctions among the sorts of activities engaged in—ranging from agriculture and mineral extraction to manufacturing, services, and information flows—provide a useful way to compare and contrast one place with another. Finally, there are many different people and entities involved in development-related activities, each having their own particular influence on the process.

A Brief History of Development

The term *development* emerged in the post–World War II era when scholars as well as government and policy officials, particularly those in the United States, became concerned that the standard of living in many regions of the world was quite low. By standard of living, they were referring to such things as literacy rates, infant mortality, life expectancy, and poverty levels. They realized that, in general, a low standard of living was most prevalent in countries whose economies were based primarily on subsistence agriculture (see Chapter 8). These officials and scholars believed that introducing new technologies and skills would enable these countries to develop more productive forms of agriculture, and that *industrialization*—the transformation of raw materials into commodities—would follow. These new economic initiatives would then lead to a higher standard of living. Transforming the economic structure of a region in order to raise the standard of living is what at that time was known as development. The term *development* referred to a process of both economic and social transformation, and provided a way of categorizing different regions of the world.

Stages of Economic Growth   One of the earliest and most influential models of the process of economic development was suggested by the economist W. W. Rostow in 1962. Rostow posited that economic development was a process that all regions of the world would go through and experience in similar ways. His model assumed that regions would progress in a linear fashion through defined **stages of economic growth,** which he compared to an airplane pulling out of the gate, taxiing along the runway, taking off, and ultimately flying high. The first stage, which he called a "traditional" economy, is one that is based on agriculture and has limited access to or knowledge of advanced technologies—in other words, the economy is undeveloped. This first stage would be followed by a series of other stages characterized by increasing technological sophistication and the introduction of industrialization. Rostow's final stage, called the "age of high mass consumption," is one that he saw as characteristic of economies in the United States and much of western Europe. In this final stage of development, industrialization has gained such momentum that goods and services are widely available and most people can accumulate wealth to such a degree that they no longer have to worry about meeting basic needs. According to Rostow, the reason some countries were developed while others were not was that they were simply at different stages along the same flight path. Eventually, however, all nations would reach the "cruising altitude" of development as defined by high levels of mass consumption.

Rostow's stages of economic growth model has been criticized by scholars for a variety of reasons. While the simplicity of the model is appealing, its assumptions about the world often don't match up with reality. For example, the model suggests that countries and regions proceed along the path to development in isolation from one another. But, as we will see in this chapter, different countries and their economies are interlinked in complex ways; for example, if one country's economy is based primarily on producing goods and services, it needs other regions and economies to supply its food and raw materials. In addition, the model assumes that all economies will develop without obstacles from other countries, and we know that this is rarely true, since, for example, some countries may deliberately try to block economic development in other countries that they think will become their competitors.

Indeed, some critics of Rostow's approach claimed that developed countries—those enjoying high levels of mass consumption—depended on the exploitation and impoverishment of other regions in order to achieve their success. The cheap natural and human resources of less developed regions were exploited in order to permit more developed regions to enrich themselves. In other words, underdevelopment of some places is an active process that goes hand in hand with the development of others.

Finally, the end goal of Rostow's development model—consuming lots of material goods—may not be considered as desirable as other goals, such as political freedom or decent working conditions.

World Systems Theory An alternative to Rostow's model is world systems theory, developed by the economic historian Immanuel Wallerstein. **World systems theory** categorizes world

history as moving through a series of socioeconomic systems, culminating in the modern world system, which is our current interdependent, capitalist world economy rooted in nation-state economies. World systems theory divides the world into three regions: core, periphery, and semi-periphery. **Core** world regions have the most advanced industrial and military technologies, complex manufacturing systems, and external political power and the highest levels of wealth and mass consumption. Historically, the first core region emerged in England and northern Europe with the Industrial Revolution (discussed later in the chapter). Through the course of the nineteenth century, other areas became core regions, notably the European settler colonies in Australia and North America. Japan, whose government oversaw a massive industrialization and modernization effort in the 1900s, also became a core region. The **periphery** is comprised of places that exhibit characteristics opposite those of the core: they have relatively little industrial development, simple production systems focused mostly on agriculture and raw materials, and low levels of consumption. Together these concepts form the **core–periphery model,** a spatial model of the interrelations between core regions where economic, political, and cultural powers are centered, and peripheral regions that are dependent upon and serve the needs of the core. Finally, Wallerstein posited the existence of semi-peripheral regions. The **semi-periphery** exhibits a mix of characteristics of both the core and periphery and plays a key role in mediating politically and economically between them. Countries of the semi-periphery may or may not be in a transitional stage toward either the core or the periphery. Examples include Mexico, Brazil, and South Korea.

Unlike the stages of economic growth model, world systems theory suggests that countries do not inevitably march through a set of development stages. Rather, countries are part of a global system of interdependencies that defines their roles in it. Their movement into or out of the core or periphery is dependent largely on long-term historical shifts in economic and political power, rather than on their position on a scale of preordained stages. World systems theory is dynamic, recognizing that the status of a place can change over time.

However, world systems theory has been criticized, primarily due to its global scale of analysis. As a sweeping, centuries-spanning explanation of the world economy, it is a necessarily broad brush, blotting out some of the complexities of development that do not neatly fit the model. For example, as you will learn later in this chapter, there are large geographic areas within core regions that have socioeconomic profiles more similar to the periphery than to the core. Development geography is all about identifying and analyzing such complex spatial realities.

Many Worlds There have been other ways of using notions of development to categorize world regions. For instance, during the mid-twentieth century, the world was seen as being divided

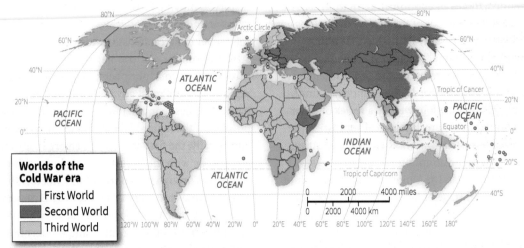

FIGURE 9.1 The three worlds of the Cold War era. During the middle decades of the twentieth century, when the world order was shaped by the geopolitical tensions among the liberal democracies and communist nations, the world was divided into those nations aligned with these two positions and those nonaligned nations known as the Third World. As the Cold War progressed, the terms First and Third World became synonymous with wealthy and poor nations, respectively.

into the First, Second, and Third Worlds (**Figure 9.1**). The **First World** represented the world's wealthy, democratic, capitalist nations, while the **Third World** was made up of poor countries. Because the so-called Cold War was the principal frame for power from the end of World War II in 1945 to the fall of the Soviet Union in 1991, these labels also had political meanings (see Chapter 6). The **Second World** consisted of those countries that were ruled by communist or socialist governments and had central or command economies, rather than capitalist ones. The Third World nations were not aligned with either the First or Second World.

After 1991 the terms *First, Second,* and *Third World* had primarily economic connotations, along with the **Fourth World,** those regions considered to be so poor that they existed beyond the reach of contemporary society.

In the 1970s, West German Chancellor Willy Brandt proposed what came to be known as the Brandt Line as a way of understanding the world. The **Brandt Line** divided the countries of the world into two regions: wealthy and poor (**Figure 9.2**). Notice that wealthy countries tend to be in the Northern Hemisphere, while poorer ones are in the Southern Hemisphere.

By the 1980s, however, many places in the so-called Third World had made significant economic advances, and the Cold War had lost some of its traction as an organizing framework with which to understand power in the world. Thus, the division of the globe into different regions, by the political and economic criteria represented by the four worlds, or the "have" versus "have-not" distinction demarcated by the Brandt Line, fell out of fashion. Today, you can find divisions of the globe that focus

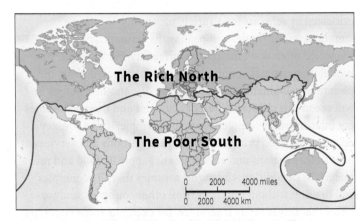

FIGURE 9.2 The Brandt Line. This line divides the wealthy nations of the North from the poor nations of the South. It provided another way of conceptualizing world regions from a development perspective.

on its more contemporary developmental aspects—for instance, highlighting the world's fast-emerging economies, or those with access to resources such as technology, which distinguish the current haves from have-nots (see Figure 2.21).

A Contemporary Approach to Development In this chapter, we use the terms *developing* and *developed* to refer to different regions of the world, but we don't assume, as the Rostow model does, that there is a normal and linear progression from one stage of development to another. We use the term **developing** to refer to regions whose economies include not only a good deal of subsistence activities but also some manufacturing and service activities. In these regions, people are often unable to accumulate wealth, since they produce just enough, or at times *not* enough, food and other resources for their own immediate needs. These places therefore are considered relatively poor. We use the term **developed** to refer to those regions of the world whose economies are based more on manufacturing, services, and information. These regions are considered wealthier because people living there are better able to accumulate resources. The landscape of development is a complex one, as individual cities, regions, and countries can contain a mixture of developed as well as underdeveloped places.

Measuring Development

As we noted in the introduction to this chapter, there exist a variety of ways to measure development. Development has traditionally been understood as primarily about wealth accumulation; therefore, the classic measures of development have involved assessing indicators of economic strength. More recent understandings of development have emphasized the fulfillment of basic needs, quality of life, and important but difficult-to-quantify states such as happiness.

Economic Measures of Development The most common measure of development is **Gross Domestic Product,** or **GDP.** Gross Domestic Product totals the value of all goods and services produced in a country over a specific period, usually a year. The map in **Figure 9.3** shows the vast disparities in GDP from country to country.

What do those differences really mean? In fact, not much can be inferred from GDP numbers except that some countries have economies that are far larger—and thus produce more total goods and services—than the economies of other countries. For instance, Brazil's economy, as measured by the total value of goods and services produced annually, is the ninth largest in the world. It is approximately 10 times the size of New Zealand's economy. Is the average Brazilian 10 times richer than the average New Zealander? Ten times more productive? Not so fast. Comparing Brazil's GDP to New Zealand's is hardly fair. New Zealand's population is only about 2 percent of Brazil's. So it is understandable that far more populous Brazil can produce more total goods and services in a year than New Zealand.

How then can we compare apples to apples—in this case, the share of GDP that corresponds to each Brazilian or New Zealand resident? To do that, we need to divide GDP by the national population, which gives the **GDP per capita.** GDP per capita provides a crude way to compare how wealthy or poor the "average" Brazilian or New Zealander may be. In our example, New Zealand's GDP per capita—which in 2016 was measured at $39,899—was much larger than Brazil's at $8,649. **Figure 9.4** shows GDP per capita figures globally. Now you can begin to assess how different nations stack up against one another in terms of their per capita economic production, bearing in mind that GDP per capita figures cannot demonstrate how incomes are actually distributed across national populations. Brazil, for instance, is notoriously unequal in its income distribution. (For a measure of income inequality, see **Figure 9.5.**)

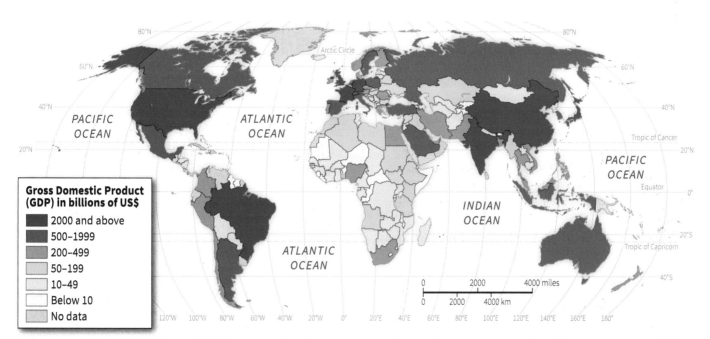

FIGURE 9.3 Gross Domestic Product. GDP measures the total value of all of the goods and services produced in a country annually. This map illustrates the regional patterns of economic output.

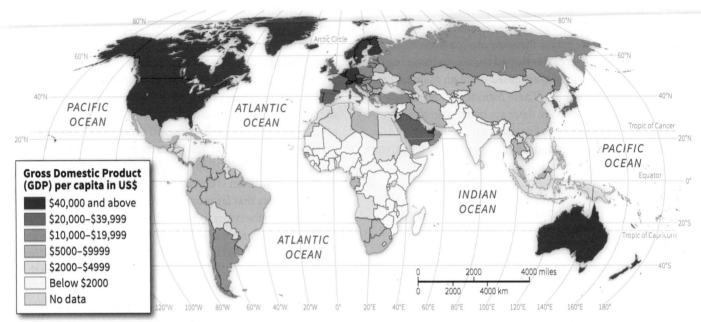

FIGURE 9.4 Gross Domestic Product per capita. GDP per capita is calculated by dividing a nation's GDP by its total population. On this map you can begin to discern areas of poverty and wealth. This statistic cannot, however, capture how equally (or unequally) income is actually distributed among people on the ground.

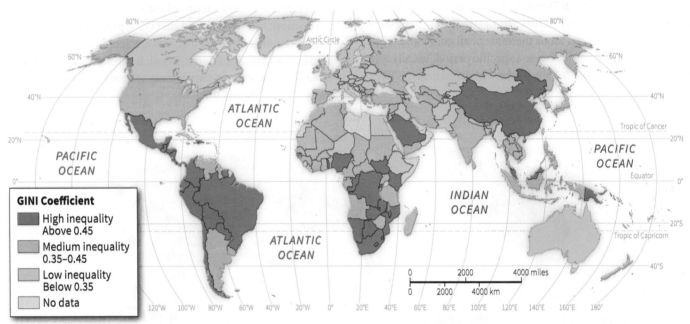

FIGURE 9.5 Income inequality. The Gini coefficient (named after the Italian sociologist who developed it) provides a numerical measure of income inequality. Zero represents perfect income equality; thus, those nations that score lowest are the most equal with respect to income distribution, while the high scorers are more unequal. Not all nations regularly gather the data necessary to calculate this statistic.

Development as Quality of Life It is certainly true that without sufficient economic resources, societies are unable to provide for their citizens. But—to put it crudely—is money all there is to development? Or, to turn this question around, does having monetary resources necessarily mean that societies will choose to spend those resources to improve the well-being of their populations?

Development geographers and others who study such questions largely agree that development is indeed about more than just money. Development is, at its heart, a much more encompassing idea: it is about enhancing individual and societal quality of life. To paraphrase the great Indian leader Mahatma Gandhi, development is about each human being's ability to reach his or her full positive potential. In a material sense, we can understand development as the fulfillment of basic needs for food, clean water, adequate shelter, and clothing. Some understandings of basic needs add education, health care, and sanitation to this list. **Figure 9.6** shows a world map of the **Human Development Index,** or **HDI,** a traditional measure of how well people's basic needs are being met, while **Figure 9.7** takes a closer look at well-being in the United States.

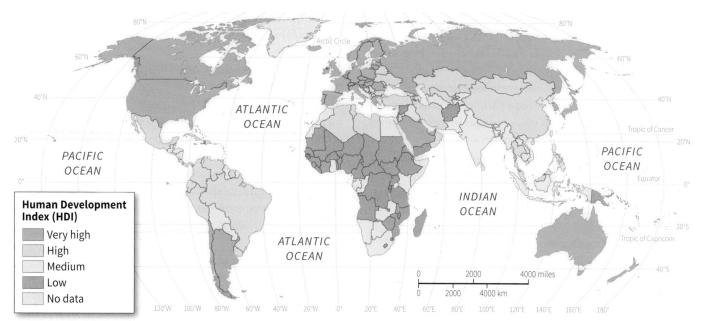

FIGURE 9.6 Human Development Index. The HDI was developed by the United Nations to categorize nations into tiers according to the well-being of their citizens. The HDI is a composite index derived from three separate statistics for life expectancy, education, and income.

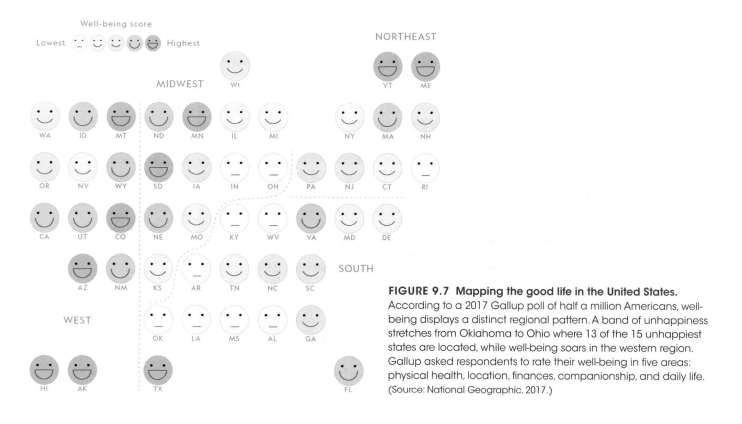

FIGURE 9.7 Mapping the good life in the United States. According to a 2017 Gallup poll of half a million Americans, well-being displays a distinct regional pattern. A band of unhappiness stretches from Oklahoma to Ohio where 13 of the 15 unhappiest states are located, while well-being soars in the western region. Gallup asked respondents to rate their well-being in five areas: physical health, location, finances, companionship, and daily life. (Source: National Geographic, 2017.)

Although access to a certain baseline amount of economic resources is necessary in order to fulfill basic needs, merely having the financial means to provide them is no guarantee that basic needs will, in fact, be met for all members of society. The principal allocators of such resources—in most nations, that means governments, corporations, and nongovernmental organizations (NGOs) such as churches and foundations—may choose to direct economic resources to other areas. For instance, purchasing military equipment and funding armies may be a higher priority than making sure basic human needs are met. Or some form of prejudice—for instance, discrimination by gender, religion, age, race, sexuality, or ethnicity—may exist, such that only some members of society have their basic needs met while others systematically do not. For their part, individuals may simply not have sufficient economic resources to meet their own basic needs or those of their dependent family members. And, as with larger entities such as governments or corporations, individuals may choose to direct their economic resources in ways that do not fulfill their basic needs—for instance, forgoing medical care in order to gamble.

The Sustainable Development Goals In 2000, the United Nations established eight Millennium Development Goals, specifying targets for all nations to achieve by 2015 in areas such as promoting gender equality and empowering women, eradicating extreme poverty and hunger, and achieving universal primary education. Some countries reached all eight goals, while others achieved none. In 2015, the United Nations adopted the Sustainable Development Goals, a collection of 17 goals intended to end poverty, protect the planet, and achieve prosperity by 2030 (**Figure 9.8**). Like the Millennium Development Goals, these are ambitious and not all countries will be able to achieve the stated targets. However, the Sustainable Development Goals do reflect a transition, on the part of international development agencies, away from the purely economic understanding of development, and toward a more holistic approach encompassing human and environmental well-being.

Development, Freedom, and Happiness Thinking beyond the material, social, and infrastructural dimensions of development for a moment, what other conditions might be important for a place to be considered truly developed? Development economist Amartya Sen has proposed that freedom lies at the heart of development. Sen intends for freedom to be interpreted as encompassing political and civic freedoms, as well as freedom from the vulnerabilities (such as illiteracy or early death) associated with poverty. Geographers might add mobility to this notion—in other words, the freedom to move about without restrictions.

In the 1970s, the king of Bhutan coined the term *Gross National Happiness index* to counter what he viewed to be an overly economy-centered approach to development taken by Western nations. Instead, he proposed that happiness was also an important part of development, and that, in fact, it

FIGURE 9.8 U.N. Sustainable Development Goals. These 17 goals are intended to further a sustainable development agenda that will, in the view of the United Nations, "transform our world" by the year 2030. (Source: United Nations.)

FIGURE 9.9 Bhutan: The happiest place on Earth? From this photo you might guess this to be the case. However, Bhutan has been criticized for some unhappy conditions, including the oppression of ethnic minority populations, pervasive poverty, and unemployment. Some claim Bhutan's Gross National Happiness index is merely a clever marketing ploy intended to lure tourists to visit the tiny mountain nation. (Keren Su/China Span/Alamy)

could be measured. Not surprisingly, the nation of Bhutan, while poor, ranked very high on the Gross National Happiness index. Cultural and ecological diversity, good governance, and psychological well-being are some of the elements included in the Gross National Happiness index. These values resonate

within Bhutan's Buddhist culture (see Chapter 7), while not being exclusive to Buddhism. It is important to note, however, that Bhutan's government has been sharply criticized for its oppression of ethnic minorities, which brings into question its "happiest place" designation (**Figure 9.9**).

The idea of happiness as an indicator of development has gained some traction. In 2012, the United Nations commissioned a "Global Happiness Report," issued annually on March 20th—also known as the International Day of Happiness. In 2017, the report ranked 155 countries by their happiness level (**Figure 9.10**). The world's happiest countries enjoy high levels of trust, shared purpose, generosity, and good governance.

As you can see, development geography has a firm foundation in economic well-being, but it is hardly limited to monetary measures.

Categorizing Types of Economic Activity by Sector

In order to fully understand the differences and similarities between regions, it is helpful to distinguish the types of economic activities that characterize them. In general, scholars divide types of economic activities into four broad categories, or sectors. **Primary sector** activities involve extracting natural resources from the Earth. Fishing, farming, hunting, lumbering, oil extraction, and mining are examples of primary industries. **Secondary sector** activities process the raw materials

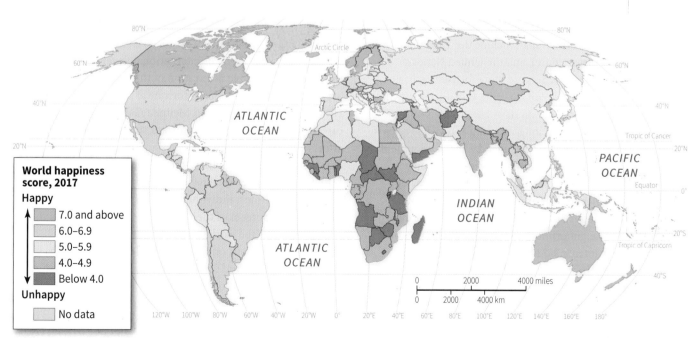

FIGURE 9.10 World happiness, 2017. Roughly 3000 respondents in each of 155 countries were asked to rate their lives, with "0" being the worst possible life and "10" being the best possible life. The results display distinct regional patterns. (Source: Helliwell, J., Layard, R., & Sachs, J. (2017). *World Happiness Report 2017*, New York: Sustainable Development Solutions Network.)

extracted by primary industries, transforming them into more usable forms. Ore is converted into steel, logs are milled into lumber, and fish are processed and canned. *Manufacturing* is a more common way of referring to secondary industries. Manufacturing activities are found throughout the world, but, as we will discuss later in this chapter, they tend to cluster in particular areas because of favorable circumstances such as cheap power sources or available labor.

In parts of the world where people import the bulk of their manufactured products, economic activities are dominated by services. Services make up the **tertiary sector,** which refers to all the different types of work necessary to move goods and resources around and deliver them to people. So wide is the range of services that some geographers find it useful to distinguish three different types: transportation/communication services, producer services, and consumer services. Finally, the fastest-emerging set of economic activity is occurring in the **quaternary sector.** The quaternary sector includes intellectual and informational activities.

Development Actors

Now that you have some idea of the various definitions and approaches to development, it is logical to ask who "does" development and what kinds of activities count as development? Traditionally, there have been several major actors in the field of development, including international organizations, nations, corporations, and others.

International bodies specifically charged with some aspect of economic development, many of them emerging from the aftermath of World War II, include the World Bank, the International Monetary Fund, and the United Nations Development Programme. Others are more focused on basic needs fulfillment, such as the World Health Organization, which targets disease eradication, or Oxfam, which provides hunger relief. Such bodies have played long-standing and visible roles in international development by setting agendas for development, providing expertise, and coordinating policies and people. Some churches and religiously affiliated NGOs also provide development assistance in cash or in kind. Sometimes their assistance is offered internationally, but it can also be locally targeted at specific neighborhoods or other local places.

National governments, both individually and working together, are also significant development players. Groups of countries, particularly wealthy ones such as member nations of the Organisation for Economic Co-Operation and Development (OECD) or the Group of 7 (G-7), have been very influential in shaping the course of economic development through promoting international trade. Individual nations also have played decisive roles thanks to their economic might as well as

their political influence. Though the United States gives more total foreign aid than any other country in the world—around $31 billion per year—other countries dispense much more when measured as a share of the national economy. In 2015, the United States gave only 0.17 percent of its national income in direct foreign aid—one of the lowest percentages among the wealthy nations—while top donor Sweden gave 1.4 percent. The United States has been criticized for channeling much of its international development assistance toward promoting its foreign policy agenda and supporting military and political allies around the world (compare **Figures 9.11** and **9.12**).

Corporations can be significant development agents as well. The World Economic Forum recently observed that Walmart ranks as the 10th largest economic entity in the world, just behind Canada in 9th place. In other words, if Walmart were a country, it would be economically bigger than Australia, British Petroleum would be larger than Switzerland, Toyota would outrank India, and Apple would come in ahead of Belgium. Of course, corporations are not countries, and in some important ways the comparison is an unfair one. Yet the analogy does a good job of highlighting just how economically powerful corporations can be. Some corporations are noted for their development-related investments, which are framed as an aspect of corporate social responsibility. Walmart, for instance, donated $40 million to hurricane relief efforts in Texas, Florida, and Puerto Rico in 2017. While corporate leaders argue that national governments and NGOs have until now done little to alleviate poverty and provide for basic needs, their critics accuse corporations of being interested in donating solely in order to grow their market share and brand loyalty in impoverished places.

Individuals associated with corporations can also become quite active in the arena of international development. Perhaps the most recognizable illustration of this is Bill Gates, who amassed a multibillion dollar fortune as cofounder of Microsoft. Now the head of the Bill and Melinda Gates Foundation, he directs his energy and wealth toward supporting development efforts aimed at alleviating poverty and disease and promoting education globally. Together with billionaire investor Warren Buffett, in 2010 they established The Giving Pledge, whereby America's wealthiest individuals and couples pledge to give more than 50 percent of their wealth toward philanthropic efforts "to make the world better" (**Figure 9.13** on page 322). Here, criticisms focus on the individuals' motivations for giving; namely tax breaks, and increasingly the opportunity to shape public policy through political donations.

Despite the criticisms, all of these development actors are visible, recognized, and legitimate participants in the development enterprise. But it is important to know that there are development actors and activities that are not sanctioned or recognized. For instance, Mexico's drug lords consider

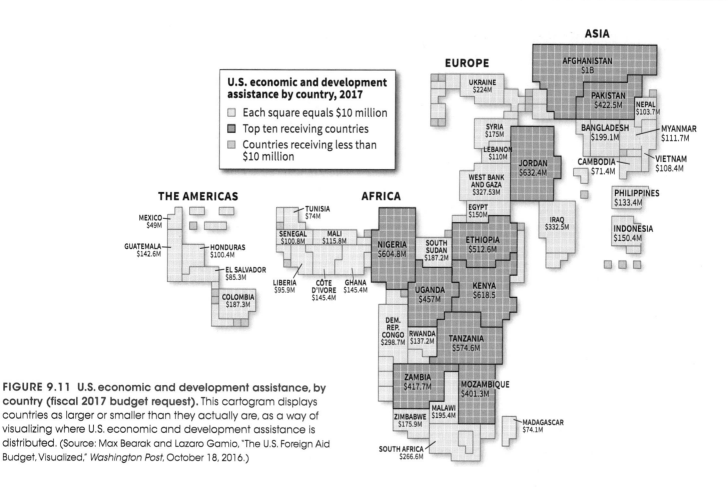

FIGURE 9.11 U.S. economic and development assistance, by country (fiscal 2017 budget request). This cartogram displays countries as larger or smaller than they actually are, as a way of visualizing where U.S. economic and development assistance is distributed. (Source: Max Bearak and Lazaro Gamio, "The U.S. Foreign Aid Budget, Visualized," *Washington Post*, October 18, 2016.)

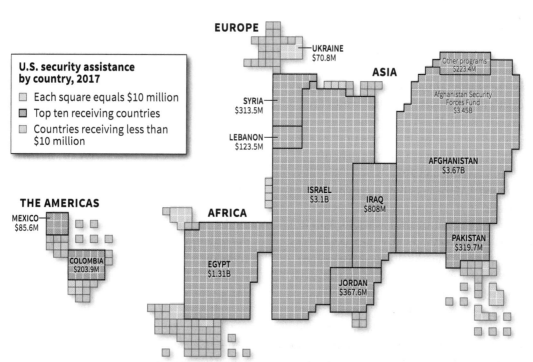

FIGURE 9.12 U.S. security assistance, by country (fiscal 2017 request). This cartogram displays countries as larger or smaller than they actually are, as a way of visualizing where U.S. security assistance is distributed. (Source: Max Bearak and Lazaro Gamio, "The U.S. Foreign Aid Budget, Visualized," *Washington Post*, October 18, 2016.)

FIGURE 9.13 The Giving Pledge. Bill and Melinda Gates visited the village of Jamsaut, near Patna, India, as part of The Giving Pledge campaign. (Aftab Alam Siddiqui/AP Photo)

themselves to be the primary source of development aid to regions of the country that have been neglected by the Mexican government for many years. In some states, such as Sinaloa, drug cartels have long based their activities both as producers of locally grown marijuana and heroin, and as traffickers for these drugs and for cocaine from South America. Drug lords are often the only investors concerned with building local schools, parks, and infrastructure in rural areas. In 2012 one Sinaloa cartel kingpin, Joaquín "El Chapo" Guzmán, was ranked number 63 on *Forbes*'s list of the world's most powerful people; he was later removed from the list because his assets were deemed too difficult to verify.

Indeed, the story that such cartels tell the local people who grow and traffic narcotics on their behalf is that they are merely taking from the wealthy, drug-addicted gringos and giving back to Mexico's forgotten poor. In other words, drug lords paint themselves as Robin Hood figures enacting social justice on behalf of the downtrodden. It is a powerful narrative, told in quasi-religious terms through the patron saint of narcotraffickers, Jesús Malverde, and in the lyrics of so-called narcocorridos—traditional Mexican ballads (*corridos*) singing the praises of drug lords.

It is important to emphasize that acknowledging how drug cartels and kingpins have channeled some of their wealth into local development activities in the places where they operate is not the same as condoning these activities. Indeed, Mexico's drug cartels are engaged in some of the most violent criminal behavior ever seen, seriously threatening the rule of law. Yet it is worth asking whether the sorts of criticisms leveled at the drug lords are in some ways similar to those directed at legitimate development actors. More broadly, is development just about helping people, or is it also about controlling them? Finally, and importantly for you as students of geography, what is the future role of powerful and wealthy countries as the world's

primary development actors? The emergence of development alliances among countries that were formerly the recipients of development aid, discussed under Globalization on page 331, provides yet another example of how the contemporary regional geographies of development are changing.

Mobility

9.2 Analyze the relationships between transportation and development.

The movement of people, goods, and ideas from place to place has rarely been symmetrical. Instead, some places are net exporters of people, goods, and ideas, while others import them. This has led to the enrichment of some places at the expense or impoverishment of others. In this section, we consider aspects of how the movement of people, goods, and ideas has both driven, and been driven by, development. We start by considering the importance of transportation and transportation-related infrastructure.

Transportation and Industrialization

Development, industrialization, and urbanization are profoundly connected. Development as we know it—both of the narrow economic sort and as the broader attainment of basic needs and loftier human aspirations—could not have occurred without industrialization (i.e., the shift from primary to secondary sector activities) or the existence of cities (see Chapter 10). And it is the movement of goods, people, and ideas from one place to another—transportation, in other words— that provides perhaps the most important element linking development, industrialization, and urbanization.

The **Industrial Revolution** saw a radical transformation in economic activities and processes. The Industrial Revolution began in England in the early 1700s. Machines, new power sources, and novel chemical processes replaced human hands in the making of products, initially in the cotton textile industry. In addition, many aspects of how people lived their lives in places were altered, including concepts of time, money, work, home, and gender roles.

The Industrial Revolution both encouraged and was furthered by the development of new forms of transportation. Traditional wooden sailing ships gave way to steel vessels driven by steam engines, and later railroads became more prevalent. The need to move raw materials and finished products from one place to another cheaply and quickly was the main stimulus that led to these transportation breakthroughs. Without them, the impact of the Industrial Revolution would have been minimized.

Once in place, the railroads and other innovative modes of transport associated with the Industrial Revolution fostered additional cultural diffusion. Ideas and, in fact, the Industrial Revolution itself spread more rapidly and easily because of this efficient transportation network.

Great Britain maintained a virtual monopoly on the Industrial Revolution well into the 1800s. Indeed, the British government actively tried to prevent other countries from acquiring the various inventions and innovations that

distinguished the Industrial Revolution. After all, they gave Britain an enormous economic advantage and contributed greatly to the growth and strength of the British Empire. Nevertheless, this technology finally spread beyond the bounds of the British Isles, with continental Europe feeling the impact first. In the last half of the nineteenth century, the Industrial Revolution took firm hold in the coalfields of Belgium, Germany, and other nations of northwestern and central Europe. The growth of railroads in Europe provides a good index of the spread of the Industrial Revolution there. The United States began rapid adoption of this new technology around 1850,

GEOGRAPHY @ WORK

Matthew Toro

Director of Map, Imagery, and Geospatial Services, Arizona State University

Education: MA Geography, University of Miami

BA Geography and International Relations, Florida International University

Courtesy of Matthew Toro.

Q. *Why did you major in geography and decide to pursue a career in the geography field?*

A. I wanted to understand the way the world works, and no other field seemed to offer the sort of comprehensive, multidimensional, multiscale answers afforded by the space-based perspective of geography. Other disciplines I explored seemed to look at one or a few isolated components of social or "natural" reality. Geography was able to interweave seemingly unrelated human and geophysical systems, helping me see more completely the way the world system operates. I'll never succeed at comprehending the world in its infinite complexity, but my geographic perspective gives me an insight that I'm confident no other field can match.

Q. *Please describe your job.*

A. I run a library-based mapping center. It's a place for learning about GIS and cartographic resources (data, maps, satellite and aerial imagery, etc.), as well as various geospatial software tools that can bring those resources to life in powerful ways. It's a place where people can explore how to integrate spatial analysis, map interpretation, and 21st century cartography into their work. We collect and process hundreds of geospatial datasets, digitize and georeference thousands of aerial photos and maps, create engaging

educational programming, and conduct original research. Outside academia, I consult on various data analysis and geo-visualization projects.

Q. *How does your geographical background help you in your day-to-day work?*

A. Geography exposes the world as a set of processes rather than a fixed, immutable reality. That's empowering, as it provides a framework for changing those processes in more sustainable ways. My professional life involves multiple roles: from getting deep into the technicalities of data analysis, to GIS project management, to overseeing the operations of a geospatial technology center. Geography endowed me with indispensable critical thinking and information processing skills to effectively jump among those roles.

Q. *In terms of employment, what advice do you have for students considering a career in the geography field?*

A. Exploit geography's multi- and transdisciplinary nature to the fullest. Gain exposure to multiple theoretical approaches and applied methods, while specializing in a few that lend useful skills for the real economy. Be entrepreneurial and ambitious, and don't be afraid to market yourself. Appreciation for geographers—and how we think and what we can do—is growing tremendously across all sectors.

followed half a century later by Japan, the first major non-Western nation to undergo full industrialization. In the first third of the twentieth century, industry and modern transport spilled over into Russia and Ukraine.

Transportation and the Colonial Legacy

As you have already seen in previous chapters, Europe's colonization of large parts of the world from the fifteenth century through the early decades of the twentieth century has shaped much of today's cultural geographies. That colonization was, in great measure, what allowed Europe to become wealthy and industrialized—developed—and also led to the progressive impoverishment and underdevelopment of the colonized areas.

The physical transportation of raw materials from the site of their extraction from the Earth to places where they were processed into manufactured goods was an essential aspect of industrialization. Before the advent of the steam engine in the late 1700s, for instance, the running water of streams and rivers drove the mechanical textile looms of northwestern England. However, once the steam engine came into use, textile factories were no longer geographically tied to locations with running water. They could be moved to cities where labor was cheaper and where transportation of their finished product to market—in ships and locomotives also powered by steam—was more centralized and efficient.

Thanks then to the steam engine and other transportation innovations, colonizer and colonized became drawn together in a geographically ever-wider and increasingly complex relationship. Raw materials such as cotton could be brought into England's mill towns from as far away as India to be spun, woven into fabric, dyed, and made into clothing. These finished goods were then reexported to the colonies, to be sold at far higher prices than the original raw cotton had commanded.

All these activities involved transportation from one point to another, with many stops along the journey from raw material to final consumer. Transportation technologies and infrastructure (such as roads, railways, and ports) played a central role in the relationship between colonizer and colonized. Thus, it should not come as a surprise that struggles for independence were often waged on and around the means of transportation. In many colonial places, **nationalization**—the government takeover of the transportation infrastructure and fuel sources from private foreign ownership—was a crucial element. For example, Japan nationalized its railways in 1906, Mexico nationalized its petroleum industry in 1938, and Egypt nationalized the Suez Canal in 1956. Airlines, banks, and a wide variety of natural resources have also been subject to nationalization as a part of decolonization and independence movements.

Today, many formerly colonized countries have inherited the transportation infrastructure that was built during the colonial era. Because colonies were seen largely as sources of raw materials, the primary focus of colonial-era infrastructure was moving extracted products—lumber, ores, minerals, and agricultural products—to ports and transporting them to factories in the cities of England, France, Germany, Belgium, Spain, and other colonial power centers to be processed into finished goods. Latin America and Africa, in particular, have transportation infrastructures oriented toward the coasts and port cities. They do not, however, necessarily move people and goods about efficiently within the region.

Places formerly colonized by the same European power are also not likely to be connected to each other. Long-distance air travel to African nations tends to make at least one stop in a European capital city en route, rather than flying direct. And you may find it cheaper and even quicker to fly from one African city to a European capital first in order to get to a second African city, rather than flying directly within Africa. In other words, even modern transportation infrastructure such as airports and flight routes may still today be oriented toward transporting people to and from former colonial powers.

It is important to note, however, that the development of transportation networks and infrastructure predates colonialism. For instance, the roadways of the Incan and Mayan societies in the Americas were noted for their complexity and high quality. The sixteenth-century Spanish conquerors of Peru observed the Incan road system to be better than the Roman roads at the time (**Figure 9.14**). Other remarkable roadways include those built by imperial China, roads in the indigenous U.S. Southwest, trans-regional roadways across the Sahara Desert, and the Silk Road through Asia. The major difference between precolonial and colonial-era transportation infrastructure was their orientation. Precolonial transportation routes focused inward, on moving people and goods across interiors and connecting places to each other, whereas colonial-era infrastructure for the most part was concerned with getting people and goods to sea and river ports as expediently as possible.

Major regional differences exist today in the relative importance of the various modes of transport. In Russia and Ukraine, for example, highways are not very important to industrial development; instead, railroads—and, to a lesser extent, waterways—carry much of the transport load. Indeed, Russia still lacks a paved transcontinental highway. In the United States, by contrast, highways reign supreme, while the railroad system has declined in importance. Western European nations rely on a balance among rail, highway, and waterway transport.

Even some countries that we would today consider to be "developed"—for instance, Australia, Canada, and

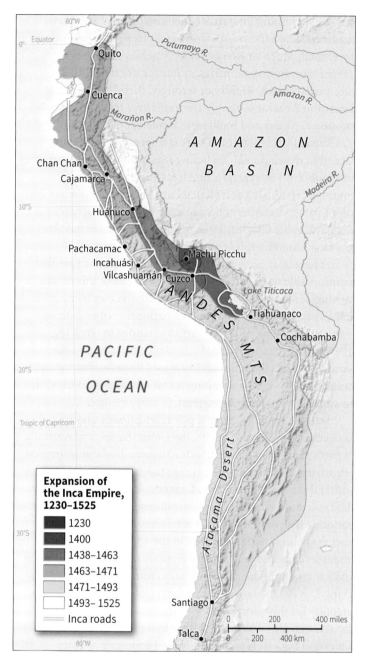

FIGURE 9.14 Incan road system. Much as with the Roman Empire, the expansion of the Incan Empire in South America in the fifteenth century c.e. depended on an extensive system of roads, bridges, tunnels, and causeways. The Incas did not use wheeled vehicles, so the roads were intended for human and animal foot traffic.

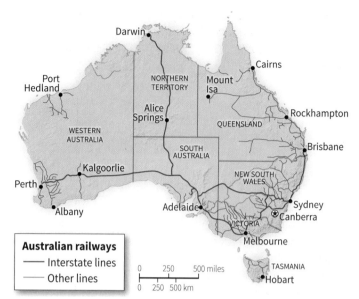

Australian railways
— Interstate lines
— Other lines

FIGURE 9.15 Australia's railway network. The country's railways are concentrated along the length of the eastern coastline, and just one line bisects the nation north–south and one east–west. This is a wonderful illustration of a colonial-era, extraction-oriented transportation infrastructure.

Russia—developed rail systems and to a lesser extent roads that are primarily focused on moving raw materials to port and then on to European colonial powers. **Figure 9.15** shows a map of the present-day Australian passenger rail system. Notice how the geographic focus of the routes is on the coastal areas, and that interior and transcontinental railways are few and far between.

Australia is coping with an additional colonial transportation legacy as well. The gauges (or widths) of railway tracks differ from state to state. Before the unified nation of Australia came into being in 1901, the continent consisted of six British colonies. Each developed its own railway system and, for a variety of personal and political reasons, chose different railway gauges. In 1848 the secretary of state for the colonies in London, Earl Grey (yes, the man associated with the tea that bears his name), advised the colonial governors to adopt a uniform railway gauge, the English Standard Gauge. Three of the colonies did so, while the others later adopted different gauges or a mixture of gauges.

Before unification, this system of differing gauges was not foreseen to be a problem because the colonial governors could not envision the need to connect their railways to one another across the continent. Rather, their focus was on moving people and raw materials to the port cities and back and forth from Europe. Transcontinental freight or passenger traffic had to transfer lines as it crossed from one colony to another. Thus, the Australian railway network became a mixture of narrow, standard, and broad-gauge tracks. Significant personnel and resources had to be devoted to so-called break-of-gauge activities in places where routes of different gauges became connected to one another. Several initiatives to convert to a standard railway gauge nationwide have been undertaken, but the work is slow, politically complicated, and expensive (**Figure 9.16**).

FIGURE 9.16 Dual-gauge railroad tracks near Melbourne, Australia. Multiple gauges, or widths between railroad tracks, are a colonial legacy in many nations. While it is costly and politically difficult to adopt a standard gauge, it is also costly as well as inefficient to operate a system of different gauges. (Paul Souders/Worldfoto)

Transportation Haves and Have-Nots

As the Australian railroad gauge mismatch problem illustrates, modernizing transportation infrastructure can be a costly and difficult undertaking. Yet having a modern transportation infrastructure is vital to the ability of a place to develop. For example, the port of San Francisco declined in importance relative to the port of Oakland just across the bay for this very reason. The local authorities were unwilling or unable to undertake the difficulties and expense of modernizing San Francisco's port facilities to accommodate containerized shipping, whereas Oakland's decision makers took the plunge and modernized that port's infrastructure (see Chapter 10).

Because of the colonial legacy, many of the world's underdeveloped places have an inadequate transportation infrastructure. Old and dilapidated, mismatched, failing to connect places to one another in an efficient fashion,

and often running on outdated technologies, these ports, roadways, and rail systems prevent a place from becoming developed. Much of Africa, Latin America, and the former communist countries in Europe are facing this very predicament. Rightly or wrongly, their leaders have determined that it is simply too expensive to demolish existing facilities and build new, modern ones.

Typically the poor have the least access to everything related to development, including transportation infrastructure and mobility. For example, in Africa 90 percent of the continent's land area and 80 percent of the population are 100 kilometers (62.1 miles) or more away from the seacoast or a navigable river. Given Africa's colonial legacy of extraction-oriented economies, this means that the majority of Africans do not have easy access to transportation infrastructure that would enable mobility. As we have seen already, this makes moving about from place to place within the continent difficult and costly, and inhibits the sort of regional cooperation and integration that would allow African nations to work together to establish a stronger global economic presence. In other words, Africa's colonial mobility legacy impedes the region's development potential, making it that much more difficult for an already poor place to progress.

Yet we have also seen that places which are today thought of as developed—Russia, Canada, the United States, and Australia, in particular—have similarly extraction-focused transportation infrastructures. This is true because these countries (or regions within them) were themselves former colonies. And, at least in the United States, a persistent unwillingness to modernize the nation's transportation infrastructure has and will continue to hurt the country economically. On the other hand, formerly underdeveloped areas that have emphasized the modernization and expansion of their transportation infrastructure, including many East Asian places, have become global hubs of economic mobility for this very reason.

We can also drill down deeper into the structure of mobility haves and have-nots within individual countries or cities.

Singapore: A Transportation Infrastructure Reboot Some formerly underdeveloped places have prioritized building a state-of-the-art transportation infrastructure. Many Asian nations fall into this category. Fifty years ago, for example, Singapore was an overcrowded, housing- and transportation-poor city-state situated at the southern the tip of the Malay Peninsula. Since the 1970s, the Singaporean government has prioritized transportation infrastructure expansion and modernization as a development strategy to position Singapore as a regional Asian as well as global hub of economic exchange.

Singapore expanded and modernized its deep-water port facilities, the second busiest in the world after Shanghai, China. Today, one-fifth of the entire world's container traffic and half of

FIGURE 9.17 Singapore's subway station. Singapore's MRT system efficiently moves more than half the population of the city-state about on a daily basis. (ROSLAN RAHMAN/AFP/Getty Images)

FIGURE 9.18 Japan's maglev train. This maglev (short for "magnetic levitation") train is a test version that Japan and other nations hope to adopt for long-distance transportation in the near future. Maglev transportation uses electromagnets to levitate and float the train forward along a guideway. The result is a fast, smooth, quiet, efficient, and clean ride when compared to conventional wheeled trains. Traveling at speeds of over 310 miles (499 kilometers) per hour, this train is still in the prototype and testing stage, and is scheduled to connect Tokyo and Nagoya in 2027. (Tomohiro Ohsumi/Bloomberg via Getty Images)

global crude oil shipments pass through the port of Singapore. Singapore's airport facilities were moved from a residential area to reclaimed land away from the urban population. Its new Changi Airport was expanded with the addition of terminals and runways, and is today an ultramodern facility handling, on average, over a million passengers per week.

Finally, Singapore's Mass Rapid Transit (MRT) rail system, built in 1987, is a model of cleanliness, safety, and efficiency (**Figure 9.17**). The MRT claims to transport 2.9 million people per day around a city-state of roughly 5.6 million inhabitants. That's more than half of the entire population moving about the city on mass transit every day.

The investment has clearly paid off, though Singapore's small size arguably makes such undertakings easier. Singapore's government is criticized for being highly restrictive with respect to the civil liberties allowed their population: when a course of national investment is set, there is not much room for meaningful protest. The Chinese government is making similar investments in transportation infrastructure, on a much larger scale. Japan's transportation infrastructure leads the world in high-tech innovation (**Figure 9.18**). South Korea as well is rapidly building a modern national transportation infrastructure. East Asia is "on track," literally and figuratively speaking, to lead the world in transportation.

The United States: A Transportation Infrastructure Left Behind The United States provides an interesting example of a wealthy country whose aging transportation infrastructure is fast becoming a serious detriment to the well-being of its citizens and a drag on economic growth. Think about recent "infrastructure incidents" in your local news: a collapsed bridge,

airport delays arising from congestion, dangerously crowded subway cars, railway accidents due to damaged track or mis-routed signals, or unsafe roadways closed to traffic (**Figure 9.19**). Why does the United States have so many problems with its transportation infrastructure?

FIGURE 9.19 The U.S.'s aging transportation infrastructure. This image shows a collapsed bridge section of Interstate 5 near Mount Vernon, in Washington State. This dramatic infrastructure failure is one of many such incidents across the United States of late. Unfortunately, such catastrophic infrastructure failures are becoming more common. (Stephen Brashear/Getty Images)

FIGURE 9.20 Deteriorated roadways. The image on the left is from Los Angeles, California, and the one on the right was taken in South Africa. The similarity in road conditions underscores the fact that differences between developed and underdeveloped places aren't always hard and fast, especially with respect to the quality of infrastructure. (Left: Jonathan Alcorn/Bloomberg via Getty Images; Right: EggImages/Alamy)

As in many developing regions, the transportation infrastructure of the United States is generally old, outdated, and not adequately maintained (**Figure 9.20**). Most of the country's railways, airports, seaports, bridges, and roadways were built decades, if not centuries, ago. The American Society of Civil Engineers publishes a Report Card for America's Infrastructure. The 2017 report awarded an overall grade of D+ to the transportation infrastructure of the United States, estimating that some $4.5 trillion would need to be invested by the year 2025 to update it to a state of good repair worthy of a B grade.

The World Economic Forum ranked America's infrastructure 23rd worldwide, between Spain and Chile. Compared to northern Europe, the infrastructure for U.S. roads, railways, ports, and air transport is ranked "mediocre." Traffic congestion, delays, and fatalities are significantly higher when compared to western Europe. Compared to those in Europe, U.S. railways are slow and delays common. When it comes to air travel, a 1950s-era ground tracking system forces flights to take inefficient routes to maintain contact with controllers, airports are overbooked, and cumbersome security measures cause passenger delays. The United States' inland waterways and rivers, what the American Society of Civil Engineers calls "the hidden backbone of our freight network . . . inland marine highways," received a grade of D, which the ASCE defines as "poor, at risk." Aging, unrepaired, and inadequate facilities cause an average of 49 percent of vessels throughout this system to experience delays.

The deplorable state of the U.S. transportation infrastructure poses economic costs and real dangers to its users. The United States is today the world's largest economy, but as **Figure 9.21** shows, it spends far less as a percentage of GDP than other nations on infrastructure. This stands in direct contrast to a sustained history of large-scale transportation infrastructure construction in the United States, with the nation's canal system constructed in the late 1700s, the transcontinental railway linking the East and West coasts created in the late 1800s, and the interstate highways built in the mid-1900s. What happened?

For one thing, Americans have insisted on low automobile and fuel taxes, whereas in most countries these are a principal source of funding for building and maintaining roadways. Poor regional planning and political squabbling often lead to reluctance to fund construction and repairs. Budget crises at all

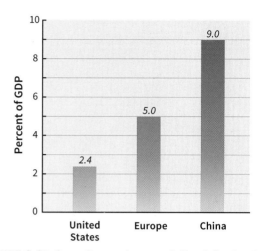

FIGURE 9.21 Spending on transportation infrastructure. This graph depicts wide variations in the proportion of GDP that is devoted to building and repairing transportation infrastructure, such as bridges, roads, waterways, airports, and railroad facilities.

levels—federal, state, and local—mean less public investment in transportation infrastructure, particularly in a deficit-averse environment. And solutions that would privatize these costs, such as tolls, taxes, and fees levied on users, are unpopular. Only one thing is clear: the political gridlock preventing the prioritization of investment to repair and improve the U.S. transportation infrastructure is resulting in literal gridlock on the nation's roads, railways, waterways, and airports.

Nicaragua: The Hovering Elite Often the elite residents of a place—whether a city, a country, or a world region—have access to different and better (safer, quicker, more comfortable) transportation options than the less well-off residents. This discrepancy is an example of how development geographies, in general, tend to be uneven, with the benefits accruing disproportionately to those who are already relatively well-off. Uneven development has spatial implications as well, with some places becoming progressively better connected, wealthier, and more modern while others grow isolated, impoverished, and dilapidated.

If we consider the scale of most cities, the way that uneven development typically plays out is in the creation of enclaves or areas of relative wealth that exist side-by-side with areas of relative poverty. You might recall driving through a city where luxurious gated communities are located right next to impoverished ghettos or *barrios*. (See also Chapter 3, pages 110–111, Fig. 3.34). In this case, spatial differences in development are very clear on the ground. Travel through such cities is seldom a straightforward exercise of moving from point A to point B, because roadways frequently bypass the poorer

neighborhoods, while walls and fences make it impossible to transit freely through wealthier ones. As we discuss in Chapter 10, this compromises the notion of public space, which so many have regarded as central to the exercise of democracy.

Research from Managua, Nicaragua, demonstrates how uneven development in some metropolises has become verticalized. Dennis Rodgers found that "rather than fragmenting into an archipelago of self-sustaining and isolated islands of wealth within a wider sea of poverty, urban space has undergone a process whereby a whole layer of the metropolis has been disconnected from the general fabric of the city." Wealthy Managuans attempt to escape from rising crime rates by going up, rather than further out into the suburbs. Managua's urban elites inhabit what Rodgers calls a "fortified network," moving back and forth from high-rise apartments, office towers, and shopping and leisure spaces located above the poverty, insecurity, and pollution found in the street-level areas of the city. Roadways connecting elite urban spaces to each other were modernized so that elite car traffic could move at higher speeds across the urban space. Roundabouts were constructed to reduce carjackings at intersections. Use of these roads presupposes ownership of a private vehicle and the ability to pay tolls. Pedestrians face significant dangers simply trying to cross multilane high-speed roadways: Rodgers reports that pedestrian fatalities constitute fully 40 percent of all traffic fatalities in Managua.

The next time you move through a large city, look around and see if you can find evidence of the verticalization of privilege. More and more cities around the globe look like Managua, where the urban elites have literally detached their lives from the social and spatial fabric of the city, hovering above it. Indeed, globalization in many areas—economic, technological, and even of celebrity—has reshaped places and their relationships to other places in ways that bring into question some of the fundamental assumptions of development.

Globalization

9.3 Explain how formerly poor places are today significant agents of development.

As we have already seen in this chapter, development today looks a lot different than it did right after World War II. In some very important ways, economic, technological, cultural, and even media-related transformations have brought different actors and sites into prominence. These transformations have fundamentally reoriented the landscape of global development geographies.

Postdevelopment and the Rise of the Global South

Were today's less developed places always poor, isolated, and largely engaged in primary sector activities? Not necessarily. As you have learned already, and will see again in the next chapter on urban geographies, many civilizations now considered "underdeveloped" were once hubs of economic activity, technological innovation, cultural diversity, and political dominance.

Consider the arc of Islamic influence, stretching from Rabat in northern Africa to Manila in Southeast Asia. As former university president William Frawley describes, "There is a fifteen-century-old line of faith and business that runs straight from Morocco to the Philippines and that has drawn Muslims and workers of need to countries of religion and economic prosperity." Beginning in the seventh century C.E., this 7600-mile (12,300-kilometer) sweep of Islamic influence constituted a vital corridor of economic, cultural, and political activity—an arc of development. It flourished in a world where Europe had languished for centuries in the Dark Ages, and the so-called New World was populated by indigenous societies that were themselves centers of innovation and prosperity (see also Chapter 10, pages 358–359).

The stark division between "developed" and "underdeveloped" places discussed earlier under the Region theme has itself been largely redrawn. In reality, very few of the world's places are impoverished, isolated, and predominantly agricultural any more. Individual countries, like China, or entire regions, like sub-Saharan Africa or Southeast Asia, that did fit this profile as recently as 30 years ago have witnessed the transformation of their economic, political, technological, and cultural systems. Even within countries that may still be less developed as a whole, there are pockets of accelerated urbanization, innovation, and connection to global networks that make these locales look, feel, and act wealthy, contemporary, and plugged-in (**Figure 9.22**).

By the same token, places that several decades ago unquestionably occupied the ranks of the wealthy, connected, and modern—in other words, developed countries and regions—have slipped. Earlier, we discussed the United States' crumbling transportation infrastructure and noted the high toll placed on economic productivity by this situation. Europe, too, has seen its seemingly secure First World status called into question by debt, financial crisis, an aging workforce, fears of terrorism, conflicts over immigration, and racist violence. In addition, places within these countries and regions, once centers of economic, technological, political, and cultural dynamism, have fallen on hard times.

The city of Detroit is a perfect example. Once a world leader in the auto industry and the United States' fourth largest city, Detroit was home to secure jobs and was seen as a cultural embodiment of quintessentially midwestern American values and lifestyle.

FIGURE 9.22 Songdo International Business City. This new development is located on reclaimed land along South Korea's Incheon Waterway, near the capital city of Seoul. It includes ultramodern residential, commercial, education, leisure, and health care facilities. Forty percent of the area is devoted to green spaces such as the park you see here. Songdo is a "smart city" representing the cutting edge of urban design: sustainable, green, and wired. (Ann Hermes/The Christian Science Monitor via Getty Images)

In 2013 Detroit declared bankruptcy, the largest U.S. city ever to do so. It has lost over half of its population since its heyday in 1950, and with it the tax base needed to pay for city services and maintenance. Financial mismanagement and the burden of health care and pensions for city workers have further impoverished Detroit. Detroit's urban landscape reflects this downward spiral in the appearance of the built environment (**Figure 9.23**).

FIGURE 9.23 Detroit's deteriorated cityscape. This abandoned house is one of many in Detroit. The city estimates that one-fifth of its housing stock is abandoned and blighted. (Peter Baker/Getty Images)

Some scholars see these changes as evidence of **postdevelopment**: the idea that "development" was simply a way to categorize most of the world as needing the assistance of a few wealthy nations. However, as scholars of postdevelopment argue, development as we knew it—in the form of monetary aid, investments, expertise, and projects—was seldom successful. Instead, development in reality constituted a modern form of **imperialism,** whereby wealthy nations ruled poor nations by controlling their economic, political, and cultural systems. The age of development, they argue, is over.

Whether or not one agrees with this statement, it is clear that places formerly dismissed as underdeveloped have emerged as major players in a global world. The term **Global South** has largely replaced Third World when referring to the countries of Latin America, Africa, and most of Asia. The term also represents an attempt to move beyond the negative connotations of "Third World," recognizing that much of the world's growth—economic, population, and urban—is happening in these countries, and that powerful contributions and alliances are emerging among them without the involvement of traditionally wealthy nations.

South–South cooperation is a major dimension of the dynamism occurring within the Global South. Several countries traditionally not part of the developed nations—Brazil, China, India, South Africa, Malaysia, and Indonesia among them—have large populations and growing economies. In many respects, they are the new centers for technical and economic development vis-à-vis their poorer neighbors. These countries are emerging as major global lenders, investors, researchers, educators, and employers (**Figure 9.24**). A report by the World Bank finds that South–South trade, for instance, has grown on average by 13 percent every year since 1995. They are also standing shoulder to shoulder with the world's former political powerhouses, demanding a voice in issues like global climate change, energy use, and the world's political organization.

Technology and the Global South

Technology has long played a central role in economic development. Earlier in this chapter, we discussed Europe's transformation from an agricultural to an industrial society, beginning in the eighteenth century. Recall that technological developments, particularly in powering machinery for manufacturing goods and transporting them, were the key to this transformation. Throughout the centuries that followed, and today as well, revolutions in transportation and communications technologies continue to play a huge role in economic development. Small to medium-sized enterprises in countries of the Global South are generating low-cost solutions to common problems there. Tata Motors, one of whose best-selling products is shown in Figure 9.23, is just one example. In another example, the One Laptop per Child initiative aims to facilitate children's engagement with education by distributing low-cost, rugged, low-power, and connected laptops to students in the Global South, with the stated goal of "collaborative, joyful, and self-empowered learning" (**Figure 9.25**). Other innovations are occurring in the fields of energy, environment, public health, education, and transportation.

FIGURE 9.24 India's Tata Nano. Billed as "the world's cheapest car," the Nano, designed and manufactured by Tata Motors, is a good example of technology developed by and for the Global South. Retailing for around $3700, the Nano is lightweight, compact, simply designed, and fuel-efficient. (BAY ISMOYO/AFP/Getty Images)

FIGURE 9.25 One Laptop per Child. This initiative supplies rugged, low-cost, low-power, connected laptops to over 3 million children and teachers in 60 countries, including the United States. The objective is to empower children in the Global South to engage in their own education, and to learn, share, and create together. (OLPC, Inc.)

Mobile phones provide another field of South–South technological innovation and exchange. Over 5 billion people on the planet have access to a mobile phone, and usage in the Global South has already surpassed that in wealthy northern countries. Mobile phones are arguably the most ubiquitous modern technology. In some poorer regions of the world, more people have access to a mobile phone than to a bank account, electricity, or even clean water. Smartphone apps are used by farmers, fishers, and traders to monitor weather conditions, track schools of fish, and follow market prices. Health care providers use smartphones to remotely diagnose ailments and to text patients medication reminders. Smartphones are used by migrants to send money home, by citizens to monitor elections, and by workers to apply for jobs.

All these projects attempt to close the digital divide—the rift between those who benefit from easy access to the Internet and, more importantly, the vast storehouse of knowledge available online, and those who do not have easy access (see Chapter 2, page 58 (Fig. 2.21), and The Video Connection). The digital divide is a contemporary aspect of the long-standing gap between development's haves and have-nots. Those who do not have easy access to the Internet are disadvantaged to the extent that they cannot readily participate in today's globally connected world. Though Internet access is growing rapidly, about half of the world's inhabitants do not use the Internet (**Figure 9.26**). Across all places, those who are wealthier, younger, and better educated have higher rates of Internet usage. Even within

wealthy countries, some segments of the population—the elderly, disabled people, and the poor—are far less likely to go online than are the young, healthy, and wealthy.

As Internet access continues to expand in the Global South, it is worth considering how the Internet's content and delivery (and language; see Chapter 4) might be reshaped. For instance, since 2000 the Middle East has been the most active world region in terms of growth in online use. Yet less than 1 percent of Internet content is in Arabic. How might growing numbers of Arabic-speaking Internet users reshape the online environment?

Gender, Globalization, and Development

The traditional development models formulated after World War II were gender-blind. They did not distinguish between women and men in their understanding of how development unfolds in a place. In other words, development was supposed to affect everyone equally. Beginning in the 1970s, development geographers and others demonstrated that development is, in fact, a highly uneven process with respect to gender. Feminists in wealthy northern countries were at that time concerned that the labor market in the developed world had systematically left women behind. Women were underpaid, unemployed or underemployed, and marginalized into certain sectors of the

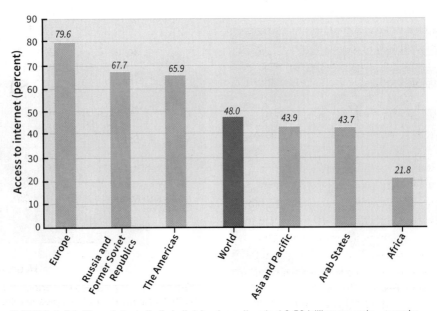

FIGURE 9.26 The global digital divide. An estimated 3.58 billion people—nearly half the world's population—are estimated to be online. As you can see from this figure, that proportion varies from region to region.

▶ The Video Connection

Wiring the Amazon
◈ Watch at 🅜 SaplingPlus

The small Amazonian village of Palestina, Peru, is featured in this video. The population of the village has dwindled over time to 65 residents. The complicated logistics of bringing digital connectivity to the village, via telephone and cell phone service and free laptops for schoolchildren, is explored. Alongside the technical difficulties of bringing modern communications to the village, the cultural changes that happen are highlighted.

Thinking Geographically

1. This video depicts the "digital divide." While this concept can be abstract, how does being on the other side of the digital divide play out in the day-to-day lives of the residents of Palestina?

2. There are many ways of being connected or disconnected. Describe the ones that apply to the village portrayed in this video.

3. As the village gains intermittent access to the Internet and to telephone, then cell phone, technology, the role of the village's younger residents becomes more important. In what ways does their position in society shift?

---

economy, such as teaching and nursing. The situation, feminists speculated, was likely to be the same in countries that were just then undergoing the process of development.

Evidence shows that women in the developing as well as the developed world earn less than their male counterparts, have fewer employment choices, and are far more likely to lose their jobs during economic downturns. In some societies in the developing world, there are cultural prohibitions against women working for wages outside the home. These women are entirely dependent on their fathers, brothers, husbands, and sons for economic support. Research conducted in many areas of the developing world has demonstrated that men are more likely to spend their wages on alcohol and leisure activities, whereas women tend to use money for food, education, and health care (see also Subject to Debate).

Women's economic vulnerability often translates into vulnerabilities in the other key arenas of development, among them health, housing, safety, and education. The logical solution, according to this understanding, is to more fully include women in the development process, particularly in paid employment outside the home. Women's inclusion in the labor market would reduce their economic dependence on men; in addition, it would raise overall economic productivity. One study estimated that $12 trillion would be added to global GDP by 2025 if women's equality were advanced. Yet such a solution overlooks several important factors.

First, many women are already involved in the development process, even if they do not participate in paid employment outside the home. All of the labor performed by women in the home for free, including child care, meal preparation, in-home health care, and cleaning, subsidizes the world economy. This is true everywhere, not just in the Global South. As **Figure 9.27** shows, women the world over perform the bulk of unpaid work.

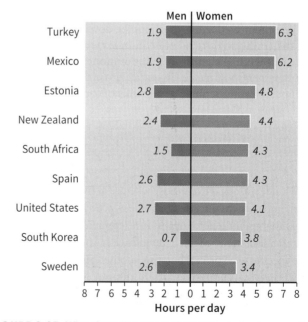

FIGURE 9.27 **Who does the chores?** Women everywhere perform the bulk of unpaid labor—chores, in other words—in the form of grocery shopping, food preparation, laundry, cleaning, and caring for children, the ill, and the elderly. This leaves less time for women to engage in paid work outside the home or in leisure activities.

SUBJECT TO DEBATE Microfinance and Development

In the 1970s, an economics professor by the name of Moham-med Yunus was troubled. On the one hand, he was teaching textbook economic theory, while on the other hand, he was sur-rounded by the poverty and hunger of so many inhabitants of his native Bangladesh. One thing that stood out to Professor Yunus was the difficulty faced by poor people, particularly women, in securing credit. Banks didn't want to lend to them because they had no assets or credit history. The loan sharks who did lend to women demanded exorbitant interest rates in return.

In 1983 Yunus founded the Grameen ("Village") Bank. The Grameen Bank lends only to those without assets or credit, and requires loan recipients to educate their children and establish savings. Yunus himself initially loaned a total of $27, which was enough to erase the debts of 42 individuals. Most of the micro-loan recipients are women or groups of women, who—far from being credit risks—have very high loan repayment rates. The loans, which average $100, are used to establish small businesses and to lift families out of poverty. In 2006 Yunus was awarded the Nobel Peace Prize; at that time, over 7 million individuals had received loans through the Grameen Bank.

Why do microfinance efforts like the Grameen Bank focus on women? Research and experience have shown that poor women in the Global South by and large prioritize expenditures on family well-being, including health care, nutrition, and edu-cation. Poor men, on the other hand, often spend money on short-term pursuits that do not enhance family well-being, such as leisure activities and alcohol. In addition, women are far more likely than men to repay loans on time and in full and less likely to default.

Microfinance efforts throughout the developing world chal-lenge the traditional focus of economic development, which emphasizes lending significant sums of money to organizations and governments to finance large-scale projects. Women are typ-ically not well represented in the ranks of formal development organizations or national governments. In some ways, you might think of microfinance as a postdevelopment undertaking.

Continuing the Debate

Consider the debates surrounding microfinance, and ask your-self these questions:

- Do you believe that women are truly better credit risks than men? Why or why not?
- Microloans are for such small amounts of money—can they really make a difference?
- Are microfinance initiatives under way where you live?

Thanks to bank loans, Maryam was able to buy and repair rickshaws and create her own company. She employs eleven drivers. (John van Hasselt/Sygma via Getty Images)

Second, women—even those who are actively participating in paid employment outside the home—face systematic barriers that men do not. For instance, women often confront discrimination, harassment, and violence in the labor market and in specific workplaces, but also in other public spaces as well as inside their homes. Prevalent cultural associations of women with motherhood and the spaces of domesticity (see also Chapter 3, page 93 including Fig. 3.15) mean that women who work outside the home are seen as transgressing into male-dominated spaces. In order to force these women back into their homes and mothering roles, men—and other women as well as public policy and the media—subtly (and not so subtly) retaliate in a variety of ways.

Because such scenarios occur throughout the world, you are probably familiar with some of these retaliatory tactics. For instance, women moving through public spaces such as city streets and on public urban transit are subjected to catcalls, groping, and sexual assault. Working Indian women have been raped on public transit and in city spaces. Since the 1990s, hundreds of female factory workers in northern Mexico have been murdered as they traveled to and from their jobs.

Women's mobility is restricted in other ways as well. In Saudi Arabia, for instance, women have been legally forbidden to drive automobiles, a prohibition only recently overturned. Many societies expect women to dress in restrictive clothing that limits their mobility and comfort in the name of modesty. You might think of high-heeled shoes as an example of this tactic. Even the persistently lower pay for women the world over who hold the same jobs as men can be seen as a tactic to discourage women who work. In the United States, women working full-time, on average, make 76 percent of what their male counterparts are paid; in South Korea and Japan, women earn less than 70 cents for every dollar earned by men. In many ways, women are still second-class citizens, a reality that hinders progress in development by holding back fully one-half of humanity.

Celebrities and Development

Technology is not just good for solving logistical problems in poor areas, such as bringing electricity to remote villages or tracking the height and weight of children. In the development field, technology has also been used to create communities of concern and action that are truly global in scope. Many times, celebrity figures from the world of music, film, and sports are the faces of such communities. This is not a new development. In 1940, for instance, silent movie star Charlie Chaplin's first talking film, *The Great Dictator,* condemned Hitler, Nazi Germany, and fascism. Film star Jane Fonda was a very vocal participant in the antiwar movement of the 1960s. Singer Johnny Cash performed at California's Folsom State Prison in 1968, in part to protest conditions there.

The Ethiopian famine of 1984, however, heralded the advent of contemporary high-profile celebrity-led development efforts televised worldwide. The 1985 Live Aid charity concert, assembled by the Boomtown Rats' front man Bob Geldof, was the most visible of these early efforts (**Figure 9.28**). Geldof had been deeply affected by coverage of the famine, which claimed more than 1 million lives. Geldof assembled a superstar musical lineup, including Bono, Mick Jagger, Elton John, Madonna, Paul McCartney, and B. B. King, performing in one of two shows broadcast via satellite technology. Phil Collins actually performed in both events—first in London and then, after flying across the Atlantic, to perform again with the reunited superband Led Zeppelin in Philadelphia. Each performer had no more than 17 minutes on stage, and the performances were interspersed with film clips depicting the famine and calls by Geldof to donate money to the famine relief efforts.

MTV estimated that 1.4 billion of the world's then 5 billion inhabitants watched, and that at one point

FIGURE 9.28 Live Aid concert. Organized by Bob Geldof in 1985, Live Aid was at the time the biggest and most ambitious concerts ever staged. (Michael Ochs Archives/Getty Images)

95 percent of the world's television sets were tuned in to the concert. More than $200 million was raised. Keep in mind that this huge participation occurred in the mid-1980s, when cell phones, the Internet, and Twitter did not exist. Indeed, the 1981 birth of MTV—and the music video—can itself be seen as a technological innovation facilitating massive charity rallies such as Live Aid. Broadcasts like this one allowed virtual participation, leading to a widening of the concerned community away from just the face-to-face audience at live concerts. MTV thrived on the repeated playing of music video clips, and those shot at Live Aid were soon iconic, such as U2's performance of "Sunday Bloody Sunday." These performances resonated with the career development of this era's pop musicians, which would become increasingly video-based rather than linked solely to live performances.

Since then, there have been any number of celebrities associated with causes in the Global South. Actors, musicians, and television personalities make up a large portion of such celebrity ranks: Angelina Jolie, Madonna, George Clooney, Oprah Winfrey, Akon, and Kim Kardashian have all engaged in social justice charity work in Africa, for instance. To be sure, the wealth, mobility, and cultural visibility of celebrities can go a long way toward bringing worldwide attention to events that might otherwise remain under the digital radar. These individuals leverage their fame to raise millions of dollars for development-related causes, sometimes far more effectively than the traditional economic development agencies officially charged with funding development efforts in the Global South.

Nevertheless, there is also widespread criticism of celebrity development efforts. Most media celebrities are not trained economists, climate change scientists, medical doctors, educators, or social workers. In some ways, they are playing a role—that of development expert—they are not really qualified to play. A few celebrities acknowledge this and have teamed up with international development organizations or experts (**Figure 9.29**). Others, however, seem oblivious to their limitations, as can be seen in the example of Madonna's failed $3.8 million Malawian school project, apparently a victim of financial mismanagement.

Critics have noted that celebrity development efforts tend to portray the Global South and its residents as passive victims of calamity who are just waiting for wealthy Westerners to rescue them. Writes Andrew M. Mwenda, "So some [celebrities] come to save orphans, others to defend human rights, feed the hungry, treat the sick, educate our children, protect the environment, end civil wars, negotiate aid, and promote family planning lest we overproduce ourselves. It is almost as if Africans cannot do anything by themselves and need a combination of Mother Teresa and Santa Claus to survive." Critics of celebrity development figures suggest that their efforts are more about attempting to enhance the individual celebrity's brand, rather than reflecting any true concern with living people and places. In all likelihood, celebrity development efforts are not entirely naïve and self-serving undertakings. In some ways, they underscore a long-standing and problematic relationship between the world's haves and have-nots.

FIGURE 9.29 Celebrity and international development. U.S. pop star Madonna interacts with Malawian children at Mkoko Primary School in 2013. It is one of the schools Madonna's Raising Malawi organization has built jointly with the U.S. organization BuildOn. (AMOS GUMULIRA/AFP/Getty Images)

As we have seen, technological, cultural, and political transformations on a global scale will certainly continue to shape and reshape how development is practiced and who benefits.

Nature–Culture

9.4 Recognize the role that nature has played over time as concepts and practices of development have changed.

As we have seen already in this chapter, the extraction of natural resources from the Earth, and their subsequent transformation through manufacturing into products that people need and want, have for centuries been the engine of economic development. Nature, in other words, is transformed into (material) culture. Nature is also transformed symbolically into culture. Specifically, "pristine" natural environments (see Figure 9.9, for example) are cultural images often positively associated with poorer places on the planet. It is as if a loss of purity, both of the natural world and of man's place in nature, is the primary trade-off of economic development. Some scholars would go so far as to say that we have quietly agreed to accept the ongoing destruction of the planet in exchange for living comfortably today, for most economic activity creates serious ecological problems. Thus, nature–culture is a very important theme to explore with respect to development geography.

Renewable Resource Crises

Human beings have a long history of depleting the natural world around them in the name of progress. Deforestation, for example, has been a problem since the beginning of human civilization. Ancient humans used fire to burn down trees and drive woodland animals out of the forest in order to more easily hunt them. Neolithic-era agriculture involved clearing forest land for cultivation (see World Heritage Site, pages 338–339). The ancient Greeks faced problems with the silting of their harbors as the lands cleared for domestic crop cultivation eroded, a scenario also experienced by the ancient Romans and medieval Europeans. Cities were often abandoned when their ports became unusable. Europe's heavy involvement in the transatlantic slave trade, warfare, and voyages of exploration to distant lands resulted in the destruction of northern forests in order to build wooden ships for these activities.

It might seem that only finite resources are affected by industrialization, but even renewable resources such as forests and fisheries are endangered. The term **renewable resources** refers to those primary sector products that can be replenished naturally at a rate sufficient to balance their depletion by human use. But that balance is often critically disrupted by the effects of industrialization. So, while deforestation is a long-standing practice, the Industrial Revolution drastically increased the magnitude of the problem. Today, the Earth Policy Institute has identified Latin America as the region experiencing the greatest loss of forest cover, thanks primarily to tree clearance in the Amazon to make way for cattle ranches. Indeed, we are witnessing the rapid destruction of one of the last surviving great woodland ecosystems, the tropical rain forest. The most intensive rain-forest clearing is occurring in Indonesia and Brazil. Although trees represent a renewable resource when properly managed, too many countries are, in effect, mining their forests at a rate faster than they can be replaced. Even when forests are replaced by scientifically managed "tree farms," as is true in most of the developed world, ecosystems are often destroyed. Natural ecosystems engender plant and animal diversity that cannot be sustained under the monoculture of commercial forestry.

Similarly, overfishing has brought a crisis to many ocean fisheries, a problem compounded by the pollution of many of the world's seas. On average, people consume four times as much fish as they did in 1950, reaching an all-time high of 45 pounds (just over 20 kilos) per person annually in 2016. More than 85 percent of the world's fish stocks are compromised by overfishing. A BBC report suggests that "large areas of seabed in the Mediterranean and North Sea now resemble a desert—the seas having been expunged of fish." Commercial fishing methods are now depleting the world's tropical waters as well, with a 40 percent drop in tropical fish catch expected by 2050. Some good news was reported in the North Sea, though: since 2009 the size of the spawning stock of cod reached a level that scientists agreed was sustainable. Experts credited an array of conservation techniques for the turnaround. As with tree farms, fish farming has significant drawbacks, including pollution of the surrounding waters from the antibiotic and fecal waste of the concentrated fish farm populations, along with a lack of nutrients in the meat when farmed fish are fed vegetarian diets (see the discussion of aquaculture in Chapter 8).

Global Climate Change

Virtually all climate scientists now agree that we have entered a phase of **global climate change** (sometimes mistakenly called **global warming**) caused by human activity and, most particularly, by the greatly increased amount of carbon dioxide (CO_2) produced by burning fossil fuels (see Subject to Debate, Chapter 1). Fossil fuels—coal, petroleum, and natural gas—are burned to create the energy that powers the world's factories, and as that burning has increased, so has the amount of CO_2 released into the atmosphere. As **Figure 9.30** shows, the 10 hottest years on record have all occurred since 1998, with the most recent years being the warmest. Rising sea surface temperatures melt arctic ice. Melting ice raises ocean levels, threatening to obliterate coastal areas and even entire countries, such as the island nation of Maldives. Thanks to global climate change, some areas of Earth are hotter, wetter, drier, or colder than normal. As displayed in **Figure 9.31** on page 340, these changes have led to disruptions in agriculture, weather, tourism and leisure activities, and human health.

To begin to mitigate the effects of the environmental crises caused by global climate change, 38 industrialized countries

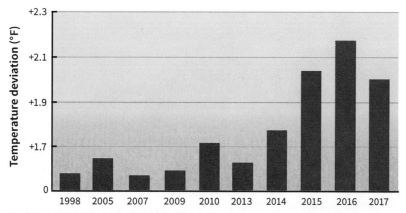

FIGURE 9.30 Ten hottest years globally. This chart shows the temperature deviation, in Fahrenheit, from a baseline in the early industrial period (1881–1910). This shows how much warming has occurred since the beginning of industrialization. Note that the warmest years are also the most recent ones.

World Heritage Site: Neolithic Flint Mines at Spiennes

Two hundred and ninety-seven of the UNESCO World Heritage Sites predate the modern era, and are considered ancient places having outstanding universal value to humankind. The Neolithic flint mines at Spiennes is one such place. Located near the city of Mons, modern-day Spiennes is a small village in Wallonia, a French-speaking region in southern Belgium. Spiennes dates

back to the middle Neolithic period (4400–3500 B.C.E.) in central Europe. Associated with the ancient settlement above ground is a vast complex of mine shafts from the same period.

- The mines span an area of over 100 hectares (247 acres) and represent the largest concentration of early primary sector activity in Europe.

- The mine shafts are among the deepest ever sunk anywhere to retrieve flint.

- The area was actively mined from 4400 to 2000 B.C.E., though there is some evidence of human activity prior to that—as early as 6000 B.C.E.

TECHNOLOGY AND DEVELOPMENT: The Neolithic era, also known as the New Stone Age, was a pivotal one in human history. It saw the domestication of crops and animals, and coincided with the development of early cities (see Chapter 10). The tools developed during this period were made of stone—the Metal Ages would come later. Flint proved to be an exceptionally useful material, as it can be worked into a very sharp edge through a process known as "knapping" and polishing. The first flint tools were hand-held; later, wooden handles were added to make axes.

- As discussed in the Nature–Culture theme for this , chapter, the Neolithic period saw the rise of agriculture and the related clearance of vast areas of forested land across Europe and the Mediterranean. The felled timber was used in the construction of houses and boats. The axes—considered to be the Swiss army knives of their time—were used to fell trees as well as for hunting and mining. Many were constructed of the flint mined at Spiennes.

- As you know from reading this chapter, industrial activities are an important component of development. Mining is a primary sector activity because it involves the extraction of raw material from the Earth. Large flint nodules were extracted from underground at Spiennes using flint picks, deer antlers, and cow bone.

TOURISM: Some ancient World Heritage Sites, such as Stonehenge in England and the Pyramids at Giza, Egypt, are visited by thousands of tourists annually. Others, such as the Neolithic flint mines at Spiennes, are by contrast off the beaten path. Until quite recently, the mines had hardly been marked and were difficult for tourists to find. The mine shafts remain almost entirely untouched, except for some that have been fortified to accommodate the visits of tourists.

- Only a few hundred tourists visit the Neolithic flint mines at Spiennes annually, often on a day trip that is part of a broader tourist itinerary in the area.

- A modern visitor center, opened in 2015, provides information and guided tours of the mines.

- Guides lead tourists down into a mine shaft, approximately 26 feet (8 meters) deep, to explore a section of the underground labyrinth.

- The area above ground is littered with tailings, which are leftover scraps of flint generated during the knapping process thousands of years ago.

(Museum of London/Heritage Images/Getty Images)

- The Neolithic era was a turning point in the development of technology for human progress, which is why the flint mines at Spiennes are designated as an Industrial Heritage Site.
- The mine complex is the earliest and largest example of primary sector activity in Europe.
- The Neolithic flint mines at Spiennes were designated a World Heritage Site in 2000.

http://whc.unesco.org/en/list/1006

(Photo Guy Focant © AWaP-SPW-be)

FIGURE 9.31 Effects of global climate change. The effects of climate change vary from place to place. This map is rendered from an interactive map produced by *National Geographic,* which can be found at http://environment.nationalgeographic.com/environment/global-warming/gw-impacts-interactive/.

signed the Kyoto Protocol in 2001. This document, originally discussed in Kyoto, Japan, in 1997 and renewed in Durban, South Africa, in 2011, binds these countries to reducing emissions of greenhouse gases so that their future levels of emissions will be less than their 1990 levels. The United States agreed to this protocol in 1997, but has since changed its position and is no longer a party to the agreement, claiming that it would cause economic setbacks and does not do enough to curtail the high emission rates of poor nations. It is true that the large emerging economies in the Global South discussed earlier are also adding quickly to the world's greenhouse gas emissions, and these nations must be central partners in any agreement to address global climate change. In 2016, the Paris Agreement went into effect, binding its member countries to work together to limit the rise in global temperatures. In 2017, President Donald Trump withdrew the United States from the Paris Agreement, leaving the United States as the planet's sole nonparticipant nation in this important climate agreement.

The Environmental Consequences of Powering Industrialization

The extraction of natural resources from the Earth, and their transformation into the myriad consumer goods we now rely on, depend on the provision of energy to run factories and transport finished products. Consumers the world over have grown accustomed to heated and cooled dwellings, artificial lighting, and running water, among other amenities. The agricultural

activities needed to feed the world's more than 7 billion inhabitants require plenty of power as well.

Renewable Energy Sources So-called **alternative energy** sources—that is, alternatives to fossil fuels—have long been pursued in an attempt to provide cleaner, cheaper, renewable power sources. These alternatives include hydroelectric, solar, wind, geothermal, nuclear, biofuel, and "clean" coal. However, the expense involved in building, running, and maintaining sites for harnessing and distributing these power sources; their intermittent nature in some cases, such as solar and wind; and the potential hazards associated with some of them have deterred their widespread adoption. As a result, the world still relies heavily on fossil fuels to power economic activities; 80 percent of the world's energy, in fact, still comes from fossil fuels. Not coincidentally, powerful oil, gas, and electrical industry groups have spent millions of dollars to lobby politicians, especially in the United States, to reject policies favorable to alternative energy. The advent of hydraulic fracturing (*fracking*)—which uses pressurized liquid to fracture rock in order to extract natural gas and petroleum—is the latest example of how the United States has, as a nation, largely chosen to continue to extract fossil fuels rather than switch to other fuel sources.

As the emerging nations of the Global South become increasingly important producers of goods and services, how they choose to fuel their economic growth will be of paramount importance. As we have already seen, South–South cooperation in alternative environmental technologies is an important growth area for these emerging economies. It is entirely possible that the

FIGURE 9.32 Europe's renewable energy island. Samsø, located off the coast of Denmark, is a model of alternative energy in Europe. This photo depicts some of the island's wind-powered turbines, which supply 100 percent of the 4200 residents' electrical needs. Solar power and wood chips are used to provide heating. In all, Samsø removes more carbon dioxide from the atmosphere than it contributes. (Pep Roig/Alamy Stock Photo)

FIGURE 9.33 China's deteriorated air quality. This photo was taken at Beijing's Tiananmen Square during the winter, when air contamination from burning coal is at its worst. During the harsh northern winters, people routinely burn traditional (not "clean") coal—a highly contaminating fuel source—to keep warm. (Lintao Zhang/Getty Images)

next wave of alternative energy ideas will emerge from technical innovations developed in the Global South. Lesotho exports water to South Africa, while Paraguay, Bhutan, and Mozambique are all net exporters of electricity to neighboring countries.

Some nations in wealthier northern regions of the world are also noted for their use of renewable energy. Iceland, the world's largest volcanic island, has abundant geothermal and hydro power. Iceland's electricity is clean: it is produced entirely by renewable energy. Samsø, an island off the coast of Denmark, has become a wind power showcase for the world. Hot water is produced by wood chips and straw provided by local farmers, as well as by solar panels. As a result, Samsø can boast its status as carbon-neutral and a net exporter of clean electricity (**Figure 9.32**). Nations the world over—North and South—are in the vanguard of developing alternative energy sources to fuel their activities.

China's Harmful Air Quality China provides an example of the negative environmental, as well as political, consequences of rapid industrialization. For several decades, China has followed the same course, energy-wise, as the now wealthy nations of the North did to fuel their industrial development. Specifically, China's factories, as well as households, have been fueled by burning coal, a notably polluting fuel source. In addition, China's cities are clogged with vehicular traffic. In 2006 China's greenhouse

gas emissions exceeded those of the United States for the first time, and the gap has continued to accelerate since then. An "environmental census" conducted in late 2017 revealed that the sources of pollution in the country had increased by more than 50 percent since the first census, conducted in 2010. China is now the world's biggest emitter of pollution.

The resulting air contamination has caused regular shutdowns of businesses, public places such as schools, and transportation venues from bus service to international airports in cities such as Shanghai and Beijing (**Figure 9.33**). Not only is human health severely affected by the thick soup of particulate matter given off by cars, buses, household heating and cooking, and industry; visibility is also reduced, so much so that drivers cannot see across a roadway, or a pilot down a runway to safely land a plane. On a particularly bad day, the level of fine particulate matter—those particles small enough to lodge deep inside human lungs—as measured in the northeastern city of Harbin can be up to 40 times higher than the international safety standard set by the World Health Organization.

China's air (and water and ground) pollution is the result of its rapid, sustained rate of economic growth, which has averaged 10 percent annually since the late 1970s. This growth has lifted many Chinese out of poverty, and has transformed the country's cultural landscape from a primarily rural to a primarily urban one. Yet the economic toll when business and transportation activities must be halted due to contamination is steep, as are the long-term costs to human health. Urban residents there must limit their time outdoors, and many wear masks when they do venture outside. China's tight control over dissent means that protests against polluters are shut down, information on air quality is censored, and social media is of limited effectiveness in pulling together a national antipollution movement. Despite the government's attempts to control information and contain protest, and its repeated pledges to reduce smog in the winter months, pollution is poised to become China's primary political flashpoint.

Recognizing the environmental, health, and political consequences of this environmental degradation, China's government declared a "war against pollution" in 2013. New coal plants have been barred and existing plants have been ordered to curb emissions, while major cities have restricted the number of cars allowed on the roads, and residents of Beijing had their coal-powered heaters removed. Since then, air quality has improved markedly in China's largest cities.

The link between human economic activities and other aspects of well-being, including planetary health, is particularly visible—and vulnerable—where natural resources are concerned. These economic activities have also left distinct marks on the cultural landscape.

Cultural Landscape

9.5 Describe how development has shaped landscapes.

It is difficult to look at any particular landscape and not see some evidence of economic development. Factories, shopping malls, housing developments, industrial agriculture, and cities—all provide examples of landscapes that emerge over the course of development. There is literally no place on Earth that has not been touched in some way by human activity.

Cultural Landscapes of Resource Extraction

Primary industries produce perhaps the most drastic transformations. The resulting landscapes contain slag heaps, clear-cut commercial forests, strip-mining and open-pit mining scars, and forestlike clusters of oil derricks (**Figure 9.34**). The destruction of nature is less apparent in other types of primary industrial landscapes. The fishing

FIGURE 9.34 Mining landscape. This image shows a conveyor moving ore from an open pit mine to a processing plant. Such bleak landscapes are common in areas of primary industry, which involves extracting natural materials from the Earth. (Thomas_Moore/iStock/Getty Images)

FIGURE 9.35 A fishing village in Newfoundland. Primary industrial landscapes can be pleasing to the eye. (JG Photo/Shutterstock)

villages of Portugal and Newfoundland continue to attract tourists even though the fisheries are depleted (**Figure 9.35**). In still other cases, efforts are made to restore the preindustrial landscape or to repurpose it into something more attractive. Examples include the establishment of artificial new grasslands in old strip-mining areas of the American Midwest, and the creation of recreational ponds in old mine pits along interstate highways.

Cultural Landscapes of Industry and Services

Among the most obvious features of the landscapes of secondary sector activities (manufacturing) are factory buildings. Early to mid-nineteenth-century industrial landscapes are easy to identify, because the technologies that ran these factories were reliant on water power. Given this locational requirement, these mill complexes were most often built in rural areas. Entire communities, usually with housing for the workers, thus had to be constructed along with the mills. These mill towns dotted the landscape of England, Scotland, and New England, the site of the first industrial developments in the United States. Later—as water power was supplanted by steam, coal, and then electric power—factories were located near urban areas, taking advantage of the housing supply already there and the proximity to a large consumer base. In the second half of the twentieth century—with the development of the interstate highway system and the trucking industry, as well as the switch to high-tech industries such as electronics—industrial landscapes took on a different form. Factories began to move to industrial parks at interstate exchanges, and their architecture began to resemble other types of mass-produced architecture.

Service industries, too, produce a cultural landscape. Its visual content includes elements as diverse as high-rise bank buildings, fast-food restaurants, gasoline stations, strip malls, and the concrete and steel webs of highways and railroads. Producer services related to financial activities, such as legal services, trade, insurance, and banking, traditionally were located in high-rise buildings in urban centers, but with suburbanization they have taken on a nonurban form. Some consumer services, particularly retailing, have created distinctive and, within the American context at least, socially important landscapes. Shopping malls are now dominant features of the North American suburb, and they often serve as catalysts for suburban land development, in effect creating entirely new landscapes, all geared toward consumption. As more and more of this activity moves online, it is interesting to contemplate what sorts of landscapes will be created.

Emerging Cultural Landscapes of the Internet

Indeed, the quaternary sector landscapes of intellectual and informational activities—including the development of technology that allows the other sectors to conduct business in an online environment—are notoriously difficult to see. Yet these landscapes do exist. Named after a place that founder Jeff Bezos thought to be "exotic and different," Amazon.com

FIGURE 9.36 Amazon fulfillment center. Amazon has 140 such fulfillment centers, employing tens of thousands of people and utilizing robots to fill and ship online orders. Centers such as this one have a huge footprint, occupying over a million square feet. (Marty Bicek/ZUMA Wire/Alamy Live News)

began life as an online bookstore. Today, you can purchase just about anything you can think of from Amazon.com. Its headquarters are located in Seattle, Luxembourg, and Belgium. Nevertheless, its physical locations also include a vast global network of mammoth warehouses and fulfillment centers, as well as software development and customer service offices (**Figure 9.36**). Much of Amazon.com's business, however, is conducted in noncommercial locations such as households, with customers ordering online in their pajamas, for instance. It also occurs in non-places. Think, for example, of Amazon's cloud computing services, or the books and video only available in digital format. The company is envisioning same-day delivery of orders via drone in some urban markets as soon as the regulatory environment will permit it (**Figure 9.37**).

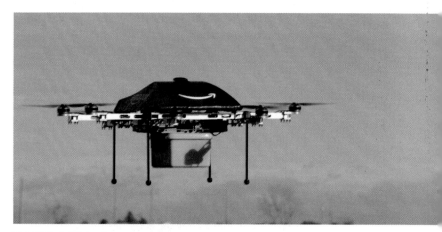

FIGURE 9.37 Delivery by drone. This drone is a prototype used by the online retailer Amazon.com. Though the service has not yet been approved by the Federal Aviation Administration, Amazon envisions using drones to deliver packages to customers' doorsteps. (AMAZON/UPI/Newscom)

Conclusion

Development has traditionally been understood in primarily economic terms, as the transition from a state of poverty to one of prosperity. Though it is true that economic activities have profoundly reshaped the world, it has long been recognized that development is about more than merely the pursuit of wealth. Quality of life, as measured by the provision of basic human needs and the attainment of happiness, is also an important aspect of development. Contemporary scholars of development geography recognize that today's world can no longer be divided into a wealthy "here" and a poor "there." Rather, a postdevelopment perspective emphasizes that poor neighborhoods and regions, where the basic indicators of an adequate quality of life are missing, can be found within some of the wealthiest nations. Additionally, the rise of the Global South underscores the power that formerly poor nations are gaining, not just economically, but as innovators and financiers of tomorrow's development efforts.

Chapter Summary

9.1 Discuss the history of development, the diverse ways to measure it, and how development has shaped, and reshaped, regions.

- Theories of development, dating from the mid-twentieth century, conceptualize development as a set of stages (Rostow's stages of economic growth), or power relationships between regions (Wallerstein's world systems theory).
- Development can be measured and mapped by many statistics, including economic ones (Gross Domestic Product, GDP per capita) as well as measures that include quality of life indicators (Human Development Index, Gross National Happiness).
- Economic activities are categorized by sectors: primary (extractive activities), secondary (industry), tertiary (services), and quaternary (information).
- The people and entities that support, fund, and enact development-related activities—development actors—include various international bodies, national governments, corporations, nongovernmental organizations (NGOs), and wealthy individuals.

9.2 Analyze the relationships between transportation and development.

- Technological innovations in manufacturing and transportation made the Industrial Revolution possible.
- The transportation infrastructures, especially roadways and railroads, of the formerly colonized nations in Africa, Latin America, and Asia were developed to serve the needs of European colonizers, and are not particularly suited to the contemporary needs of former colonies.
- Transportation haves and have-nots—in other words, those places that have invested in developing and maintaining modern, efficient transportation infrastructures, and those that have not done so—are an important aspect of development geographies.

9.3 Explain how formerly poor places are today significant agents of development.

- The rise of the so-called Global South points to the reality that nations and regions formerly dismissed as "underdeveloped" are emerging as powerful and innovative forces in today's world, while nations and regions typically viewed as wealthy and developed are not uniformly so.
- Technological innovations arising in the Global South are reshaping development geographies and closing the digital divide between those who have access to the Internet and smartphones, and those who do not.
- Development plays out unevenly by gender the world over, with women experiencing restrictions to their mobility, violence, and lower salaries compared to men.
- Technology is used to create communities of concern, enabling celebrities to become major actors in the development enterprise.

9.4 Recognize the role that nature has played over time as concepts and practices of development have changed.

- Development efforts have utilized natural resources, in some cases depleting or polluting them.

- Activities associated with development, specifically the burning of fossil fuels to provide power, have led to global climate change that is reshaping the world's map and impacting human activities.
- Decisions made by the nations and regions of the Global South about how to power their development efforts will greatly impact future global climate change.

9.5 Describe how development has shaped landscapes.

- Primary sector activities, which involve extracting raw materials from the Earth, often (but not always) create aesthetically unpleasant landscapes.

- Industrial and service sector activities have given rise to distinctive cultural landscapes marked by factories and commercial buildings such as fast-food restaurants and shopping malls.
- Internet commerce connects households, warehouses, and novel delivery methods in new ways, as well as creating ephemeral cultural landscapes such as "the cloud."

Key Terms

stages of economic growth A development model associated with the economist W.W. Rostow, positing that all nations would inevitably pass through the same sequence of stages of economic growth on their way to becoming wealthy, consumerist, and developed (page 313).

world systems theory A development model associated with Immanuel Wallerstein, categorizing world history as moving through a series of socioeconomic systems, culminating in the modern world system, which is our current, interdependent, capitalist world economy, rooted in nation-state economies (page 313).

core Regions having the most advanced industrial and military technologies, complex manufacturing systems, external political power, and the highest levels of wealth and mass consumption (page 313).

periphery countries Nations exhibiting characteristics opposite those of the core: they have relatively little industrial development, simple production systems focused mostly on agriculture and raw materials, and low levels of consumption (page 313).

core–periphery model A spatial model associated with world systems theory, of the interrelations between core regions where economic, political, and cultural powers are located, and peripheral regions that are dependent on and serve the needs of the core (page 313).

semi-periphery countries Countries exhibiting a mix of characteristics of both the core and periphery, and playing a key role in mediating politically and economically between them (page 313).

First World A term used to describe the group of wealthy, democratic, capitalist nations across the world (page 314).

Third World A term used to describe the world's poor nations as a group. During the Cold War era, these were also the nations that were neither communist nor socialist; in other words, they were not aligned with either the First or Second World (page 314).

Second World A term used during the Cold War era to describe the group of countries that were communist or socialist in their governmental philosophy, and had centrally planned rather than free-market economic systems (page 314).

Fourth World A term used to describe the world's poorest nations (page 314).

Brandt Line Division of the countries of the world into two regions: wealthy and poor. The Brandt Line is named after former West German Chancellor Willy Brandt, and was a concept commonly applied in the 1970s through the 1980s (page 314).

developing Refers to regions that are considered relatively poor, because they produce just enough, or at times not enough, food and other resources for their own immediate needs (page 314).

developed Refers to those regions that are considered wealthier, because people living there are better able to accumulate resources (page 315).

Gross Domestic Product (GDP) The total value of all goods and services produced in a country over a specific period, usually a year (page 315).

GDP per capita The mathematical result of dividing a country's GDP by its national population (the Latin *per capita* means "per head") (page 315).

Human Development Index (HDI) A numerical value devised by the United Nations Development Programme that is used to measure how well basic needs are being met. It is a composite index, derived from three areas: life expectancy, education, and income (page 317).

primary sector The set of economic activities that involves extracting natural resources from the Earth (page 319).

secondary sector Commonly referred to as manufacturing, this is the set of economic activities that involve processing the raw materials extracted by primary sector activities into usable goods (page 319).

tertiary sector The set of economic activities that refers to all the different types of work necessary to move goods and resources around and deliver them to people; in other words, service activities (page 320).

quaternary sector Intellectual and informational economic activities (page 320).

Industrial Revolution Beginning in England in the early 1700s, the Industrial Revolution saw the rapid transformation of the economy through the introduction of machines, new power sources, and novel chemical processes. These replaced human hands in the making of products, initially in the cotton textile industry (page 322).

nationalization Government takeover of transportation infrastructure, fuel sources, and industries from private foreign ownership, occurring in many previously colonized nations in the twentieth century (page 324).

postdevelopment The idea that traditional approaches to "development"—in the form of monetary aid, investments, expertise, and projects—were ineffective, merely providing a way to categorize most of the world as needing the assistance of a few wealthy nations (page 331).

imperialism A relationship whereby wealthy nations dominate poor ones by controlling their economic, political, and cultural systems (page 331).

Global South A term that has largely replaced "Third World" when referring to regions of Latin America, Africa, and most of Asia, in recognition of the fact that much of the world's dynamism, growth, and power resides in these places (page 331).

South–South cooperation Trade, technological innovation, and other forms of exchange and assistance that occur between the larger and wealthier nations of the Global South, and poorer nations (page 331).

renewable resources Those primary sector products that can be replenished naturally at a rate sufficient to balance their depletion by human use (page 336).

alternative energy Alternatives to traditional fossil fuels, including hydroelectric, solar, wind, geothermal, nuclear, biofuel, and "clean" coal (page 340).

electronic waste Used electronics, such as mobile phones, computers, office equipment, television sets, and refrigerators, which can no longer be used for their intended purpose and are discarded. Also known as e-waste (page 349).

Practice at Sapling Plus

Read the interactive e-Text, review key concepts, and test your understanding.

Story Map. Explore life expectancy in various countries.

Web Map. Examine the geography of development issues.

Doing Geography

ACTIVE LEARNING: Development by the Numbers

Hans Rosling (1948–2017) was a Swedish doctor, statistician, and showman. He was known for, among other achievements, his many TED talks, which are both entertaining and informative. In them he explores economic and social development, demonstrating that the idea of a "developed" and a "developing" world is today overly simplistic. Rather, different regions, countries, and local places have changed vastly since the mid-twentieth century, when—as you now know—much of the classical thinking on development was done. Rosling uses simple props, such as boxes he bought at the Swedish furniture store IKEA, to make his points. His presentations are supported by development data, which is projected using his signature Gapminder software, which shows graphically how a data set evolves over time. In this activity, you will explore Rosling's work.

More guidance at **Sapling**Plus.

EXPERIENTIAL LEARNING: The Where and Why of What You Wear

The textile and garment industries are a large component of some countries' secondary sector economic activities. China, Vietnam, Bangladesh, and Sri Lanka, for instance, are several of the world's leading garment producers. Textile and garment industries have historically been very mobile, moving production facilities to locations that suit manufacturers, often because of the low cost of labor. In today's global economy, that tendency is even more pronounced, with manufacturers outsourcing many aspects of production to complete different tasks in various parts of the world. It's likely, then, that the clothes and shoes you are wearing right now were designed in one place, woven into fabric in another place, and assembled in yet another.

This exercise is about tracing the "origins" of the clothes and shoes you are wearing and asking "Why?" The cost of labor is important, but it certainly isn't the only factor in determining where garments and shoes are made. In manufacturing, other factors include the locations of markets, state policies, and international treaties, such as NAFTA. Sometimes it is the lack of policies—or lack of enforcement—that manufacturers find attractive. This can lead to unsafe working conditions, as seen in Bangladesh in 2013 when a factory building collapsed, killing 1200 garment workers.

For clothing and shoes, another important factor is fashion. Styles change often, and manufacturers need to be able to change their production quickly. This would lead companies to locate manufacturing facilities close to their main markets, which might mean in the United States or Canada.

Steps to Tracing the Origins of What You Wear

Step 1: Locate as many labels as you can from the shoes and clothing you are wearing.

Step 2: Read the labels carefully and create a list of the places mentioned in the "Made in" section of the labels.

Step 3: Locate those places on a map. This step alone should give you a good sense of the global nature of this industry.

Now that you have a better understanding of the global nature of the clothing industry, consider these questions:

- Why do you think certain items are manufactured in particular places?

- What are the similarities in these countries? What are the differences?

Consider each of the locational factors we've mentioned—labor, markets, state policies, trade treaties, consumer trends—and try to generalize the why of what you wear.

Workers in Bangladesh (*top*), produce clothing that more than likely will end up in the stores where this young woman (*bottom*) is shopping. (Top: MUNIR UZ ZAMAN/AFP/Getty Images; Bottom: KidStock/Blend Images/Getty Images)

SEEING GEOGRAPHY

Mapping the Life of a Smartphone

Have you ever wondered where your smartphone started its life, and where it will go when it dies?

This photo montage depicts some of these places. To begin with, no smartphone would be truly cool without so-called rare earth elements. Yttrium, lanthanum, terbium, cerium, and five other elements are used to manufacture a smartphone's screen, speakers, and vibration unit. The properties of rare earth elements are what make your phone "light, bright, and loud," according to Cecilia Jamasmie. Ninety-five percent of rare earth minerals are mined in China's Inner Mongolia region, both because these elements are concentrated there but also because of the environmental damage caused by the techniques used to extract them, which are not allowed in other places.

The actual assembly of your smartphone occurs in a factory. Foxconn is the largest manufacturer, making smartphones as well as other electronic devices. Foxconn's largest factory, shown here, is located in Longua, a district in Shenzen, China. No one knows exactly how many employees labor here, but it is estimated that nearly half a million people work at this factory. In fact, this place is commonly referred to as "Foxconn City." Workers sleep in dormitories, shop in stores, and engage in leisure activities, all located on the factory premises. Foxconn has been criticized for harsh labor practices that result in employee suicides, unrest, and exposure to dangerous conditions. The stress of meeting production demands when, for instance, a new iPhone version is released can be overwhelming. Literally millions of new phones can be sold in a single weekend.

What happens to old smartphones? According to a Google survey, 51 percent end their lives in a drawer or closet. Some are refurbished and resold, while others are too old or broken for that. These turn into **electronic waste**. Some phones or components are sent to places like Manila, in the Philippines, where children disassemble them to scavenge copper wire, and then burn the rest. Burning e-waste is an increasing health hazard in places such as the one shown here.

Top to bottom: China's Mongolia mining region; Foxconn workers assembling smartphones; using the latest iPhone; children sifting through electronic waste in Manila, Philippines. (From the top: Ren Junchuan/Xinhua/1010292141/CHINE NOUVELLE/SIPA/Sipa Press; Ym Yik/EPA/REX/Shutterstock; KEIZO MORI/United Press International (UPI)/Newscom; CHERYL RAVELO/REUTERS/Newscom)

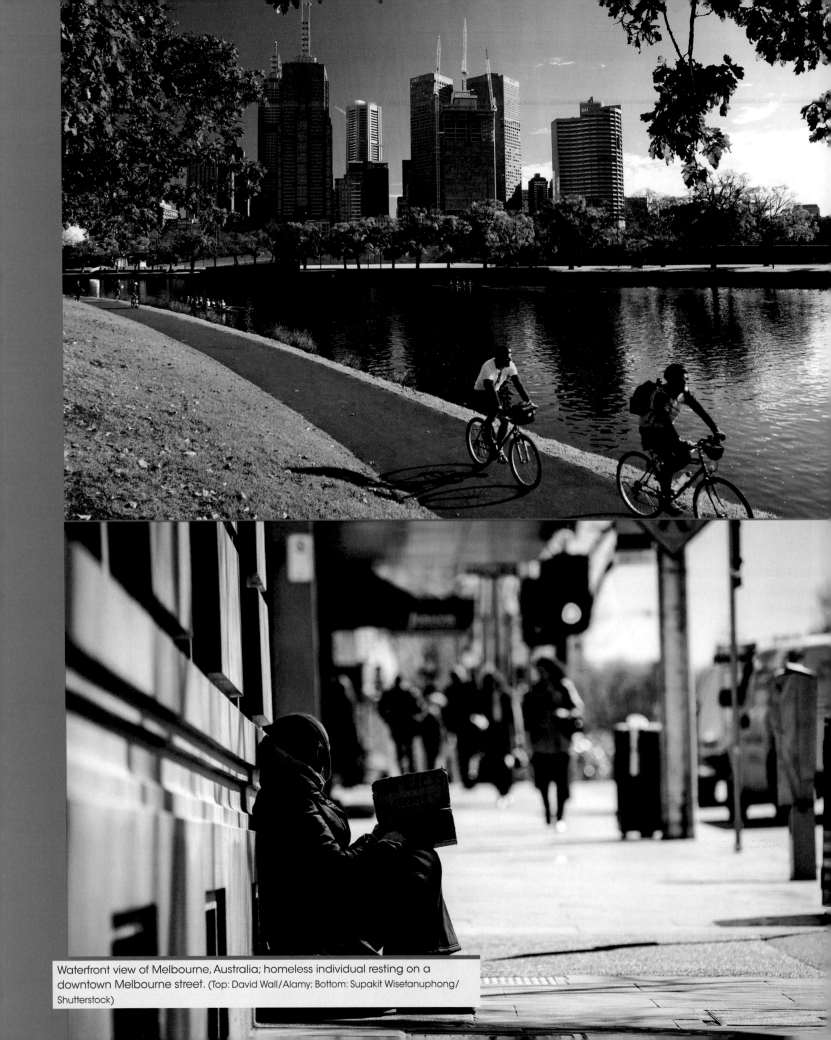

Waterfront view of Melbourne, Australia; homeless individual resting on a downtown Melbourne street. (Top: David Wall/Alamy; Bottom: Supakit Wisetanuphong/Shutterstock)

chapter 10

Urban Geography

A WORLD OF CITIES

What makes a city truly livable?

Think about these photos and question as you read.
We will revisit them in Seeing Geography on page 387.

Learning Objectives

10.1 Understand the historical and contemporary regional patterns of urbanization.

10.2 Evaluate the ways that migration has shaped the growth of cities, how cities have grown and changed, and how humans get around within cities.

10.3 Explain the notion of global cities and some of the problems associated with them.

10.4 Recognize the ways that human and natural worlds interact in cities, and the effects on the health and well-being of those living in cities.

10.5 Interpret an urban landscape, and identify the trends shaping cities today and into the future.

(David Wall/Alamy)

Imagine the 2 million years that humankind has spent on Earth as a 24-hour day. In this framework, settlements of more than a hundred people came about only in the last half hour. Towns and cities emerged only a few minutes ago. Yet it is during these "minutes" that we see the rise of civilization as we know it. *Civitas*, the Latin root word for *civilization*, was first applied to settled areas of the Roman Empire. Later it came to mean a specific town or city. *To civilize* meant literally "to citify."

In this chapter, we apply the five themes of human geography to consider overall patterns of urbanization, learn how urbanization began and developed, and discuss the differing forms of cities in the developing and developed worlds. We will see the many ways that migration has shaped the world's urban population, as well as how mobility within cities is facilitated or hindered. The spectacular rise of the so-called global city, along with its universally recognizable form but also the many social problems that go along with it, will be considered. The interaction of built environments and natural elements and processes that occur in cities is discussed. Finally, we turn to processes currently shaping cities, as well as how you might "read" urban landscapes from a cultural geography perspective.

stand at 70 percent (**Figure 10.1**). The World Bank estimates that cities grow by a combined total of 3 million people each week. The cultural geography of the world will change dramatically as we become a predominantly urban people.

We all know from our own travels, locally or internationally, that some regions contain lots of cities while others contain relatively few, and that the size, population density, and look and feel of those cities can vary greatly. How then can we begin to understand the location, distribution, and size of cities? We start with a consideration of global patterns of urbanization, examining the general distribution of urban populations around the world. A quick look at **Figure 10.2** reveals differing patterns of **urbanized population**—the percentage of a nation's population living in towns and cities—around the world. For example, the countries of Europe, North America, Latin America, and the Caribbean have relatively high levels of urbanization, with approximately 75 percent of each country's population living in urban areas. The nations of Africa and Asia, on the other hand, are less urbanized, with many countries in these regions having 40 percent or less of their population residing in urban areas. How do geographers explain these varying regional patterns of urbanization?

Region

10.1 Understand the historical and contemporary regional patterns of urbanization.

Today, just over half—54 percent—of human beings live in cities. By the year 2050, the United Nations estimates that this figure will

Patterns and Processes of Urbanization

According to United Nations estimates, almost all the worldwide population growth in the next several decades will be concentrated in urban areas, with the cities of the less developed regions accounting for most of that increase (**Figure 10.3**). The reasons

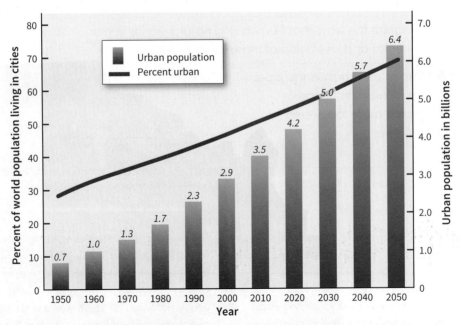

FIGURE 10.1 People living in cities. This graph depicts the changing percentage and total number of people living in urban areas across the century spanning 1950 to 2050. (Source: United Nations, Department of Economic and Social Affairs, Population Division.)

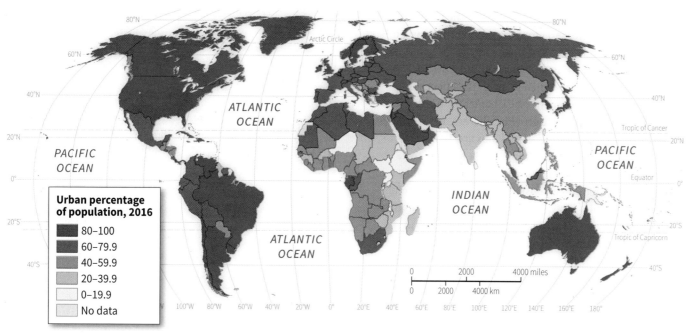

FIGURE 10.2 Urbanized population in the world. Just over half—54 percent—of the world's inhabitants live in cities. (Source: United Nations, Department of Economic and Social Affairs, Population Division.)

FIGURE 10.3 The world's largest cities in 1975, 2000, and 2025. Large cities are located in almost every world region. However, Asia and Africa will be home to the large cities of the future. (Source: Population Reference Bureau.)

for this urban population growth and its uneven distribution around the world vary, as each country's unique history and society presents a slightly different narrative of urban and economic development. Making matters more complex is the lack of a standard definition of what constitutes a city. Consequently, the criteria used to calculate a country's urbanized population differ from nation to nation. Using data based on these varying criteria would result in misleading conclusions. For example, the Indian government defines an urban center as an area having 5000 inhabitants, with an adult male population employed predominantly in nonagricultural work. In contrast, the U.S. Census Bureau defines a city as a densely populated area of 2500 people or more, and South Africa counts as a city any settlement of 500 or more people. It is important to remember, then, that an international comparison of urbanized population data can be made only by taking into account the varying definitions of a city.

Urban growth comes from two sources: the migration of people from rural areas to cities (see also the section on Mobility) and the higher natural birth rates of these recent migrants (see Chapter 3). People move to the cities for a variety of reasons, most of which relate to the effects of uneven economic development (see Chapter 9). Cities are often the centers of economic growth, whereas opportunities for land-ownership or farming-based jobs are, in many countries, rare. Because urban employment is unreliable, many migrants continue to have large numbers of children in order to construct a more extensive family support system. Having a larger family increases the chance that someone in the household will get work. The demographic transition to smaller families comes later, when a certain degree of security is ensured. Often, this transition occurs as women enter the workforce (see Chapter 3).

The Rise of Cities

Humans have not always lived in cities. Indeed, for most of our existence we humans have lived in small, nomadic groups of 25 to 50 hunter-gatherers. Such groups eventually became sedentary and formed into small villages, but these were nothing like what you would recognize as a real city today.

In the Middle East, where the first cities appeared, a network of permanent agricultural villages developed about 10,000 years ago. These farming villages were modest in size, rarely with more than 200 people, and were probably organized on a kinship basis.

Although small farming villages predate cities, it is wrong to assume that a simple change in size took place whereby villages slowly grew, first into towns and then into cities. Neolithic settlements that date from around 9000 B.C.E. (Jericho, in the modern-day Palestinian territories on the West Bank) and 7500 B.C.E. (Çatalhöyük, in modern-day Turkey) are considered by some archaeologists to be the earliest real cities (**Figure 10.4**). Others do not believe these qualified as true cities, based on their small populations and lack of apparent social stratification.

Urban settlements dating to around 4000 B.C.E. in Mesopotamia (the area between the Tigris and Euphrates rivers in modern-day Iraq) were without doubt true cities (**Figure 10.5**). Small by current standards, Mesopotamian cities covered 0.5 to 2 square miles (1.3 to 5 square kilometers), with populations that rarely exceeded 30,000. Nevertheless, the densities within these cities could easily reach 10,000 people per square mile (4000 per square kilometer), which is comparable to the densities in many contemporary cities.

By that time, agricultural technologies in Southwest Asia permitted larger settlements. True cities, then and now, differ

FIGURE 10.4 Ancient urban dwelling. A group of archaeologists uncovers the interior of a small home in the ancient Anatolian city of Çatalhöyük, revealing a living room and small bedrooms. Excavations have revealed that people kept their homes scrupulously clean as very little trash has been uncovered. Occupants entered and exited their homes through holes in the roof. Since there was no space between houses, people got from house to house by walking on the rooftops as if they were streets. (Images & Stories/Alamy)

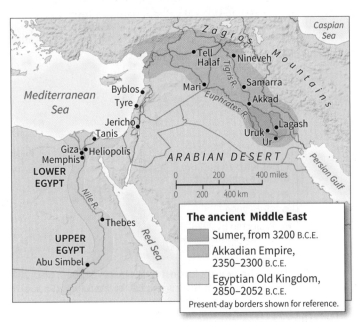

FIGURE 10.5 Early Mesopotamian and Egyptian cities. Note the location of these earliest cities: almost all are situated along rivers.

qualitatively from agricultural villages. All the inhabitants of agricultural villages were involved in some way in food procurement—tending the agricultural fields or harvesting and preparing the crops. Cities, however, were more removed, both physically and psychologically, from everyday agricultural activities. Food was supplied to the city, but not all city dwellers were involved in obtaining it.

Two elements were necessary for this dramatic social change: the creation of an **agricultural surplus** and the emergence of **socioeconomic stratification**. Surplus food, which is a food supply larger than the everyday needs of the agricultural labor force, is a prerequisite for supporting nonfarmers—people who work at administrative, religious, military, or handicraft tasks. Socioeconomic stratification, involving the hierarchical arrangement of society through the emergence of distinct socioeconomic classes, facilitates the collection, storage, and distribution of resources through well-defined channels of authority that can exercise control over goods and people. A society with these two elements—surplus food and an organized means of storing and distributing it—was prepared for urbanization.

One way to understand the transition from village to city life is to model the development of urban life, assuming that a single factor is the trigger behind the transition. The question that geographers and other scholars of urban life ask is: "What activity could be so important to an agricultural society that its people would be willing to give some of their surplus to support a social class that specializes in that activity?"

The **hydraulic civilization model**, developed by Karl Wittfogel, assumes that the development of large-scale irrigation systems was the primary driver of urbanization, and that a class of technical specialists were the first urban dwellers. Irrigation of agricultural crops yielded more food, and this surplus supported the development of a large non-farming population. A strong, centralized government backed by an urban-based military could expand its power into the surrounding areas. Farmers who resisted the new authority were denied water. The power of the elite class was further enhanced by the increased need for organizational coordination to ensure continued operation of the irrigation system. Labor specialization developed. Some people farmed; others worked on the irrigation system. Still others became artisans, creating the implements needed to maintain the system, or administrative workers in the direct employ of the urban elite's court.

Although the hydraulic civilization model fits several areas where cities first arose—China, Egypt, and Mesopotamia—it cannot be applied to all urban hearths. In parts of Mesoamerica, for example, an urban civilization blossomed without widespread irrigated agriculture and therefore without a class of technical experts.

Urban Hearth Areas

The first cities appeared in distinct regions, such as Mesopotamia, the Nile River valley, Pakistan's Indus River valley, the Yellow River valley of China, Mesoamerica, and the Andean highlands and coastal areas of Peru. These are called **urban hearth areas** (**Figure 10.6**).

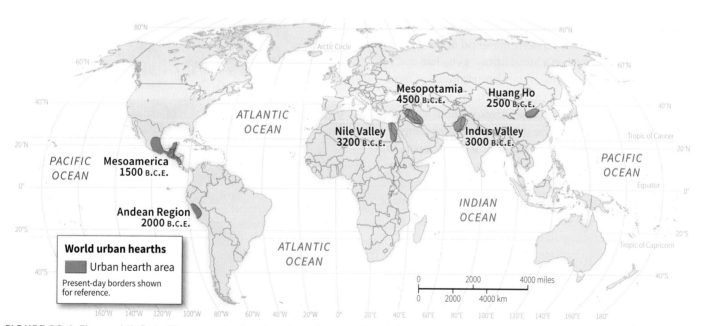

FIGURE 10.6 The world's first cities arose in six urban hearth areas. The dates shown are conservative figures for the rise of urban life in each area. For example, some scholars would suggest that urban life in Mesopotamia existed by 5000 B.C.E. Ongoing discoveries suggest that urban life appeared earlier in each of the hearth areas and that there are probably other hearth areas—in West Africa, for example (see also the World Heritage Site, the Great Zimbabwe National Monument described in Chapter 1).

Scholars have referred to these first cities in Mesopotamia and elsewhere as **cosmomagical cities**, cities that are arranged spatially according to religious principles. The spatial layouts of cosmomagical cities are similar in three important ways. First, great importance was given to the city's symbolic center, which was also believed to be the center of the known world. It was therefore the most sacred spot and was frequently identified by a vertical structure of monumental scale that represented the point on Earth closest to the heavens. This symbolic center, or **axis mundi**, took the form of the ziggurat (a massive structure having the form of a terraced step pyramid with receding levels) in Mesopotamia, the palace or temple in China, and the pyramid in Mesoamerica (**Figure 10.7**). Often this elevated structure, which usually served a religious purpose, was close to the palace or seat of political power and to the granary (a storehouse for grain). These three structures were often walled off from the rest of the city, forming a symbolic center that both reflected the significance of these societal functions and dominated the city physically and spiritually.

The second spatial characteristic common to cosmomagical cities is that they were oriented toward the four cardinal directions. By aligning the city in the north–south and east–west directions, the geometric form of the city reflected the order of the universe. It was thought that this alignment would ensure harmony and order over the known world, which was bounded by the city walls.

In all these early cities, one sees evidence of a third spatial characteristic: an attempt to shape the form of the city according to that of the universe. The arrangement of the space of the city was believed to be essential to maintaining harmony between the human and spiritual worlds. In this way, the world of humans would symbolically replicate the world of the gods. This characteristic may have taken a literal form—a city laid out, for example, in the pattern of a major star constellation. Far more common, however, were cities that symbolically approximated mythical conceptions of the universe. Angkor Thom was an early city in Cambodia that presents one of the best examples of this parallelism. An urban cluster that spread over 6 square miles (15.5 square kilometers), Angkor Thom was a representation in stone of a series of religious beliefs about the nature of the universe. Thus, the city was a re-creation on Earth of an image of the larger universe.

FIGURE 10.7 Symbolic centers of ancient cities. Top right: The Great Ziggurat of Ur was built in Mesopotamia by the Sumerians in the mid-sixth century B.C.E. Center right: The Forbidden City in Beijing, China, remains one of the best examples of the guarded, fortress-like "city within the city." Bottom right: This pyramid was at the center of the Mayan civilization at Chichén Itzá, on Mexico's Yucatán Peninsula. (Top: Everett Collection Inc/Alamy; Center: 06photo/iStock/Getty Images; Bottom: cinoby/iStock/Getty Images)

Imperialism and Urbanization

Although urban life originated at several specific places in the world, cities are now found everywhere. How did city life come to these regions? There are two possible explanations. Cities could have evolved spontaneously in different places across the globe, as populations developed new agricultural technologies allowing for a surplus and underwent socioeconomic stratification. Or contact with city dwellers, whether through trade, transoceanic voyages, or conquest, could have diffused the techniques and ideas associated with urban life to other areas.

Cities and Conquest There is little doubt that diffusion has been responsible for the dispersal of the city in historical times (**Figure 10.8**), because the city has commonly been used as a means for imperial expansion. Typically, urban life is carried outward in waves of conquest as the borders of an empire expand. Initially, the military controls the newly won lands and sets up collection points for local resources, which are then shipped back to the heart of the empire. As the surrounding countryside is increasingly pacified, the new collection points lose some of their military atmosphere and begin to show the social diversity of a city. Artisans, merchants, and bureaucrats increase in number; families appear; the native people are slowly assimilated into the settlement as workers. Finally, the process repeats itself as the empire pushes farther outward: with first a military camp, next a collection point for resources, then a

full-fledged city expressing a true division of labor and social diversity.

Such a process, however, did not always proceed without opposition. The imposition of a foreign civilization on native peoples was often met with resistance, both physical and symbolic. Expanding urban centers relied on the surrounding countryside for support. Their food was supplied by farmers living fairly close to the city walls, and tribute was demanded from the agricultural peoples living on the edges of the urban world. The increasing needs of the city required more and more land from which to draw resources. However, the peasants farming that land may not have wanted to change their way of life to accommodate the city. The fierce resistance of many Native American groups to the spread of Western urbanization is testimony to the potential power of indigenous society to defy urbanization, although the destructive long-term effects of this kind of resistance suggest that the organized military efforts of urban society were difficult to overcome.

Ancient Empires and Imperial Expansion The ancient Greek civilization, to which both Western civilization and the Western city trace their roots, expanded in part through spreading cities throughout the Mediterranean. Greek imperial cities reached as far as the north shore of Africa, Spain, southern France, and Italy. These cities were of modest size, rarely containing more than 5000 inhabitants.

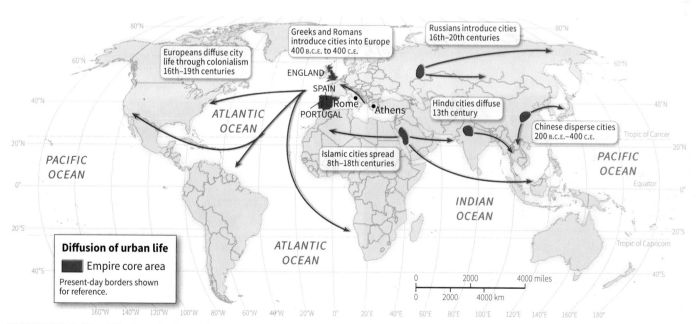

FIGURE 10.8 The diffusion of urban life. In many areas of the world, the practice of living in cities spread along with the expansion of empires.

Athens, however, may have reached a population of 300,000 in the fifth century B.C.E.

The Roman civilization supplanted ancient Greece, and was itself centered on a city: Rome. The Romans adopted many urban traits from the Greeks as well as the Etruscans, a civilization of central Italy that Rome had conquered. As the Roman Empire expanded, city life diffused farther into France, also reaching Germany, England, interior Spain, the Alpine countries, and parts of eastern Europe—areas that had not previously experienced urbanization. Most of these cities were, initially, military and trading outposts of the Roman Empire.

Fundamental to Roman cities was the grid street pattern, composed of straight streets intersecting at right angles. Indeed, transportation lay at the heart of Roman cities. Water was transported into cities from remote regions using aqueducts, many of which still punctuate the landscapes of European cities today (**Figure 10.9**). The Roman Empire itself was held together by a complicated system of roads and highways linking towns and cities. In choosing a site for a new urban settlement, the Romans made access to transportation a major consideration. By 400 C.E., the Roman Empire had declined and so too did urban life throughout much of Europe (see Chapter 3, Figure 3.32). The highway system that linked them fell into disrepair, so that cities could no longer exchange goods and ideas. Cities were either destroyed or left to decay

on their own, becoming small villages where a few hundred people eked out a living. The urban areas, along with the rest of the territory formerly under Roman rule, fell into the Dark Ages.

Cities in other parts of Europe and in other world regions, however, thrived. Spanish cities, for example, flourished as centers of learning, commerce, and governance thanks to the North African Moorish imperial presence in Iberia from the eighth through fifteenth centuries. (See the World Heritage Site described in Chapter 4—Alhambra, Generalife, and Albayzín.) Chinese cities developed their own cosmomagical features, including sacred sites connecting residents to ancestors, pagodas that functioned as axis mundi, and fabulous palaces reflecting the divine order on Earth (see Figure 10.7). In the Americas, several indigenous cultures established urban centers prior to their colonization by Spain in the fifteenth century. American cities were religious ceremonial centers, as well as control centers for the subjugation of surrounding conquered rural peoples. Some grew to a size and complexity that outpaced anything that Europeans of the time had experienced.

Tenochtitlan—the Aztec capital now known as Mexico City—was founded in 1325 on an island in the middle of Lake Texcoco, with canals providing transportation routes within the city and causeways reportedly wide enough to accommodate 10 horses leading to and from the mainland (**Figure 10.10**). By the

FIGURE 10.9 Pont du Gard, near Remoulins, France. This Roman-era aqueduct is a striking example of the infrastructure that extended throughout the Roman Empire. It was built in the first century C.E. and is one of the best-preserved aqueducts in existence. (LucynaKoch/E+/Getty Images)

FIGURE 10.10 Pre-conquest Tenochtitlan. This image, a detail from a mural by Mexican artist Diego Rivera, shows Tenochtitlan in the background. Notice how orderly, clean, and happy this painting makes this ancient city—today known as Mexico City—look. (Schalkwijk/Art Resource, NY)

time of the Spanish conquest of Mexico in 1521, the city boasted over 200,000 inhabitants, far outstripping any Spanish city at that time. Bernal Díaz del Castillo, a foot soldier in Hernán Cortés's army, described coming upon it in awe-inspired tones: "And when we saw all those cities and villages built in the water, and other great towns built on dry land, and that straight and level causeway leading to Tenochtitlan, we were astounded. These great towns and *cues* [temples] and buildings rising from the water, all made of stone, seemed like an enchanted vision. . . . Indeed, some of our soldiers asked whether it was not all a dream."

Cities of the Industrial Revolution

The agricultural innovations mentioned in the previous sections—primarily technological innovations leading to the domestication of crops and animals that allowed for the production of a food surplus—permitted the world's first true cities to develop because they facilitated support of a dense human population in one place, the rise of a nonagricultural class, and the accompanying socioeconomic stratification. The connection of agricultural innovations with the rise of cities is often referred to as the **first urban revolution**.

Cities obviously did not stop evolving, as evidenced by the many differences between early cities such as Tenochtitlan or ancient Rome and the contemporary cities you are familiar with today. Why and in what ways have cities changed over the last six millennia?

The answer to this question can be traced, in large part, to what scholars term the **second urban revolution**. Rather than being tied to technological innovations in agriculture, the second urban revolution arose as societies, initially those located in western Europe and later North America, became industrialized. At its heart, the second urban revolution developed from innovations in mining and manufacturing (see also Chapter 9). As industrialization progressed, many transformations occurred in society that fundamentally reshaped cities, both spatially and in terms of the social, cultural, political, and economic functions assumed by cities.

The transformations that culminated in capitalism had begun to reshape the cities of western Europe from the mid-sixteenth to the mid-eighteenth century. Agriculture became increasingly specialized and commercialized. Land formerly owned and worked collectively became enclosed and owned by individuals who purchased it and then paid others to work the land for them. The very landscape of the European countryside became reorganized (see Chapter 8, Seeing Geography).

Perhaps of greatest significance was how the capitalist mentality introduced a notion of urban land as a source of income. Proximity to the center of the city, and therefore to the most pedestrian traffic, added economic value to the land. Other specialized locations, such as areas close to the river or harbor, or along major thoroughfares into and out of the city, also increased land value.

In the emerging capitalist city, the ability to pay determined where one would live. The city's residential areas became segregated by economic class. The wealthy lived in the desirable neighborhoods; those without much money were forced to live in the more disagreeable parts of the city—for example, low-lying flood-prone zones, places contaminated by factory waste, or parts of the city located at higher altitudes where roadways and urban services were not easily provided (**Figure 10.11**).

In earlier urban configurations, there was no separation between one's home and one's workplace. For most people, work was organized by family units, it centered on the home, and the workday was governed by the seasons and the rising and setting of the sun. In the capitalist city, the place of work became spatially distinct from the home, such that the laborer had to travel back and forth between them. This spatial separation of work from home, of public space from private space, both reflected and helped to shape the changing worlds of men and women. In general, men generated monetary income from work outside the home and therefore became associated with moving about and occupying the public spaces of the city. Women, who primarily did domestic work, were considered the keepers

FIGURE 10.11 Early industrial-era London. This painting depicts mid-nineteenth-century London, awash in industrial air and water pollution. (Science & Society Picture Library/Getty Images)

of the private world of the home and family. They were not monetarily compensated for their labors. This association of women with private domestic space and men with public work space deepened and became more complex over the next few hundred years. (See also Chapter 3, page 93, and Chapter 9, pages 332–335.)

The center of the capitalist city—its axis mundi—was not the ziggurat, palace, or cathedral, but instead the buildings devoted to business enterprises. A downtown defined by economic activity emerged, and as industrialization progressed, it expanded and evolved into specialized districts. With the downtown devoted to commerce and industry, the upper classes moved to the outskirts of the city and built large houses that displayed their wealth.

Modern Urban Locations

The geographic criteria for city placement changed as industrialization took hold. Early cities tended to be located in places that were easily defended—on islands, hilltops, sheltered harbors, or river bends—or along trade routes (**Figure 10.12**). Industrial-era cities sought locations near sources of raw materials, such as coal, wood, or metals, used in factories. Location along trade routes continued to be important, but the parameters broadened, inasmuch as the waterways, railways, and roadways needed to accommodate increased

traffic associated with raw materials entering and finished goods exiting the sites of industrial production (**Figure 10.13**).

Geographers are, as you are no doubt aware by now, fond of using models—including maps—to represent how space is ordered in the abstract, as well as to predict and plan for its optimal ordering. Urban geographers are no exception, and they have used a wide variety of models to understand how urban space is arranged and how to guide its arrangement in better ways. **Central place theory** is the term given to a set of models that attempts to understand why cities are located where they are, as well as to help planners position cities in space most efficiently.

Central place theory is associated with Walter Christaller, a German urban geographer who first published his work on this concept in 1933. Christaller understood cities primarily as economic centers concerned with distributing goods to people, who would travel certain distances to acquire them. According to Christaller, people would travel only short distances to acquire everyday low-order goods such as food and other regularly purchased household items. However, in order to obtain specialized or costly high-order goods and services such as automobiles, the advice of a cardiologist, or to interview for a passport, he hypothesized that people would travel much longer distances. These travel distances dictated the placement of different orders of urban settlements, ranging from tiny hamlets and villages on one end, through towns and cities, and finally to regional capitals (**Figure 10.14**).

FIGURE 10.12 The classic defensive site of Mont St. Michel, France. A small town clustered around a medieval abbey, which was originally separated from the mainland during high tides, Mont St. Michel now has a causeway that connects the island to shore, allowing armies of tourists to penetrate the town's defenses easily. (RAPHO AGENCE/Science Source)

FIGURE 10.13 Pittsburgh's Golden Triangle. The Allegheny River meets the Monongahela River at the site shown in the photograph. Both rivers were early trade routes. Today, Pittsburgh's commercial center is located at their juncture. (Jose Luis Stephens/Alamy)

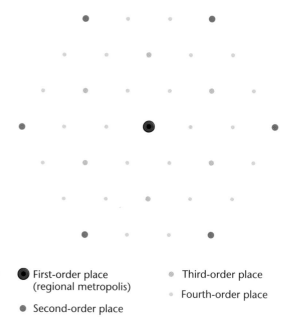

First-order place
(regional metropolis)

Third-order place

Fourth-order place

Second-order place

FIGURE 10.14 Christaller's hierarchy of central places shows the orderly arrangement of towns of different sizes. This is an idealized presentation of places performing central functions. For each large central place, many smaller places are located within the larger place's hinterland.

Because people would travel only short distances to acquire the low-order goods sold in smaller urban settlements, there were many villages and hamlets (what Christaller termed fourth-order places) and to a lesser extent towns (third-order places) located relatively close to one another. Each one could be supported by the routine traffic in low-order goods that nobody

would travel very far to purchase. In Christaller's model, smaller urban places have a lower **threshold**, the number of people required to make provision of a good feasible, and a smaller **range**, the distance people will travel to acquire a good.

Costlier and less frequent high-order purchases, however, could only be made in cities (second-order places) and regional capitals (first-order places). This is because people were willing to travel longer distances to obtain high-order goods and services (in other words, big cities have a larger range), while on the other hand, fewer urban areas could actually afford to carry these high-order goods and services because people didn't purchase them as frequently (in other words, the higher-order goods have a higher threshold).

The requirements of trade-based urban locations continue to shift to this day. For instance, San Francisco's importance as a West Coast port city was high before the advent of container ships. But once containerized cargo became the norm starting in the 1970s, San Francisco's piers were no longer needed and there was no room in the city to build the large holding lots needed for containers to go from port to rail. Oakland, San Francisco's rival city just across the bay, was quick to invest in the large cranes and other equipment needed to handle containers. Oakland was also able to accommodate containerized cargo because it filled in huge tracts of shallow bay lands, creating a massive area for the loading, unloading, and storage of cargo containers. As a result, San Francisco's importance as a port city declined, and Oakland is now the second-largest West Coast port city (**Figure 10.15**).

FIGURE 10.15 The seaport of Oakland. Cranes are loading a large container ship. Note the ample area devoted to container movement and storage at this port. Oakland is an intermodal facility, which means that containers can be loaded to and from ships, trucks, and railways. (David R. Frazier Photolibrary, Inc./Alamy)

As capitalism shifts from a focus on manufacturing goods and exchanging them for money, and toward the provision of services and information, the question of what constitutes an ideal urban location changes yet again. It is no longer as critical to be located near physical transportation routes as it once was. Indeed, an argument may be made that other sorts of networks, particularly those associated with communications technology, are just as important (if not more so) criteria for city placement. It is worth considering whether the historical legacy of where the world's major cities are located today will be outweighed by these new considerations, and whether the sites of the important cities of the future will shift accordingly.

Mobility

10.2 Evaluate the ways that migration has shaped the growth of cities, how cities have grown and changed, and how humans get around within cities.

As noted at the beginning of this chapter, we have entered an era in which more people live in cities than in rural areas—quite a historical milestone. As we saw in the preceding section, when cities first arose, only a very small portion of the population lived in them. But over time, cities began to diffuse across the globe, and established cities grew larger. In this section, we look at the process of rural-to-urban migration that has, both in the past and today, fueled the growth of cities. Some cities have grown to a very large size, either in absolute population terms or relative to the other cities around them. We will take a look at where these very large cities are located, where they will appear in the future, and what role such cities play in a country's urban hierarchy. Within urban areas, transportation that facilitates the movement of people has also shaped the size and layout of cities. In this section, we focus particularly on the advent of the automobile and its central role in shaping human mobility in and between cities.

Rural-to-Urban Migration

Increasing urbanization is caused by three related phenomena: natural population increase (see Chapter 3), rural-to-urban migration, and urban-to-urban migration. The last two phenomena can be domestic, as in a family moving from one city to another in the same country, or the result of international migration from one country to another. Although the United Nations estimates that natural increase accounts for the majority of the recent growth in urban population, it was rural-to-urban

migrations that historically brought millions of people into urban life.

As we have seen, wide-scale urbanization occurred hand in hand with industrialization. As capitalism took root and flourished, first in Europe and North America and subsequently shaping life everywhere in the world, cities allowed for the large workforces needed to run factories and to provide services of all sorts to factory owners and laborers. In the United States, rural-to-urban migration resulted in the growth of large cities such as Chicago in the nineteenth century, and in Asia large urban centers such as Tokyo and Mumbai in the twentieth century. Today, Europe and more recently North America as well as much of Latin America are highly urbanized societies, with about three-quarters of their populations living in cities. (See the World Heritage Site and Seeing Geography features in this chapter.)

China presents a particularly interesting example of rural-to-urban migration. Until the late 1970s, the country was predominantly rural, and its communist government was explicitly antiurban. Indeed, many urban Chinese were forced out of cities to work on farms during the decade of the Cultural Revolution from 1965 to 1975, under the logic that cities fostered class privilege. State policy reform after 1978 lifted restrictions on internal migration, which enabled its rural population to become more mobile. Many Chinese abandoned life in the countryside for futures in China's booming industrial cities, among them Guangzhou, Shanghai, and Beijing. Though around 56 percent of China's population is urban today, it is estimated that by 2025, 70 percent of China's population will be urban, and there will be 240 Chinese cities with a population over 1 million (**Figure 10.16**). China's "urban billion" is expected to materialize by 2030.

Today, particularly across Africa and Asia, large numbers of people still depart their rural villages and migrate to cities in search of better economic and social opportunities. The United Nations predicts that about four-fifths of the world's future urban growth will result from rural-to-urban migration in these two regions. Others will be forced from their lands due to rural conflict, natural disaster, climate change, or seizure of their land. Thus, cities can increase in size not necessarily because there is economic growth requiring urban laborers, but rather because conditions in the countryside are dismal. People leave their farms and rural villages to escape economic, political, and environmental hardships, in the hope that urban life will offer a slight improvement. Often it does. Given the new global economy, however, many of the employment opportunities available in such cities are low-skilled and low-paying manufacturing and service jobs with harsh working conditions. The result is that rural-to-urban migrants often find themselves either unemployed or with jobs that barely provide a decent living (**Figure 10.17**).

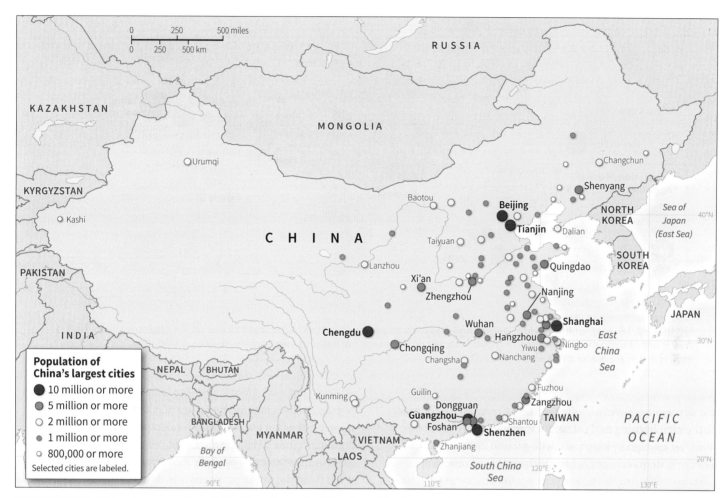

FIGURE 10.16 China has more than 100 cities of over 1 million inhabitants, a number that is expected to surpass 200 by the year 2025. Long focused on the coast, China's urban expansion is moving inland.

FIGURE 10.17 Beggar at an intersection in Islamabad, Pakistan. Begging in traffic is common in major cities around the world. Beggars can be recent immigrants to the city or people who have difficulty securing employment. (FAROOQ NAEEM/AFP/Getty Images)

When Are Cities Too Large?

Although rural-to-urban migration affects nearly all cities in the developing world, the most visible cases are the extraordinarily large settlements we call **megacities**, those having populations of over 10 million. **Figure 10.18** shows the world's 30 largest cities, a majority of which are located in the developing world. This is a major change from 50 years ago, when the list was dominated by Western, industrialized cities.

The urban population in the developing world is growing at impressive rates. Indeed, the World Bank estimates that most of the urban growth in the world today—over 90 percent of it, in fact—is occurring in the developing world. Seventy million people move into cities in the developing world each year. The world's two poorest regions, South Asia and sub-Saharan Africa, expect their urban populations to double by 2030.

And with this incredible increase in the sheer numbers of urban dwellers in the less developed regions of the world comes a large list of problems. Unemployment rates in cities of

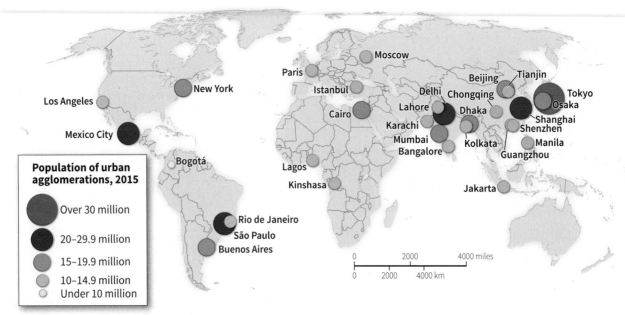

FIGURE 10.18 Map of the world's megalopoli. This map shows the locations of the world's large urban agglomerations. Notice how many of them that have arisen or grown in recent decades are clustered in Asia and Africa. (Source: United Nations, Department of Economic and Social Affairs, Population Division.)

the developing world are often over 50 percent for newcomers to the city; housing and infrastructure frequently cannot be built fast enough to keep pace with growth rates; water and sewage systems can rarely handle the influx of new people. Consequently, one of the world's ongoing challenges will be how to cope with the radical restructuring of population and culture as people in developing countries move into cities.

Urban Primacy

The destination for much of the rural-to-urban migration in the developing world is the **primate city**. This is a city that, because it is so much larger than all the others, dominates the economic, political, and cultural life of a country. Some scholars maintain that in order to qualify as a primate city, a city must be 10 times or more larger than the next largest city. With 13 million residents, Buenos Aires is an excellent example of a primate city because its population is exactly 10 times that of the next largest city, Rosario, which is the second largest city in Argentina, with a population of 1.3 million. Although many developing countries are dominated by a primate city, often a former center of colonial power, urban primacy is not unique to these countries: think of the way London, Moscow, Athens, and Paris dominate their respective countries.

These primate cities are often the ones that claim scholars' and policy makers' attention because of their dominance and large size. However, the United Nations finds that most of the world's

population lives in mid-sized cities of 500,000 or less. Indeed, the majority of urban growth today is taking place in smaller, less dominant cities. Urban planners and policy makers are beginning to focus their attention on these mid-sized cities scattered throughout the world, particularly in developing countries, in order to examine the impacts of increased urbanization.

This is true of the United States, too, where so-called legacy cities—well-known, established, large urban centers such as San Francisco, New York, Chicago, and Los Angeles—are not growing nearly as fast as mid-sized cities, particularly those located in the Sunbelt, known as "aspirational cities"—among them Raleigh, Houston, Phoenix, San Antonio, and Las Vegas. Legacy cities have simply priced all but the elites out of their housing markets. Lower-skilled people are attracted to aspirational cities because they offer employment, affordable housing, and a lower cost of living. Many of the cities on *Forbes'* list of "coolest cities" are aspirational cities. Youth, ethnic and racial diversity, and a growing slate of hip downtown amenities like coffee shops and craft breweries, combined with employment and affordability, have earned these cities this distinction (**Figure 10.19**).

Urban Transportation

Getting around inside the city, and transporting people and goods from city to city, has—for as long as cities have existed—been a major concern. You will recall that Rome was famous

FIGURE 10.19 Houston's hipster gentrifiers. These urbanites are partaking in Houston's happening cultural scene. This restaurant has more than 75 beers on tap. Houston has gone from cow-town to cool-town thanks to the influx of young gentrifiers. (RICHARD CARSON/ REUTERS)

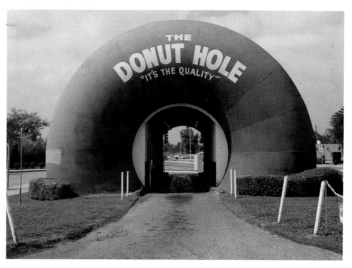

FIGURE 10.20 Drive-through architecture. This uniquely American architectural form emphasizes convenience and accessibility. As this photo shows, drive-through architecture can also be quirky. The Donut Hole drive-through is located in Los Angeles, California— ground zero for the drive-through. (Barry Winiker/Getty Images)

for its engineering marvels, including grid-patterned streets facilitating traffic flow within cities, and the remarkable network of roadways connecting cities across the Roman Empire.

Railroads, streetcars and trolleys, light rail systems, airplanes, buses, and subways have all presented innovations in urban transportation modes and infrastructure. These innovations have shaped and reshaped the layout and size of cities over time. For instance, the advent of streetcars—fixed rail systems within many U.S. cities that arose in the 1880s—led to the development of so-called **streetcar suburbs**. Streetcar suburbs allowed urban residents to live farther away from their workplaces than they could in the past, when commuting was done by foot or horse.

No innovation in transportation has had a more profound impact on cities, however, than the automobile. In the United States, Henry Ford's mass-produced Model T initiated the norm of automobile ownership among the middle classes, and by 1920 over a million had been sold. Automobiles increased personal mobility exponentially, and it has been said that Henry Ford "freed common people from the limitations of geography." Automobiles brought with them a network of roadways and, by the 1950s, interstate highways. Housing architecture was reshaped around the automobile with the addition of the now-ubiquitous garage. Drive-through commerce and the architectural forms associated with it became a hallmark of America's obsession with cars (**Figure 10.20**).

Because so much of the middle class owned a car, it became possible to locate housing further from the urban **central business district (CBD)**, as its placement was not limited to locations close to existing train lines. As long as

roads could be built to connect these settlements to the CBD, the new automobile suburbs would be viable. Cities grew ever outward in a process that came to be known as **sprawl** (**Figure 10.21**). Some U.S. cities, such as Miami, Atlanta, Las Vegas, and Phoenix, are known as **automobile cities** because they have been significantly built up recently enough that their layouts are primarily oriented around cars.

FIGURE 10.21 Aerial view of downtown Phoenix, Arizona. Phoenix is a quintessential automobile city. Note the straight streets that intersect at right angles, and the fact that it sprawls out into the surrounding desert for miles. Only with an automobile or another form of transportation such as a bus, train, or subway can an individual truly move about comfortably in such a city. The distances are simply too far to bike or walk. (NNehring/E+/Getty Images)

The Downsides to Urban Car Culture

Automobiles also brought undesirable consequences to cities. Cities became increasingly polluted with the fossil fuel residue released by the automobile's internal combustion engine, affecting the health of city residents. Safety, too, became a concern, and automobile fatalities—for both passengers and pedestrians—is the leading cause of accidental death. Traffic congestion is a major complaint, and many people squander time and money idling in traffic as they make their way to work, commuting from far-out suburban residences or nearby towns to workplaces (**Table 10.1**). In addition, the norm of individual commuters who spend a great portion of their day in a car has led to social isolation and inactivity. Some urban and suburban roads no longer have sidewalks, which has made pedestrian traffic—and the face-to-face social encounters seen by many sociologists to be at the heart of the urban experience—all but disappear.

In fact, there seems to be a growing disenchantment with automobiles, particularly among young people in Europe and the United States. French teenagers, for example, are less likely to hold a driver's license: in 1983, 35 percent of 18- and 19-year-olds did, but by 2010 only 19 percent did. The numbers for U.S. teens show a similar decline, from 46.2 percent of 16-year-olds in 1983 to 24.5 percent in 2014. Researchers speculate that concerns over the costs of owning, operating, insuring, and maintaining a car have driven this decline, as has a disenchantment among today's youth with the negative environmental impacts of automobiles along with a growing use of online technology as the preferred way to connect with friends, work, and shop.

This phenomenon has led to a rise in alternative modes of urban transportation. So-called millennials—people now in their teens and twenties—are opting to bike or take public transportation rather than to drive or even own cars. Car and bike rental services have sprung up in cities across the globe (**Figure 10.22**). For a nominal fee, a user can avoid the cost and hassle of owning a bike or car, and simply rent it for the time needed, or subscribe for a monthly fee. In the United States, reservations and payments are often made through smartphones and the user simply parks the vehicle at a rental station when done. In Europe, the process is even more flexible, as drivers locate the nearest available car using their smartphone and simply pick up the vehicle wherever in the city the previous user parked it. As a result, fewer people own cars, and fewer families own more than one car, which reduces the need for parking spaces and curtails pollution.

FIGURE 10.22 Bike rental. This woman is unlocking a Citi Bike in New York City's Greenwich Village, a service for which she pays an annual membership fee. When she is done riding, she can simply dock the bike at any Citi Bike kiosk. Bike rental and sharing services are helping to reduce the congestion, contamination, and dangers associated with automobile traffic. (Courtesy of Patricia L. Price.)

TABLE 10.1 Top Ten Worst Commutes in the United States

Rank	City	Average Daily Commute (in minutes)
1	San Jose, CA	39
2	Jersey City, NJ	38
3	Long Beach, CA	36
4	Oakland, CA	35
5	New York, NY	34
6	Mesa, AZ	33
7	Aurora, CO	32
8	Arlington, TX	32
9	Chicago, IL	30
10	Philadelphia, PA	29

Source: U.S. Census Bureau, American Community Survey, 2016.

Globalization

10.3 Explain the notion of global cities and some of the problems associated with them.

Many of the globalizing forces we have already discussed in this book—the integration of international economies, the interaction of different peoples, and the reshaping of culture—are centered in urban life. In other words, cities are the places where one can see both the multitude of benefits of globalization and the numerous pitfalls that it can bring. In this way, all the world's cities can be considered global: they are places being shaped by the new global forces.

Global Cities

Narrowly defined, **global cities** are those cities that have become the control centers of the global economy—the sites of major decisions about the world's commercial networks and financial markets. Rather than merely being big cities, global cities house a concentration of multinational and transnational corporate headquarters, international financial services, media offices, and related economic and cultural services. According to sociologist Saskia Sassen, who coined the term "global city" in 1991, there were at that time only three such cities operating at this level: New York City, London, and Tokyo. These cities had become, in many ways,

the headquarters for a global economy, and they form the top level of a hierarchical global system of cities.

Yet an argument could certainly be made that cities have been, for at least six thousand years, a globally relevant mode of human dwelling in the world. To be sure, there were not as many cities back then, nor were they as large as today's megacities, but—as we have seen—cities in the ancient world were certainly powerful places. As industrialization progressed, cities became larger, more powerful centers of decision making, and through rural-to-urban migration they affected the lives of greater numbers of people.

Contemporary definitions of what constitutes a global city focus on the possession of certain characteristics. Among them are economic might, political power, information exchange, and research and development activity, as well as less tangible qualities such as influence, engagement, the ability to attract talented individuals, diversity, cultural experience, accessibility, and environmental quality. Being labeled a global city is desirable, and therefore many urban officials lobby to make the list of top global cities and—unsurprisingly—the number of lists has proliferated. Yet the same 20 or so cities appear on most of the lists, though their exact place on a certain list may vary (**Figure 10.23**).

Urban geographers have broadened the analysis of global cities in order to also understand the next tier of cities within the urban hierarchy—those that contain a large percentage of producer service firms (law, accounting, advertising, financial services, consulting) that are international in scope. In this framework, cities are categorized by the number and type of transnational firms they house—not only headquarters but

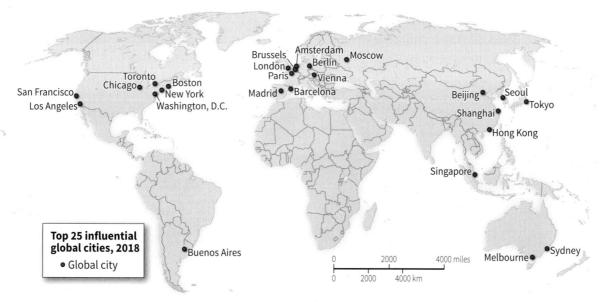

FIGURE 10.23 Top 25 global cities. This map shows the location of the world's top 25 global cities according to the annual A.T. Kearney Global Cities Report. That list used the following factors to determine global city status: business activity, human capital, information exchange, political engagement, and cultural experience. Note the world regions that have the most global cities, as well as those having none. (Source: Data from the A.T. Kearney 2018 Global Cities Report.)

This World Heritage Site encompasses a portion of the city of Rio de Janeiro, Brazil. Specifically, it includes the natural elements located within the urban boundaries that have fundamentally shaped the city of Rio: the peaks of the Tijuca National Park, the Botanical Gardens established in 1808, and the hills surrounding Guanabara Bay. These distinctive urban landscapes have been celebrated artistically in paintings, poetry, song, and literature.

- From 1763 to 1960, Rio de Janeiro was the capital of Brazil. In an effort to decentralize the national population away from the southeastern coast, the capital was relocated to Brasília, in the country's interior (see Figure 5.27).

- Stretching from the mountains to the sea, this World Heritage Site is bounded by the best vantage points for appreciating the interaction of nature and culture.

- As with many Latin American urban landscapes, Rio's is one of expensive high-rise housing located immediately adjacent to *favelas* (slums).

CULTIVATING NATURE: The word Carioca refers to a resident of Rio de Janeiro. It is derived from the indigenous Tupi word for "houses of the whites," referring to the Portuguese settlements built along the river in the sixteenth century C.E. Today, Carioca indicates both a dialect of Brazilian Portuguese, as well as the friendly, relaxed, and cosmopolitan personality associated with Rio de Janeiro. Carioca culture is defined by its outdoor living, framed by the beaches and mountains, but also the streets and parks. Urban planners have conscientiously highlighted the city's dramatic local natural features, and have enhanced them through landscaping and constructed elements of the urban environment.

- This World Heritage Site is home to the largest urban forest on the planet: the Tijuca Forest. It is but a remnant of the Atlantic forest ecosystem, which once extended along the length of Brazil's coastline. From the founding of Rio in 1565 through the mid-1800s, trees were felled for timber and fuel for the expanding city. In 1860 Emperor

(JW/Radius Images/Getty Images)

Pedro II ordered the replanting of the forest with native species, acknowledging that the forest was essential for the city's environmental and cultural well-being.

- Although it is known for its natural features, this site also includes low-income urban neighborhoods. With their tendency to be built on the city's steep, forested hillsides, favelas such as Santa Marta put contemporary environmental pressures on the natural landscape.

TOURISM: Frequently referred to as Cidade Maravilhosa (the Marvelous City), Rio de Janeiro is the country's top destination for leisure travelers. It is widely described as one of the most beautiful cities in the world.

- Rio has been a global tourist attraction since the early 1900s, when it constructed one of the world's first aerial cable car lines to the top of a monolithic mountain called Sugar Loaf.

- Regardless of their religious affiliation, tourists from around the globe are keen to visit the world's largest statue of Jesus, located on Corcovado Mountain in the Tijuca Forest.

- Rio de Janeiro hosted the World Cup soccer games in 2014 and the 2016 Summer Olympics. Brazil hosted 6.6 million foreign visitors in 2016. The city's infrastructure is currently being upgraded to handle the expected influx of tourists.

- As with any large city, street crime is a concern. As of early 2014, assaults, theft, and muggings had increased dramatically in Rio in advance of the World Cup soccer games and again with the 2016 Olympic Games.

(Alex Robinson/Getty Images)

- Rio de Janeiro is known as one of the world's most beautiful cities, in great part because of how planners have showcased its dramatic natural features.
- This city is home to the world's largest urban forest.
- The Carioca Landscapes between the Mountains and the Sea were designated a World Heritage Site in 2012.

whc.unesco.org/en/list/1100

(Christopher Pillitz/Getty Images)

also regional and national offices. This analysis has identified globally and regionally dominant cities and those that are major participants in the new global economy. Geographers Yefang Huang, Yee Leung, and Jianfa Shen refer to these as international cities—places that are significant because they are centers of the new international economy. Their analysis allowed them to identify degrees of internationalization based on the number and locations of the international firms each city contained. The result is a very interesting list organized into classes of international cities, dominated by six in Class A (London, New York, Hong Kong, Tokyo, Singapore, and Paris), followed by 10 in Class B and 44 in Class C. In this way, we can begin to understand globalization as a multidimensional process that has impacts on many cities previously left unexamined.

The Globalization of Urban Wealth and Poverty

Whatever the precise parameters used to rank them, the interconnected and powerful nature of global cities is what makes them stand out from all other urban centers. And with the rise of the global city comes the rise of the global cosmopolitan class, consisting of those individuals who, in the words of geographers Sam Schueth and John O'Loughlin, "belong to the world." Rather than being especially identified with one country, the global cosmopolitan moves with ease from place to place.

Ease of movement is usually only possible when one possesses wealth, and global cities are home to a great many wealthy individuals. Indeed, the presence of significant numbers of high net worth individuals is one criterion for inclusion as a global city, because this indicates a spatial concentration of economic power (**Figure 10.24**). And, though London tops the list when it comes to concentration of high net worth individuals, the super-wealthy—those with over $30 million in net assets—may be found nearly worldwide.

The tastes and needs of the global cosmopolitan class have shaped the global city in particular ways. One of these is the prevalence of highly secure residential enclaves known as **gated communities**. Cities throughout the world are home to residential enclaves that are more or less privately governed: they have their own services such as security and landscaping, their own oversight board that sets and enforces policies, and their own amenities such as clubhouses, golf courses, and swimming pools. Some communities have walls or fences around them, use surveillance technology to track entry and exit, and restrict access through gates to those who live there or have been approved for temporary entry, such as the guests of residents. Others do not have obvious barriers setting them apart from the surrounding urban areas. Regardless, according to geographer Renaud Le Goix and urban planner Chris Webster, "The issue is not really the gate as such, it is the fragmentation of the urban governance realm into micro-territories."

Approximately 15 percent of the housing stock in the United States is privately governed, though it is much more

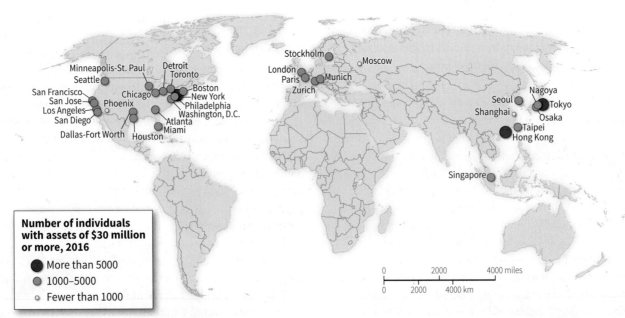

FIGURE 10.24 Leading global cities of the super-rich. This map of the world's top 30 global cities shows urban clusters of HNWIs (high net worth individuals). HNWIs are those who possess more than $30 million in net assets. There are around 173,000 HNWIs globally, which amounts to 0.002 percent of the world's population. One-third of them reside in the United States, and together the world's HNWIs hold around $20 trillion in global wealth. (Source: http://www.citylab.com/work/2013/03/global-cities-super-rich/4951/.)

FIGURE 10.25 South African gated community. This "security village" in Johannesburg is fortified with high walls, guarded gates, and electric fencing. South Africa's well-to-do, concerned about urban crime, are driving a huge and growing security industry in the country. (Eden Breitz/Alamy)

FIGURE 10.26 Poverty amidst plenty. This trash-strewn vacant lot is located in Los Angeles's Watts neighborhood. Though Los Angeles is home to the glitterati of Hollywood and Beverly Hills, it also has poor neighborhoods characterized by higher than average numbers of gangs and households headed by single parents. (Julie Dermansky/Getty Images)

common for newer housing. And while it may be tempting to assume the gated community is a U.S. export, the researchers note it is an urban form that has arisen independently in cities for diverse reasons (**Figure 10.25**). Most of the reasons, however, can be traced to rising income inequality and fear. The wealthy in global cities have responded to this situation by attempting to shut out the have-nots, both from their residential spaces and from their urban governance processes. In some ways, then, the urban wealthy have opted out of participation in the global city.

Scholars of urban poverty have traditionally focused on the specific ways that poverty is manifest in urban environments located in developing regions. But recent economic hardship in Europe and North America has affected cities in these regions in many ways. Unemployment, deterioration of housing and urban infrastructure, homelessness, and other manifestations of poverty in the urban landscape are quite visible (**Figure 10.26**). You have encountered many examples thus far in this textbook of how urban poverty can be a problem in wealthy places as well as poor ones (see, for instance, Figures 3.35 and 3.36 in Chapter 3 and Figures 9.21 and 9.24 in Chapter 9). So, as with the rise of global cities and the global cosmopolitan class discussed above, it is important to bear in mind that urban poverty is not limited to the developing world but is rather a global phenomenon. Most global cities have pockets of both extreme wealth and extreme poverty.

However, in cities of the developing world, the combination of large numbers of migrants and widespread unemployment leads to overwhelming pressure for low-rent housing in ways not frequently seen in cities in wealthier regions. Governments have rarely been able to meet these needs through housing projects, so one of the most characteristic landscape features of cities in the developing world has been the spontaneous construction of impoverished housing in areas known as **shantytowns**.

Shantytowns, also called favelas, barriadas, or squatter settlements, are often simply referred to as slums—areas of degraded, precarious, inadequate, and often illegal housing (see also Chapter 3, pages 110–111. The United Nations defines a slum household as a cohabiting group that lacks one or more of the following conditions: an adequate physical structure that protects them from extreme climatic conditions, sufficient living area such that no more than three people share a room, access to a sufficient amount of water, access to sanitation, and secure habitation or protection from forced eviction. Over 1 billion people in the world live in slums: that is one out of every seven people on Earth. In greater Manila alone, for instance, scholars estimate that almost half of the city's 19 million inhabitants live in slum conditions within shantytowns.

In his book *Planet of Slums*, urban theorist Mike Davis speculates that the proportion of humanity living in slums will increase sharply in the next few decades, to 2 billion or more, constituting a potentially volatile political group. Yet historically, shantytowns simply constitute the margin of urban growth in developing world cities. Because governments do not or will not expend the resources to build housing for the poor, the poor are forced to undertake this task themselves.

Shantytowns usually begin as collections of crude shacks constructed from scrap materials; gradually, they become increasingly elaborate and permanent. Paths and walkways link houses, vegetable gardens spring up, and often water and electricity are bootlegged into the area so that a common tap or electrical outlet serves a number of houses. At later stages, municipal governments install paved roads, electricity, and sewage systems and typically initiate some process—usually timed with the municipal elections cycle—to grant residents legal occupancy to shantytown lands. Thus, the city proper eventually takes most shantytowns under its wing, and in many instances these supposedly temporary settlements become permanent parts of the city, functioning as regularized and recognized urban neighborhoods. Some former slums in Mexico City, for example, have become so established that they are now attracting the wealthy and undergoing a process of gentrification (page 378).

Given their similarities across the globe, it is fair to ask whether global cities are converging in terms of how they look and "feel" to their residents (**Figure 10.27**). Is there, in other words, an emerging global city model? How much does it really matter if you are part of the global cosmopolitan class in Melbourne or Mumbai? Or poor in Shanghai or São Paulo? This question is far too complex and subjective to be addressed appropriately here, but it is certainly worth considering.

FIGURE 10.27 Global high-rise landscapes. These photos were shot in Toronto, Dubai, Taipei, and Kuala Lumpur (Clockwise from top left: naibank/Moment/Getty Images; Iain Masterton/Getty Images;SeanPavonePhoto/iStockphoto/Getty Images; Martin Puddy/Stone/Getty Images)

Nature-Culture

10.4 Recognize the ways that human and natural worlds interact in cities, and the effects on the health and well-being of those living in cities.

At first glance, cities seem totally divorced from the natural environment. What possible relationships could the shiny glass office buildings, paved streets, and high-rise apartment complexes that characterize most cities have with forests, fields, and rivers? In this section, we examine several different ways to think about the interactions between urban life and its natural setting. We start by exploring the climate of cities, particularly the vegetation, weather, and hydrology unique to urban environments. We then turn to an examination of the reciprocal relationship that exists between increasing urbanization and global environmental problems such as climate change and vulnerability to natural disasters, and how cities are becoming increasingly resilient in the face of these challenges.

Urban Weather and Climate

Cities alter virtually all aspects of local weather and climate. Temperatures are higher in cities, rainfall increases, the incidence of fog and cloudiness is greater, and levels of atmospheric pollution are much higher.

The causes of these changes are no mystery. Because cities cover large areas of land with streets, buildings, parking lots, and rooftops, about 50 percent of the urban area is a hard surface. Rainfall is quickly carried into gutters and sewers, so that little standing water is available for evaporation. Because evaporation removes heat from the air, when moisture is reduced, evaporation is lessened and air temperatures are higher.

Moreover, cities generate enormous amounts of heat. This comes not just from the heating systems of buildings but also from the heat generated by automobiles, industry, and even human bodies. The hard surfaces so prevalent in urban areas not only repel water, they also retain heat very effectively. This results in a large mass of warmer air sitting over the city, called the **urban heat island** (**Figure 10.28**). The urban heat island causes yearly temperature averages in cities to be 3.5°F (2°C) higher than in the countryside; during the winter, when there is more city-produced heat, the average difference can easily reach 7°F to 10°F (4°C to 5.6°C).

Urbanization also affects precipitation. Because of higher temperatures in the urban area, snowfall is about 5 percent less than in the surrounding countryside. However, rainfall can be 5 to 10 percent higher in cities as well as areas downwind from them. The increased rainfall results from two factors: the large number of dust particles in urban air and the higher city temperatures. Dust particles are a necessary precondition

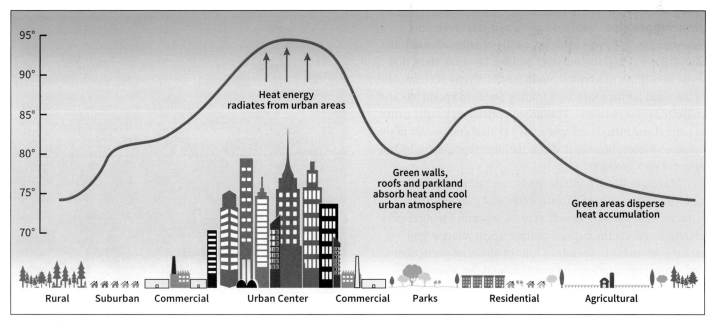

FIGURE 10.28 Urban heat island. Cities experience higher than normal air temperatures, thanks primarily to the heat-retaining and -deflecting properties of urban building materials. Cooling these overheated buildings stresses urban power grids, particularly in the summer. In addition, it is expensive and contaminates the atmosphere.

for condensation, offering a nucleus around which moisture can collect. An abundance of dust particles, then, facilitates condensation. That is why fog and clouds are usually more frequent around cities.

Urban Hydrology Residential areas are usually the greatest consumers of water in urban areas. Water consumption can vary, but generally each person in the United States uses from 80 to 100 gallons (303 to 379 liters) per day in a residence. Residential demand is greater in drier climates as well as in middle- and high-income neighborhoods. Higher-income groups usually have a larger number of water-using appliances, such as washing machines, dishwashers, and swimming pools.

Not only is the city a great consumer of water; it also alters water runoff patterns in a way that increases urban flooding. Urbanization can increase both the frequency and the magnitude of flooding because cities create large impervious areas where vegetation has been replaced with pavement, and water cannot soak into the earth. Instead, precipitation is converted into immediate runoff. It is forced into gutters, sewers, and stream channels that have been straightened and stripped of vegetation, which results in more frequent high water levels than may be found in a comparable area of rural land. Furthermore, the time between rainfall and peak runoff is reduced in cities; there is more lag in the countryside, where water runs across soil and vegetation into stream channels and then into rivers. So, because of hard surfaces and artificial collection channels, runoff in cities is concentrated and immediate, leading to urban flooding.

Urban Vegetation Until a decade ago, it was commonly believed that cities were made up mostly of artificial materials: asphalt, concrete, glass, and steel. Studies, however, show that about two-thirds of a typical North American city is composed of trees and herbaceous plants (mostly weeds in vacant lots and cultivated grass in lawns). This urban vegetation, usually a mix of natural and introduced species, is a critical component of the urban ecosystem because it affects the city's topography, hydrology, and meteorology.

More specifically, urban vegetation influences the quantity and quality of surface water and groundwater; reduces wind velocity as well as turbulence and temperature extremes; affects the pattern of snow accumulation and melting; absorbs thousands of tons of airborne particulates and atmospheric gases; and offers habitat for mammals, birds, reptiles, and insects, all of which play some useful role in the urban ecosystem. Furthermore, urban vegetation influences the propagation of sound waves by muffling much of the city's noise; affects the distribution of natural

and artificial light; and, finally, is an extremely important component in the development of soil profiles that, in turn, control hillside stability.

Our urban settlements are still closely tied to the physical environment. Cities change these natural processes in profound ways, and we must understand these disturbances in order to make better decisions about adjustments and control.

Cities and Environmental Vulnerability

As we examined earlier (see also Chapter 9), industrialization goes hand in hand with urbanization, so the environmental impacts of industrialization are often found in cities. But urbanization generates its own set of environmental impacts: supplying enough energy, food, and water to large concentrations of people puts an array of stresses on the natural environment. Scholars refer to the extent of these varied impacts of urban areas on the environment as the **urban footprint**. For example, Las Vegas is one of the fastest-growing urban areas in the United States, yet it is located in a desert (**Figure 10.29**). The city's primary water source is the Colorado River, but in more ways than one the costs of delivering that water are extremely high. It is expensive to construct the dams and infrastructure necessary to move the water into the city, and the environmental damage to the surrounding region has been very costly. The dams alter the flows of water through the Colorado

FIGURE 10.29 Desert suburb. This suburb of Las Vegas extends into the desert. Providing water to its residents is costly and causes environmental damage. (Cameron Davidson/Getty Images)

Valley, harming fish and disrupting aquatic life cycles, and the energy required to divert the water to the city leads to higher sulfur dioxide emissions, which have been shown to contribute to global climate change. Las Vegas's urban footprint, in other words, is large.

Environmental Effects The environmental effects of increased urbanization—the urban footprint—can also spread beyond the immediate stresses caused by higher concentrations of population in a place. Urban living often encourages rising levels of consumption, as new urbanites gain access to better jobs and acquire more disposable income. This growing demand for things like different foods can impact areas far from urban centers. For example, as the United Nations report *Unleashing the Potential of Urban Growth* suggests, tropical forests in Tabasco, an area 400 miles (644 kilometers) from Mexico City, have been transformed into cattle-grazing areas in response to urbanites' demands for meat. In a second example, a major factor contributing to the deforestation of the Amazon is the increased demand for soybeans from the newly urbanizing regions of China as well as the urbanites of the United States, Japan, and Europe.

Urbanization, however, does not necessarily imply environmental degradation (see Subject to Debate on page 377). For example, the higher densities of population in cities can be seen as a form of sustainable growth. Half the population of the world—the urban half—lives on approximately 3 percent of the Earth's landmass. The concentration of people in cities, therefore, opens up other areas that can be protected and left relatively free from human use. Many countries in the developing and developed worlds are working on local and regional planning projects that, on the one hand, maintain urban boundaries and prevent sprawl, and, on the other hand, protect natural environments outside urban areas from the sorts of environmental degradations we have just described. These sorts of urban environmental conservation projects will become increasingly important in future years.

Natural Disasters If you follow the news, you are surely aware of the natural disasters that seem to strike urban areas with a certain vengeance: tornados, floods, hurricanes, landslides, and earthquakes are just some of them. For the most part, there is nothing particular to cities as such that makes them more vulnerable to natural disasters. Yet in urban areas, with their dense concentrations of people, disasters are more destructive because so many lives and livelihoods are at risk. In addition, since many cities are located near rivers or coasts, they often lie in the direct paths of disasters such as hurricanes and tsunamis.

Making matters even worse, scientists have recently noted an increase in the number of natural disasters, attributable in part, they believe, to global climate change (**Figure 10.30**). Sea level rise and storm surges have particularly affected cities because, as we just noted, so many of them are located near coastlines. Rising temperatures have also led to extended summer heat waves that affect city residents. This combination of increasing urbanization and a growing number of natural disasters means that more and more people's lives are being impacted adversely. It is most often poor people who are the most affected, since they tend to live in structures that are not well built and in sections of the city that are more vulnerable. In cities in the developing world, shantytowns are frequently situated on steep hillsides or poorly drained areas, making these areas far more vulnerable to landslides, flooding, and other natural disasters.

The International Red Cross and Red Crescent Societies have termed this disparity in vulnerability to disaster the **urban risk divide**: as the world's population becomes increasingly concentrated in large cities, disasters and disaster risk become an urban phenomenon. Some cities are better prepared for disasters when they strike—a component of resilience, discussed in the next section—while others, whether for lack of funding or concern on the part of government officials, are not. Compare, for instance, the human toll taken by two major earthquakes in 2010: in Chile, the death toll from an 8.8-magnitude quake was in the hundreds, while in Haiti, the toll from a quake of slightly less magnitude was over 200,000 dead and more than 1 million homeless.

The wealthy, however, are not immune to natural disasters. Hurricane Sandy, which hit the coastline of New Jersey and New York in October 2012, disproportionately affected somewhat wealthier individuals. The storm caused $62 billion in damages and loss, with 375,000 homes destroyed, subway tunnels flooded, and business activities curtailed. (Also see Figure 3.26.) Without a doubt, global climate change and its urban impacts are a worldwide concern.

Urbanization, Sustainability, and Resilience

How have cities—specifically, urban residents and governing bodies—responded to the negative environmental impacts associated with their presence and growth and the challenges raised by natural disasters, particularly global climate change? As with many geographic questions, the answer depends on where you look. Some cities have been leaders in creatively confronting environmental challenges and reducing risk, while others have lagged behind. You might be surprised by which

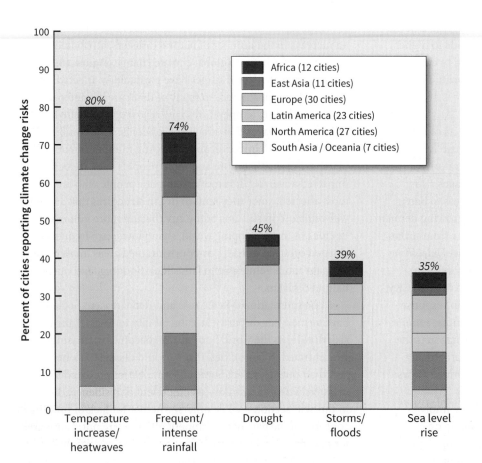

FIGURE 10.30 Urban risks arising from climate change. In the 110 cities surveyed by the Carbon Disclosure Project (CDP), these are the top five reported physical risks posed by climate change. (Source: http://www.cdpcities2013.net/doc/CDP-Summary-Report.pdf, p. 5.)

areas of the world make the top of the list in the area of disaster readiness (**Figure 10.31**). As a region, Latin American cities lead the pack, with 95 percent of the region's cities engaging in some form of climate change adaptation planning, while U.S. cities lag behind all other places at 59 percent. Why is the United States so far behind everyone else in this area? The reasons are varied and can differ from city to city. Generally, however, the conversation about global climate change is a relatively recent one, coming at a time when many U.S. cities are facing severe budgetary shortfalls and even declaring bankruptcy. In addition, there are still many local as well as national policy makers in the United States who, for political or ideological reasons, do not believe that global climate change is a reality.

Cities can take some measures that attempt to mitigate, or lessen, the impact of climate change on natural systems, humans, and the urban built environment. Mitigation attempts to reduce the extent of climate change by decreasing the emission of greenhouse gases into the environment by cities. Such mitigation measures can include restrictions on urban industry regarding pollutant emissions, shifting to low-carbon and renewable fuels, and reforestation to reduce atmospheric carbon dioxide levels.

The very architecture of buildings can also be reconfigured in ways that mitigate climate change. For instance, the installation of so-called green roofs and green walls provides a

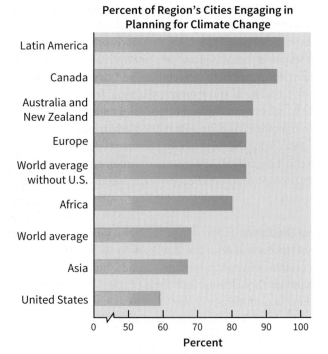

Percent of Region's Cities Engaging in Planning for Climate Change

FIGURE 10.31 Climate change adaptation planning. This chart shows the percentage of a region's cities that are planning for climate change. The world average of cities engaging in some planning, including U.S. cities, is 68 percent; without U.S. cities, the percentage rises to 84. (Source: Carmin et al., 2012.)

SUBJECT TO DEBATE Can Urbanization Be Environmentally Sustainable?

The fact that more than half of the world's population now lives in cities rather than rural areas has refueled the debate over the environmental impacts of increasing urbanization. On the one hand, increasing urbanization can be seen as efficient. Cities concentrate human populations at one point in space, which leaves surrounding lands available for conservation, agriculture, or other low-density uses. Cities create economies of scale because more people can benefit from the urban infrastructure—transportation, power supply routes, water servicing, and so on—allowing these services to be delivered at a lower cost per person. On the other hand, these same concentrations of population lead to greater demands on resources that must be brought in from outside of the city, such as water, energy, and food. City residents, with their relative wealth, tend to demand more and better resources, and such consumer demands, in turn, affect large swaths of countryside outside the city that are recruited to meet these demands. Are cities, then, sustainable?

First, we have to think about what *sustainable* actually means, and this issue itself is subject to debate! For our purposes, **sustainable urbanization** is the creation of a situation whereby a society can meet the needs of contemporary urban dwellers for water, food, and shelter, while not damaging the ability of future urban dwellers to meet their needs. It was only in the late nineteenth and twentieth centuries, with industrialization (see Chapter 9), that cities in Europe and North America grew rapidly in size, adding stress to the environment. That stress has been further exacerbated by the recent and rapid growth of cities in the developing world. And it has been this recent and rapid growth that has caused alarm among policy makers and scholars.

On one side of the debate are those who focus on the adverse environmental impacts of large concentrations of people. Not only do cities have a direct impact on the land, in terms of the amount of space they occupy and the pollution they create; they also have indirect impacts, since urban dwellers typically have higher incomes and expectations in terms of material satisfaction than do rural dwellers, which places additional stress on the environment. More consumer demand for items such as meat, wood, and metals from city dwellers in, say, Vancouver, can have adverse environmental impacts throughout Canada and other parts of the Asia-Pacific Rim.

Those who are optimistic about creating a more sustainable future with increasing urbanization argue that higher-density human settlements are better for the environment in the long run than less dense settlements. Even with the current growth rate of cities in terms of size and number, recent estimates based on satellite data show that urban settlements occupy only 2.8 percent of the total land on Earth. Urban spatial expansion per se does not appear to be a major environmental stressor. Urban density actually helps maintain fragile ecosystems by keeping them free from human interference. Proponents of this side of the debate argue that policies geared toward dispersing the world's population away from big cities are misguided. The real problem, they argue, rests with unsustainable forms of production and consumption, and it is on managing these issues that the world's attention should be focused.

Continuing the Debate

The debate over the environmental sustainability of urbanization is certain to grow in intensity as the world's population continues its urban course. After all, many, many people are drawn to urban life. Given what you have read here, consider the following questions:

- Can an increasingly urban world become sustainable? How?

- Does your city have a sustainability plan? If so, how might you learn more about this plan and its feasibility?

- How vulnerable is your city to environmental or man-made threats? Is your city among those that have begun to engage in resilience planning (see the Nature–Culture section in this chapter)?

Aerial view of New York City. Because of the high density of its population and its mass transportation system, New York City is considered to be a relatively "green" city. Urban parks, like Central Park shown here, often provide important sanctuaries for flora and fauna in the heart of the city. (JGI/Daniel Grill/Getty Images)

surface that does not retain heat (thereby addressing the urban heat island phenomenon), and absorbs water (addressing rapid runoff) and carbon dioxide. In addition, these surfaces calm urban noise levels and provide a pleasant environment for urban humans, as well as refuge for urban wildlife (**Figure 10.32**).

Yet in many ways, the conversation on climate change is shifting from a focus on mitigation or even **sustainability**— attempting to preserve conditions as close as possible to what they have been in the past—to include the notion of resilience. **Resilience** involves the ability to better predict future trends rather than looking backward to how things used to be. Resilience involves an acceptance of climate change as the

new normal, and an emphasis on how to most appropriately respond to threats—both human (e.g., terrorism) and natural—in order to minimize risk to human lives, livelihoods, and property.

How cities respond to climate change will be a defining feature of life on Earth in the coming decades.

Cultural Landscape

10.5 Interpret an urban landscape, and identify the trends shaping cities today and into the future.

As you probably already know from your local, regional, or international travels, the look and "feel" of cities varies from place to place. Some cities have bustling downtowns where most of the residents work and perhaps live, while others are characterized by nodes of activities clustered on the suburban fringe; some cities are built around a set of historical buildings with political or religious significance, while the centers of other cities are dominated by new skyscrapers housing financial offices. Finally, some cities' streets are laid out in a gridlike pattern, while in other cities streets are a crisscrossed jumble without any apparent order.

How can we make sense of this range of urban landscapes? In Doing Geography on page 386, you can try your hand at reading the urban landscape in the way that cultural geographers do. Here, we will discuss some of the major trends reshaping the look and feel of cities across the globe: gentrification, efforts to create more livable cities, and movements to "take back the city" from developers who have increasingly privatized formerly public urban spaces.

Gentrification

Beginning in the 1970s, urban scholars began to observe what seemed to be a trend opposite to suburbanization. This trend, called **gentrification**, was the movement of upper-middle-class people into deteriorated areas of city centers (see also Chapter 2). Today, some older suburbs are becoming gentrified, as well as some neighborhoods that began their lives as shantytowns (as we saw earlier with respect to Mexico City; **Figure 10.33**). Gentrification often begins in an inner-city residential district, with gentrifiers moving into rundown housing or even spaces that didn't start out as housing—for example, warehouses. These places are more affordable than newer suburban housing and are often closer to workplaces or mass transit lines. The infusion of new capital into these neighborhoods usually results in

FIGURE 10.32 Green walls. The city of Sydney, Australia, has developed a "Green Roofs and Walls" initiative in an attempt to encourage more beautiful and environmentally friendly urban surroundings. (James D. Morgan/Getty Images)

FIGURE 10.33 Global gentrifiers. Condesa is a neighborhood in Mexico City—probably not a place you associate with gentrification. Yet several Mexico City neighborhoods, some that started their lives as lower-income settlements, have become targets of gentrification because of their desirable locations. (Eye Ubiquitous/Nick Bonetti/Alamy)

higher property values, and this, in turn, often displaces existing residents who cannot afford the higher prices. The displaced residents are disproportionately poor, immigrant, and racialized minority populations. Their displacement opens up even more housing for gentrification, and the gentrified district continues its spatial expansion.

Commercial gentrification usually follows residential gentrification, as new patterns of consumption are introduced into the inner city by the middle-class newcomers. Pedestrian shopping districts with expensive boutiques and art galleries attract cultural consumers, and bars and restaurants catering to this new urban middle class provide entertainment and nightlife for the gentrifiers (see Figure 10.19).

The speed with which gentrification has proceeded in many of our cities, and the extent of changes—in the urban landscape, but also culturally, economically, and politically—that accompany gentrification are causing dramatic shifts in the urban fabric. What factors have led to gentrification?

Economic Factors Some urban scholars look to broad economic trends to explain gentrification. Throughout the post–World War II era in North America and Europe, most investments in metropolitan land were made in the suburbs; as a result, land in the inner city was devalued. By the 1970s, many home buyers and commercial investors found land in the city much more affordable, and a better economic investment, than in the higher-priced suburbs. This situation brought capital into areas that had been undervalued and thus accelerated the gentrifying process.

In addition, most Western countries have been experiencing **deindustrialization**, a process whereby the economy is shifting from one based on secondary industry to one based on the service sector. This shift has led to the abandonment of older industrial districts in the inner city, including waterfront areas. Many of these areas are prime targets of gentrifiers, who convert the waterfront from a noisy, commercial port area into an aesthetic asset. In Buenos Aires, for example, one of the new gentrified neighborhoods is Puerto Madero, an area that was once home to docking facilities and a wholesale market (**Figure 10.34**). The shift to an economy based on the service sector also means that the new productive areas of the city are dedicated to white-collar activities. These often take place in relatively clean and quiet office buildings, contributing to a view of the city as a more livable environment.

Political Factors Many metropolitan governments in the United States, faced with the abandonment of the central city by the middle class and therefore with the erosion of their tax base, have enacted policies to encourage commercial and residential development in downtown areas. Some policies provide tax breaks for companies willing to locate downtown; others furnish local and state funding to redevelop central-city residential and commercial buildings.

At a more comprehensive level, some larger metropolitan areas have devised long-term planning agendas that target certain neighborhoods for revitalization. Often this is

FIGURE 10.34 Puerto Madero, Buenos Aires. These expansive dock facilities in Buenos Aires have been converted by the city and private entrepreneurs into a center of nightlife for the city, complete with clubs, restaurants, and shopping. New condominiums and hotels are now under construction. (Courtesy of Mona Domosh.)

accomplished by first condemning the targeted area, thereby transferring control of the land to an urban-development authority or other planning agency. Such areas are frequently older residential neighborhoods originally built to house people who worked in nearby factories, which have usually been torn down or transformed into lofts or office space. The redevelopment authority might locate a new civic or arts center in the neighborhood. Public-sector initiatives often lead to private investment, thereby increasing property values. These higher property values, in turn, lead to further investment and the eventual transformation of the neighborhood into a middle- to upper-class gentrified district.

Sexuality and Gentrification Gentrified residential districts are often correlated with the presence of a significant gay and lesbian population (Chapter 2, pages 51–53). It is fairly easy to understand this correlation. First, the typical suburban life tends not to appeal to people whose lifestyle may be regarded as different and whose community needs frequently diverge from those of people living in the child-centric suburbs. Second, gentrified inner-city neighborhoods provide access to the diversity of city life and amenities that often include gay cultural institutions. In fact, the association of urban neighborhoods with gays and lesbians has a long history. For example, urban historian George Chauncey has documented gay culture in New York City between 1890 and World War II, showing that a gay world occupied and shaped distinctive spaces in the city, such as neighborhood enclaves, gay commercial areas, and public parks and streets.

Yet, unlike this earlier period, when gay cultures were often forced to remain hidden, the gentrification of the postwar period has provided gay and lesbian populations with the opportunity to reshape entire neighborhoods actively and openly. Indeed, many geographers have explored how gays and lesbians played very influential roles in the gentrification of certain neighborhoods, which have become heavily identified with these populations. Other geographers have focused on black gentrifiers in New York's Harlem and Chicago's South Side neighborhoods. Both areas of work underscore that it is not always affluent whites who are the primary movers and shakers in gentrification. More recently, however, urban geographers have pointed out that while these contributions are important, it is vital to remember that gentrification is mostly driven by and for affluent white urban populations.

The Costs of Gentrification Gentrification often results in the displacement of lower-income people who are forced to leave their homes because of rising property values. This displacement can have serious consequences for the city's social fabric.

Because many of the displaced people come from disadvantaged groups, gentrification frequently contributes to racial and ethnic tensions. Displaced people are routinely forced into neighborhoods more peripheral to the city, a trend that only adds to their disadvantages. In addition, gentrified neighborhoods usually stand in stark contrast to surrounding areas where investment has not taken place, thus creating a very visible reminder of the uneven distribution of wealth within cities.

The success of a gentrification project is usually measured by its appeal to an upper-middle-class clientele. Indeed, the so-called hipster gentrifiers—artists, musicians, youth, and other creative types initially attracted to these lower-cost neighborhoods—are often displaced by later waves of wealthier incomers who find the "cool factor" of such up-and-coming neighborhoods irresistible. Richard Campanella refers to the final wave as "*bona fide* gentry, including lawyers, doctors, moneyed retirees, and alpha professionals from places like Manhattan or San Francisco." Thus, gentrification ultimately gravitates toward the suburban notion of residential homogeneity, eliminating what many people consider to be a great asset of urban life: its diversity and heterogeneity.

The Livable City Probably for as long as humans have lived in cities, there have been some who have dedicated themselves to contemplating exactly what makes city life so attractive, pondering as well how we might go about making it better. Some five thousand years ago, urbanites in Mesopotamia, Egypt, and the Indus River valley planned their cities to have straight streets that intersected at right angles and devised drainage systems. Since then, we have attempted to invent better—more beautiful and more efficient—ways to house, employ, and entertain city residents; to convey people and the things they need within cities and between them; and to fortify cities for defense from attack.

What makes a city livable? In many ways, the answer to that question depends on who you ask. Certainly, beautiful surroundings—natural and man-made—are important. Clean air and water, public transit, freedom from crime, historical importance, accessibility, affordability—all of these factors and more are also vital components of making a city not just livable, but truly great. Indeed, making the list of "the world's most livable cities" is something many cities would like to achieve; you might not be surprised to learn then that many such lists exist. And given that there is no real way to measure and thereby quantify many of the positive aspects of cities listed above, a lot of subjectivity goes into compiling these lists. In 2017, for instance, Melbourne, Australia, was at the top of the Economic Intelligence Unit's "Most Livable Cities" list (see also Seeing Geography, page 378). Yet other listings place

Vancouver, Canada, in the top spot. All of the lists, however, skew notably toward Australia, New Zealand, Canada, and continental Europe.

Refocusing cities on people seems to be at the heart of movements to make cities more livable. Moving away from individually owned and operated automobiles as the drivers, literally and figuratively, of our cities is part of the solution (**Figure 10.35**). As we discussed earlier in this chapter, more and more cities are adopting bike- and car-sharing programs and encouraging mass transit, carpooling, biking, and walking as inexpensive and less polluting ways to get around.

As our reliance on the automobile decreases, cities can be reshaped. No longer dependent on paved roadways, parking lots and garages, and long hours commuting alone in cars across sprawling urban jungles, people are freer to spend time in their neighborhoods. Space becomes available for meeting places such as plazas, parks, and pedestrian areas. With fewer automobiles, city spaces become safer for bicyclists and pedestrians. Danish architect Jan Gehl describes the multiple activities that city residents can engage in when they inhabit more livable city spaces: "purposeful walks from place to place, promenades, short stops, longer stays, window shopping, conversations and meetings, exercise, dancing, recreation, street trade, children's play, begging and street entertainment."

The Right to the City

Throughout this chapter, we have noted that the processes that change cities—their layout, demographic profile, ecology, and what it means to live in them—affect people differently. Almost invariably, immigrants, racialized minorities, and the impoverished experience negative consequences stemming from urban transformations. These are the people who are displaced when neighborhoods gentrify. They are most at risk from natural disasters and climate change. They have the least access to transportation, housing, and urban services.

We have also seen, for instance in the case of gated communities, that some processes at work in cities are explicitly aimed at excluding disadvantaged people from access to certain parts of the city. To provide an additional example, think about the last time you visited a shopping mall. Did you see a sign like the one pictured in **Figure 10.36**? Though it may come as a shock to many, shopping malls are, in fact, not public spaces. While they may look open to the public—indeed, some open-air shopping malls are not enclosed by any physical barrier such as walls or fences—most shopping malls are owned by developers and function as private spaces. The mall developers look out for the business interests of their tenants—the stores renting space in the mall. When groups of youths, or homeless individuals, use shopping malls for social gatherings, to take care of daily

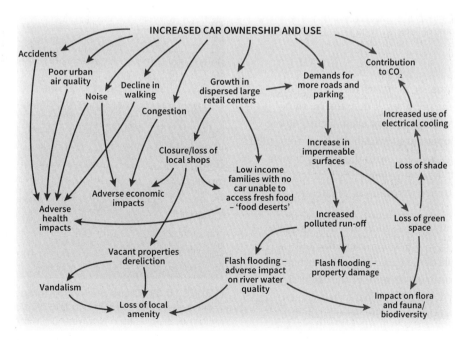

FIGURE 10.35 Cars and urban consequences. This flowchart, adapted from one produced by the government of the United Kingdom, shows the various negative environmental and social outcomes of increased car ownership and use.

these spaces are becoming private property, and developers can then decide who stays and who goes. Some beaches, for instance, have been purchased by hotel and condominium developers who then restrict access to residents and guests only. Surveillance technologies such as drones and cameras may be used to watch people in public spaces and have them removed based on what is deemed threatening or disruptive behavior. The spaces themselves may be configured in ways that make them inaccessible to certain people or for certain purposes. This area of urban planning goes by the name of "crime prevention through environmental design." For instance, benches in waiting areas for public transit are often constructed in ways that deter people from sleeping on them (**Figure 10.37**).

Are our cities made more or less livable by such measures? What is lost when access to city spaces is curtailed? What is gained? Who benefits, and who loses out?

FIGURE 10.36 Code of conduct. Though you may think of shopping malls as public places, they are not. Some malls have adopted parental escort policies or parental patrols, and ban groups of unaccompanied minors from occupying the mall during certain days or times. (Courtesy of Patricia L. Price.)

needs like sleeping or bathing, or as venues for begging, they are swiftly escorted from the premises. If they return, such individuals can be cited for trespassing or even disturbing the peace. Why? As one developer put it, his Albany, New York, mall "had become a babysitting service . . . with thousands of teen mall rats roaming around." Rambunctious crowds of teens at the food court and theater entrance were seen as frightening away older, wealthier shoppers.

But what about the truly public spaces of cities: the parks, pedestrian promenades, plazas, libraries, sidewalks, and beaches? Are certain people denied access to public urban spaces? The answer, unfortunately, is yes (see the discussion of public space in Chapter 1, page 12). What urban scholar Henri Lefebvre called the **right to the city**, specifically the right of all urban residents to be in and move through public urban places, is becoming increasingly curtailed. More of

FIGURE 10.37 Crime prevention through environmental design. These benches, located in a public area outside a halfway house for former drug users in New York City's Flatiron district, have either bars or curved surfaces to deter sleeping. The benches are referred to as "bum proof." (Courtesy of Patricia L. Price.)

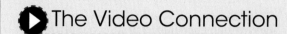

The Video Connection

"Vic Invades": The Life of an Urban Explorer
Watch at **Sapling Plus**

This video follows the adventures of Vic, a Brooklyn youth engaged in exploring the subway tunnels and skyscrapers of New York City. UrbExers enter urban spaces deemed off limits, both to photograph them and for the thrill of discovery.

Thinking Geographically

1. Vic considers himself to be, first and foremost, an artist. At one point in the video he remarks that the city is "one big canvas." Is what Vic and his associates are doing indeed art, trespassing, or a combination of the two?

2. How do UrbExers like Vic understand Henri Lefebvre's concept of the right to the city?

3. Geographers have, historically, been interested in the exploration of space and documenting that exploration for the purposes of ownership, conquest, and imperial expansion. Can Vic and his associates be viewed through this lens? How or how not?

Conclusion

The first cities arose as new technologies—particularly the domestication of plants and animals—facilitated the concentration of people, wealth, and power in a few specific places. This transformation from village to city life was accompanied by increased social stratification. Although the first cities developed in specific hearths, urban life has now diffused worldwide, and the majority of the world's population has become urban.

The future of cities is an evolving question. Focused urban planning measures might alleviate many present-day ills, but long-range hope lies with endeavors to make cities more livable and resilient.

Chapter Summary

10.1 Understand the historical and contemporary regional patterns of urbanization.

- Levels of urbanization, and the growth rate of the urbanized population, vary from one world region to another.
- The world's first cities developed around 10,000 years ago in Mesopotamia, in connection with the creation of an agricultural surplus and the emergence of socioeconomic stratification.
- The layout of early cities mirrored religious or mythical principles.
- Cities diffused throughout the world along with conquest and imperial expansion.
- The Industrial Revolution in western Europe and North America led to specialization in how urban land was used, and increased spatial segregation of the urban population.
- The criteria for situating cities have changed since ancient times, shifting from site-specific factors such as defense and trade routes, toward location with respect to market factors.

10.2 Evaluate the ways that migration has shaped the growth of cities, how cities have grown and changed, and how humans get around within cities.

- Rural-to-urban migration has been, and will continue to be, the major driver of a regional demographic shift away from the countryside and toward cities.
- Very large cities—either in absolute terms, or relative to the other cities—present particular challenges.

- Innovations in transportation have profoundly reshaped cities.
- Urban car culture is giving way, in an increasing number of cities, to alternate modes of transportation, such as bike sharing and rental.

10.3 Explain the notion of global cities and some of the problems associated with them.

- Global cities are not simply large urban centers; they are also powerful regional and national drivers of progress.
- A global cosmopolitan class shapes cities in specific ways—for instance, in the prevalence of gated communities.
- The urban poor are found in all cities, shaping them in particular ways—for example, in the existence of shantytowns at the urban periphery.

10.4 Recognize the ways that human and natural worlds interact in cities, and the effects on the health and well-being of those living in cities.

- Cities create their own distinct weather and climate patterns, as manifest in increased heat, water consumption and runoff, and the role of urban vegetation.
- The environmental impact of cities on surrounding areas can be seen as both beneficial and detrimental.
- How cities are affected by, and respond to, global climate change can be viewed through the lenses of sustainability and resilience.

10.5 Interpret an urban landscape, and identify the trends shaping cities today and into the future.

- Cities and urban populations are constantly undergoing demographic, cultural, political, and economic shifts.
- The move to make cities more livable centers on improving the lived experience of cities—in particular, refocusing on people and away from reliance on automobiles.
- A variety of subtle and not-so-subtle practices shape who can go where in the city.

Key Terms

urbanized population The proportion of a country's population living in cities (page 352).

agricultural surplus The amount of food grown by a society that exceeds the demands of its population (page 355).

socioeconomic stratification The hierarchical arrangement of society through the emergence of distinct socioeconomic classes (page 355).

hydraulic civilization model Hypothesizes that urban civilization arose from large-scale crop irrigation (page 355).

urban hearth areas Regions in which the world's first cities evolved (page 355).

cosmomagical cities Types of cities that are arranged spatially according to religious principles; characteristic of very early cities (page 356).

axis mundi The symbolic center of cosmomagical cities, often demarcated by a large vertical structure (page 356).

first urban revolution The connection between agricultural innovations and the rise of the world's first true cities (page 359).

second urban revolution The connection between industrial innovations and the rise of capitalist cities (page 359).

central place theory A set of models designed to explain the spatial distribution of urban centers (page 360).

threshold In central place theory, the size of the population required to make the provision of goods and services economically feasible (page 361).

range In central place theory, the average maximum distance people will travel to purchase a good or service (page 361).

megacities Particularly large urban centers (page 363).

primate city A city of large size and dominant power within a country (page 364).

streetcar suburbs The extension of urban residential areas along streetcar lines in the late nineteenth century (page 365).

central business district (CBD) A dense cluster of offices and shops located at the city's most accessible point, usually its center (page 365).

sprawl The tendency of cities to grow outward in an unchecked manner, which is particularly notable in automobile cities (page 365).

automobile cities Cities whose spatial layout—in terms of both extent and form—is dictated by the near ubiquity of individual automobile ownership (page 365).

global cities Cities that are control centers of the global economy (page 367).

gated communities Highly securitized residential enclaves that are more or less self-governing (page 370).

shantytowns Precarious and often illegal housing settlements, usually made up of temporary shelters and located on the outskirts of a large city (page 371).

urban heat island A mass of warm air generated and retained by urban building materials and human activities; it sits over the city and causes urban temperatures to be greater than those of surrounding areas (page 373).

urban footprint The spatial extent of the impacts of urban areas on the natural environment (page 374).

urban risk divide As the world's population becomes increasingly concentrated in large cities, disasters and disaster risk become an urban phenomenon (page 375).

sustainable urbanization The creation of a situation whereby a society can meet the needs of contemporary urban dwellers for water, food, and shelter, while not damaging the ability of future urban dwellers to meet their needs (page 377).

sustainability The ability to use resources in a way that does not deplete them over the long term (page 378).

resilience The ability to recover quickly from adversity (page 378).

gentrification The displacement of lower-income residents and economic activities by higher-income residents and activities; frequently associated with a restoration of buildings in deteriorated areas of the city (page 378).

deindustrialization The decline of industry, accompanied by a rise in service and information sectors of the economy (page 379).

right to the city The notion that all urban residents, not just the privileged, should be able to access city spaces and have a voice in how the city is shaped and used (page 382).

Practice at SaplingPlus

Read the interactive e-Text, review key concepts, and test your understanding.

Story Map. Explore patterns of land use in large cities.

Web Map. Examine issues related to urban geography.

Doing Geography

ACTIVE LEARNING:
Adapting to Climate Change

For this exercise, you will explore some of the topics discussed in the Nature–Culture section of this chapter. Specifically, you will take a closer look at how global climate change is playing out in a place you have lived and how your campus is—or is not—dealing with climate change. Some places have done next to nothing to adapt to climate change, while others are incredibly proactive. What should the locations you have chosen be doing to prepare?

More guidance at 🅜 SaplingPlus

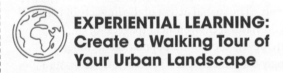

EXPERIENTIAL LEARNING:
Create a Walking Tour of Your Urban Landscape

Even without extensive training in geography, most of us are avid "readers" of our landscapes. We drive through residential neighborhoods and make judgments about the types of people who live there; we walk downtown and, finding that the streets grow narrower, surmise that they were built a long time ago, before automobile traffic was a factor; we marvel at how high buildings rise into the sky, knowing that those with offices at the top make "top" dollar. But reading landscapes can be tricky business. What we see in front of us may not always be as it appears, nor built when and for whom we think. This exercise, then, is about interpreting your own urban landscape and about understanding the limitations of that interpretation.

Your goal in this exercise is to create a walking tour of part of your city that can be used by tourists (or anyone, really) and that is both informative and engaging. In other words, the idea is to get people to look around them, think, and make connections—but also to understand the limitations of "just looking." Your walking tour should include a map, and the information you provide for each location can be in the form of a document, audio file, or video. To do this exercise, follow the steps below.

Steps to Reading Your City

Step 1: Pick your place. Decide on a route that takes people through a particular part of your city—it could be downtown, an ethnic landscape, a commercial area, or a local neighborhood. Almost anywhere will work. After reading this chapter, you should have a fairly good idea of the types of urban landscapes that are of interest to you. Create a map showing the area of your tour. You can draw a map yourself, or download a map of the area.

Step 2: Walk the route, marking places where you want people to stop, look, and think, and taking pictures or video if necessary.

Step 3: Investigate the places you marked. You should be able to use the information in this chapter and elsewhere in the book to do a good bit of the investigative work, but the specifics of your city will require more information: archival research, perhaps informal interviews, or something else altogether. At each point, you should ask yourself: What can I learn from looking at this landscape and what would I never have guessed?

Step 4: Based on what you have learned, compose your tour, providing information and questions for each particular stop on the route.

Finally, try out your "self-guided walking tour." Then swap tours with a classmate, and provide one another feedback. Revise your tour as needed.

Observing the urban landscape. (Starcevic/iStock/Getty Images)

SEEING GEOGRAPHY

Melbourne, Australia

What makes a city truly livable?

Melbourne, Australia, has again been voted by the Economist Intelligence Unit "The World's Most Livable City." As you have learned, "livability" is a subjective term. In fact, many lists of "most livable cities" exist; they all focus on similar aspects of urban life and cities: grand architecture and abundant green spaces, including accessible public parks, plazas, and promenades; nearby natural features such as mountains or beaches; plentiful opportunities, whether economic, cultural, or educational; and finally, a sense of safety, family-friendliness, and community.

Melbourne, Australia, certainly ticks all of these boxes. If you were to visit Melbourne, you might see these firsthand: beaches for swimming and water sports, playing fields for sporting events, fresh food at farmers' markets, endless restaurants and cafés, public gardens and parks, and dramatic architecture. It is little wonder that Melbourne has topped the list of the world's most livable cities for several years. Does everyone who lives in Melbourne have access to all of the amenities that make it number one? By now, you may be thinking "probably not"—and you are correct. While some aspects of life are free and available to all—for instance, experiencing a vibrant sunrise—others cost money. Zoos, theaters, concerts, restaurants, and sporting events: all of these require a ticket to enter. And still other places are not as accessible as they may seem. Public beaches, for instance, may be experienced as places of harassment, not enjoyment, for women in Australia's notoriously masculinist culture.

In 2016, the homeless population was deemed by the government to be at emergency levels. Thanks to rising housing costs, growing income inequality, and increasing ranks of drug users, more and more people in Melbourne are "sleeping rough"— an Australian term for sleeping out-of-doors. At least 30 homeless encampments dotted the city's central district, and Melbourne boosters became concerned that the begging, littering, and drug use would tarnish the city's world-class image. In 2017, a massive clearance project removed the homeless from the city center in advance of the Australian Open tennis tournament.

Homelessness complicates the image of "livability" that Melbourne and other cities wish to project. Social justice issues, such as discrimination, poverty, inequality, and lack of affordable housing and other basic needs, are always part of the urban landscape, no matter how appealing a city may appear. As human geographers, it is our job to look closely at cities as one of humanity's greatest, but also most problematic, inventions.

Waterfront view of Melbourne, Australia; homeless individual resting. (Top: David Wall/Alamy; Bottom: Ray Warren Australia/Alamy)

Appendix

THE LANGUAGE AND MEANING OF MAPS

The late cultural geographer Barney Nietschmann used to tell his students at the University of California at Berkeley, "The map is the starting point, not the ending point for geographers." He meant that the main practice of human geography is not making maps. (Map making is a technical field called cartography.) Rather, human geographers use the spatial arrangement of social and natural phenomena to formulate and answer questions about relationships, including causal relations. Maps are the common language for pursuing these questions.

For our purposes in *Contemporary Human Geography*, **maps** are two-dimensional representations of three-dimensional spatial phenomena. Cartographers have agreed on certain graphic conventions that are shared by all maps. These conventions allow cartographers to represent a wide range of spatial phenomena such as movement, qualitative and quantitative variation across space, boundaries, location, and so on, in a way that is recognizable to the trained observer. Maps, therefore, are a type of text and, as with all texts, literacy in the medium is required. We will start with some of the basic conventions and then move on to discuss how different mapping choices can provide different answers.

Basic Map Conventions

Most maps have a scale. **Map (or cartographic) scale** is the distance on the map in relation to distance in actual space. For example, 1 inch on the map might equal 1 mile of distance in actual space. Scale on a map can be represented by a bar, as we see in **Figure 5.20**, or it can be represented as a numerical ratio, as in 1/100,000 or 1:100,000.

Most maps also have a **map legend**. Legends are essential for reading maps, for they provide a key to the meaning of **map symbols**. Map symbols are graphic figures that represent any number of real-world phenomena. For example, the legend for **Figure 6.24** has line symbols representing different types of boundaries (e.g., the extent of Soviet influence) by using different styles and colors of lines. Dots are used to represent specific locations (in this case, sites of U.S. military conflict involvement). Dots are so widely understood to indicate specific location that they often are not included in map legends.

Finally, all maps have an **orientation**, which refers to the compass direction of the top of the map. Many map legends include an arrow indicating north so that the reader knows the map's orientation. The north arrow is essential in maps used for way finding, but less so for maps used for other purposes. Figure 6.24 does not have an arrow, because orientation is not particularly important to the map's purpose. Established convention dictates that maps are oriented to the north. As with dots, a northward orientation is so widely understood that a north arrow is often not included in legends.

Thematic Maps

Maps are broadly categorized as either **baseline maps** (also called general maps) or **thematic maps**. Baseline maps provide basic information on coastlines, country boundaries, and so forth. Thematic maps emphasize a specific phenomenon or process, such as transportation, migration, and agricultural production. Most maps in *Contemporary Human Geography* are thematic.

The theme of **Figure 5.11**, for example, is returning refugees who had been displaced by armed conflict. The theme requires that a fixed, two-dimensional map symbol somehow represent the mass movement of people in space. Arrow-tipped lines are a common map symbol for showing movement, with the arrow end indicating movement direction and the flat end indicating the origin. In Figure 5.11, the arrow symbols could convey other information in addition to origin and direction. For example, the width of the line can indicate quantity, but in this case the number of displaced people is represented numerically.

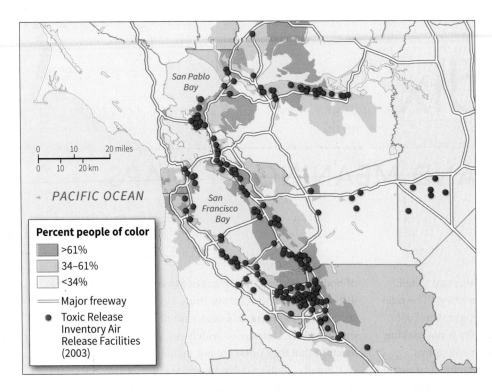

FIGURE 5.20 Industrial air pollution and minority neighborhoods. (Page 171.)

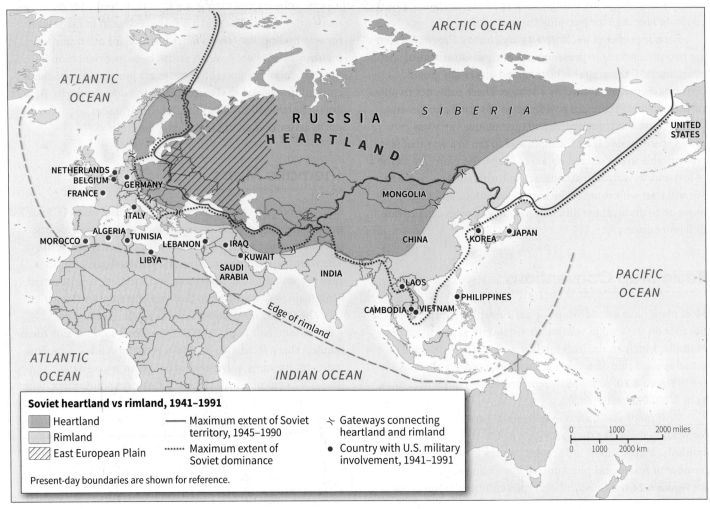

FIGURE 6.24 Heartland versus rimland in Eurasia. (Page 209.)

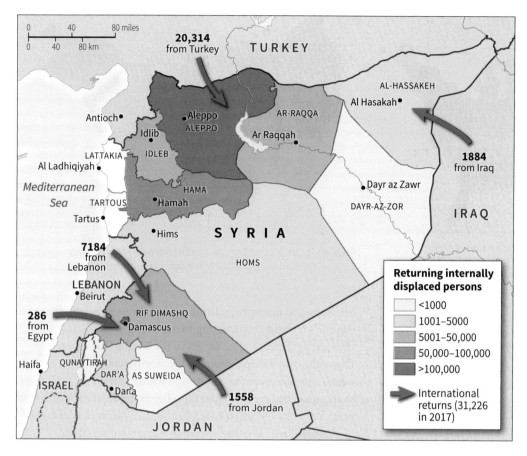

FIGURE 5.11 Syrian refugees and internally displaced persons returning home. (Page 161.)

Note in the case of Figure 5.11 that the quantitative measure (number of people) is represented numerically, so the size of the lines remains the same, while the qualitative information (locations of origin and destination) changes. Returning refugees are shown as returning from countries, rather than returning from specific towns or cities. This reflects the way displaced people's identities shift when they cross international borders. They become simply "repatriated Syrians" and their particular home city or region becomes largely irrelevant to their new classification.

As we see in these examples, simple symbols can carry a wealth of information, and the more closely we look, the more we are able to discern patterns and relationships and the more we are sparked to inquire more deeply. We are also increasingly able to read the limitations of the symbols and to think about what maps hide as well as reveal.

Let's look at **Figures 6.10** and **6.12** to further illustrate this last point. These are common thematic maps known as **choropleth maps**. A choropleth map uses data aggregated over a predefined geographic area, typically a political designation such as a county, province, or state. Often colors are used to represent differences in values in different areas. In Figure 6.10, red represents a majority vote for the Republican Party and blue represents a majority vote for the Democratic Party.

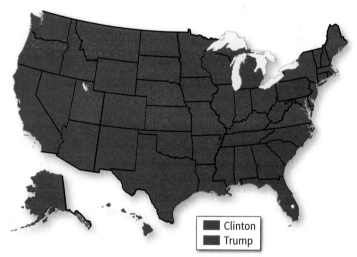

FIGURE 6.10 2016 presidential election results. (Page 196; courtesy of Mark Newman, University of Michigan.)

As noted in Chapter 6, the story of political party divisions within the United States changes depending on which geographic unit we decide to use. If we aggregate data at the state level, as in Figure 6.10, we tell one sort of story. If we aggregate data at the county level, as in Figure 6.12, the story of political difference changes by becoming more

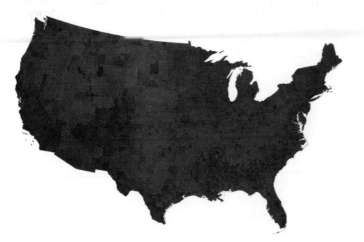

FIGURE 6.12 Purple America. (Page 197; courtesy of Mark Newman, University of Michigan.)

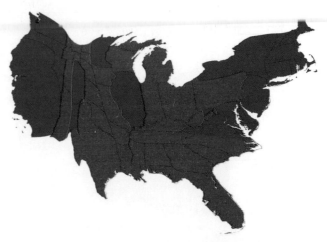

FIGURE 6.11 Cartogram of the 2016 presidential election results. (Page 196; courtesy of Mark Newman, University of Michigan.)

nuanced and less starkly defined. In addition, the coloring technique changed from a color binary in Figure 6.10 to a color progression in 6.12. Progressive color shading is better able to represent the subtle geographic variation in voting patterns.

Another type of thematic map used in *Contemporary Human Geography* is the **cartogram**. A cartogram is a map that results from using some thematic variable—the number of people voting Republican or Democrat in the case of **Figure 6.11**—rather than geographic area or distance. (*Note:* Map scale is no longer relevant for such maps.) Once again, we see that the story of electoral geography in the United States changes when we represent the data differently through different mapping choices.

Equipped with these basic map-reading skills, we are ready to use maps as tools for formulating and answering questions about spatial relations. Take another look at Figure 5.20 on page 171, which accompanies our discussion of environmental racism in Chapter 5. The problem of environmental racism is fundamentally a spatial problem, the scope of which only becomes evident through mapping. Early researchers of the problem noticed that many toxic facilities were located near minority communities in the United States. Based on these observations, they asked what would happen if we created a map that included the location of toxic facilities and census data on residence by race. Maps such as 5.20 were the result, clearly demonstrating a close spatial relationship between minority neighborhoods and pollution sources.

World Map Projections

All cartographers, from ancient times to the present, have faced the same difficult choices when representing the spherical Earth on the flat surface of a paper map. Such representations

are known as **map projections**, and there are many different ways to go about constructing them. Every method, however, necessarily distorts some aspect of the Earth's true surface. Depending on a map's intended use, cartographers must choose which characteristics to distort and which to hold true, whether shape, area, distance, or direction. For example, if the projection holds compass direction true, then surface distance is distorted.

We will see why this is so in our examination of four common world map projections. For each example we will use projections based on an equatorial aspect, meaning the cartographer has placed the equator at the center of the map's horizontal axis. As with other mapping decisions, the choice of aspect is limitless. We could, for example, present a series of projections using a polar aspect. The equatorial aspect, however, is the one used most commonly in world maps.

The Mercator projection, named for the sixteenth-century cartographer Gerardus Mercator, is a common world map projection that maintains true shapes (**Figure A1**). Such true-shape maps are known as **conformal projections**. At every point, the actual shapes of the world's land masses are accurately represented. This makes conformal maps useful for navigation, because the lines connecting points on the map represent the true compass direction. The Mercator map, therefore, became widely adopted during Europe's Age of Exploration and was later adopted as a standard projection in geographic education. However, in the Mercator projection, area (size) is truly represented only along the equator and becomes increasingly distorted as we move poleward.

Some educators have argued that a map designed for navigation is not the best way to represent the world. For example, it greatly distorts the size of Europe and North America, making each appear much larger than Africa. In fact, Africa is more than three times as large as the lower-48 United States. The Peters projection, named for the German historian Arno Peters,

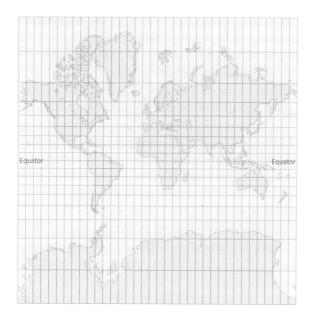

FIGURE A1 Mercator projection. The poleward distortion of land area size is particularly noticable in the depiction of Antartica at the bottom of the map.

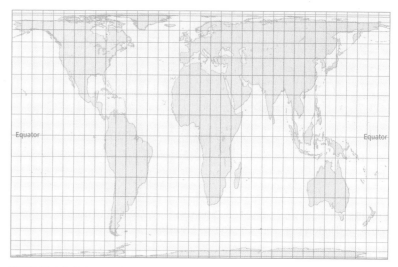

FIGURE A2 Peters projection. Here the relative sizes of land areas are accurate, but their shapes are greatly distorted.

who popularized it in the 1970s, maintains true area to correct this distortion (**Figure A2**). This type of projection is known as an **equal-area projection**. Peters, supported by many educators and institutions, asserted that the Mercator projection also distorts our understanding of the world, making the Northern Hemisphere appear naturally more dominant and significant than it is. An equal-area projection corrects this misconception. In doing so, however, the Peters projection greatly distorts shape.

Another equal-area map is the Goode projection, named for American geographer and cartographer John Paul Goode (**Figure A3**). Goode's solution to the problem of shape distortion was to avoid the restrictions of the rectangular map and create "interruptions" in the map's continuity, thereby reducing the distortion of shape. This projection became known as the

"orange-peel" map because it resembles a flattened rind from a peeled orange. It is the projection used throughout the popular *Goode's World Atlas* and is often adopted for world maps comparing national-level data (e.g., economic and demographic statistics).

The Mercator versus Peters projection debate came to symbolize two polarizing political positions, one celebrating Euro-American historical accomplishments and the other critically assessing Euro-American colonialism and imperialism. The Robinson projection, named for American geographer and cartographer Arthur Robinson, has been offered as a compromise solution to the problem of distortion, in both the cartographic and political senses (**Figure A4**). The Robinson projection is neither conformal nor equal-area, but rather attempts to create the most visually appealing representation by keeping all types of distortion relatively low over most of the map. The Robinson projection is increasingly replacing the Mercator projection in

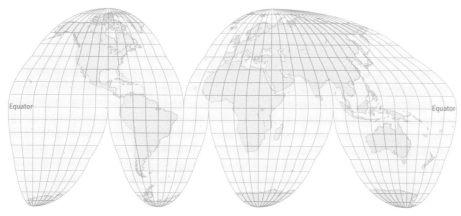

FIGURE A3 Goode projection. The problem of holding both size and shape true is solved by creating cartographic "interruptions."

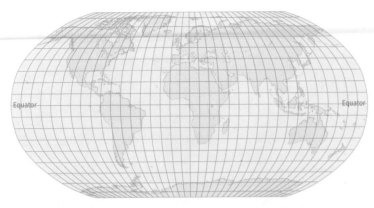

FIGURE A4 Robinson projection. Distortion is minimized but not eliminated in this popular compromise solution.

geography classrooms. In *Contemporary Human Geography* we use the Robinson projection for world maps.

We can see, now, how maps can be powerful analytical tools, not just handy aids for finding the nearest pizza. The environmental racism map (Figure 5.20), for example, both answered the question of scope and raised new questions about how structural racism operates geographically. Armed with a basic understanding of how information can be represented spatially in the form of maps, we are better able to interpret the message of maps, sometimes even finding undisclosed ideologies or political agendas. Just as importantly, we can begin to ask our own questions—sometimes based on everyday observations, sometimes in an effort to solve difficult real-world problems—and to discover important spatial relations that only mapping can reveal.

Key Terms

map Two-dimensional representations of three-dimensional spatial phenomena (page A-1).

map scale (cartographic scale) The distance on a map in relation to the distance in actual space (page A-1).

map legend A visual explanation of what the symbols on a map mean (page A-1).

map symbol Graphic figure that represents any number of real-world phenomena (page A-1).

orientation Refers to the compass direction of the top of a map (page A-1).

baseline map A map that provides basic geographic information such as coastlines, rivers, roads, and country boundaries (page A-1).

thematic map A map that emphasizes a specific phenomenon or process, such as transportation, migration, and agricultural production (page A-1).

choropleth map A map that uses data aggregated over a predefined geographic area, typically a political designation such as a county, province, or state (page A-3).

cartogram A map that results from using a thematic variable, such as population data, rather than geographic area or distance (page A-4).

map projection A method for representing the spherical earth on a flat surface (page A-4).

conformal projection A map projection, such as the Mercator projection, that preserves the true shape of the Earth's features (page A-4).

equal-area projection A map projection, such as the Peters projection, that preserves the true size of the Earth's features (page A-5).

Glossary

absolute distance   The precise measurement of physical space between one point on the Earth's surface and another, as in the case of the longitude and latitude system (page 59).

absolute location   The precise position of anything on (or below) the surface of the Earth (page 14).

absorbing barrier   A barrier that completely halts diffusion of cultural innovations (page 22).

acculturation   An ethnic group's adoption of enough of the ways of the host society to be able to function economically and socially (page 153).

adaptive strategies   The unique way in which each culture uses its particular physical environment to provide the necessities of life—food, clothing, shelter, and safety (page 101).

aerial photography   A remote sensing technique using cameras mounted on fixed-wing aircraft to systematically photograph the Earth's surface from directly overhead (page 15).

agribusiness   A firm engaged in all parts of farming operations, from cultivation to storage, transport, processing, and marketing. The term can also refer to the entire enterprise of commercial, industrial, large-scale, and mechanized agriculture (page 285).

agricultural surplus   The amount of food grown by a society that exceeds the demands of its population (page 355).

agriculture   The cultivation of crops and rearing of livestock to produce food, animal feed, drink, and fiber (page 268).

agroforestry   Farming systems that combine the growing of trees with the cultivation of agricultural crops (page 65).

alternative energy   Alternatives to traditional fossil fuels, including hydroelectric, solar, wind, geothermal, nuclear, biofuel, and "clean" coal (page 340).

amenity landscape   Landscapes with attractive natural features such as forests, scenic mountains, or lakes and rivers that have become desirable locations for retirement or vacation homes (page 72).

Anatolian hypothesis   A theory of language diffusion holding that the movement of Indo-European languages from the area in contemporary Turkey then known as Anatolia followed the spread of plant domestication technologies (page 123).

animal geography   A subfield of human geography that emphasizes the active participation of nonhuman species in nature–culture interactions, investigating how the human and nonhuman act together to transform space, place, and landscape (page 66).

animistic religion   The belief that souls or spirits exist not only in humans but also in animals, plants, rocks, natural phenomena such as thunder, geographic features such as mountains or rivers, and other entities of the natural environment (page 239).

Anthropocene   The proposed designation for a new geological epoch beginning around 1800 C.E. in which the dominant influence on the Earth's natural systems is human activities (page 30).

anti-Semitism   Hostility toward or persecution of Jews (page 165).

aquaculture   The cultivation and harvesting of aquatic organisms under controlled conditions (page 276).

Arctic Circle   The latitude line of 66 degrees, 34 minutes north (page 212).

assimilation   The blending of an ethnic group into the host society, resulting in the loss of distinctive ethnic traits (page 153).

automobile cities   Cities whose spatial layout—in terms of both extent and form—is dictated by the near ubiquity of individual automobile ownership (page 365).

axis mundi   The symbolic center of cosmomagical cities, often demarcated by a large vertical structure (page 356).

barrio   An impoverished, urban, Hispanic neighborhood (page 157).

baseline map   A map that provides basic geographic information such as coastlines, rivers, roads, and country boundaries (page A-1).

bilingualism   The ability to speak two languages fluently (page 120).

biodiversity   Biodiversity—a blending of biology and diversity—refers to the variety and variability of life on Earth as measured at various scales, including diversity among individuals, populations, species, communities, and ecosystems (page 61).

biofuel   A fuel derived from organic wastes or plant material (page 297).

borderland   A region straddling both sides of an international boundary where national cultures overlap and blend to varying degrees (page 188).

boundary   A precisely demarcated line that marks both the limits of a territory and the division between territories (page 188).

Brandt Line   Division of the countries of the world into two regions: wealthy and poor. The Brandt Line is named after former West German Chancellor Willy Brandt, and was a concept commonly applied in the 1970s through the 1980s (page 314).

Buddhism   Monotheistic religion based on the teachings of Prince Siddhartha Gautama (also known as Buddha or "the awakened one"), which posits that the root of suffering is desire (page 238).

buffer state   Independent but politically and economically weak country lying between the borders of two more powerful, potentially belligerent empires or countries (page 210).

cadastral survey   A systematic documentation of property ownership, shape, use, and boundaries (page 37).

carrying capacity   The maximum number of people that can be supported in a given area (page 83).

cartogram   A map that results from using a thematic variable, such as population data, rather than geographic area or distance (page A-4).

central business district (CBD)   A dense cluster of offices and shops located at the city's most accessible point, usually its center (page 365).

central place theory   A set of models designed to explain the spatial distribution of urban centers (page 360).

centrifugal force   Any factor that disrupts the internal unity of a country (page 190).

centripetal force   Any factor that supports the internal unity of a country (page 190).

cereal grains   Seeds that come from a wide variety of grasses cultivated around the world including wheat, maize, rice, millet, sorghum, rye, barley, and oats (page 270).

cereal–root crop farming   A diversified system of agriculture based on the cultivation of cereal grains, root crops (e.g., potatoes, yams), and the rearing of herd livestock (page 270).

Christianity   Monotheistic proselytic faith of Christians, related to Judaism and to Islam, whose principles are rooted in the teachings of Jesus as recorded in the Old and New Testaments of the Bible (page 233).

choropleth map   A map that uses data aggregated over a predefined geographic area, typically a political designation such as a county, province, or state (page A-3).

circulation   An ongoing set of multidirectional movements of people, ideas, or things that have no particular center or periphery (page 22).

colonialism   Centuries-long process of European countries competing among themselves to conquer and control other peoples and their territories across the world (page 188).

colonization   The appropriation, control, and occupation of a territory by a geographically distant power (page 6).

commercial agriculture   Farming oriented exclusively toward the production of agricultural commodities for sale in the market (page 268).

community-supported agriculture (CSA)   A direct-to-consumer marketing arrangement whereby the farmers are guaranteed buyers for their produce at guaranteed prices and consumers receive fresh food directly from the producers (page 291).

concentrated animal feeding operation (CAFO)   The practice of confining livestock, including cattle, sheep, turkeys, chickens, and hogs, at high densities in cages only large enough to allow the animal's body to grow and to accommodate equipment for feeding and refuse removal (page 284).

concentration   The degree to which landscape elements are grouped (page 36).

conformal projection   A map projection, such as the Mercator projection, that preserves the true shape of the Earth's features (page A-4).

Confucianism   Derived from the teachings of Chinese philosopher K'ung Fu-tzu, this has bureaucratic, ethical, and hierarchical overtones and is more of a way of life than a religion (page 238).

consumer nationalism   A situation in which local consumers favor nationally produced goods over imported goods as part of a nationalist political agenda (page 60).

contact conversion   The spread of religious beliefs by personal contact (page 241).

contagious diffusion   A type of expansion diffusion in which cultural innovation spreads by person-to-person contact, moving wavelike through an area and population without regard to social status (page 21).

contract farming   An arrangement between an independent farmer and an agribusiness company in which the farmer receives all the inputs from the company needed to produce a crop under conditions specified by the company in exchange for a guaranteed buyer and price for the crop (page 286).

conventional agriculture   Used in contrast to organic agriculture, this is another term for farming that depends on manufactured synthetic inputs, GMO seeds, and other such industrial practices (page 296).

convergence hypothesis   A hypothesis holding that cultural differences among places are being reduced by improved transportation and communications systems, leading to a homogenization of popular culture (page 60).

cool chain   The system that uses new refrigeration and food-freezing technologies to keep farm produce fresh in climate-controlled environments at every stage of transport from field to retail grocers and restaurants (page 285).

core   Regions having the most advanced industrial and military technologies, complex manufacturing systems, external political power, and the highest levels of wealth and mass consumption (page 313).

core–periphery model   A spatial model associated with world systems theory, of the interrelations between core regions where economic, political, and cultural powers are located, and peripheral regions that are dependent on and serve the needs of the core (page 313).

cornucopians   Those who believe that science and technology can solve resource shortages. In this view, human beings are our greatest resource rather than a burden to be limited (page 99).

corporate consolidation   When a business grows by amalgamating many companies into one large company through mergers and acquisitions (page 287).

cosmomagical cities   Types of cities that are arranged spatially according to religious principles; characteristic of very early cities (page 356).

creole   A language derived from a pidgin language that has acquired a fuller vocabulary and become the native language of its speakers (page 120).

cult   An unconventional belief system, often focusing on a person (page 229).

cultural ecology   A field of study that uses concepts borrowed from biological ecology to study culture as a system that facilitates human adaptation to the environment (page 26).

cultural landscape   The built forms that cultural groups create in inhabiting the Earth—farm fields, cities, houses, and so on—and the meanings, values, representations, and experiences associated with those forms (page 31).

cultural maladaptation   Poor or inadequate adaptation that occurs when a group pursues an adaptive strategy that, in the short run, fails to provide the necessities of life or, in the long run, destroys the environment that nourishes it (page 169).

cultural practices   The social activities and interactions—ranging from religious rituals to food preferences to clothing—that collectively distinguish group identity (page 2).

cultural preadaptation   A complex of adaptive traits and skills possessed in advance of migration by a group, giving it survival ability and competitive advantage in occupying the new environment (page 168).

cultural simplification   The process by which immigrant ethnic groups lose certain aspects of their traditional culture in the process of settling overseas, creating a new culture that is less complex than the old (page 169).

culture   A dynamic process through which humans create, learn, and change shared behaviors, beliefs, values, meanings, and practices (page 2).

culture hearth   A focused geographic area where important innovations are born and from which they spread (page 240).

dairying   A farming system that specializes in the breeding, rearing, and utilization of livestock (primarily cows) to produce milk and its various by-products, such as yogurt, butter, and cheese (page 275).

death rate   The number of deaths per year per 1000 people (page 85).

deindustrialization   The decline of industry, accompanied by a rise in service and information sectors of the economy (page 379).

demographic transition   The movement from high birth and death rates to low birth and death rates (page 88).

density   The amount per unit area of a given landscape element, such as houses (page 36).

depopulation   A decrease in population that sometimes occurs as the result of sudden catastrophic events, such as natural disasters, disease epidemics, and warfare (page 107).

desertification   The process whereby once-fertile land is converted to desert either through climate variation, human activities, or a combination of the two (page 295).

developed   Refers to those regions that are considered wealthier, because people living there are better able to accumulate resources (page 315).

developing   Refers to regions that are considered relatively poor, because they produce just enough, or at times not enough, food and other resources for their own immediate needs (page 314).

dialect   A distinctive local or regional variant of a language that remains mutually intelligible to speakers of other dialects of that language; a subtype of a language (page 119).

diaspora culture   Ethnic, racial, and national population concentrations of people displaced and geographically scattered from their homelands. Such displaced groups often maintain strong social and economic ties to their homelands (page 9).

Diaspora   When written with a capital "D," *Diaspora* refers to the forced dispersal of the Jews from Palestine in Roman times and the resulting wide spatial reach of contemporary Jewish populations across the globe. When written with a lowercase "d," *diaspora* refers to the dispersal across places of people who have a common religious, ethnic, or national background (page 232).

diffusion   The pattern by which a phenomenon—people, ideas, technologies, or preferences—spreads from one location through space and time (page 20).

digital divide   A pattern of unequal access to advanced information technologies produced by socioeconomic inequalities and measured at scales ranging from the individual to countries and world regions (page 57).

dispersed settlement   A pattern where individual buildings are scattered apart from one another (page 36).

domesticated animal   Animals that depend on people for food and shelter and that differ from wild species genetically, physically, and behaviorally as a result of human-controlled breeding and frequent human contact (page 281).

domesticated plant   A plant that is bred, planted, protected, cared for, and used by humans (page 279).

domestication   The multigenerational process through which humans selectively breed, protect, and care for individuals taken from populations of wild plant and animal species to create genetically distinct species, known as domesticates (page 279).

double-cropping   The farming practice of planting and harvesting the same parcel of land twice per year (page 269).

doubling time   The amount of time, in years, that it will take for a given population to double in size (page 99).

ecofeminism   A philosophy and a political movement. As a philosophy, ecofeminism suggests that women—as traditional child bearers, gardeners, and nurturers of the family and home—are more in touch with the rhythms of nature than men, and are therefore better stewards of the Earth. From this arose a movement in the 1970s, identifying a parallel between how women are discriminated against by a masculinist society, and how nature—often portrayed as a feminine entity—is similarly exploited, degraded, and oppressed. (pages 64 and 252).

ecology   A biological science that studies the complex relationships among living organisms and their physical environments. Also used when referring to the biophysical conditions of a particular region or place (page 26).

ecotheology   The study of the influence of religious belief on habitat modification (page 251).

effective sovereignty   The idea that states' power to effectively enforce or ignore sovereignty claims irrespective of territorial boundaries varies in time and from country to country (page 203).

electoral geography   A subfield of political geography that analyzes the geographic character of political preferences and how geography can shape voting outcomes (page 193).

electronic waste   Used electronics, such as mobile phones, computers, office equipment, television sets, and refrigerators, which can no longer be used for their intended purpose and are discarded. Also known as e-waste (page 349).

enclave   A piece of territory surrounded by, but not part of, a country (page 190).

environmental determinism   The belief that the physical environment is the dominant force shaping cultures and that humankind is essentially a passive product of its physical surroundings (page 26).

environmental justice   General term for the movement to redress environmental discrimination (page 172).

environmental perception   The process whereby people act based on how they perceive their environment, rather than on how it actually is (page 27).

environmental racism   The targeting of areas where ethnic or racial minorities, immigrants, and/or poor people live with respect to environmental contamination or failure to enforce environmental regulations (page 171).

environmental refugees   People who are displaced from their homes due to severe environmental disruption (page 104).

equal-area projection   A map projection, such as the Peters projection, that preserves the true size of the Earth's features (page A-5).

ethnic cleansing   The forced removal of an ethnic group by another ethnic group to create ethnically homogeneous territories (page 200).

ethnic flag   A readily visible marker of ethnicity on the landscape (page 172).

ethnic group   A group of people who share a common ancestry and cultural tradition, often living as a minority group in a larger society (page 153).

ethnic homelands   Sizable areas inhabited by an ethnic minority that exhibits a strong sense of attachment to the region and often exercises some measure of political and social control over it (page 153).

ethnic islands   Small ethnic areas in the rural countryside; sometimes called folk islands (page 154).

ethnic neighborhood   A voluntary urban community where people of like origin reside by choice (page 156).

ethnic religion   A religion identified with a particular ethnic or tribal group; does not seek converts (page 229).

ethnic separatism   Separation of an ethnically distinct group to form a politically autonomous region within an existing state, or to secede and form a new nation-state (page 191).

ethnic substrate   The geographical residue of a formerly ethnic population that has now been absorbed into the cultural mainstream (page 155).

ethnoburb   A suburban ethnic neighborhood, sometimes home to relatively affluent immigration populations (page 158).

ethnographic boundary   A political boundary that follows some cultural border, such as a linguistic or religious border (page 190).

ethnolect   A dialect spoken by a particular ethnic group (page 128).

Eurocentric   Using the historical experience of Europe as the benchmark for all cases (page 90).

European Union (EU)   A political, economic, and social union of 28 independent countries in Europe that promotes the free movement of people, goods, services, and capital among its members (page 205).

exclave   A part of national territory separated from the main body of a country by the territory of another country (page 190).

exclusive economic zone (EEZ)   The zone of resource control established by the UN Convention on the Law of the Sea extending 200 nautical miles (370 kilometers) out from the mean low-tide line of a country's coastline (page 212).

expansion diffusion   The spread of innovations within an area in a snowballing process, so that the total number of knowers or users becomes greater and the area of occurrence grows (page 20).

extensive agriculture   Crop cultivation and livestock rearing systems that require low levels of labor or capital input relative to the areal extent of land under production (page 268).

fair trade   A third-party certification program that supports good crop prices for Third World farmers and environmentally sound farming practices (page 291).

farm villages   Clustered rural settlements of moderate size, inhabited by people who are engaged in farming (page 106).

farmers' market   A setting (ranging from a few stalls in the street to covered enclosures extending a few city blocks) for farmers to sell their produce directly to consumers (page 291).

farming systems   Populations of farms that share a similar resource base, including climate, water availability, slope, topography, and soil quality, and a similar set of practices including crop and livestock types, property ownership, farm size, temporal and spatial patterns of cultivation and animal rearing, and the use of machinery or irrigation (page 268).

farmstead   The center of farm operations, containing the house, barn, sheds, and livestock pens (page 106).

federal state   An independent country that disperses significant authority among subnational units (page 193).

feedlot   A fenced enclosure used for intensive livestock feeding that serves to limit livestock movement and associated weight loss (page 274).

feng shui   The practice of harmoniously balancing the opposing forces of nature in the built environment by choosing environmentally auspicious sites for locating houses, villages, temples, and graves (page 248).

first urban revolution   The connection between agricultural innovations and the rise of the world's first true cities (page 359).

First World   A term used to describe the group of wealthy, democratic, capitalist nations across the world (page 314).

folk architecture   Building styles of folk cultures constructed in a local manner and style, without the assistance of professional architects or blueprints, using locally available raw materials (page 66).

folk culture   Rural, unified, largely self-sufficient groups that share similar customs, material practices, and ethnicities (page 5).

food security   When all people, at all times, have physical and economic access to sufficient safe and nutritious food to meet their dietary needs and food preferences for an active and healthy life (page 275).

food sovereignty   The principle that all people have a right to sufficient, healthy, and culturally appropriate food and that control over food production should remain in local farmers' hands rather than those of multinational corporations (page 290).

foodborne outbreaks   Illnesses that occur when two or more people consume the same contaminated food or drink (page 289).

foodways   Customary behaviors associated with food preparation and consumption (page 178).

forced migration   Situations where people are involuntarily compelled, by either disasters or social conditions, to leave their homes (page 24).

formal region   A geographical area inhabited by people who have one or more cultural traits in common (page 17).

Fourth World   A term used to describe the world's poorest nations (page 314).

friction of distance   The inhibiting effect of distance on the intensity and volume of most forms of human interaction (page 21).

frontier   A region at the margins of state control and settlement, an idea that can be traced to ancient notions of the limits of civilization and the savage world beyond (page 188).

functional region   A geographic area that is defined by some economic, political, cultural, or social function (page 18).

fundamentalism   A movement to return to the founding principles of a religion, which can include literal interpretation of sacred texts, or the attempt to follow the ways of a religious founder as closely as possible (page 230).

Gaia hypothesis   The theory that there is one interacting planetary ecosystem, Gaia, that includes all living things and the land, waters, and atmosphere in which they live; further, that Gaia functions almost as a living organism, acting to control deviations in climate and to correct chemical imbalances, so as to preserve Earth as a living planet (page 252).

gated communities   Highly securitized residential enclaves that are more or less self-governing (page 370).

GDP per capita   The mathematical result of dividing a country's GDP by its national population (the Latin *per capita* means "per head") (page 315).

gender roles   The set of generally held expectations around the behaviors of men and women in different cultural and historical contexts (page 93).

gendered space   Spaces that are classified by social norms and practices as predominantly either male or female (page 2).

generic toponym   The descriptive part of many place-names, often repeated throughout a culture area (page 138).

genetic engineering   A technology that artificially modifies, manipulates, or recombines an organism's DNA (page 285).

genetically modified organisms (GMO)   Living organisms, including crops and livestock, that are produced through genetic engineering (page 285).

genocide   The systematic killing of members of a racial, ethnic, religious, or linguistic group (page 166).

gentrification   The displacement of lower-income residents and economic activities by higher-income residents and activities; frequently associated with a restoration of buildings in deteriorated areas of the city (page 378).

geographic information system (GIS)   A computer software application that is used to collect, store, retrieve, and visually present geospatial data (page 16).

geographic marginalization   A government-led process whereby a population is forced to migrate to remote, often rugged and unproductive lands, far from political and economic centers (page 49).

geometric boundary   A political border drawn in a regular, geometric manner, often a straight line, without regard for environmental or cultural patterns (page 190).

geopolitics   A term originally coined to describe the influence of geography and the environment on political entities (page 208).

gerrymandering   The practice of manipulating voting district boundaries to favor a particular political party, group, or election outcome (page 193).

ghetto   Traditionally, an area within a city where an ethnic group lives, either by choice or by force. Today in the United States, the term typically indicates an impoverished African American urban neighborhood (page 156).

global cities   Cities that are control centers of the global economy (page 367).

global food system   The agro-industrial complex, organized at the global scale, that encompasses all elements of growing, harvesting, processing, transporting, marketing, consuming, and disposing of food for people (page 283).

Global Positioning System or GPS   A constellation of satellites orbiting the Earth, which transmit radio signals earthward to receiving devices, such as smartphones, locating them precisely in time and three-dimensional space (page 15).

Global South   A term that has largely replaced "Third World" when referring to regions of Latin America, Africa, and most of Asia, in recognition of the fact that much of the world's dynamism, growth, and power resides in these places (page 331).

globalization   Processes of economic, political, and social integration that operate on a global scale and have collectively created ties that make a difference to lives around the planet (page 24).

grain farming   An intensive, highly mechanized commercial farming system that specializes in the production of cereal grains, the most globally important of which are rice, maize (corn), and wheat (page 272).

Great Migration   The twentieth-century movement of 6 million African Americans from the rural southern United States to the cities of the midwestern and northeastern states (page 23).

green revolution   The U.S.-supported development of high-yielding varieties (HYVs) of cereal crops and accompanying agricultural technologies for transfer to economically less developed countries (page 282).

greenhouse effect   The global warming trend caused by the increase of greenhouse gases such as carbon dioxide (CO_2) (page 30).

greenhouse gas (GHG)   A compound in the atmosphere, such as the carbon dioxide (CO_2) released from fossil fuel combustion, that absorbs and traps heat energy near the Earth's surface (page 30).

Gross Domestic Product (GDP)   The total value of all goods and services produced in a country over a specific period, usually a year (page 315).

Hajj   The so-called Fifth Pillar of Islam is an annual religious pilgrimage to the city of Mecca. Able-bodied Muslims who can afford the journey are encouraged to perform the Hajj at least once in their lifetime (page 244).

heartland   The interior of a sizable landmass, removed from maritime connections; in particular, the interior of the Eurasian continent (page 209).

heartland theory   A 1904 theory of Harold Mackinder that the key to world conquest lay in control of the continental interior of Eurasia (page 209).

heirloom seeds   Locally bred seeds of crop varieties (also called heritage seeds) that for marketing reasons are not suitable to the global food system and are open pollinated rather than produced through hybridization (page 290).

herbicide   A common type of pesticide that is designed to kill or inhibit the growth of unwanted plants (i.e., weeds) that compete with crop plants (page 284).

hierarchical diffusion   A type of expansion diffusion in which innovations spread from one important person to another or from one urban center to another, bypassing other persons or rural areas (page 20).

Hinduism   Polytheistic ethnic faith and the parent religion of Buddhism, closely tied to India and its ancient culture (page 237).

Human Development Index (HDI)   A numerical value devised by the United Nations Development Programme that is used to measure how well basic needs are being met. It is a composite index, derived from three areas: life expectancy, education, and income (page 317).

human geography   The study of the relationships between people and the places and spaces in which they live (page 2).

hunter-gatherers   Culture groups (e.g., the Inuit of arctic Greenland, Alaska, and Canada) that gain a livelihood by hunting wild game, fishing where possible, and gathering edible and medicinal wild plants (page 277).

hydraulic civilization model   Hypothesizes that urban civilization arose from large-scale crop irrigation (page 355).

iconoclasm   The destruction of statues, monuments, plaques, and other religious or political landscape elements (page 177).

imperialism   A relationship whereby wealthy nations dominate poor ones by controlling their economic, political, and cultural systems (page 331).

independent invention   A cultural innovation developed in two or more locations by individuals or groups working independently (page 20).

indigenous culture   A culture group made up of the original inhabitants of a territory. The term commonly refers to subjugated cultures distinct from national cultures that are rooted in colonial domination (page 6).

indigenous technical knowledge (ITK)   Highly localized knowledge possessed by indigenous culture groups about environmental conditions and sustainable land-use practices (page 62).

Industrial Revolution   Beginning in England in the early 1700s, the Industrial Revolution saw the rapid transformation of the economy through the introduction of machines, new power sources, and novel chemical processes. These replaced human hands in the making of products, initially in the cotton textile industry (page 322).

industry concentration   A process whereby an increasing proportion of production is controlled by decreasing numbers of companies (page 287).

infant mortality rate   The number of infants per 1000 live births who die before reaching one year of age (page 86).

intensive agriculture   Crop cultivation and livestock rearing systems that use high levels of labor or capital relative to the size of the landholding (page 268).

intercropping   The farming practice of planting multiple crops together in the same clearing (page 268).

interdependence   Relations between regions or countries of mutual, but not necessarily equal, dependence (page 25).

internal migration   Human migration that occurs within the borders of a country (page 23).

internally displaced persons (IDPs)   Persons or groups that have been forced to flee their homes due to conflict, natural disaster, or persecution but who remain within the borders of their home country (page 161).

international migration   Human migration across country borders (page 22).

irredentism   The political claim to territory in another country based on ethnic affiliations and historic borders (page 191).

irrigated agriculture   Farming that relies on the controlled application of water to cultivated fields (page 291).

Islam   Monotheistic proselytic faith of the Muslims, whose principles are established in the word of Allah as conveyed by the Prophet Mohammed and recorded in the Qur'an (or Koran) and practiced in the Five Pillars of the faith (page 236).

isogloss   The spatial border of usage of an individual word or pronunciation (page 126).

Judaism   Monotheistic ethnic faith of the Jews, and the parent religion of Christianity, whose principles are rooted in the Torah (page 232).

Kurgan hypothesis   A theory of language diffusion holding that the spread of Indo-European languages originated with animal domestication in the central Asian steppes and grew more aggressively and swiftly than proonents of the Anatolian hypothesis maintain (page 123).

land reclamation   The term used in reference to efforts to drain land inundated with either fresh or salt water (page 291).

language   A mutually agreed-upon system of symbolic communication that has a spoken and usually a written expression (page 118).

language family   A group of related languages derived from a common ancestor (page 120).

language hotspots   Those places on Earth that are home to the most unique, misunderstood, or endangered languages (page 133).

latitude   The degree of distance north or south from the equator, which is 0 degrees, as far as the poles, which are at 90 degrees (page 14).

leisure landscape   Landscapes planned and designed primarily for entertainment and tourism, such as theme parks or ski and beach resorts (page 72).

LGBT district   Urban neighborhoods inhabited or frequented by a disproportionately large number of lesbian, gay, bisexual, and transgendered people and identifiable by a geographic concentration of "gay-friendly" public spaces and establishments such as bookstores, bars, theaters, and restaurants (page 52).

linear settlement   A pattern where buildings are arranged in a line, often along a road or river (page 36).

lingua franca   An existing, well-established language of communication and commerce used widely where it is not a mother tongue (page 120).

linguistic refuge area   An area protected by isolation or inhospitable environmental conditions in which a language or dialect has survived (page 135).

livestock fattening   An intensive system of animal feeding, utilizing fenced enclosures to finish livestock, mostly cattle and hogs, for slaughter and processing for the market (page 274).

livestock ranching   The practice of using extensive tracts of land to rear herds of livestock to market as meat, hides, or wool (page 274).

locavores   People who dedicate themselves to slow-food diets and to obtaining as much of their nutrition as possible from local farmers (page 290).

longitude   The degree of distance east or west of the prime meridian (Greenwich Meridian), which is 0 degrees, to the line exactly on the opposite side of the Earth, which is 180 degrees (page 14).

Malthusian   Those who hold the views of Thomas Malthus, who believed that overpopulation is the root cause of poverty, illness, and warfare (page 99).

map   Two-dimensional representations of three-dimensional spatial phenomena (page A-1).

map legend   A visual explanation of what the symbols on a map mean (page A-1).

map projection   A method for representing the spherical earth on a flat surface (page A-4).

map scale (cartographic scale)   The distance on a map in relation to the distance in actual space (page A-1).

map symbol   Graphic figure that represents any number of real-world phenomena (page A-1).

mariculture   A form of aquaculture oriented to the farming of saltwater species such as shrimp, oysters, marine fish, and more (page 276).

market gardening   A small-scale farming system of one to a few acres, usually farmer-owned and managed, that produces a mixture of vegetables and fruits for mostly local and regional markets (page 271).

material culture   All physical, tangible objects made and used by members of a cultural group, such as buildings, furniture, clothing, food, artwork, musical instruments, and so forth (page 4).

megachurch   Large Protestant church structures, which have large congregations (2000–10,000 members) and utilize business models to tailor their spaces and services to their congregation's needs (page 253).

megacities   Particularly large urban centers (page 363).

melting pot   The belief, no longer widely held, that North America is a crucible of sorts, where ethnic differences will eventually blend together and disappear (page 153).

microaggressions   The subtle, indirect, or unintentional insults, exclusions, or small acts of violence that marginalized people encounter as they go about their daily lives (page 154).

migration   The mass movement of a population between different areas of the Earth's surface (page 22).

mobility   The relative ability of people, ideas, or things to move freely across space (page 20).

monoculture   The cultivation of a single commercial crop on extensive tracts of land (page 285).

monotheistic religion   The worship of only one God (page 229).

nation   A community of people bound to a homeland and possessing a common identity based on a shared set of cultural traits such as language, ethnicity, and religion (page 188).

national culture   The controversial idea that citizens possess a set of recognizable values, behaviors, and beliefs—often including the same ethnic and linguistic traits—that express the core culture of each modern nation (page 6).

nationalism   The sense of belonging to and self-identifying with a national culture (page 188).

nationalization   Government takeover of transportation infrastructure, fuel sources, and industries from private foreign ownership, occurring in many previously colonized nations in the twentieth century (page 324).

nation-state   An independent country dominated by a relatively homogeneous cultural group (page 188).

natural boundary   A political border that follows some feature of the natural environment, such as a river or mountain ridge (page 190).

natural hazard   A physical danger present in the environment, such as floods, hurricanes, volcanic eruptions, or earthquakes, that may be perceived differently by different peoples (page 27).

nature–culture   How values, beliefs, perceptions, and practices have ecological impacts, and how, in turn, ecological conditions influence perceptions and practices (page 26).

neo-Malthusians   Modern-day followers of Thomas Malthus (page 99).

node   A central location in a functional region where functions are coordinated and directed (page 18).

nones   Refers to those who do not belong to any organized religious faith (page 246).

nonmaterial culture   The wide range of beliefs, values, myths, and symbolic meanings passed from generation to generation of a given society (page 5).

nucleated settlement   A pattern of buildings clustered around a center (page 36).

organic agriculture   The production of crops and livestock using ecological processes, natural biodiversity, and renewable resources rather than relying on industrial practices and synthetic inputs (page 296).

orientation   Refers to the compass direction of the top of a map (page A-1).

orthodox religions   Strands within most major religions that emphasize purity of faith and are not open to blending with other religions (page 230).

paddy rice farming   A system of wet rice cultivation on small, level fields bordered by impermeable dikes that are flooded with 4–6 inches (10–15 centimeters) of water for about three-quarters of the growing season (page 269).

paganism   A term used to refer to rural, polytheistic non-Christians during the Roman empire; today *paganism* denotes an unconventional religion (page 241).

pastoralism   A system of breeding and rearing herd livestock (also called pastoral nomadism or nomadic herding) by moving them over expansive areas of open pasturelands (page 272).

pattern   The consistent or characteristic spatial arrangements of objects or phenomena (page 3).

peasants   Small-scale farmers who own their fields, rely chiefly on family labor, and produce both for their own subsistence and for sale in the market (page 270).

periphery countries   Nations exhibiting characteristics opposite those of the core: they have relatively little industrial development, simple production systems focused mostly on agriculture and raw materials, and low levels of consumption (page 313).

permeable barrier   A barrier that permits some aspects of an innovation to diffuse through it but weakens and retards continued spread; an innovation can be modified in passing through a permeable barrier (page 22).

pesticide   An industrially manufactured chemical that kills or repels the animals, insects, or plants that can damage, destroy, or inhibit the growth of crops (page 284).

physical environment   All aspects of the natural physical surroundings, such as climate, terrain, soils, vegetation, and wildlife (page 3).

pidgin   A composite language consisting of a small vocabulary borrowed from the linguistic groups involved in commerce (page 119).

pilgrimages   Journeys to sacred places or places of religious importance (page 243).

place   Locations that acquire meaning and value through people's subjective experiences of them (page 11).

placelessness A feeling resulting from the standardization of the built environment, increasingly on a global scale, that diminishes regional variation and eliminates the unique meanings associated with specific locations (page 7).

plantation Large landholding devoted to capital-intensive, specialized production of a single tropical or subtropical crop for the global marketplace (page 270).

polyglot Involving many languages (page 122).

polytheistic religion The worship of many gods (page 229).

popular culture The modern material and symbolic practices associated with the rise of mass-produced, machine-made goods and the invention of long-distance communication technologies, such as radio and television (page 7).

population density A measurement of population per unit area (e.g., people per square mile) (page 82).

population explosion The rapid, accelerating increase in world population since about 1650, and especially since 1900 (page 98).

population geography The study of the spatial and ecological aspects of population, including distribution, density per unit of land area, fertility, gender, health, age, mortality, and migration (page 82).

population pyramid A graph composed of back-to-back bars showing the age and sex composition of a population (page 91).

possibilism The theory that the environment presents constraints and opportunities for cultural innovation, but the culture that develops in a particular environment ultimately depends on human agency (page 26).

postdevelopment The idea that traditional approaches to "development"—in the form of monetary aid, investments, expertise, and projects—were ineffective, merely providing a way to categorize most of the world as needing the assistance of a few wealthy nations (page 331).

primary sector The set of economic activities that involves extracting natural resources from the Earth (page 319).

primate city A city of large size and dominant power within a country (page 364).

prime meridian The line of 0 degrees longitude running through Greenwich, England. Also called the Greenwich Meridian (page 14).

proselytic or universalizing religions Religions that actively seek new members and aim to convert all humankind (page 229).

protracted refugee situation (PRS) Results when people are born and mature into adulthood in refugee camps with no promise of becoming citizens in a new country and little hope of returning to their home country (page 201).

proxemics The study of the size and shape of people's envelopes of personal space (page 115).

public space Geographical areas that are owned and controlled by a government agency and legally accessible to all citizens (page 12).

push and pull factors Unfavorable, repelling conditions (push factors) and favorable, attractive conditions (pull factors) that interact to affect migration and other elements of diffusion (page 96).

quaternary sector Intellectual and informational economic activities (page 320).

race A classification system that is variously understood as arising from genetically significant differences among human populations, or from visible differences in human physiognomy, or as a social construction that varies across time and space (page 151).

racial profiling Treating a person as a criminal suspect because of his or her race, ethnicity, nationality, or religion (page 164).

racialize To understand people or practices through the lens of race (page 153).

racism The belief that human capabilities are determined by racial classification and that some races are superior to others (page 153).

range In central place theory, the average maximum distance people will travel to purchase a good or service (page 361).

reapportionment The process by which the 435 seats in the U.S. House of Representatives are divided proportionately by population among the 50 states following every U.S. census (page 193).

redistricting The process of drawing new boundaries for U.S. congressional districts to reflect the population changes since the previous U.S. census (page 193).

refugees Persons or groups that have been forced to flee their home country. Many nations recognize as legitimate only **political refugees**—those who are at risk of harm from their own government due to the political beliefs they hold, their ethnicity or race, or because of political conflict in their home country. However, people can become **economic refugees,** driven to migrate by economic necessity; or **environmental refugees,** who are involuntarily displaced from their homeland due to natural disaster (pages 96 and 161).

region A geographical unit based on a set of common characteristics or functions (page 17).

regional trading blocs Multi-country agreements that reduce or eliminate customs duties and import tariffs to promote the freer flow of goods and services across international borders (page 205).

relative distance The space between one point on the Earth's surface and another measured by changes in time and cost of movement (page 59).

relative location The position of one place, person, or object in relation to the position of another place, person, or object (page 15).

religion A social system involving a set of beliefs and practices through which people seek harmony with the universe and attempt to influence the forces of nature, life, and death (page 228).

religious extremism An intolerant expression of religious fundamentalism, resulting in violence (page 230).

relocation diffusion   The spread of an innovation or other phenomenon that occurs with the bodily relocation (migration) of the individual or group responsible for the innovation (page 20).

remote sensing   The collection of information about the Earth's surface using devices located at distance, generally high above (page 15).

renewable resources   Those primary sector products that can be replenished naturally at a rate sufficient to balance their depletion by human use (page 336).

resilience   The ability to recover quickly from adversity (page 378).

return migration   The phenomenon of migrants returning to their place of origin after long-term residency elsewhere (page 24).

right to the city   The notion that all urban residents, not just the privileged, should be able to access city spaces and have a voice in how the city is shaped and used (page 382).

rimland   The maritime fringe of a country or continent; in particular, the western, southern, and eastern edges of the Eurasian continent (page 209).

sacred spaces   Areas recognized by a religious group as worthy of devotion, loyalty, esteem, or fear to the extent that they become sought out, avoided, inaccessible to nonbelievers, or removed from economic use (page 256).

salad bowl   A contemporary approach to racial and ethnic differences in North America, which views them as mixing together, but never entirely blending and disappearing (page 153).

satellite imagery   Information about the Earth's surface gathered from sensors mounted on orbiting satellites that record in both the visible and non-visible portions of the electromagnetic spectrum (page 16).

satellite state   A nominally independent country, but politically, militarily, and economically directly controlled by a more powerful state (page 210).

seasonal migration   Often associated with crop harvest periods; migrants move according seasonal changes in weather (page 24).

second urban revolution   The connection between industrial innovations and the rise of capitalist cities (page 359).

Second World   A term used during the Cold War era to describe the group of countries that were communist or socialist in their governmental philosophy, and had centrally planned rather than free-market economic systems (page 314).

secondary sector   Commonly referred to as manufacturing, this is the set of economic activities that involve processing the raw materials extracted by primary sector activities into usable goods (page 319).

sect   A subgroup holding somewhat different religious beliefs from the main religion of which it is a part (page 235).

self-determination   The freedom of culturally distinct groups to govern themselves in their own territories (page 188).

semi-periphery countries   Countries exhibiting a mix of characteristics of both the core and periphery, and playing a key role in mediating politically and economically between them (page 313).

sense of place   The subjective human emotions, both negative and positive, evoked by a particular place. A shared sense of place is commonly a key component in defining a culture (page 11).

settlement pattern   The spatial arrangement of buildings, roads, towns, and other features that people construct while inhabiting an area (page 36).

sex ratio   The numerical ratio of females to males in a population (page 91).

shamanism   An animistic belief system wherein a shaman serves as an intermediary between people and spirits (page 239).

shantytowns   Precarious and often illegal housing settlements, usually made up of temporary shelters and located on the outskirts of a large city (page 371).

Shinto   An animistic faith that was once the state religion of Japan and has long been blended with Buddhism and Confucianism; Shinto adherents worship ancestors and *kami*, or spirits, that inhabit natural objects such as waterfalls and mountains (page 239).

slang   Words and phrases that are not part of a standard, recognized vocabulary for a given language but that are nonetheless used and understood by some of its speakers (page 128).

slash-and-burn agriculture   A cultivation method (also called swidden or shifting cultivation) that involves cutting small plots in forests or woodlands, burning the cuttings to clear the ground and release nutrients, and planting in the ash of the cleared plot (page 268).

slow food   A global consumer movement and an international organization promoting local food production and traditional cuisines and foodways (page 290).

small island developing states   Small island countries whose economies rely mostly on tourism and ocean fisheries (page 212).

socioeconomic stratification   The hierarchical arrangement of society through the emergence of distinct socioeconomic classes (page 355).

South–South cooperation   Trade, technological innovation, and other forms of exchange and assistance that occur between the larger and wealthier nations of the Global South, and poorer nations (page 331).

sovereignty   The right of individual states to control political and economic affairs within their territorial boundaries without external interference (page 187).

space   In the spatial science tradition, an abstract, geometric property that can be scientifically modeled through the application of universal laws (page 10).

spatial model   Simulations of social life, often represented in mathematical terms, that abstract a limited set of variables from real-world situations in search of universal laws and regularities in spatial patterns (page 10).

spatial scale   The geographic extent of an area that is under investigation, generally determined by the problem being studied and therefore also called scale of analysis (page 10).

sprawl   The tendency of cities to grow outward in an unchecked manner, which is particularly notable in automobile cities (page 365).

stages of economic growth   A development model associated with the economist W. W. Rostow, positing that all nations would inevitably pass through the same sequence of stages of economic growth on their way to becoming wealthy, consumerist, and developed (page 313).

state   Independent political units with a centralized authority that makes claims to sole legal, political, and economic jurisdiction over a bounded territory. Often used synonymously with "country" (page 187).

stateless nation   A large ethnic group whose population is divided by one or more international boundaries (page 192).

stateless people   Displaced people, often long-term refugees from armed conflict, who are denied a nationality and therefore deprived of basic rights such as education, health care, employment, and freedom of movement (page 201).

stepwise migration   Human migration conducted in a series of stages (page 23).

stimulus diffusion   A type of expansion diffusion in which a specific trait fails to spread but the underlying idea or concept is accepted (page 21).

streetcar suburbs   The extension of urban residential areas along streetcar lines in the late nineteenth century (page 365).

subculture   A group of people with distinct norms, values, and material practices that differentiate them from the dominant culture surrounding them (page 9).

subsistence agriculture   Food production mainly for consumption by the farming family and local community, rather than principally for sale in the market (page 268).

subsistence economies   Rural economies that have few or no external inputs and are oriented primarily toward food production for local consumption, rather than production for sale on the market (page 62).

supranational organization   Political bodies of international integration that nation-states establish in cooperation with their neighbors for mutual political, military, economic, or cultural gain (page 205).

supranationalism   When a collection of nation-states and their citizens relinquish some sovereign rights to a larger-scale political body that exercises authority over its member states (page 205).

sustainability   The ability to use resources in a way that does not deplete them over the long term (page 378).

sustainable agriculture   The production of food and fiber in a manner that does not diminish the future capacity of the land, harm the environment, or damage human health (page 296).

sustainable urbanization   The creation of a situation whereby a society can meet the needs of contemporary urban dwellers for water, food, and shelter, while not damaging the ability of future urban dwellers to meet their needs (page 377).

symbolic landscapes   Landscapes that express the values, beliefs, and meanings of a particular culture (page 33).

syncretic religions   Religions, or strands within religions, that combine elements of two or more belief systems (page 229).

synthetic fertilizer   A nutrient needed for plant growth, most commonly nitrogen, phosphorus, and potassium, that is industrially manufactured using petroleum by-products (page 284).

Taoism   Both an established religion and a philosophy, Taoism emphasizes the dynamic balance depicted by the Chinese yin-yang symbol, and is centered on the "three jewels of Tao"—humility, compassion, and moderation (page 239).

territorial seas   The zone of territorial sovereignty extending 12 nautical miles (22 kilometers) out from the mean low-tide line of a country's coastline (page 212).

territoriality   A learned cultural response, rooted in European history, that produced the external bounding and internal territorial organization characteristic of modern states (page 188).

tertiary sector   The set of economic activities that refers to all the different types of work necessary to move goods and resources around and deliver them to people; in other words, service activities (page 320).

thematic map   A map that emphasizes a specific phenomenon or process, such as transportation, migration, and agricultural production (page A-1).

Third World   A term used to describe the world's poor nations as a group. During the Cold War era, these were also the nations that were neither communist nor socialist; in other words, they were not aligned with either the First or Second World (page 314).

threshold   In central place theory, the size of the population required to make the provision of goods and services economically feasible (page 361).

time–distance decay   The decrease in acceptance of a cultural innovation with increasing time and distance from its origin (page 21).

time-space convergence   The phenomenon whereby the time taken to travel between places is progressively reduced through the introduction of new transportation technologies (page 59).

toponym   A place-name, usually consisting of two parts—the generic and the specific (page 138).

total fertility rate (TFR)   The number of children the average woman will bear during her reproductive lifetime (15–44 or 15–49 years of age) (page 84).

tourism   A modern activity that arose with the introduction of rail, sea, and air mass transportation and involves people traveling from their homes for the purposes of business, leisure, entertainment, or recreation (page 53).

township and range system   The land survey system created by the U.S. Land Ordinance of 1785, which divides most of the country's territory into a grid of square-shaped "townships" with 6-mile (9.6-kilometer) sides (page 214).

transculturation The notion that people adopt elements of other cultures as well as contributing elements of their own culture, thereby transforming both cultures (page 153).

transhumance A herding practice in which herders move with their livestock seasonally in search of the best forage for the animals as pasture conditions change (page 273).

transnational migration The movement of groups of people who migrate continuously between countries, thereby maintaining ties to both their countries of origin and their countries of destination (page 24).

transnationalism The experience of living and working in more than one country that can produce new and distinct cultural identities. Sometimes used synonymously with globalization to refer to economic, political, or cultural processes operating across international borders (page 207).

truck farming A scaled-up version of market gardening, with larger acreages, less diversity of crops, and oriented toward more distant markets (page 271).

unitary state An independent state that concentrates power in the central government and grants little or no authority to subnational political units (page 193).

urban agriculture The practice of growing fruits and vegetables in small private plots or shared community gardens within the confines of a city (page 275).

urban footprint The spatial extent of the impacts of urban areas on the natural environment (page 374).

urban hearth areas Regions in which the world's first cities evolved (page 355).

urban heat island A mass of warm air generated and retained by urban building materials and human activities; it sits over the city and causes urban temperatures to be greater than those of surrounding areas (page 373).

urban risk divide As the world's population becomes increasingly concentrated in large cities, disasters and disaster risk become an urban phenomenon (page 375).

urbanized population The proportion of a country's population living in cities (page 352).

vernacular region A culture region perceived to exist by its inhabitants, based in the collective spatial perception of the population at large and bearing a generally accepted name or nickname (such as "Dixie" for the southern United States) (page 19).

vertical integration When a single firm is engaged at multiple stages of the production process (page 285).

working landscape Landscapes of rural areas that are shaped and maintained by the type of productive activity that predominates there, such as cattle ranching or various forms of agriculture (page 72).

World Heritage Sites Places, including buildings, cities, forests, lakes, deserts, and archaeological ruins, that the UN's International Heritage Programme judges to possess outstanding cultural or natural importance to the common heritage of humanity (page 32).

world systems theory A development model associated with Immanuel Wallerstein, categorizing world history as moving through a series of socioeconomic systems, culminating in the modern world system, which is our current, interdependent, capitalist world economy, rooted in nation-state economies (page 313).

zero population growth A stabilized population created when an average of only two children per couple survive to adulthood. Eventually, the number of deaths equals the number of births (page 84).

References

Agnew, John. 1998. *Geopolitics: Re-Visioning World Politics.* London: Routledge.

Agnew, John. 2011. *Globalization and Sovereignty.* Lanham, Md.: Rowman & Littlefield.

Agnew, John, and Luca Muscarà. 2012. *Making Political Geography.* 2nd ed. Lanham, Md.: Rowman & Littlefield.

Airriess, Christopher. 2002. "Creating Vietnamese Landscapes and Place in New Orleans." In Kate A. Berry and Martha L. Henderson (eds.), *Geographical Identities of Ethnic America: Race, Space, and Place,* pp. 228–254. Reno: University of Nevada Press.

Anderson, Kay, and Fay Gale. 1999. *Cultural Geographies.* Melbourne: Pearson Education.

Bassett, Thomas, and Alex Winter-Nelson. 2010. *The Atlas of World Hunger.* Chicago: University of Chicago Press.

Basso, Keith H. 1996. *Wisdom Sits in Places: Landscape and Language Among the Western Apache.* Albuquerque: University of New Mexico Press.

Bebbington, Anthony. 1996. "Movements, Modernizations, and Markets: Indigenous Organizations and Agrarian Strategies in Ecuador." In R. Peet and M. Watts (eds.), *Liberation Ecologies: Environment, Development, Social Movements,* pp. 86–109. London: Routledge.

Bebbington, Anthony. 2000. "Reencountering Development: Livelihood Transitions and Place Transformations in the Andes." *Annals of the Association of American Geographers* 90(3): 495–520.

Blaikie, Piers, and Harold Brookfield. 1987. *Land Degradation and Society.* London: Methuen.

Blaut, James M. 1977. "Two Views of Diffusion." *Annals of the Association of American Geographers* 67: 343–349.

Bullard, Robert D. (ed.). 1993. *Confronting Environmental Racism: Voices from the Grassroots.* Boston: South End Press.

Campanella, Richard. 2013. "Gentrification and Its Discontents: Notes from New Orleans." NewGeography, March 1. http://www.newgeography.com/content/003526-gentrification-and-its-discontents-notes-new-orleans.

Carmin, JoAnn, Nikhil Nadkarni, and Christopher Rhie. 2012. *Progress and Challenges in Urban Climate Adaptation Planning: Results of a Global Survey.* Cambridge, Mass.: MIT Press.

Carney, Judith. 2001. *Black Rice: The African Origins of Rice Cultivation in the Americas.* Cambridge, Mass.: Harvard University Press.

Clevinger, Woodrow R. 1938. "The Appalachian Mountaineers in the Upper Cowlitz Basin." *Pacific Northwest Quarterly* 29: 115–134.

Clevinger, Woodrow R. 1942. "Southern Appalachian Highlanders in Western Washington." *Pacific Northwest Quarterly* 33: 3–25.

Dalby, Simon. 2002. "Environmental Governance." In R. Johnston, P. Taylor, and M. Watts (eds.), *Geographies of Global Change: Remapping the World,* pp. 427–440. London: Routledge.

Diamond, Jared. 2002. "Evolution, Consequences and Future of Plant and Animal Domestication." *Nature* 418: 700–707.

Eliade, Mircea. 1987 [1957]. *The Sacred and the Profane: The Nature of Religion.* Willard R. Trask (trans.). San Diego: Harcourt.

Faria, Caroline. 2010. "Contesting Miss Sudan." *International Feminist Journal of Politics* 12(2): 222–243.

Faria, Caroline. 2011. "Staging a New South Sudan in the USA: Men, Masculinities, and Nationalist Performance at a Diasporic Beauty Pageant." *Gender, Place, and Culture* 20(1): 87–106.

Foote, Kenneth E. 1997. *Shadowed Ground: America's Landscapes of Violence and Tragedy.* Austin: University of Texas Press.

Ford, Clark. "Early World History: Indo-Europeans to the Middle Ages." http://www.public.iastate.edu/~cfford/342worldhistoryearly.html.

Foster, Jeremy. 2008. *Washed with Sun: Landscape and the Making of White South Africa.* Pittsburgh: University of Pittsburgh Press.

Freidberg, Susanne. 2001. "Gardening on the Edge: The Conditions of Unsustainability on an African Urban Periphery." *Annals of the Association of American Geographers* 91(2): 349–369.

Frey, William H. 2001. "Micro Melting Pots." *American Demographics* (June): 20–23.

Glassie, Henry. 1968. *Pattern in the Material Folk Culture of the Eastern United States*. Philadelphia: University of Pennsylvania Press.

Global Education Project. 2014. *Earth: A Graphic Look at the State of the World*. "Fisheries and Agriculture." http://www.theglobaleducationproject.org/earth/fisheries-and-aquaculture.php.

Griffin, Ernst, and Larry Ford. 1980. "A Model of Latin American City Structure." *Geographical Review* 70: 397–422.

Guthman, Julie. 2004. *Agrarian Dreams: The Paradox of Organic Farming in California*. Berkeley: University of California Press.

Hägerstrand, Torsten. 1967. *Innovation Diffusion as a Spatial Process*. Allan Pred (trans.). Chicago: University of Chicago Press.

Hall, Edward T. 1966. *The Hidden Dimension*. Garden City, N.Y.: Doubleday. This is the classic study of proxemics, conducted by an anthropologist. Hall argues that culture, above all else, shapes our criteria for defining, organizing, and using space.

Harner, John, and Bradley Benz. 2013. "The Growth of Ranchettes in La Plata County, Colorado, 1988–2008." *Professional Geographer* 65(2): 329–344.

Hehman, Eric, Jessica K. Flake, and Jimmy Calanchini. 2018. "Disproportionate Use of Lethal Force in Policing Is Associated with Regional Racial Biases of Residents." *Social Psychological and Personality Science* 9:4. https://doi.org/10.1177/1948550617711229.

Heynen, N., 2010. Cooking Up Non-Violent Civil Disobedient Direct Action for the Hungry: Food Not Bombs and the Resurgence of Radical Democracy. *Urban Studies* 47(6): 1225–1240.

Hollander, Gail. 2005. "The Material and Symbolic Role of the Everglades in National Politics." *Political Geography* 24(4): 449–475.

Hollander, Gail. 2008. *Raising Cane in the 'Glades: The Global Sugar Trade and the Transformation of Florida*. Chicago: University of Chicago Press.

Hopkins, Jeffrey. 1990. "West Edmonton Mall: Landscape of Myths and Elsewhereness." *Canadian Geographer* 34: 2–17.

Hopkins, Terence K., and Immanuel Maurice Wallerstein. 1982. *World-Systems Analysis: Theory and Methodology*. Thousand Oaks, Calif.: Sage Publications.

Huang, Yefang, Yee Leung, and Jianfa Shen. 2007. "Cities and Globalization: An International Perspective." *Urban Geography* 28: 209–231.

Hyndman, Jennifer, and Wenona Giles. 2011. "Waiting for What? The Feminization of Asylum in Protracted Situations." *Gender, Place & Culture: A Journal of Feminist Geography* 18(3): 361–379.

IDRC (International Development Research Centre). 2004. http://web.idrc.ca/en/ev-1248-201-1-DO_TOPIC.html.

ISAAA. 2016. International Service for the Acquisition of Agri-Biotech Applications website. http://www.isaaa.org.

Jackson, Peter. 2004. "Local Consumption Cultures in a Globalizing World." *Transactions of the Institute of British Geographers* NS 29: 165–178.

Jamasmie, Cecilia. 2013. "Infographic: The Periodic Table of Smartphones." Mining.com, February 5, http://www.mining.com/infographic-the-periodic-table-of-smartphones-56390/.

Johnston, R. J., Peter Taylor, and Michael Watts (eds.). 2002. *Geographies of Global Change: Remapping the World*. 2nd ed. Oxford: Blackwell.

Kniffen, Fred B. 1965. "Folk Housing: Key to Diffusion." *Annals of the Association of American Geographers* 55: 549–577.

Knopp, Lawrence, and Michael Brown. 2008. "Queering the Map: The Productive Tensions of Colliding Epistemologies." *Annals of the Association of American Geographers* 98(1): 40–58.

Latrimer Clarke Corporation Pty Ltd. http://www.altapedia.com.

Le Goix, Renaud, and Chris J. Webster. 2008. "Gated Communities." *Geography Compass* 2(4): 1189–1214.

Light, Duncan. 2012. *The Dracula Dilemma: Tourism, Identity and the State in Romania*. Farnham, U.K.: Ashgate.

Massey, Doreen. 1994. *Space, Place, and Gender*. Minneapolis: University of Minnesota Press.

Mackinder, Halford J. 1904. "The Geographical Pivot of History." *Geographical Journal* 23: 421–437.

Mitchell, Don. 1996. *The Lie of the Land: Migrant Workers and the California Landscape*. Minneapolis: University of Minnesota Press.

Mitchell, Katharyne. 1998. "Fast Capital, Race, and the Monster House." In R. George (ed.), *Burning Down the House: Recycling Domesticity*, pp. 187–212. Boulder, Colo.: Westview Press.

Mitchell, Katharyne. 2002. "Cultural Geographies of Transnationality." In K. Anderson, M. Domosh, S. Pile, and N. Thrift (eds.), *Handbook of Cultural Geography*, pp. 74–87. London: Sage.

Motzafi-Haller, Pnina. 1997. "Writing Birthright: On Native Anthropologists and the Politics of Representation." In Deborah E. Reed-Danahay (ed.), *Auto/Ethnography: Rewriting the Self and the Social*, pp. 195–222. Oxford: Berg.

Mwenda, Andrew M. 2013. "Madonna and Africa's 'Celebrity Saviors.'" CNN, April 17, http://www.cnn.com/2013/04/16/opinion/madonna-charity-africa-mwenda/.

Nietschmann, Bernard. 1973. *Between Land and Water: The Subsistence Ecology of the Miskito Indians, Eastern Nicaragua.* New York: Seminar Press.

Newman, M. E. J. 2016. "Maps of the 2016 U.S. Presidential Election Results." University of Michigan Department of Physics and Center for the Study of Complex Systems.

Nietschmann, Bernard. 1979. "Ecological Change, Inflation, and Migration in the Far West Caribbean." *Geographical Review* 69: 1–24.

O'Loughlin, John, Frank Witmer, Thomas Dickinson, Nancy Thorwardson, and Edward Holland. 2007. "Preface and Map Supplement." Special Issue: The Caucasus: Political, Population, and Economic Geographies. *Eurasian Geography and Economics* 48(2): 127–134.

Ormrod, Richard K. 1990. "Local Context and Innovation Diffusion in a Well-Connected World." *Economic Geography* 66: 109–122.

Ó Tuathail, Gearóid. 1996. *Critical Geopolitics.* Minneapolis: University of Minnesota Press.

Pillsbury, Richard. 1998. *No Foreign Food: American Diet in Time and Place.* Boulder, Colo.: Westview Press.

Pountain, Chris. 2005. "Varieties of Spanish." Queen Mary School of Modern Languages. http://www.qmul.ac.uk/~mlw058/varspan/varspanla.pdf.

Pulido, L., and Peña, D. 1998. "Environmentalism and Positionality: The Early Pesticide Campaign of the United Farm Workers Organizing Committee 1965–71." *Race, Class, and Gender* 6(1): 33–50.

Raitz, Karl B. 1987. "Place, Space and Environment in America's Leisure Landscapes." *Journal of Cultural Geography* 8(1): 49–62.

Ramadan, Adam. 2008. "The Guests' Guests: Palestinian Refugees, Lebanese Civilians, and the War of 2006." *Antipode* 40(4): 658–677.

Rashtogi, Nina Shen. 2010. "Is Kosher Better for the Planet?" *Washington Post,* February 2. http://articles.washingtonpost.com/2010-02-02/opinions/36839736_1_kosher-meat-halal-meat-kosher-products.

Rehder, John B. 2004. *Appalachian Folkways.* Baltimore: Johns Hopkins University Press.

Relph, Edward. 1976. *Place and Placelessness.* London: Pion.

Relph, Edward. 1981. *Rational Landscapes and Humanistic Geography.* New York: Barnes & Noble.

Roberts, Sam. 2010. "Listening to (and Saving) the World's Languages." *New York Times,* April 28. http://www.nytimes.com/2010/04/29/nyregion/29lost.html?pagewanted=all.

Rocheleau, Diane, Barbara Thomas-Slayter, and Esther Wangari (eds.). 1996. *Feminist Political Ecology: Global Issues and Local Experiences.* New York: Routledge.

Rodgers, Dennis. 2004. "'Disembedding' the City: Crime, Insecurity and Spatial Organization in Managua, Nicaragua." *Environment and Urbanization* 16: 113–123.

Rose-Redwood, Reuben, and Derek Alderman (eds.). 2011. "Special Thematic Interventions Section: New Directions in Political Toponymy." *ACME: An International E-Journal for Critical Geographies* 10(1): 1–41. https://www.acme-journal.org/index.php/acme/issue/view/67. A collection of the latest geographic scholarship on political place-names.

Rostow, W. W. 1962. *The Process of Economic Growth.* New York: W. W. Norton.

Sack, Robert D. 1986. *Human Territoriality: Its Theory and History. Studies in Historical Geography,* No. 7. Cambridge, U.K.: Cambridge University Press.

Sack, Robert D. 1992. *Place, Modernity, and the Consumer's World: A Rational Framework for Geographical Analysis.* Baltimore: Johns Hopkins University Press.

Sassen, Saskia. 1991. *The Global City: New York, London, Tokyo.* Princeton, N.J.: Princeton University Press.

Sauer, Carl O. 1952. *Agricultural Origins and Dispersals.* New York: American Geographical Society.

Sauer, Jonathan D. 1993. *Historical Geography of Crop Plants.* Boca Raton, Fla.: CRC Press.

Sayre, Nathan. 2005. *Working Wilderness: The Malpai Borderlands Group Story and the Future of the Western Range.* Tucson: Rio Nuevo Publishers.

Schueth, Sam, and John O'Loughlin. 2008. "Belonging to the World: Cosmopolitanism in Geographic Contexts." *Geoforum* 39(2): 926–941.

Sen, Amartya. 1999. *Development as Freedom.* New York: Alfred A. Knopf.

Shortridge, Barbara G., and James R. Shortridge (eds.). 1998. *The Taste of American Place: A Reader on Regional and Ethnic Foods.* Lanham, Md.: Rowman & Littlefield.

Singh, Rana P. B. 1994. "Water Symbolism and Sacred Landscape in Hinduism." *Erdkunde* 48: 210–227.

Sivanandan, A. 2001. "Poverty Is the New Black." *The Guardian,* August 17. http://www.theguardian.com/politics/2001/aug/17/globalisation.race.

Sparke, Matthew. 2013. *Introducing Globalization: Ties, Tensions, and Uneven Integration.* Oxford: John Wiley & Sons.

Stavans, Ilan. 2003. *Spanglish: The Making of a New American Language.* New York: Rayo (an imprint of HarperCollins).

Ter-Ghazaryan, Diana K. 2013. "'Civilizing the City Center': Symbolic Spaces and Narratives of the Nation in Yerevan's Post-Soviet Landscape." *Nationalities Papers: The Journal of Nationalism and Ethnicity* 41(4): 570–589.

Toal, Gerard, and John Agnew. 2002. "Introduction: Political Geographies, Geopolitics and Culture." In K. Anderson, M. Domosh, S. Pile, and N. Thrift (eds.), *Handbook of Cultural Geography,* pp. 455–461. London: Sage.

Trachtenberg, Stephen Joel, Gerald B. Kauvar, and E. Grady Bogue. 2013. *Presidencies Derailed: Why University Leaders Fail and How to Prevent It.* Baltimore: Johns Hopkins University Press.

Tsuda, Takeyuki. 2006. "When Minorities Migrate: The Racialization of Japanese Brazilians in Brazil and Japan." In Lola Romanucci, George De Vos, and Takeyuki Tsuda (eds.), *Ethnic Identity: Problems and Prospects for the Twenty-First Century,* pp. 208–232. Lanham, Md.: AltaMira Press.

Tweed, Thomas. 2008. *Crossing and Dwelling: A Theory of Religion.* Cambridge, Mass.: Harvard University Press.

von Thünen, Johann Heinrich. 1966. *Von Thünen's Isolated State: An English Edition of Der Isolierte Staat.* Carla M. Wartenberg (trans.). Elmsford, N.Y.: Pergamon.

Walker, Peter, and Louise Fortmann. 2003. "Whose Landscape? A Political Ecology of the 'Exurban' Sierra." *Cultural Geographies* 10: 469–491.

Warf, Barney, and Morton Winsberg. 2010. "Geographies of Megachurches in the United States." *Journal of Cultural Geography* 27(1): 33–51.

Worthen, Molly. 2012. "One Nation Under God?" *New York Times Sunday Review,* December 22. http://www.nytimes.com/2012/12/23/opinion/sunday/american-christianity-and-secularism-at-a-crossroads.html?pagewanted=all&_r=0.

Index

Note: Page numbers followed by f indicate figures, those followed by t indicate tables, and those followed by m indicate maps.